OBJECT DETECTION, COLLISION WARNING & AVOIDANCE SYSTEMS VOLUME 2

PT-133

Edited by

Ronald K. Jurgen

Published by
SAE International
400 Commonwealth Drive
Warrendale, PA 15096-0001
U.S.A.
Phone (724)776-4841
Fax (724)776-0790

For permission and licensing requests contact:

SAE Permissions
400 Commonwealth Drive
Warrendale, PA 15096-0001-USA
Email: permissions@sae.org
Fax: 724-776-3036
Tel: 724-772-4028

2009 06 08

Global Mobility Database®

*All SAE papers, standards, and selected
books are abstracted and indexed in the
Global Mobility Database.*

For multiple print copies contact:

SAE Customer Service
Tel: 877-606-7323 (inside USA and Canada)
Tel: 724-776-4970 (outside USA)
Fax: 724-776-0790
Email: CustomerService@sae.org

ISBN-13 978-0-07680-1810-3
Library of Congress Catalog Card Number: 2006939315
SAE/PT-133
Copyright © 2007 SAE International

SAE Order No. PT-133

Printed in USA

OBJECT DETECTION, COLLISION WARNING & AVOIDANCE SYSTEMS
VOLUME 2

Other SAE books in this series:

Adaptive Cruise Control
(Product Code: PT-132)

Electronic Braking, Traction, and Stability Controls, Volume 2
(Product Code: PT-129)

Multiplexing and Networking, Volume 2
(Product Code: PT-128)

Automotive Software
(Product Code: PT-127)

Electronic Engine Control Technologies, Second Edition
(Product Code: PT-110)

Sensors and Transducers, Second Edition
(Product Code: PT-105)

Electric and Hybrid Electric-Vehicles
(Product Code: PT-85)

On- and Off-Board Diagnostics
(Product Code: PT-81)

Electronic Transmission Controls
(Product Code: PT-79)

Multiplexing and Networking
(Product Code: PT-78)

Electronic Steering and Suspension Systems
(Product Code: PT-77)

Electronic Braking, Traction, and Stability Controls
(Product Code: PT-76)

Navigation and Intelligent Transportation Systems
(Product Code: PT-72)

INTRODUCTION

New Sensors Are Key

Highly sensitive sensors and powerful microprocessors will be the enablers of the "sensitive" automobile with predictive driver assistance systems. With the availability of appropriate sensors, new vehicle systems will range from warning systems to systems with vehicle interaction. These conclusions are expressed by the authors of SAE Paper No. 2004-01-1111 in this book, PT-133, which has 51 papers covering object detection, collision warning, and collision avoidance. None of those papers is included in PT-70, the first volume of 32 papers on these topics.

The papers in the first section in this book are on object detection and cover a variety of topics, including parking aids, target tracking with cameras, sensor combinations, blind spot detection, imager chips, and lane tracking. Collision warning papers cover lane and road departure warning, sensor fusion, and intersection collision warning. Collision avoidance papers cover front- and rear-end crash avoidance, automatic collision avoidance systems, braking systems for collision avoidance, and driver-vehicle interface requirements. The fourth and final section presents three papers covering future needs in a variety of areas in order to further progress in object detection, collision warning, and collision avoidance.

Among the many insights to be found herein are the following:

- "Improved signal processing and threat assessment algorithms would enhance FCW (forward crash warning) alert efficiency by recognizing slower lead vehicles transitioning from the path of the host vehicle to out of its path... Proceeding with further FCW enhancement activities may depend on successful results (driver satisfaction, units sold, and positive safety impact) from short-term deployment and good market penetration levels... Vehicle to vehicle communications are suggested to improve the forward-looking sensing capability of FCW for long-term deployment plans... This enhancement would call upon lead vehicles to transmit information about their dynamic state to following vehicles, given wider deployment of FCW in the vehicle fleet." (2006-01-0573)

- "On the supplier side, virtually every Tier One supplier who trades in automotive electronics is developing some ITS systems. In Europe it's Continental AG, Siemens VDO Automotive and Valeo. In Japan, the lead is with Denso, Aisin and Hitachi, but many others are also active. The U.S. effort is led by Delphi with the foregoing European and Japanese firms also very active.

"If you want to see the future face of new automotive technology in general and ITS in particular, watch Mercedes, Toyota and Honda. These companies have little fear of technological risk—in fact, they seem to revel in it—because at the end of the day it brings them image and, in the long term, image brings customers, revenue and profits." (2006-01-0095)

- "This paper has addressed the needs of precrash sensing as they relate to vehicles approaching objects in the path of travel. An important extension to this is expanding the coverage region so that a precrash system could potentially predict objects approaching from the sides such as at an intersection.

  "Finally an algorithm must be developed that relates telemetric data of an object to the probability of impact. This involves not only generating significant simulation, but also real world testing to ensure that the simulations are representative." (2005-01-3578)

- "No matter how you approach it, the future of automotive safety systems is certainly an integrated, vehicle systems-level approach.

  "Safety technology roadmaps are beginning to look alike. Collision avoidance sensors and occupant recognition sensors employ basically the same technologies. The same can be said about vehicle dynamic control sensors and vehicle crash sensors, as well as distributed safety architectures and distributed mobile multimedia architectures.

  "Safety features and functions are blending. The best way to protect an occupant is to avoid the accident. Subsystem information can and should be shared (vehicle dynamic state estimation information, occupant information, airbag status, scene information, etc.). Subsystem blending can enhance vehicle (and integrated safety) performance (e.g., a collision threat can mute a radio and cell phone, etc.)... As a result, a system approach to integrated safety is driving our future developments." (2000-01-0346)

*　*　*　*　*　*　*　*　*

This book and the entire Automotive Electronics Series are dedicated to my friend Larry Givens, a former editor of SAE's monthly publication, *Automotive Engineering International.*

Ronald K. Jurgen, Editor

TABLE OF CONTENTS

INTRODUCTION

OBJECT DETECTION SYSTEMS

COLLISION WARNING SYSTEMS

COLLISION AVOIDANCE SYSTEMS

FUTURE TRENDS AND NEEDS

OBJECT DETECTION SYSTEMS

Driver Performance Research Regarding Systems for Use While Backing

Charles A. Green and Richard K. Deering
General Motors

ABSTRACT

General Motors has pursued research to develop systems intended to assist drivers in recognizing people or objects behind them when they are backing, and this paper summarizes results from this research. We are currently working with ultrasonic rear parking assist systems, rear radar backing warning systems, and rear camera systems, which are briefly described and their utility for assisting drivers in recognizing people or objects behind them discussed. Our research on driver performance with a prototype long range backing warning system found that audible and visual warning combinations may not be effective in warning distracted drivers about unexpected objects. Driver expectancy is thought to play a significant role in this result. However, further research found drivers were more likely to notice an unexpected obstacle behind their vehicle with a prototype rear view video camera system compared to ultrasonic rear parking assist and trials that had no system. These results suggest that rear view video camera systems may provide limited benefit in some backing scenarios. While parking assist systems may assist drivers with parking tasks, they are not designed to be warning systems and may not effectively warn drivers of unexpected obstacles. We continue to pursue work with rear view video camera systems including integration with object detection sensors. We are also investigating automatic vehicle braking and haptic warning strategies for long range backing warning systems.

INTRODUCTION

The purpose of this paper is to summarize our recent research directed towards the development of systems intended to assist drivers in recognizing people or objects behind them when they are backing. A recent study by NHTSA (NHTSA, 2004) estimates approximately 120 deaths and over 6,000 injuries (mostly minor) annually that are associated with backing incidents. Children under 5 years of age and older persons over the age of 70 are over represented in the fatality statistics.

PARKING AIDS

Several automobile manufacturers currently offer backing/parking aids designed to assist drivers while parking. These systems can include side view mirrors which rotate downward when the vehicle is placed into reverse, rear view cameras, and ultrasonic rear parking assist systems. All of these systems are designed to help the driver locate known fixed objects that are behind the vehicle and near the bumper, in order to help them more precisely guide the vehicle. They are not intended to function as collision warning or avoidance systems.

Rotating side view mirrors and rear view parking cameras were designed to support parking tasks, not as warning devices. Both of these systems are passive, relying on the driver to gain information from their use, but otherwise to operate the vehicle in a safe manner. Both of these devices are not designed to be the exclusive source for the driver for visual information, as they have limited fields of view. They are supplemental in nature, designed to provide additional information.

ULTRASONIC REAR PARKING ASSIST

Ultrasonic Rear Park Assist (URPA) was designed to provide supporting information regarding parking-related obstacles (such as walls, large poles, and other vehicles) to the driver to help them maneuver their vehicle while parking. URPA utilizes a "chime", the same as used in many seatbelt reminders, as its auditory indicator. URPA was not designed to prevent collisions involving pedestrians or vehicles in motion, and as such the chime utilized was not designed to be a time critical warning. The range of the URPA system is less than 2 m. A driver moving at greater than ~5 kph would have very little time to stop in response to the

already focused on those indications and they were already intending to stop -- as they would be if parking. This range supports the driver who is parking, not the driver who is backing.

The URPA system was designed to detect larger poles and parking barriers, of greater than 7.5 cm in diameter with a length of 1.0 meter; it was not designed to detect curbs and other lower road features for which indications may not be helpful when parking. Thus the system does not detect obstacles lower than 25 cm or obstacles directly below the bumper or under the vehicle. In addition, smaller or thinner objects or pedestrians may not be detected. Thus, our instructional materials for URPA include the following warning: "If children, someone on a bicycle, or pets are behind your vehicle, URPA won't tell you they are there. You could strike them and they could be injured or killed. Whether or not you are using URPA, check carefully behind your vehicle and then watch closely whenever you back up."

BACKING WARNING SYSTEMS

Backing warning systems are intended to alert drivers to the presence of unexpected or unseen objects behind their vehicles. To be effective, such a system would include a warning designed to capture the attention of the driver with sufficient advance notice to allow the driver to stop or otherwise avoid the object. Analysis of this situation as a simple two body non-linear vehicle motion problem, incorporating a suitable delay for driver perception-reaction time, would suggest that a simple sensing and warning system should be effective in this situation. However, a recently completed study we sponsored on the effectiveness of backing warnings showed surprisingly low effectiveness for the warnings tested (Llaneras, Green, Chundrlik, Altan, and Singer, 2004). Driver expectancy is thought to play a significant role in this result.

Driver performance testing was conducted in an open parking lot using two instrumented vehicles, both equipped with a prototype backing warning system. This system utilized a highly accurate research-grade sensor not suitable for production. The system was intended to stimulate the driver to brake hard to a stop in response to a provided warning. A variety of approaches for presenting warning information to the driver were investigated using a surprise trial methodology. These warnings all utilized a visual component of blinking LEDs near the rear window (which could be seen in the inside rear view mirror or with direct glances towards the rear of the vehicle) along with several different auditory warnings including those successfully utilized in past forward collision warning work (Kiefer, 1999). Warning timing was optimized from a prior study involving alerted backing.

All drivers in this study were trained on the integrated parking assist capability built-into this prototype backing warning system, which shared the visual display and auditory system. Further, a portion of the drivers in the study were provided specific training on the backing warning functionality. Due to drivers' inherent vigilance,

in order for the surprise obstacle to be missed allowing the warning to trigger it was necessary to distract drivers from the backing task by asking them to monitor a small video screen adjacent to the rear window.

The study found that for those trials where the driver was successfully distracted (approximately two-thirds of trials) and a warning was issued, only 13% of drivers avoided hitting the obstacle (five of 39); over 87 percent of the drivers collided with the obstacle following the warning. While many drivers who experienced the warning (68%) demonstrated precautionary behaviors in response to the warning by covering the brake, tapping the brake, or braking (44% braked), the level of braking was generally not sufficient to avoid colliding. Thus, although the data provide some evidence that the warnings were influencing driver behavior, warnings in this context were not reliably inducing drivers to immediately brake to a stop.

The Llaneras et al. backing warning study data further suggest that knowledge and experience with the backing warning system may not significantly improve the situation (driver compliance and immediate response to the warning). Specific training on the warning system was provided to eight drivers; but only one of these drivers avoided the obstacle. In all cases, drivers reported that they did not expect there to be any obstacle in their path. Many also reported searching for an obstacle following the warning, but since they "didn't see anything" they continued to back. These observations suggest that expectancy is a powerful determinant, guiding driver perception and behavior. Although warnings in this study appeared to orient some drivers to search for an obstacle and/or take precautionary action (reduce speed, etc.), they did not necessarily lead drivers to brake hard in response to the warning. Many drivers appeared to want direct sensory confirmation of the existence of an object before initiating immediate avoidance behaviors. Similar behavioral results to this backing warning study were found by Lee et al. (2002) in the context of driver responses to warnings from a Rear-End Collision Avoidance System (RECAS). These researchers found that the primary effect of a warning was to redirect driver attention, rather than triggering an immediate response by drivers. However, unlike a forward collision warning situation, where drivers can simply look out the forward view and quickly detect an in-path threat, the detection of a rear obstacle by drivers presents a difficult challenge.

REAR CAMERA SYSTEMS

Any type of camera system which can provide the driver additional visibility has the potential for assisting the driver in avoiding an incident. We have investigated a number of issues relating to the human factors engineering of a rear camera system intended for drivers' use while parking. A key driver performance issue for camera systems involves what field-of-view they provide to the driver – that is, how far behind the vehicle should the rear camera image provide a view, and view of what portions of the vehicle or surrounds could be useful if included in the rear camera image. Also involved in the field-of-view decisions is the issue of image distortion, what types of distortion are acceptable for the rear camera image. Other key issues involve minimum size and resolution for the rear camera image display, and acceptable locations for the rear camera image display. Lastly, appropriate indications for the display of the rear camera image (e.g. transmission into reverse gear), and appropriate instructions and warnings on the use of the rear camera system have been addressed.

We have also sponsored external research studies (i.e. performed by a contracted research supplier) into driver performance with a prototype rear camera system. This prototype system utilized a single color camera with a wide angle lens mounted in the center of the liftgate and a display mounted in the "center stack" of the vehicle similar to a navigation system display. As part of one study, drivers' performance in noticing (and avoiding) an obstacle unexpectedly placed behind their vehicle prior to their beginning a backing maneuver was examined (McLaughlin, Hankey, Green and Kiefer, 2003). The main focus of the study was on parking behaviors and compared the rear camera system and ultrasonic rear parking assist together and separately with traditional parking methods (i.e. neither system). As an additional investigation, an unexpected obstacle technique was utilized near the end of the session to examine any potential obstacle avoidance performance differences that could result from the presence or absence of these parking assistance systems.

After the last of a number of parking trials, a ruse was used where while the experimenter spoke to an individual outside the vehicle, the vehicle was blocked from forward travel by two objects (cone and folding chair) while a plastic pylon was (unbeknownst to the participant) placed behind the vehicle. The participant was then told he or she could return to the building (where the study procedure had begun). At this time, the participant would either back into the pylon or detect it.

The ruse was set-up and executed successfully for 29 of the 32 study participants. The number of participants exposed to the ruse for the traditional, ultrasonic rear parking assist (URPA), rear camera system (RV), and URPA + RV conditions was seven, seven, nine, and six, respectively. Because so many participants were hitting the obstacle, its size was increased with the addition of a large white Styrofoam board after the first 11 participants. In sum, 24 of the participants hit the obstacle leaving five who avoided hitting the obstacle. Of the five participants who did not hit the obstacle:

three saw the obstacle using the RV (two in the RV condition, one in the URPA + RV condition), one saw the obstacle in their mirror (in the URPA + RV condition), and one saw the obstacle out the back window (in the RV condition). A χ^2 test indicates that participants in conditions when RV was present were less likely to hit the pylon than participants in conditions where RV was not available (with $\alpha=0.05$).

We have also sponsored a second external research study of driver performance with a prototype rear camera system. In this study each participant parked with the rear camera system and available ultrasonic rear park assist (URPA) more than 30 times including practice trials. Again, while it was not the central focus of the study, a "ruse" technique was utilized near the end of the session. After a final trial wherein the participant was directed to pull into a perpendicular parking space where the vehicle's nose was at a curb, the participant was directed to place the vehicle in Park. While the participant was filling out paperwork, a compatriot experimenter placed a 22 cm wide by 1.2 m tall object three feet behind their vehicle. When the participant completed the paperwork, they were asked to drive back to the building. As the participant placed the vehicle in Reverse, both URPA and the rear camera system became active.

In this study, a flashing symbol was in some cases overlaid upon the rear camera imagery in the approximate location of the ruse object, and the ruse object was either placed towards the left, centered, or towards the right edge of the rear bumper. While there were no statistically significant effects of either the presence or type of symbol or the location of the ruse object, it was noteworthy that 31 of the 48 participants who experienced this ruse, or 65%, noticed and successfully avoided the obstacle. Some reasons why the participant success rate in avoiding the obstacle may have been measured greater in the second study as compared to the first may be the greater experience participants had with the camera system in the second study, or it could be due to the larger number of ruse trials available in the data set.

CONCLUSION

These results suggest that rear view video camera systems may provide limited benefit in some backing scenarios, while parking assist systems may not effectively warn drivers of unexpected obstacles. Our current research related to backing is proceeding along several lines in an attempt to overcome the issues observed in our prior work, particularly related to backing warning effectiveness. Research has recently been concluded relating to the integration of obstacle warnings and camera displays. Research is ongoing to determine the feasibility and appropriate implementation of automatic braking of the vehicle to provide avoidance of obstacles at lower speeds. Research is also ongoing to investigate a tactile backing warning.

REFERENCES

1. Kiefer, R., LeBlanc, D., Palmer, M., Salinger, J., Deering, R., and Schulman, M. (1999). Development and Validation of Functional Definitions and Evaluation Procedures for Collision Warning/Avoidance Systems. Report DOT-HS-808-964. NHTSA, U.S. Department of Transportation.
2. Lee, J.D., McGehee, D.V., Brown, T.L., and Reyes M.L. (2002). Collision Warning Timing, Driver Distraction, and Driver Response to Imminent Rear-End Collisions in a High-Fidelity Driving Simulator. *Human Factors*, Vol. 44 (2), 314-334.
3. Llaneras, R.E., Green, C.A., Chundrlik, W., Altan, O.D., and Singer, J.P. (2004). Design and Evaluation of a Prototype Rear Obstacle Detection and Driver Warning System. *Human Factors* (in press).
4. McLaughlin, S.B., Hankey, J.M., Green, C.A., and Kiefer, R.J. (2003). Driver Performance Evaluation of Two Rear Parking Aids. Proceedings of the 2003 Enhanced Safety Vehicle Conference.
5. NHTSA (2004). Data Collection Study: Deaths and Injuries Resulting from certain Non-Traffic and Non-Crash Events. Vehicle-Generated Carbon Monoxide, Vehicle Backing, Vehicle Heat (Weather Induced), Power Windows. A Continuation of the Study of Non-Traffic and Non-Crash Motor Vehicle-Related Safety Issues Focusing on 1998 Death Certificates and Other Sources Containing Relevant Data and Information. NHTSA, U.S. Department of Transportation.

2006-01-1294

OptiVeo: A Vision-Based Platform for Driving Assistance

Patrice Reilhac, Julien Rebut, Joël Lelevé and Benoist Fleury
Valeo - Driving Assistance Domain

ABSTRACT

Most of the Driving Assistance functions require sensors that are looking ahead of the vehicle. Among these sensors, cameras, already providing Lane Departure Warning and Night Vision, will play a major role. Since it won't be possible to multiply the number of sensors located behind the windscreen, it is then crucial to perform with cameras as many detections as possible. We describe in this paper some algorithmic methods for performing camera based tunnel and fog detection.

INTRODUCTION

Pioneered by Lane Departure Warning and Night Vision Systems for front vehicle applications and by Park Assist Systems at the rear, embedded video cameras promise to play a major role in new Driving Assistance applications in the next decade.

Watching front, side and rear scenes, extracting information as diverse as pedestrian location or fog density, cameras address all types of Driving Assistance applications: night vision, obstacle detection, environment detection, parking assistance, automatic windscreen wiping and finally lighting related applications.

The potential functionalities of such sensors can be gathered into four main topics (Fig.1):

- Infrastructure Detection

- Night Vision

- Obstacle Detection

- Lighting and Wiping Automation

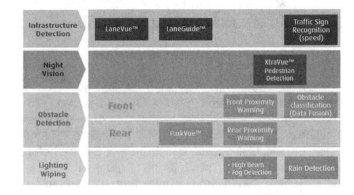

Figure 1 – Vision-based Application roadmap

Most of these vision-based applications are studied at Valeo within the framework of the OptiVeo™ project.

Valeo is currently providing IR-based RLT - for Rain, Light and Tunnel - sensors. These sensors, located behind the upper part of the windscreen, provide ambient light measurement and Tunnel Detection for Low beam automation and secondarily Rain detection for Wipers automation. On the other hand, Valeo is also producing front cameras, also located in the vicinity of the interior mirror, for Lane Departure Warning (Fig.2).

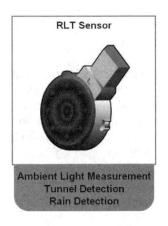

Figure 2 – Available systems

It is then obvious that it would be of a great benefit to integrate Low Beam and High Beam automation into this camera and to even extend its functionality to Fog detection and Rain detection (Fig.3).

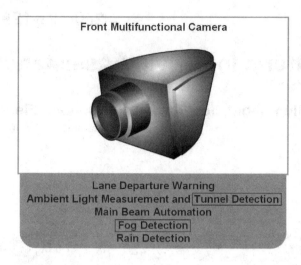

Figure 3 – Front multifunctional camera

Considering automatic switching of lighting devices, camera based systems must reproduce at least the performances of existing sensors. Among these functionalities, one can list the anticipated tunnel detection which allows to switch on low beams before entering into tunnels, what is of a great interest, for instance, with Xenon lamps whose ramp-up is about one second (Fig.4).

Figure 4 – Tunnel detection

We will describe in a first part how to perform Tunnel detection with a camera based system.

Among lighting lamps, the fog lamps have still no system for automatic activation. But to perform such automation, one needs to build a system able to detect the occurrence of fog (Fog.5) whatever is the ambient light condition. Moreover, in the case where such a system is able to evaluate the visibility distance, one can easily imagine how this information would be useful for other applications.

Figure 5 – Fog detection

Anticipating the decrease of visibility distance would allow to warn the driver, what would certainly reduce the number of accidents by fog condition; knowing that they represent 1% of the total number of accidents but are twice more fatal.

One can also point out that different algorithms using a same video signal would take benefit of working in collaboration. Indeed, the visibility distance evaluation would give to the other algorithms a good assessment of the relevance of the video information. In the case of a deep fog, the visibility distance would fall down and objects detection algorithm or infrastructure detection could warn the driver that they are no more active.

We will describe in a second part how to perform a Fog Detection and a Visibility Distance Measurement by using a video sensor at daylight and at night

TUNNEL DETECTION

For optimum lighting at the entrance to a dimly-lit tunnel or on bright days, it is appropriate to switch on dipped headlights in anticipation. However, the system as described above, does not provide for such anticipation, and specific processing therefore needs to be implemented.

At a speed of 130 km/h and assuming a 1-second response time for switching on Xenon headlamps, a tunnel must be detected at least 36 m before the vehicle enters.

Detection of the tunnel involves tracking the dark areas in the image [1]. In fact, a tunnel is a dark area of the image (low grey level ~0) featuring a roughly rectangular shape. Also, as the vehicle approaches the tunnel, the area of the corresponding dark zone is strictly increasing. After a sufficiently large number of appearances, it is then possible to establish the presence of a tunnel.

The image of a road scene includes zones of low interest (Fig6). In fact the bottom third of the image represents the road and the top third represents the sky. The area of interest is therefore defined as the central third of the image.

Figure 6 - Zone of interest

The dark areas are then extracted by segmentation. To do this we use thresholding with hysteresis which provides better results than simple thresholding for a slightly longer computing time. The dark areas in such case are correctly closed and located.

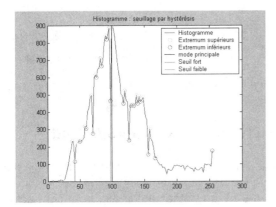

Histogram

High threshold

Low threshold

Thresholding with hysteresis

Figure 7 – Picture Thresholding

As stated previously, the area of the tunnel increases as the vehicle approaches it. To check this expanding area, the dark zones must be tracked with respect to time from one image to the next.

Space and time tracking enables the displacement of the dark source to be monitored from one image to the next. To do this, it is essential to be able to apply time correlation between the objects detected in the image at t=2 with those detected in the image at t=1 and so on. This correlation may be achieved for example by minimising the Euclidean distance. In such case, if R_i represents the regions detected in image 1 and R_j represents the regions detected in image 2, we can establish a matrix of distances:

$$
\begin{array}{c}
\ \ \ \ \ \ \ \ \ \ R_j \\
R_i\ \
\begin{array}{cccc}
d_{11} & d_{12} & \ldots & d_{1n} \\
d_{21} & d_{22} & \ldots & d_{2n} \\
\ldots & \ldots & \ldots & \ldots \\
d_{m1} & d_{m2} & \ldots & d_{mn}
\end{array}
\end{array}
$$

where $\quad d_{11} = \sqrt{\left(x_{R1} - x_{R2}\right)^2 + \left(y_{R1} - y_{R2}\right)^2}$

This matrix of distances enables allocation of a source R_j to the source R_i which will minimise the distance d_{ij} (Fig.8).

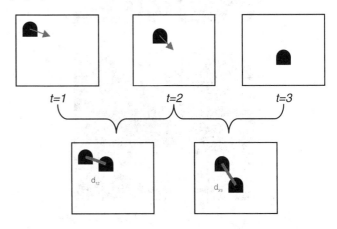

Figure 8 – Frame to frame object distance

Thus, the space and time tracking enables the elaboration of a history of the appearance of objects. According to the acquisition frequency and from a certain number of appearances, an object becomes a potential tunnel. To confirm the presence of a tunnel, it is simply required to confirm that its area is indeed increasing (Fig.9).

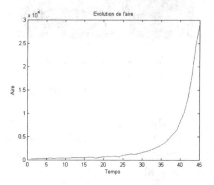

Figure 9 - Dark area (tunnel) growth.

The figure below (Fig.10) shows some results. In a straight line, a tunnel is detected at a distance of more than 150 m.

Figure 10 – Tunnel detection (snapshots)

As we have seen, the shutter speed is controlled automatically by the camera according to the ambient lighting. This integration time can then be used to confirm the presence of a tunnel, to detect the presence of a tunnel if it has failed to be detected by image processing (operation similar to a single photodiode) but especially it can be used to switch off the headlamps at the tunnel exit. The headlamps need to be extinguished about 15 seconds after leaving the tunnel. As is shown in the next figure (Fig.11), the integration time is very short in daylight conditions and increases significantly when the vehicle is in a tunnel.

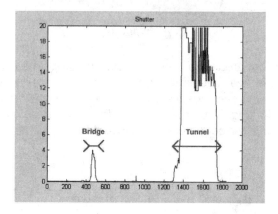

Figure11 - Integration time according to driving conditions

VISIBILITY DISTANCE MEASUREMENT

By optimising the lighting to driving conditions, automatic front fog lamp activation provides the same advantages for drivers as any other automatic front light activation. Rear fog lamps automatic activation would also provide outstanding advantages since this would anticipate the decreasing visibility distance and warn following drivers in time. Furthermore, it would, as well, turn off the fog lamps as soon as they are not required, preventing them from dazzling drivers behind.

Unfortunately, the influence of the fog on the recorded video is not the same at daytime and at night. Therefore, we need to consider dedicated algorithm for these two driving situations.

FOG DETECTION AT DAYLIGHT

The light beams passing through fog are diffused in all directions by the water droplets, thus reducing visibility. When the driver looks at the road, the objects contained within his field of view transmit or reflect light energy which penetrates the eye and illuminates the retina. This distribution of light is then analysed by the brain in terms of spatial frequencies. The following analogy can therefore be put forward : fog can be considered as an optical filter which modifies the visual signal by attenuating the frequency contrast. It is therefore possible to characterise the fog by a frequency contrast operator.

To do this, we begin by searching for the homogeneous zones of the image [1],[2]. In fact, a standard road scene generally contains two homogeneous zones, corresponding to the sky and the road. In fog, these zones tend to disappear to form a single region which is also homogeneous. To identify these homogeneous zones, we construct the correlation matrices.

In the example below (Fig.12), the homogeneous zones are the sky and the road.

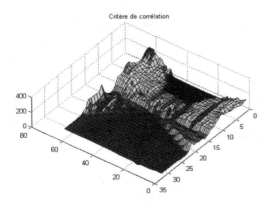

Figure 12 – Homogeneous zones

From this we can deduce (Fig.13) the correlation image of the homogeneous zones.

Figure 13 – Correlation image

Calculation of the parameters for these textured regions enables the detection of the presence or otherwise of fog. If this first stage confirms the presence of fog, the visibility range is then estimated by analysing the point of inflection of a vertical profile of the image passing through the centre of gravity of the homogeneous region (Fig.14).

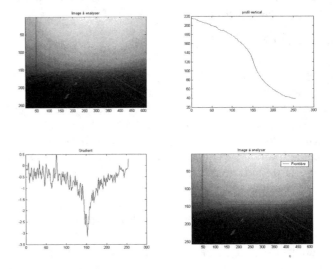

Figure 14 – Visibility distance evaluation

Below are some examples (Fig.15) of fog detection and visibility range measurement.

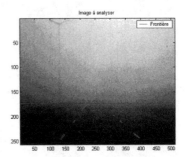

Figure 15 – Visibility distance evaluation (snapshots)

As soon as fog is detected and the visibility range falls below a predefined threshold, the fog lamps can be automatically activated.

FOG DETECTION AT NIGHT

At night, the former fog detection algorithm is no more applicable because of a lack of visibility of the scene. Indeed, the road is only lit by the car headlamps but the backlighting of fog droplets prevents the camera from recording in a proper way the road texture. We will then analyze the halation generated, being considered that its shape or geometrical characteristics are linked to the fog density i.e. the visibility distance.

To do so, the algorithm runs the following steps:

1) Shape extraction of the halation by thresholding the recording.

2) Thanks to tangential method, determination of the ellipse which is mostly corresponding to the halation. Evaluation of the ellipse parameters: 'a', 'b' and centre location.

3) Visibility distance evaluation deduced from the position of the point of inflection of a vertical profile of the image passing through the highest point of the ellipse.

On the two examples here bellow (with and without fog), one sees the contour extractions of the thresholding, the approximated ellipses and the point clouds of 'a' and 'b' values (Fig.16):

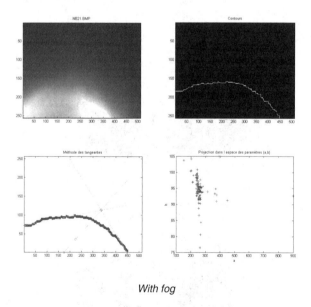

With fog

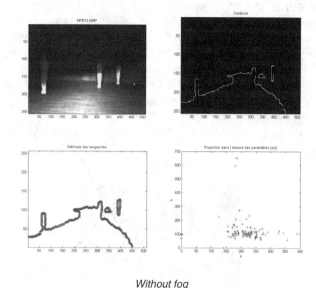

Without fog

Figure 16 – Visibility distance evaluation at night

We see that, despite scale differences, the two point clouds are significantly different. This would allow a rough estimation of the visibility distance. As said previously, we then improve the accuracy of the evaluation by locating point of inflection of a vertical profile of the image passing through the highest point of the ellipse.

On the other hand, we can see on the second example (without fog) that the ellipse calculation eliminates the distortions due to obstacles – such as pedestrians – in the field of view. Nevertheless, the spread of 'a' and 'b' values determines the quality of the measurement and gives the algorithm the possibility to reject wrong images.

This evaluation of visibility distance has to comply with:

- the different patterns of beams the future AFS regulation (Adaptive Front Lighting System) will allow (town lighting, motorway lighting).

- the combination of fog lamps with low-beam.

- the low-beam orientation when the cars are equipped with Bending Light systems.

In the first case, the algorithm will get from the CAN the typology of the active beam and will adjust its parameter according to this.

In the second case, we have checked that the addition of fog lamps to low-beam does not change significantly the shape of the low-beam.

In the last case, field tests have shown the robustness of the system towards orientation of the low-beam.

CONCLUSION

We have seen how Dipped Beam headlights and Fog Lamps can be properly activated through tunnel and fog detection performed by a single common front camera. But many other functions associated with lighting could be enhanced by the use of vision based detection.

Indeed, any system which leads to the automation of driver tasks, function enhancement or the provision of additional but relevant information to the driver will enhance driving safety. Vision Based systems and Front and Rear Lighting have numerous uses in Driving Assistance applications. We have listed some of those applications. Interfaces between these systems and future functionalities, are limited only by the imagination of development engineers.

REFERENCES

[1] J.Rebut, B.Fleury. Vision Based Systems for Lighting Automation. 6th International Symposium on Automotive Lighting, Darmstadt University of Technology.

[2] J.Lelevé, J.Rebut. Fog Lamp Automation with Visibility Sensor, the next step of lighting automation. Elektronik im Kraftfahrzeug – VDI 12.Internationales Kongress, Baden-Baden

2006-01-1158

Moving Obstacle Detection from Moving Platforms

Shima Rayej, Karl Murphy and Alberto Lacaze
Robotic Research

ABSTRACT

Developing robust algorithms for moving obstacle detection is a priority for autonomous ground robotic systems. This paper compares various techniques used for detecting moving objects from static and moving platforms and introduces two novel LADAR-based approaches for solving the problem of moving obstacle detection. Video based techniques including: codebook background subtraction, feature tracking and optical flow, and structure from motion are computationally more expensive than LADAR-based approaches and more susceptible to environmental factors, such as level of brightness in the image, as well as clutter and occlusion in scenes. In addition, LADAR-based techniques outperform video based ones when the objects to be detected are visually indistinguishable from the background, which is the case in many battlefield operations. Furthermore, most video based approaches, except for structure from motion, are unable to recover information about the moving targets or obstacles and make numerous assumptions about the input image to achieve their goals. The two novel LADAR based approaches presented, one image based and the other map based, are computationally inexpensive, provide low false alarm rates, and have the ability to easily recover various characteristics of the moving objects, such as speed and direction of travel as well as height and volume information of the obstacles.

INTRODUCTION

Accurately detecting moving objects from a mobile platform is critical to the success of autonomous robotic systems in dynamic real world environments. Detection can be performed from static or dynamic sensors. While moving obstacle detection from static platforms is most applicable to sentry operations, obstacle detection from moving platforms has a wider breadth of applications, including autonomous navigation, but is inherently more complex to solve. This research is of particular importance for autonomous battlefield operations as part of the Future Combat System which requires navigation in urban centers and motor pools with high levels of civilian presence and traffic. Major challenges involve accurately detecting moving objects at high speeds with

noisy INS solutions and shorter learning times, as well as, in situations where soldiers are indistinguishable from their surroundings, for instance when wearing camouflage gear or lying low in grassy areas. In this paper, we describe state-of-the-art algorithms for static and dynamic moving obstacle detection using video and LADAR sensing mechanisms, as well as the advantages and disadvantages of each algorithm.

MOVING OBSTACLE DETECTION

SENSORS DRIVING THE RESEARCH

Moving obstacle detection can be performed using video, LADAR, or Radar. Although video-based approaches are most common in literature, LADAR-based algorithms are less susceptible to scenarios where the moving objects are visually indistinguishable from the background. But the low cost, weight, and power consumption of cameras make them preferable over LADARs for some situations. Radar is used mainly for vehicle detection. The next sections will detail how these sensors are used for detection on static and moving platforms.

VIDEO BASED APPROACHES USING STATIC SENSORS

Codebook Background Subtraction

Object detection using static sensors is most applicable to stationary surveillance operations. A simple approach for detecting moving objects in an image from a static sensor is to first learn the background scene and subsequently "subtract" the new image from the reference background scene to obtain differences between the two. These differences in turn represent non-stationary objects in the scene.

The Codebook (CB) background subtraction algorithm developed at the University of Maryland is based on such an approach. The algorithm first builds a compressed form of the background model by clustering sample background values at each pixel into codebooks over a long image sequence [1]. For each pixel, it builds a codebook of sampled background values (codewords) clustered based on a color distortion metric and brightness ratio. At detection time, the CB algorithm computes the difference between the current image and background model in terms of color and brightness. If the input pixel's color distortion to some codeword is less than the detection threshold and its brightness lies within the brightness range of that codeword, the pixel is labeled as background; otherwise, it is classified as foreground [1]. Figure 1 shows the CB algorithm at work.

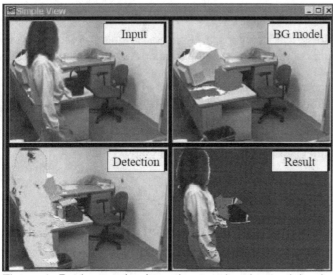

Figure 1: Background subtraction result using codebook method.

The CB algorithm has been shown to perform well for small contrast foregrounds as well as moving backgrounds [2]. This is mainly due to the algorithm's long background modeling/learning time as well as its separation of color and brightness for detection. Although the algorithm works well at detecting changes in the image, it is not designed to characterize these changes. For example, it can not easily recover the 3D direction and velocity of travel of the moving objects or the height and volume characteristics of objects without making many assumptions about the scene. This is due to two main reasons, the immobility of the sensor, as well as, the pixel-level interpretation of images. In lieu of this, CB's versatility (can be used with video or FLIR), low computational complexity, and low cost of implementation make it a suitable choice for stationary surveillance applications where detection takes precedence over characterization.

Video based approaches from mobile platfroms

Three main video-based approaches for detecting moving objects from a mobile platform are 1) stabilized codebook background subtraction, 2) feature tracking and optical flow, and 3) structure from motion. Stabilized codebook based detection using moving cameras is similar to static codebook detection preceded

by a stabilization step to account for the moving image. Optical flow and structure from motion based methods perform object tracking on a feature level as opposed to the pixel level. The optical flow method calculates the motion vector of a sequence of images based on the position of invariant feature points in successive images. Structure form motion based detection searches for moving clusters after using camera parameters and feature tracking to create a 3D model of the scene.

Stabilized codebook background subtraction

The codebook background subtraction algorithm described above for detecting moving objects from a static sensor can be extended to work for dynamic sensors. The key difference is an image stabilization step that can be performed either in software or mechanically prior to background modeling and detection. Although computationally inexpensive, the decreased learning time for building the background results in higher false alarms than the stationary sensor case.

Feature tracking and Optical Flow

Feature tracking algorithms and optical flow extract motion information from a sequence of images. The recovery of motion information is essential in image registration, or establishing correspondence between frames for moving obstacle detection. Traditional optical flow algorithms first proposed by Horn and Schunck [6] calculate the distribution of apparent velocities of the movement of brightness patterns in an image. The movement captured optically is then related to the actual movement of points on objects within view. Information about the flow field may then be used to recover the relative motion of objects to the viewer.

Closely related to the optical flow idea is the idea of tracking salient feature points between images. Feature tracking for moving obstacle detection consists of extracting feature points within an image and subsequently using correspondence algorithms to determine these points' motion vectors from one image to the next.

A critical first step is to choose invariant feature points, often "corners" that may be used to perform reliable matching between different views of an object. The topic of good features to track is discussed at length in the literature by Harris-Stephens, Shi-Tomasi, and recently David Lowe [7] in his proposed SIFT invariant features. Harris corner points are those that have high image gradients in orthogonal directions – i.e. the windows centered around the feature points are distinctive to all other windows in the immediate vicinity of the image. Figure 2 shows the Harris corner detector at work. Shi-Tomasi [8] describe techniques for locating and tracking best features in an image. Finally, David Lowe [7] describes the detection of affine-invariant, rotational-translation invariant, and scale invariant features in an image.

Figure 2: Harris corner detection method results.

Once features are located, the main focus becomes propagating features through time. The KLT algorithm [9] locates good features by examining the minimum eigenvalue of each 2x2 gradient matrix and features are tracked by using a Newton-Raphson method for minimizing the different between two image windows. The KLT tracker also allows for large displacements between features using multi-resolution tracking.

Given feature tracks, object detection from a moving platform becomes simpler due to established image correspondences. The registration of images is possible and the calculation of the warp needed to match two consecutive images is recoverable through optical flow. Subsequent clustering of features yields the motion of objects.

Although suited for structured environments, optical flow based algorithms suffer from two main setbacks that make them suboptimal for real world scenarios. First, the assumption that the grey level of all the pixels of a region remains constant between two consecutive images (brightness constancy assumption) is vital to the success of all gradient-based optical flow. Secondly, optic flow algorithms must further assume that the displacement of feature points is characterized by constant velocity. This additional constraint is due to the fact that the velocity field at each image point has two components while the change in image brightness at a point in the image plane due to motion yields only one component. The need for such assumptions makes optimal flow algorithms suboptimal for outdoor environments with varying illumination and scenes with clutter and occlusion. Approaches that are robust to motion discontinuities and illumination changes have been proposed by Kim, Martinez, and Kak [15]. Furthermore, similar to codebook based moving obstacle detection algorithms, optical flow based ones are unable to recover accurate information about the velocity and characteristics of the moving objects.

Structure from Motion

Structure from Motion is the problem of computing the 3D geometry of a scene given a sequence of 2D images

from a single camera. Single cameras provide 2D representations that are intrinsically ambiguous, since one dimension is lost is the projection from the 3D world onto a 2D imaging sensor. One way to overcome this loss is to use the dimension of time and movement of the camera. Given a series of images of a scene, typically taken by a video camera, it is sometimes possible to recover some of this lost 3-dimensional information.

Once the 2D projection of a point in the real scene has been found, its position in 3D can be assumed somewhere along the ray connecting the camera optical center and the corresponding spot in the image plane. Tracking its projections across multiple images and using triangulation allows the relatively accurate localization of the point in 3D. If extraction and correspondence can be performed for a sufficient number of points and lines and over images acquired from different directions then estimates for both the 3D locations of the features and the camera positions can be deduced. The obtained reconstruction, however, differs from the true structure by a projective transformation. Knowledge of the cameras intrinsic parameters can be used to correct for this distortion. This reconstruction method is called Structure from Motion (SfM) and is based on the process of minimizing the distance between estimated 3D structure projections and actual image measurements. Figure 3 shows the 3D surface mesh of an object created using the structure from motion algorithm.

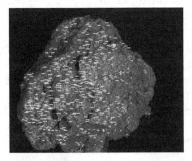

Figure 3: 3D surface mesh creation using structure from motion.

Structure from Motion (SfM) based moving obstacle detection algorithms consist of tracking features in the image domain, building a 3D representation of the world, and finally tracking features in the 3D representation by looking for moving clusters.

The first stage in the algorithm finds locations in the first image that will be good for tracking and then searches for their corresponding location in the second image using a 2D correlation-based feature tracker. The next stage is a SfM estimation that uses feature tracks to solve for the change in position and attitude (e.g. the motion) of the camera between the images and the depth to the selected features in the first image (e.g., the structure). This stage uses a non-linear least squares optimization that minimizes the distance between feature pixels by projecting the features from the first image onto the second image based on the current estimate of the scene structure and camera motion. The final stage of the algorithm uses the motion between images and the coarse structure provided by the depths to the feature tracks to efficiently generate a dense terrain map. Structure from Motion can be applied to imagery with moving objects using a multi-linear factorization approach. Vidal and Hartley have developed an approach to the problem that works for all the spectrum of affine motions: from two-dimensional and partially dependent to four-dimensional and fully independent [16]. The technique uses a combination of PowerFactorization, Generalized Principal Component Analysis (GPCA), and spectral clustering to create a purely geometric solution to the multi-frame motion segmentation problem.

The success of SfM based moving obstacle detection requires that a sufficient number of features can be identified on each moving object. Furthermore, the multi-linear factorization step is dependent on i) the photometry of the target (targets with homogeneous material or with highly reflective material will not generate a sufficient number of stable feature points), ii) the effective aperture subtended by the object (objects occupying a small portion of the visual field are equivalent to the camera having a small aperture angle), and iii) the motion undergone by each object (multi-body factorization requires each motion to be "sufficiently different" and "sufficiently rich").

Unlike codebook based and optical flow based algorithms, SfM based algorithms can easily recover the 3D direction and velocity of travel, as well as, recover height and volume characteristics of objects without making many assumptions about the scene but suffers from high computational complexity and false alarm rates.

MOVING OBSTACLE DETECTION USING LADARS

Although most of the literature focuses on detection using cameras, LADARs are more suitable for scenarios where the moving objects are visually indistinguishable from the background. In this section we will discuss current image and map-based approaches for moving obstacle detection using LADARs.

Single-line LADAR approach

CMU has developed a LADAR-based (Sick) system for detecting moving objects in city traffic by performing simultaneous localization and mapping (SLAM) and detection and tracking of moving objects (DATMO) at once [11]. The pose estimate and surrounding map generated by SLAM is used by DATMO to detect moving objects, and SLAM in turn uses the object location prediction from DATMO to filter out moving objects. Detection of moving obstacles is achieved by first converting scan data into points in the fixed world coordinate system and segmenting contiguous clusters of points into objects. Using the surrounding map and pose estimate from SLAM, moving objects are detected by finding inconsistencies between the new scan and the map. An Interacting Multiple Model (IMM) estimation algorithm tracks and predicts the motion of the moving objects using a Kalman filter to estimate object motion from the change in feature position and orientation. This step assumes that the objects are moving at constant velocity and acceleration. Multiple Hypothesis Tracking (MHT) is then applied to filter out wrong detections that may have resulted due to the roll/pitch of the vehicle and refine data association.

This system is effective in urban scenarios where the roll and pitch of the ground surface are negligible. However, when the vehicle, and thus the sensor, is subjected to roll and pitch, static objects will appear to be moving from side to side causing false detections. The constant velocity assumption also makes this algorithm impractical for some outdoor operations.

Full Frame LADAR Approach

This section presents two full frame LADAR based solutions for moving obstacle detection, one image based and the other map based.

a)LADAR Image Based

In this section we introduce a feature based tracking algorithm that models cylindrical shaped silhouettes from objects detected in a LADAR image, updates a track list based on detected features, and categorizes features based on geometric metrics. A track list of features is maintained which contains location, velocity, height, and the covariance of each feature. When a feature is detected in the LADAR image, it is checked against the track list. If the new feature does not match any in the track list, a new feature is added to the list. Tracked features are labeled as people if they are of appropriate width and height and if they move more than a meter. The minimum velocity threshold is ensures that static objects are not mistakenly classified as moving objects even though they appear to move slightly over time due to LADAR noise and errors in the vehicle's navigation solution.

Figure 4 shows a LADAR image from an ARL / GDRS XUV. The XUV was manually driven through woods while three people walked among the trees. Figure 5a show the initial features identified. Figure 5b show the features being tracked after the vehicle has moved about 20 m.

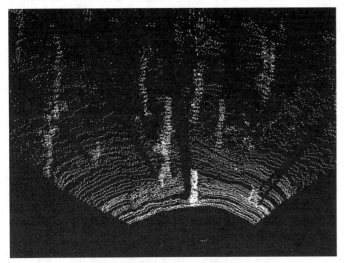

Figure 4: A LADAR image from a forested area.

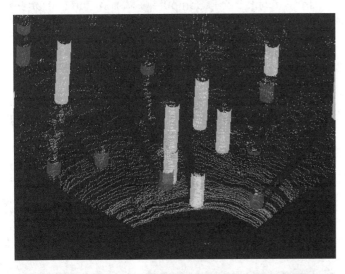

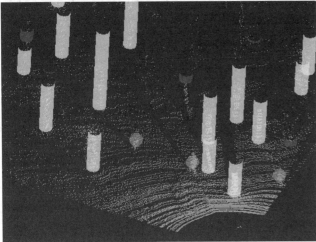

Figure 5a (top): Initial features identified
Figure 5b (bottom): Tracked features after vehicle moved 20 meters.

b) Map Based Approach

Most ground autonomous systems use a mobility map for obstacle avoidance and path planning. The mobility map represents traversability characteristics of the terrain that are used by the behavior generation system to evaluate the cost of future behaviors. In general, this map contains elevation, obstacles, ditches, roads collected by the onboard sensor as well as a-priori information. The map provides a means to probabilistically filter noisy measurements, and collect historical information on terrain outside of the range or field of view of the sensors. Since the vehicle is in motion, the inertial navigation unit (INS) is used to shift the map. Therefore, the accuracy of the map is largely dependent on the accuracy of the INS as well as the sensors being used.

In the mobility map, the detection of static features such as obstacles will match previous detections performed by the vehicle. In other words, the map is shifted using the INS solution so that these features do not "smear". However, moving obstacles present a different problem. As the features are moving throughout the map, they leave "trails" of past detections. If not erased, these trails of past detections present a strong insight into the movement of the obstacles. The fact that clear map cells change to a now detected feature is a strong indication that something has moved. The reverse is also true, if a particular feature that is consistently detected in a particular map cell is no longer detected, it provides a strong indication that something has moved.

As a person moves through the map, their leading edge will show cells changing from being clear to becoming an obstacle, and the trailing edge will show cells changing from having an obstacle to being clear. Detecting and tracking these map classification changes become surprisingly easy as the accuracy of the INS solution, and therefore of the map, increases. Figure 6 shows an example of using this methodology in a forested environment. In this case, the robotic vehicle using a GDRS Gen III LADAR traverses the environment as a person walks in front of the vehicle.

Figure 6: Map based moving obstacle detection.

CONCLUSION

The choice of whether to use a video or LADAR based technique for moving obstacle detection is heavily dependent on its application. Video based approaches are preferable when trying to minimize cost and weight, whereas LADAR-based methods are less computationally expensive and provide more accurate characterization of the moving objects. The two LADAR based techniques described in this paper are computationally inexpensive, provide low false alarm rates, and have the ability to easily recover characteristics of the moving objects. Such algorithms can be fused with detection algorithms using other means of sensing, such as FLIR, for more robust detection.

ACKNOWLEDGMENTS

This research is partially funded by FCS ANS contract DAAE07-03-9-F001.

REFERENCES

1. K. Kim et al., Background Modeling and Subtraction by Codebook Construction, 2004.
2. T. Chalidabhongse et al., A Perturbation Method for Evaluating Background, Subtraction Algorithms, 2003.
3. C.S. Kenney, M. Zuliani and B.S. Manjunath, An Axiomatic Approach to Corner Detection, UCSB Vision Research Lab, 2005.
4. H. Eltoukhy and K. Salama, Multiple Camera Tracking, Stanford University, 2002.
5. A. Bruhn, J. Weickert and C. Schnorr, Lucas/Kanade Meets Horn/Schunck: Combining Local and Global Optic Flow Methods, 2004.
6. A. Horn and B. Schunck, Determining Optical Flow, Artificial Intelligence Laboratory, MIT, 1981.
7. Lowe, Distinctive Image Features from Scale-Invariant Keypoints, University of British Columbia, January, 2004.
8. J. Shi and C. Tomasi, Good Features to Track, IEEE Conference on Computer Vision and Pattern Recognition, pages 593-600, 1994.
9. A. Tomasi and T. Kanade, Detection and Tracking of Point Features, Carnegie Mellon University Technical Report CMU-CS-91-132, April 1991.
10. C. Wang, C. Thorpe and A. Suppe, Ladar-Based Detection and Tracking of Moving Objects from a Ground Vehicle at High Speeds, Carnegie Mellon University Robotics Institute, 2003.
11. A. Steinfeld, et al., Development of the Side Component of the Transit Integrated Collision Warning System, IEEE Intelligent Transportation Systems Conference, October, 2004.
12. Johnson and L. Matthies, Precise image-based motion estimation for autonomous small body exploration, Proc. 5th Int'l Symp. On Artificial Intelligence, Robotics, and Automation in Space, June, 1999.
13. S. Soatto and R. Brockett, Optimal and suboptimal Structure from Motion, In Proceedings of the IEEE Intl. Conf. on Computer Vision and Pattern Recognition (CVPR), 1998.
14. H. Jin, P. Favaro and S. Soatto, Real-time Feature Tracking and Outlier Rejection with Changes in Illumination, Int'l Conf. on Computer Vision, pages 684-689, July 2001.
15. Y. Kim, A. Martinez and A. Kak, A Local Approach for Robust Optical Flow Estimation under Varying Illumination, 2004.
16. R. Hartley and R. Vidal, The multibody trifocal tensor: Motion segmentation from 3 perspective views, In IEEE Conf. on Computer Vision and Pattern Recognition, 2004.

CONTACT

The main author can be contacted by email at rayej@roboticresearch.com

Target Tracking by a Single Camera Based on Range-Window Algorithm and Pattern Matching

Shunji Miyahara
Visteon Japan, Ltd.

Jerry Sielagoski, Anatoli Koulinitch and Faroog Ibrahim
Visteon Corporation

ABSTRACT

An algorithm, which determines the range of a preceding vehicle by a single image, had been proposed. It uses a "Range-Window Algorithm". Here in order to realize higher robustness and stability, the pattern matching is incorporated into the algorithm. A single camera system using this algorithm has an advantage over the high cost of stereo cameras, millimeter wave radar and non-robust mechanical scanning in some laser radars. And it also provides lateral position of the vehicle. The algorithm uses several portions of a captured image, namely windows. Each window is corresponding to a pre-determined range and has the fixed physical width and height. In each window, the size and position of objects in the image are estimated through the ratio between the widths of the objects and the window, and a score is given to each object. The object having the highest score is determined as the best object. The range of the window corresponding to the best object becomes an estimated range. The pattern matching helps this algorithm when the camera image is influenced by a shadow. Since this matching adopts a warped template, it can estimate the range. This algorithm was applied to more than *4,500* real road images. It showed the range accuracy of about +/- *1* [m] and *94*% detection rate for a motorcycle, sedan, minivan, truck and bus on rural, urban and city roads. And the incorporation of the pattern matching has improved the detection rate up to *97*%. The present maximum range is *50* [m]. This algorithm is effective for the short range application like "Low speed follower".

INTRODUCTION

Vehicles with an Adaptive Cruise Control (ACC) system based on laser radar have been introduced into the Japanese market since the middle of the 1990s. At the end of the 1990s, vehicles with systems based on millimeter wave appeared in European and Japanese markets [1]. In 2004, vehicles with "Low Speed Follower (LSF)" have appeared in the Japanese market. This LSF system keeps a safe distance at low speed condition and stops the host vehicle automatically when the preceding vehicle stops. LSF may contribute to safety more than ACC and is more useful and convenient than ACC in traffic congestion. Radar is effective for range measurement, but has difficulty detecting the lateral edge of a vehicle. In the case of an offset collision, radar based LSF might not work properly because of the lateral position accuracy and limited azimuth angle information. Therefore, the author proposed a low-cost vision system capable of detecting range as well as lateral position of the preceding vehicle. The proposed system consists of only a single camera and calculates the range from a single frame of the camera image. It is based on the Range-Window Algorithm (RWA) [2]. The range-window (RW) is a virtual window in the camera image and has a fixed physical size. The RW looks different, depending on its range. If there is an object, namely a preceding vehicle, in the camera image, the ratio between the vehicle and the RW changes according to the range. The vehicle having the most adequate ratio is the most probable and the range of the RW corresponding to the vehicle becomes the estimated range. But the RWA sometimes fails to detect a vehicle when a shadow covers the vehicle. Therefore a pattern matching technique is incorporated into the RWA. Since this matching is based on the trinary image, the calculation time can be reduced. When the warped template is adopted, the range estimation can be made. The combination of the RWA and the pattern matching is expected to work complementarily and result in more robust detection.

So far there are several ways to measure range and lateral position of a preceding vehicle using a vision system. They are stereo camera [3], a motion stereo [4] and a fusion system [5-7]. Unfortunately these approaches increase the cost and/or need some help from a radar based sensor. Recently a unique approach based on a single camera was proposed [8, 9]. But it is based on the point of contact between the vehicle and the road. In a real situation, it seems difficult to identify the point regularly because of potential shadows.

Therefore the author proposed a new algorithm based on the RWA [2].

Here the RWA, the proposed pattern matching, the measuring system, the effectiveness of the matching through real data and the application of the tracking method using the RWA and pattern matching are described.

RANGE WINDOW ALGORITHM

The principle, edge-enhancement and procedure of the Range-Window algorithm (RWA) [2] are briefly described.

PRINCIPLE OF RANGE-WINDOW ALGORITHM

A basic assumption is that the host and preceding vehicles are on the same plane. Namely the road is assumed flat. The algorithm is based on the virtual/multiple windows at certain/different ranges. As seen in Fig.1, the windows are standing vertically on the ground at different ranges. The physical sizes of the windows are the same but they look different since the range is different.

Here the window of *4 × 2* [m] is chosen. The sizes of windows in the image of a camera, however, are different since their range is different as seen in Fig.2.

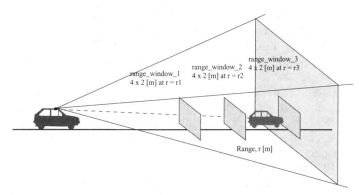

Figure 1: Principle of the Range-Window Algorithm (RWA). The windows at three ranges, r1, r2 and r3, are shown.

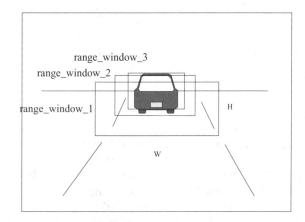

Figure 2: Appearance of the windows in a camera image. Size of the windows = W × H = 4 × 2 [m].

The object width is calculated through image processing for each window. In Fig.3 (a), the width of the object is about *30%* of the width of the window and then the estimated object width at *r1* is *1.2 [m]= 4 × 0.3*. For Fig.3 (b), the width of the object is about *45%* of the width of the window and the object width at *r2* is *1.8 [m]= 4 × 0.45*. For Fig.3 (c), the width of the object is about *80%* of the width of the window and the object width at *r3* is *3.2 [m]= 4 × 0.8*.

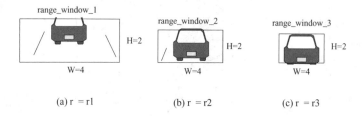

(a) r = r1   (b) r = r2   (c) r = r3

Figure 3: Object in each window.

Among the estimated object widths, *1.8* [m] for *r = r2* is the nearest to *1.7* [m], the assumed vehicle width. Then the algorithm judges that the most probable window is *(b)* in Fig.3 and then the estimated range is the range of the window, *r2*.

In making a decision, other criteria can be added. For instance, the extracted object should have the height of *1.5* or *2* [m], or be on the bottom line or ground level since a vehicle cannot fly. The larger the height is, the more probable the object should be.

EDGE ENHANCEMENT IN RWA

The original camera image (frame) is assumed monochromatic. The pre-processing makes the trinary image by edge-enhancement and binary image by a threshold. In the edge-enhancement, the vertical one is adopted.

Vertical edge-enhancement is applied to an original image and the cells in the image are classified into three groups as shown in Eq.(1). The trinary image is shown in Fig.4. Vertical Sobel filter is usually used.

$$i_{edge}(x, y) = \begin{cases} 1..........\partial i(x, y)/\partial x > d_{threshold} \\ -1.......\partial i(x, y)/\partial x < -d_{threshold} \\ 0..........otherwise \end{cases} \quad (1)$$

where $i(x,y)$, $i_{edge}(x,y)$, x, y and $d_{threshold}$ are the intensity, trinary image, lateral cell number, vertical cell number and the threshold, respectively. The threshold depends on the average or distribution of the intensity in a window.

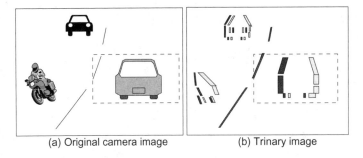

| (a) Original camera image | (b) Trinary image |

Figure 4: Trinary image based on the vertical edge-enhancement. The rectangle with dot line is the Range-Window. The trinary image in the window is used for processing in the RWA

PROCEDURE OF RANGE-WINDOW ALGORITHM

The procedure of the RWA is comprised of pre-processing including an edge-enhancement, a single window process and the estimation of the range as shown in Fig.5. The process in the single window is comprised of line-segmentation, object-segmentation, scoring and best object-in-window determination.

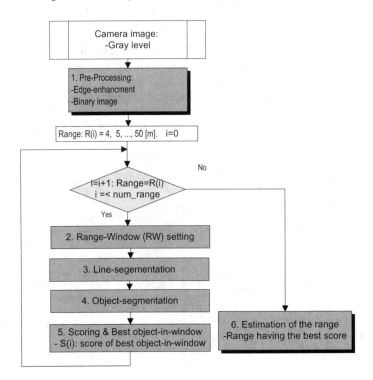

Figure 5: Flow chart of the range estimation based on the Range-Window algorithm (RWA). $R(i)$, $I=1, 2, .., N$: pre-set range of the Range-Window. $S(i)$, $I=1, 2, ..., N$: the score of best object-in-window.

Process in a single window

Line-segmentation is analyzed. The feature points, cells having +1 or -1 in the trinary image, are connected and grouped. If the cells having the same polarity are adjacent vertically, they are connected and categorized into the same line-segment. The polarity can avoid inadequate connections.

Next the object-segmentation is analyzed. An object is made from a pair of the line-segments having different polarity.

Scoring on the objects is analyzed. The scoring is based on a "vehicle model" having the following criteria.

- Width of an object
- Height and vertical position of an object
- Length and aspect ratio of line-segments
- Uniformity in the object area calculated from the binary image

The object with the highest point count is chosen as the "best object-in-window".

The score depends on the vehicle models, corresponding to a motorcycle, sedan, minivan and truck/bus. The selection of the vehicle models is based on the width and the height of the object.

Estimation of the range

After getting the best object-in-windows and its score for each RW, the object-in-window having the best score is picked as a best object and its window is identified as the most adequate window. The estimated range is that of the window. The range can be modified by the ratio between the calculated width of the object and the pre-set width. This can mitigate the error coming from the discrete ranges. Here the pitch of the ranges is selected so that the ratio of adjacent ranges might be *1.2*.

PATTERN MATCHING

In order to improve the robustness and detection rate of the RWA, a pattern matching method is added to the RWA. Here the matching method is described.

EFFECTIVENESS OF PATTERN MATCHING

In a vision based system, the shadow of a large vehicle, building, infrastructure or intensity similarity between the object and background sometimes blurs the detection of the vehicle. Usually the detection of a vision system like RWA depends on edge information at both sides of the object and tends to be sensitive to shadows or similarity. On the contrary, pattern matching is based on information in 2-dimensional space, and provides better detection under adverse condition: such as shadow and similarity. The proposed pattern matching is based on the template of the object obtained by the RWA and trinary image.

An example is shown in Fig.6. After applying the edge enhancement to an original image, the RWA extracts the object area as shown in Fig.6 (b). This object area of the trinary image becomes a template. If the vehicle enters a shadow, the left-edge of the object cannot be detected and RWA cannot identify the object any more since the RWA is based on the contour of the edges of an object. But since the pattern matching is based on all edges in the object area, the object can be identified by   partial edges. For instance, when the object comes under a shadow as shown in Fig.6 (a'), the RWA cannot work but pattern matching can identify the object. The template is compared with the matching area of an input image by shifting the area as shown in Fig.6 (b').

PATTERN MATCHING BY CROSS CORRELATION

Matching is based on the cross correlation (CC) between the template (the object area in Fig.6 (b)) and the input image (Fig.6 (b')) as shown in Eq.(2).

$$C(x,y) = \sum_m \sum_n F(m,n)I(m-x,n-y) \qquad (2)$$

where C, F and I are the cross correlation function, template and input patterns, respectively. $F \equiv \{ F(m,n); 0 \le m < M, 0 \le n < N \}$. $I \equiv \{ I(x,y); 0 \le x < X, 0 \le y < Y \}$.

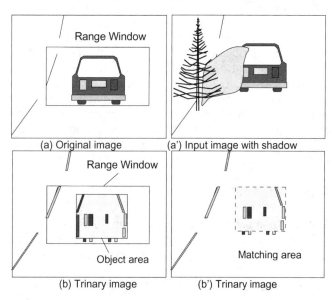

(a) Original image   (a') Input image with shadow

(b) Trinary image   (b') Trinary image

Figure 6: Pattern matching. The object area in (b) will be a template. The trinary image (b') of a new image (a') is compared with the template by shifting the template in (b'). The left part in the matching area in (b') is occluded by the shadow..

To avoid variation of edge-cells, the normalized cross correlation (NCC) is used instead of CC. The NCC, $C_n(x,y)$, is shown in Eqs.(3), (4) and (5).

$$P_F = \sum_m \sum_n F(m,n)F(m,n) \qquad (3)$$

$$P_I(x,y) = \sum_m \sum_n I(m-x,n-y)I(m-x,n-y) \qquad (4)$$

$$C_n(x,y) = \frac{C(x,y)}{\sqrt{P_F P_I(x,y)}} \qquad (5)$$

WARPED TEMPLATE FOR RANGE ESTIMATION

Matching can identify the existence of an object but cannot determine the range. Therefore a warped template is introduced. The NCC cannot express the size change of the object. The size change ratio can provide a new range. Therefore the warped template is introduced. The template is

$$\overline{F} : k \text{ columns are added/removed}$$
$$\text{from the center from } F(m,n) \qquad (6)$$

$$\overline{F} \equiv \{ \overline{F}(m,n;k); 0 \le m < M + k = \overline{M}, 0 \le n < N \} \qquad (7)$$

$\overline{F}$ and k are the horizontally warped template as shown in Fig.7 and add/delete number, respectively. $k=-1$, 0 and 1 mean to delete one column from the center of $F(m,n)$, no change and to add one zero column into the center of F, respectively. The NCC between the warped template and the input pattern, $C_n(x,y;k)$, is expressed by Eqs.(8) through (11).

$$P_{\overline{F}} = \sum_{0 \le m < \overline{M}} \sum_n \overline{F}(m,n;k)\overline{F}(m,n;k) \qquad (8)$$

$$\overline{P}_I(x,y) = \sum_{0 \le m < \overline{M}} \sum_n I(m-x,n-y)I(m-x,n-y) \qquad (9)$$

$$C(x,y;k) = \sum_{0 \le m < \overline{M}} \sum_n \overline{F}(m,n;k)I(m-x,n-y) \qquad (10)$$

$$C_n(x,y;k) = \frac{C(x,y;k)}{\sqrt{P_{\overline{F}} \overline{P}_I(x,y)}} \qquad (11)$$

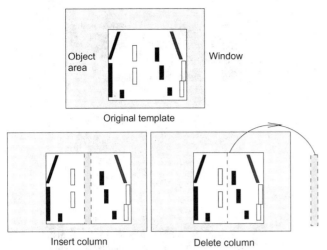

Original template

Insert column

Delete column

Figure 7: Horizontally warped template.

The new range is calculated as shown in Eq.(12). It is expressed by the product between the range of the template and the ratio, Eq.(13).

$$Range\_new = Range\_template * Ratio \qquad (12)$$

$$Ratio = \frac{N_c}{N_c + k} \qquad (13)$$

where N_c is the number of columns in the original template. The lateral position is also estimated from the displacement of the center of the warped template.

MEASURING SYSTEM

A charge-coupled device (CCD) camera and a radar are installed near the rear view mirror and the grill as shown in Fig.8, respectively. The CCD image, radar information (range, relative velocity and azimuth angle) and vehicle information (velocity and yaw rate of the host vehicle) are sampled every *0.1* [sec]. The output of the camera and radar, the velocity and yaw-rate are synchronously measured. The radar and vehicle information is not used in this RWA, but for comparison.

The CCD is monochromatic and has *320 x 240* cells. Azimuth and elevation angle widths of the camera are *48* and *35* [deg], respectively.

The measured data is used to evaluate the RWA and pattern matching. The evaluation is made off-line.

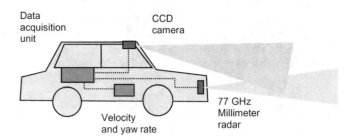

Figure 8: Measuring system by a CCD camera. Range obtained from the radar is used as a reference. Azimuth and elevation angle widths of the camera are 48 and 35 [deg], respectively.

EFFECTIVENESS OF PATTERN MATCHING

The effectiveness of the pattern matching is evaluated through real data.

PROBLEM OF RWA

The preceding vehicle is approaching the bridge as shown in Fig.9. Until *t = 31.0* [sec], the RWA works well. But it could not detect the vehicle at *t = 32.0* [sec]. Because the shadow of the bridge blurs the edge of the preceding vehicle at *t = 32.0* [sec].

TEMPLATE

The template creation is explained. The RWA determines the object area in the trinary image as shown in Figs.10 (a) and (b). The object area shown in Fig.10 (c) becomes a template.

Figure 9: Preceding vehicle is approaching the bridge. The RWA can recognize the vehicle until *t = 31.0* [sec]. But it fails in recognizing the vehicle at *t = 32.0* [sec]. The velocity between *t = 31.0* and *32.0* is *20~23* [km/h].

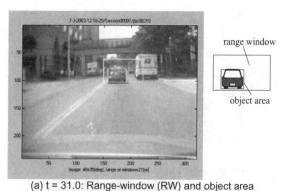

(a) t = 31.0: Range-window (RW) and object area

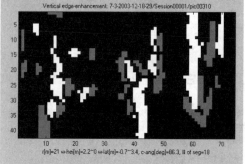

(b) Trinary image in RW

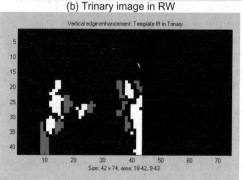

(c) Template: corresponding to the object area, subset of (b)

Figure 10: Template creation. Template is the object area of the trinary image at *t = 31.0* [sec].

APPLICATION OF PATTERN MATCHING

Object identification

Matching is applied to the image at *t = 32.0* [sec] shown in Fig.11. The 2-dimensional NCC function is shown in Fig.11 (c). Smooth function is obtained. The sharp peak is obtained for the horizontal shift as shown in Fig.11 (d), but not in the vertical shift as shown in Fig.11 (e). The peak NCC function is at the cell shift, *(x0, y0) = (1, 0)* and is about *0.45*. If the maximum NCC is larger than a threshold value, this is judged as an object. Its position is *(x0, y0)*. Usually the absolute position is evaluated by comparison with the absolute cell position of the original image shown in Fig.10 (b) since the RWA has already given the lateral position of a vehicle. The original image means one, from which the template is extracted.

(a) t =32.0 : RW and input image

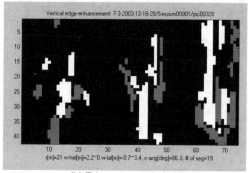

(b) Trinary image in RW

The calculation of the NCC should be confined to a position near to that of the template to avoid unnecessary calculation. Namely the shift of x-axis is small and that of y-axis should be *-1, 0* or *1* because of the continuity of the object position and the low dependence on y-axis, respectively.

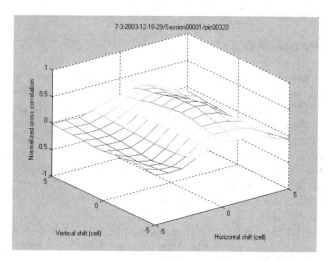

(c) Normalized cross correlation (NCC) function: x-axis and y-axis correspond to x-cell and y-cell shift, respectively.

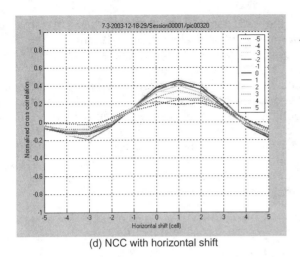

(d) NCC with horizontal shift

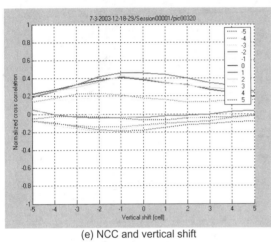

(e) NCC and vertical shift

Figure 11: Template matching. Matching is applied to he trinary image at $t = 32.0$ [sec]. The template is shown in Fig.10 (c).

Range estimation

A warped template is used for range estimation. The result is shown in Fig.12 (a). The time of the template and input is 31.0 and 32.0 [s], respectively. The NCC is the largest when the add/delete number k (warping number) is -2 and the x-shift is 2. Namely the object is farther at the input time than at template time. The distance is calculated by Eqs.(12) and (13). The non-warped ($k=0$) and warped templates ($k=2$) are shown in Fig.12 (b) and (c), respectively. Figure 12 (a) means that the warped template has larger correlation with the input image Fig.11 (b) than the non-warped one. The comparison between the RWA and the pattern matching is shown in Fig. 12 (d). The RWA cannot estimate the range at $t \geq 31.8$ [s] because of the shadow of the bridge as seen in Fig.11 (a). But the pattern matching keeps a reasonable range estimation. The pattern matching can estimate the range much more stable than the RWA even though the vehicle is partially covered by a shadow.

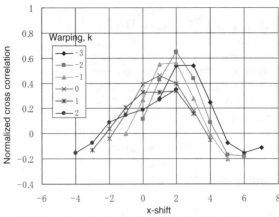

(a) NCC function with warped template, X-shift: shift of x-cells.

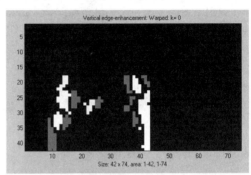

(b) Non-warped template. Warping number $k = 0$

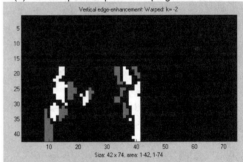

c) Warped template: Warping number $k = -2$.
The template gets smaller by 2 cell horizontally.

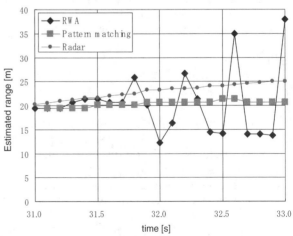

(d) Range estimation by RWA and pattern matching.
Time of the template is fixed at t = 31.0 [s].

Figure 12: Range estimation by the cross correlation with the warped template. In (d), the RWA cannot estimate the range at $t \geq 32.0$ [s] because of the shadow of the bridge as seen in Fig.11 (a). But pattern matching keeps the range estimation. The scenario is shown in Fig.9. The velocity between $t = 31.0$ and 32.0 is $20 \sim 23$ [km/h].

APPLICATION TO REAL DATA

The RWA and the tracking method, in which the RWA and pattern matching are combined, are evaluated through real data. The pattern matching cannot work alone since it needs a template. Therefore the combined method, namely the tracking method, is evaluated. A vehicle with a camera and a 77 GHz radar is run on the roads in Detroit in US. After driving, the range estimation is made by the RWA and the tracking method off-line and compared with the radar range.

TRACKING METHOD

Both the RWA and pattern matching are used in this tracking method. The template is obtained from the RWA and updated from the RWA. The range is calculated by taking by linear regression or average of the ranges of the pattern matching and RWA for the latest times after excluding the range values having large deviation. The time constant of the average is 1 [sec].

CONDITION OF RANGE-WINDOW ALGORITHM

The ranges of the RWA are 4.0, 4.2, 5.0, 6.2, 7.4, 8.8, 10.4, 12.4, 14.8, 17.6, 21.0, 24.9, 29.7, 35.3, 42.0 and 50.0 [m]. The four vehicle models are used. The size of the windows is 4 × 2.2 [m].

MEASUREMENT AND RANGE ESTIMATION

Measurements were made on the highway, urban, local and city roads. The targets are a motorcycle, sedan, minivan, truck and bus. After the measurements, the algorithm is applied to each single frame of camera images off-line. It is applied to a variety of vehicles including 30 scenarios having a vehicle (scenario-A) and 14 scenarios having no vehicle (scenario-B). The scenario-A includes 3 motorcycles, 14 sedans & others, 6 minivan, and 7 trucks & buses. And then the range estimation is applied to vehicle following tests.

RESULTS

RWA for variety of vehicles

The RWA is applied to scenario-A and scenario-B. The results are shown in Fig.13. The discrimination between the vehicle existence and non-existence was correctly made. The non-existence corresponds to "zero distance" in Fig.13. The range estimation shows the reasonable performance except for vehicles beyond the range of 50 [m].

For the 25 targets at shorter than 50 [m], the performance of this algorithm is described. The linear regression is applied to the range estimation as shown in Fig.13. The standard deviation was +/-1.2 [m] and the offset was 2.0 [m]. This offset expresses the distance between the position of the CCD camera and the radar adequately. The slope was lower than unity by 0.14. The vehicle type identification was 80% probability. The minivan was sometimes identified as a sedan or truck. The lateral position was estimated precisely as shown in Fig.14 (a). The lateral positions in the images are exactly estimated as well as the range.

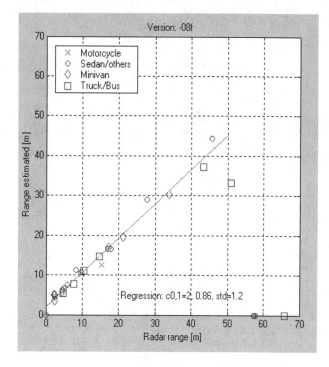

Figure 13: Comparison between the estimated and radar ranges. It is applied to 30 scenarios (scenario-A) having a motorcycle, sedan, minivan, truck and bus and 14 scenarios (scenario-B) having no vehicle. In scenario-B, the radar range is assumed zero. The vehicles in scenario-A are a motorcycle, sedan, minivan, bus and truck. The ranges of the RWA are 4.0, 4.2, 5.0, 6.2, 7.4, 8.8, 10.4, 12.4, 14.8, 17.6, 21.0, 24.9, 29.7, 35.3, 42.0 and 50.0 [m]. The solid line is the linear regression line (Estimated range [m]= 2.0 + 0.86 * Radar range [m]).

For the 14 images having no vehicle in scenario-B, this algorithm recognizes the non-existence with 100% as shown in Fig. 13. In the figure the zero range corresponds to non-existence. Some pictures in scenario-B are shown in Fig.14 (b).

(a) Scenario –A, a vehicle exists

(b) Scenario –B, no vehicle exists

Figure 14: Scenario-A and -B: Lateral position estimation is made exactly in Scenario-A. Small and large quadrilaterals are the estimated object area and the Range-Window, respectively. The non-vehicle existence is correctly identified in Scenario-B. This corresponds to the origin in Fig.13.

Tracking method (TM) for vehicle following test

The RWA and tracking method are compared through the application to vehicles running at low and middle speed in rural area, urban area and a city road. The vehicles are a motorcycle, sedan, minivan and bus. The RWA and TM are applied to a motorcycle cutting in and leaving for the adjacent lane. The result is shown in Fig.15. Although there were some estimation errors in Fig15 (a), the correlation between the estimated and radar ranges looks satisfactory. Especially the TM is much more stable than the RWA though some delay is observed. The standard deviation of the linear regression for the RWA estimation was about *0.94* [m]. The detection rates of the RWA and TM, the rate providing a reasonable accuracy, were *99%* and *100%*,

respectively. The camera images are shown in Fig.15 (b). In Fig.15 (a), the RWA and TM detected the motorcycle earlier than the radar at *t = 3* and *26* [sec] because the RWA and TM have a larger azimuth angle at short distances than the radar.

The RWA and TM are applied to a sedan in an urban area. The result is shown in Fig.16. The correlation between the estimated and radar ranges looks satisfactory as seen in Fig.16 (a). Especially the TM is much more stable and accurate than the RWA. The standard deviation of the linear regression for the RWA estimation was about *0.60* [m]. The detection rates of the RWA and TM, the rate providing a reasonable accuracy, were *97%* and *100%*, respectively. The camera images are shown in Fig.16 (b).

The RWA and TM are applied to a minivan in a city road. The result is shown in Fig.17 (a). Especially the TM is much more stable and accurate than the RWA. The standard deviation of the linear regression for the RWA estimation was about *0.99* [m] and the TM is more stable than the RWA. The detection rates of the RWA and TM, the rate providing a reasonable accuracy, were *93%* and *99%*, respectively. The camera images are shown in Fig.17 (b).

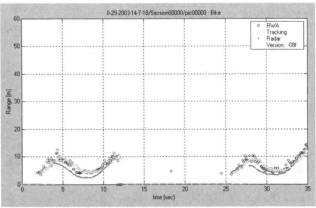

(a) Vehicle following test: motorcycle

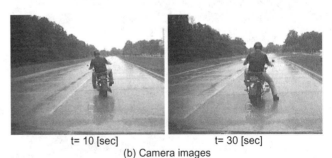

| t= 10 [sec] | t= 30 [sec] |

(b) Camera images

Figure 15: Application test to follow a low speed vehicle: motorcycle cutting in and out. The sampling of the images is every *0.1* [sec].

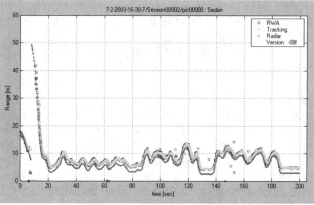

(a) Vehicle following test: sedan

t= 60 [sec]   t= 120 [sec]
(b) Camera images

Figure 16: Application test to follow a low speed vehicle: sedan. The sampling of the images is every 0.1 [sec].

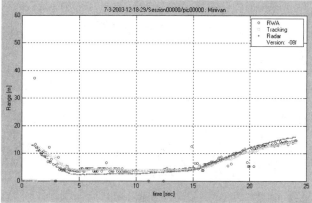

(a) Vehicle following test: minivan

t= 1.0 [sec]   t= 12 [sec]
(b) Camera images

Figure 17: Application test to follow a low speed vehicle: minivan. The sampling of the images is every 0.1 [sec].

The RWA and TM are applied to a bus in a city road. The result is shown in Fig.18 (a). Especially the TM is much more stable and accurate than the RWA. After $t = 24$ [sec], the bus is turning right and there is no preceding vehicle. Before turning, the standard deviation of the linear regression for the RWA estimation was about 0.90 [m] and the TM is more stable than the RWA.. The detection rates of the RWA and TM, providing a reasonable accuracy, were about 100%. The camera images are shown in Fig.18 (b). During turning, the RWA and TM could not work properly because there were many vertical lines on the corner of the building as seen at $t = 30$ [sec] in Fig.18 (b).

This RWA and TM were also applied to other motorcycles, sedans, minivans, trucks and buses running at low speed in highway, urban and rural roads although their results are not shown here. Overall accuracy and detection rate of about 4500 images were 1.04 [m] and 94%, respectively. The TM improved the detection rate and realized 97%.

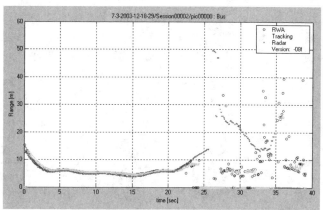

(a) Vehicle following test: Bus. The bus starts turning the right at the intersection at t= 24 [sec].

t= 0 [sec]   t= 30 [sec]: turning to the right
(b) Camera images

Figure 18: Application test to follow a low speed vehicle: Bus. The sampling is every 0.1 [sec].

The definition of the detection rate is: for radar range of less than 50 [m], it is judged as "detected" if the range difference between the RWA and radar is less than 2.5 [m]. The definition of accuracy is the standard deviation of the linear regression between the RWA and the radar for all the data in which the difference is less than 7 [m].

DISCUSSION

CALCULATION TIME OF PATTERN MATCHING

One of the problems in pattern matching is calculation time. Usually the cross correlation (CC) is calculated. 2-dimentional of the CC takes a lot of time if the gray level calculation and the search in all the area are made. Here since the trinary values, which are obtained from the RWA, are used, the time can be reduced and the size of the memory for storing the template is reduced.

Since the size of the pattern is small and the search on x-axis and y-axis in the cross correlation can be limited to a few numbers, the time is remarkably reduced.

TRINARY IMAGE OBTAINED FROM VERTICAL EDGE-ENHANCEMENT

The trinary image is calculated from the vertical edge-enhancement as shown in Eq.(1). This results in more sensitivity to the horizontal shift in the CC function than to the vertical shift as shown in Fig.11. Namely the number of the vertical shifts in the CC calculation can be reduced remarkably. This results in a short calculation time.

TEMPLATE SELECTION AND RANGE ESTIMATION

The most important thing in the pattern matching is the selection of the template. The template is obtained from the object areas of the trinary image in previous times as seen in Fig.6 (b). It is important to select a reliable area. Fortunately the RWA gives the area a score. The selection is made on the basis of:

- Higher score

- Later time

This selection provides a stable template.

The range is estimated by this matching between the warped template (Fig.12 (c)) and input image (Fig.11 (b)). The discrepancy between the ranges by the radar and pattern matching is getting larger as time proceeds if a fixed template is used. As an example, the normalized cross correlation (NCC) is shown in Fig.19 as well as the range. The discrepancy becomes larger as the correlation goes down. The range estimation should be limited by some value of the correlation. The value of $0.5 \sim 0.6$ gives a good criteria in the application here. This degradation of the NCC can be avoided by the update of a template.

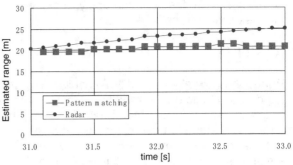

(a) Range estimation by pattern matching.
Time of the template is fixed at $t = 31.0$ [s].

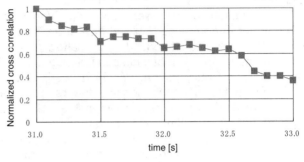

(b) Normalized cross correlation

Figure 19: Range estimation by pattern matching and Normalized Cross Correlation (NCC).

TRACKING BASED ON RWA AND PATTERN MATCHING

Unfortunately pattern matching cannot work by itself since it needs a template. In automotive applications, it is almost impossible to keep all the templates of vehicles. The template depends on the shape and color of a vehicle, the direction of the sun and the weather. But pattern matching is an effective way to identify an object with only partial information. As seen in Fig.12, the vehicle under a bridge can be detected even though the RWA cannot. On the contrary, the RWA can work by itself. It is a smart way to use the RWA and pattern matching complementarily.

The tracking method here is a simple combination of the RWA and pattern matching. The range is calculated by taking the linear regression of the ranges of the RWA and pattern matching for the latest times after excluding the range values having large deviation. The time constant of the regression is 1 [sec]. More sophisticated combination might provide a better estimation.

APPLICABLE ROAD CONDITION

Through application testing, it is judged that the tracking method consisting of the RWA and pattern matching can apply to highway, rural and urban roads for range, lateral position estimation and vehicle existence judgment. But similar to radar, during turns at intersections on city roads, the tracking method had a problem as seen in

Fig.18 (a) ($t = 24\sim30$ [sec]). It can apply to highway and city roads excluding turns.

CONCLUSION

The Range Window algorithm (RWA), which determines the range of a vehicle on the road by a single frame of a single image as well as lateral position, had been developed. The algorithm overcomes the drawback of a conventional single camera, namely the low capability of range estimation, and realizes reliable range estimates. In order to enhance the detection rate, a pattern matching technique is incorporated into the RWA. The RWA and matching can work complementarily and realize higher performance. For instance, if an image of a preceding vehicle is influenced by a shadow of a tree or infrastructure, pattern matching can help the detection of the vehicle. Usually pattern matching needs a lot of calculation time and cannot identify the range of an object. Here a trinary image is used instead of a gray image to reduce the time and a warped template is adopted to estimate the range. This RWA was applied to variety of vehicles running at low speed on rural and urban, and city roads. The range accuracy, which is calculated from the comparison with the 77GHz radar, was about *+/-1* [m] and the detection rate was *94%*. The present range from the bumper is from *2* [m] to *50* [m]. The incorporation of matching has improved the detection rate up to *97%* and has realized more stable detection. It is found that this RWA incorporated with pattern matching is promising for short range applications.

ACKNOWLEDGMENTS

The authors thank S. Fuks and T. Tiernan, Visteon Co., for the technical discussion with them.

REFERENCES

1. W. D. Jones, "Keeping Cars from Crashing", IEEE Spectrum, pp.40/45, Sept., 2001.
2. S. Miyahara, "New Algorithm for the Range Estimation by a Single Frames of a Single Camera", 2005 SAE World Congress, 2005-01-1475, April, 2005.
3. K. Hanawa and Y. Sogawa, "Development of Stereo Image Recognition System for ADA", Proceedings IEEE Intelligent Vehicles Symposium 2001, pp.177/182, 2001.
4. T. Kato, Y. Ninomiya and I. Masaki, "An Obstacle Detection Method by Fusion of Radar and Motion Stereo", IEEE Trans. Intelligent Transportation Systems, vol.3-3, pp.182/188, Sept. 2002.
5. M. Beauvais and S. Lakshmanan, "CLARK: a heterogeneous sensor fusion method for finding lanes and obstacles", Image and Vision Computing, 18, pp.397/413, 2000.
6. H. Higashida, R. Nakamura, M. Hitotsuya, K. F. Honda and N. Shima, "Fusion Sensor for an Assist System for Low Speed in Traffic Congestion Using Millimeter-Wave and an Image Recognition sensor", SAE2001, 2001-01-0800, 2001.
7. N. Shimomura, K. Fujimoto, T. Oki anf H. Muro, "An Algorithm for Distinguishing the Types of Objects on the Road Using Laser Radar and Vision", IEEE Trans. Intelligent Transportation Systems, vol.3-3, pp.189/195, Sept. 2002.
8. G. P. Stein, O. Mano and A. Shashua, "Vision-based ACC with a Single Camera: Bounds on the Range and Range Rate Accuracy", IEEE Intelligent Vehicles Symposium (IV2003), June 2003, Columbus, OH.
9. E. Dagan, O. Mano, G. P. Stein and A. Shashua, "Forward Collision Warning with a Single Camera", IEEE Intelligent Vehicle Symposium (IV2004), Parma, Italy, Jun. 2004.

2005-01-3578

Evaluation of Cost Effective Sensor Combinations for a Vehicle Precrash Detection System

John Carlin, Charles Birdsong, Peter Schuster, William Thompson and Daniel Kawano
California Polytechnic State University

Copyright © 2005 SAE International

ABSTRACT

The future of vehicle safety will benefit greatly from precrash detection – the ability of a motor vehicle to predict the occurrence of an accident before it occurs. There are many different sensor technologies currently available for pre-crash detection. However no single sensor technology has demonstrated enough information gathering capability within the cost constraints of vehicle manufacturers to be used as a stand alone device. A proposed solution consists of combining information from multiple sensors in an intelligent computer algorithm to determine accurate precrash information. In this paper, a list of sensors currently available on motor vehicles and those that show promise for future development is presented. These sensors are then evaluated based on cost, information gathering capability and other factors. Cost sensitivity is lower in large commercial vehicles than

sensitivity is lower in large commercial vehicles than in personal vehicles due to their higher initial cost and longer life span making them a good candidate for early adoption of such a system. This work forms the basis for ongoing research in developing an integrated object detection and avoidance precrash sensing system.

INTRODUCTION

Improving occupant safety has been an increasingly important area of study since the mid 1960's. Initially this work was centered on controlling post impact occupant dynamics through the use of structural modifications, seatbelts, and airbags. The next leap forward in occupant safety is precrash sensing. Figure 1 demonstrates some possible safety benefits of precrash sensing.

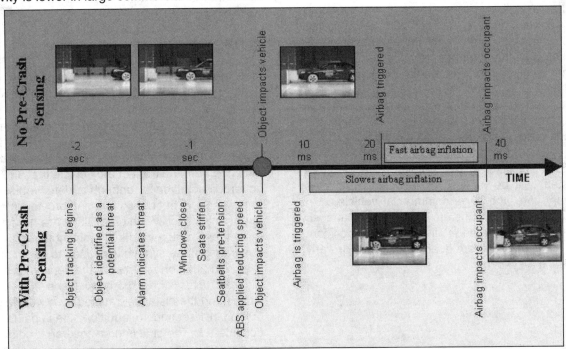

Figure 1 - Timelines for collisions with and without precrash sensing

33

The idea of using external sensors to improve vehicle safety is similarly not a new one [[1], [2]]. Ultrasonic sensors are commonly used today as parking aids on vehicles with large blind spots and radars are used in adaptive cruise control (ACC) systems to maintain safe following distance when cruise control is active.

This work will identify a possible set of requirements for precrash sensing and survey the current state of the art for a variety of sensor technologies. In comparing the sensor technologies we compare not only the capabilities of the various sensors but also the cost and overhead of including such a sensor on a vehicle. Furthermore a possible group of sensors is presented that is expected to meet these requirements.

Improving occupant safety beyond existing standards requires more information than is currently gathered by automobiles during an impact. Currently when an accident occurs the airbag sensors trigger the airbags within 10ms of the impact. The initial goal of a pre-crash system would be to bias the decision made by the impact sensors allowing for triggering based on expected impact severity. Coupled with seat belt pre-tensioning the number and severity of occupant injuries could be greatly reduced. Figure 1 shows an example impact where the precrash system triggers the airbags early, reducing the pressure required for inflation, and thereby reducing the contact force with the occupant.

The ability to predict accidents beyond the immediate event horizon has a number of other advantages. In 49% of accidents the brakes are not used at all [8]. By predicting that a collision is imminent a system could activate the brakes prior to the collision reducing the severity of the impact. Furthermore the airbags could be triggered prior to impact further reducing the force of inflation and the risk of deployment related injuries. Other safety technologies made possible with pre-crash detection include an audible alarm, automatic window closing, seat reposition and stiffening, ABS firing and eventually automatic steering to avoid the crash.

It is anticipated the initial implementation of a precrash sensing system will be on less cost-sensitive vehicles such as luxury cars and large commercial vehicles. In particular, commercial vehicles have a higher initial cost, longer life span, and greater liability from accident occurrence. This makes them good candidates for early adoption of such a system.

REQUIREMENTS OF A FRONTAL PRECRASH SYSTEM

To aid in evaluating how sensor technologies fit into the picture of precrash sensing it is important to establish what the system as a whole must be able to do.

SYSTEM ROBUSTNESS

To provide value the system must not produce false positives and must have a low occurrence of false negatives. Furthermore the system should be able to determine when it will not be effective due to road or weather conditions and inform other portions of the vehicle electronics and the driver that the system is not providing a benefit.

RESPONSE TIME

The system must be able to respond to threats in a timely manner. There are two different facets of system response time that, while related, need to be discussed separately. First there is the issue of how long the system takes to begin tracking a threat once it has entered the forward path of the vehicle. This response time will be dependant on the technical limitations of sensor covering the area in which the threat exists. The second type of system response time is related to the time required for the system to tag the object as a likely threat and notify the driver or take protective measures. This is a more difficult issue to address because while it is in part reliant on the technical limitations of system components the major factor is the accuracy of the collision prediction model used.

COVERAGE REGION

Defining the coverage region is a somewhat arbitrary task without specifically defining the vehicle and conducting an exhaustive survey of when driver warnings and other preventative measures are best performed. As a reasonable starting point we shall begin by focusing on objects in the same lane of travel as the vehicle. The distance from the front of the vehicle that should be covered should be far enough to detect objects traveling at high speed, but not so far as to provide information about objects for which predictions are uncertain. To meet these requirements the minimum region of coverage for the system is defined to be an area 3.5m wide and 30m long in front of the vehicle as shown in Figure 2. This is based on travel lane width in the United States and the distance traveled by a vehicle going 60mph in 1 second. In addition side impacts could be considered. The side impact coverage region is not discussed in this work.

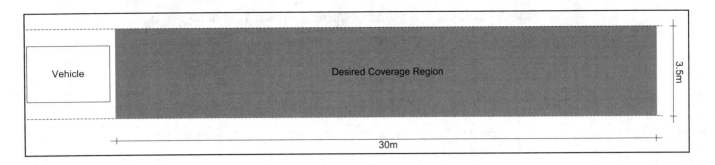

Figure 2 - Coverage Region of Suggested System

OBJECT DETECTION

A precrash sensing system must also be able to differentiate objects that are not threats to the vehicle from those that are. Table 1 gives a list to object types that should be identified and some example objects for each type.

Table 1 - Possible threat objects

Type of object	Example object
Large high mass	Tree, other vehicle, walls
Large low mass	Brush
Medium high mass	Motor cycle, cow
Medium low mass	Pedestrian
Small high mass	Road barrier, utility pole

OTHER CONSTRAINTS

Finally there are several constraints on the system. Most of them are typical for automotive electronics such as temperature, humidity, and vibration. Additionally the sensors must provide information during inclement weather. The incremental cost of such a system is a serious constraint and as a result the chosen sensors should be integrated into other, pre-existing applications such as Active Cruise Control (ACC). Sensors must be placed on vehicles in such a way that the coverage of the sensor is not severely limited due to anticipated occlusion.

POSSIBLE SENSORS

As a first step toward an integrated precrash sensing system this study will initially focus on sensors that provide telemetric data rather than classification data regarding the object of interest. In identifying possible sensors for this task it is important to understand a variety of issues including the coverage zone that each sensor is capable of providing information for, the type of information provided by each of the sensors, and the cost of the sensor. Table 2 gives a summary of these factors for each of the technologies covered in this survey for typical state of the art sensors of each type followed by a discussion of each.

ULTRASONIC SENSORS

Ultrasonic sensors have the advantage that they are already integrated into the front and rear bumpers of many vehicles for backup and parking assist. The low cost of ultrasonic sensors means they can be placed on the vehicle in such a manner as to cover any region of interest.

Backup and parking aid ultrasonic sensors work by sending out a high frequency pulse and measuring the time until the echo is received.

Individually a typical ultrasonic sensor has an aperture of roughly 45° and has a maximum range of 10m, so in practice three or four sensors are combined to cover either the front or rear bumper of the vehicle. This arrangement is made out of expedience rather than any technical limitation of the sensors themselves. Next generation sensors could use a single transmitter and combine the receivers in a phased array fashion to provide location information in a manner not dissimilar from the way many submarine sonar systems work.

LASER

Laser Imaging Detection and Ranging (LIDAR) sensors work in a manner similar to the ultrasonic sensors. The primary difference is that because of the small beam diffraction of the lasers involved, a fixed laser will not cover more than a small point directly in front of the laser. To overcome this deficiency a mirror or prism can be used to scan the beam over various angles. This means that the update rate of a LIDAR is inversely proportional to the angle of coverage.

LIDAR systems can measure distance with high accuracy. Additionally they can be made to measure object speed based on the Doppler shift of the return signal. Combined with the correct computer algorithm these systems can also provide information regarding target geometry [6].

Table 2 - Comparison Matrix of Current Sensor Technologies

	Ultrasonic	LIDAR	RADAR	Bi-Static	Vision	AIR	PIR
Cost	Low	High	High	Medium	Medium	Low	Low
Computation Overhead	Low	High	Medium	Medium	High	Low	Low
Range	3m	5m to 150m	1m to 150m	5m	Line of sight	2m	20m
Operating Conditions	Clear visibility	Clear visibility to 150m	Normal to heavy rain or snow	Normal to heavy rain or snow	Clear visibility	Normal to slight haze	Clear visibility
Commercially Available	Yes	Yes	Yes	No	Yes	Yes	Yes
Industry Acceptance	High	None	Some	None	None	None	None
Accuracy	±0.05m	±0.3m	±1.0m	±0.1m	NA	NA	NA
Update Frequency	40Hz	400Hz	10Hz	5kHz	<30Hz	NA	NA
Potential for Object Discrimination	Low	Some	Low	Low	High	None	Low
Detection Capabilities	Distance	Distance, speed, geometry	Distance, speed, cross section	Distance and radar cross section	Distance, speed, geometry, object class data	Presence	Presence
Minimum Target Size	Basketball	1" square or larger	Motorcycles and larger	Motorcycles, Pedestrians, and larger	Varies with distance	Pedestrians	Small animals

The point source nature of LIDAR systems means that they generally do not have wide coverage cones. As a result the LIDAR will have blind spots close to the vehicle the LIDAR will have blind spots as shown in Figure 7LIDAR systems rely on the ability of light to travel through whatever medium is between the source and the target. This means the presence of airborne particulates will degrade their capabilities. Military systems have been designed with enough power to overcome all but the most dense clouds of particulate, but such systems can easily damage the human eye and would be inappropriate for use on civilian vehicles.

RADAR

Classically RAdio Detection And Ranging (RADAR) has been the province of aircraft and air traffic control systems. More recently Doppler based systems have been used in traffic speed enforcement and other civilian applications. Currently radars are being used as an aid for cruise control systems to reduce driver interaction. It is these Active Cruise Control (ACC) radars that may be adapted for use in pre-crash sensing.

Radars have the capability to measure both object location and speed. ACC radars are capable of covering most of the region of interest for pre crash sensing. They have trouble detecting objects adjacent to the vehicle, however they are not as limited as LIDAR in the width of the coverage cone.

Radars can operate under nearly all practical driving conditions, although airborne particulates such as rain or snow will reduce the effective range.

BI-STATIC RADAR

Bi-static radars operate in much the same way as conventional (mono-static) radars. The difference is that the receive antenna is displaced from the transmit antenna. This means that the distance returned by such a system is actually the distance from the transmit antenna to the target and back to the receive antenna.

Mono-static radar obtains angular information by using a narrow-beam antenna pattern, and scanning the beam angle. For bi-static radar, each distance represents an ellipse rather than a circle, with the transmitter and the receiver being foci of the ellipse. As a result, an object traveling on a straight path with constant speed will appear to have acceleration. Figure 3 and Figure 4 show an example of this with simulated objects on an approach

approach path with the front of the vehicle. The closing speed is 40mph, and the transmitter-to-receiver separation distance is 1.8 meters. Speed and acceleration traces are shown for two scenarios. The first is for an impact head-on in the center of the vehicle. The second is for an approach angle 30 degrees from head-on, crossing the center line of the vehicle, and passing the front of the protected vehicle 0.5 meters outside the far antenna, or 1.4 meters from the center line of the vehicle, a near miss in the frontal crash scenario. A bi-static radar system obtains approach angle and impact-point prediction information by processing the bi-static range-ellipse data as the target approaches [9] This processing allows the bi-static radar system to predict impact speed, angle, and offset from the center-point between the two antennas.

The coverage region of bi-static radar depends upon the overlap of the transmitter beam pattern and the receiver beam pattern, which can be tailored to achieve the desired coverage region. The antenna patterns are very wide-beam, which allows the physical size of the antennas to be small, compared to the mono-static radar

system. This also allows the system to operate at lower frequencies than the mono-static radar, which makes the bi-static radar system even less susceptible to weather conditions and non-threatening clutter like brush and cardboard boxes than the higher-frequency mono-static radar systems.

A feature of the bi-static implementation is the occurrence of direct-path signal coming straight from the transmitter to the receiver without bouncing off a target. In operation, the interaction between the direct path signal and the reflected target signal provides an estimate of the target's radar cross-section, allowing the system to discriminate between a small target like a pole or motorcycle, and something large like a truck or barrier. This interaction also provides a secondary indicator of closing speed [10].

Current developments have focused upon the side-impact scenario, but the bi-static radar system is readily adaptable to the front, rear, roll-over, blind-spot, and parking aid applications.

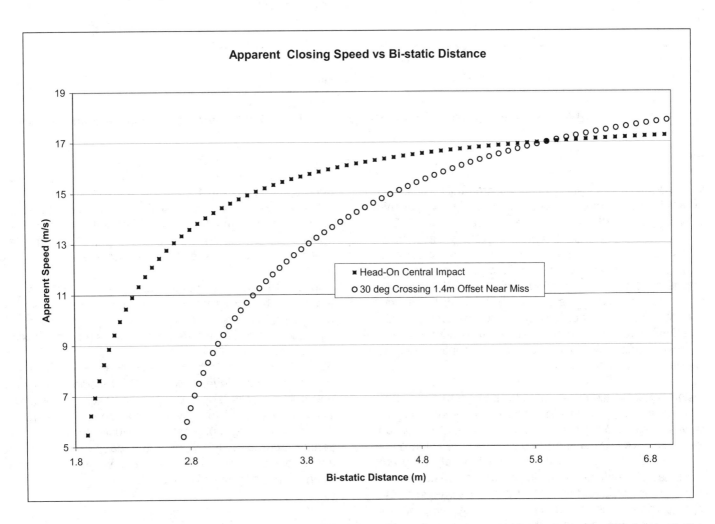

Figure 3 - Measured Speed versus Measured Distance for Two 40mph Approach Scenarios using Bi-Static Radar

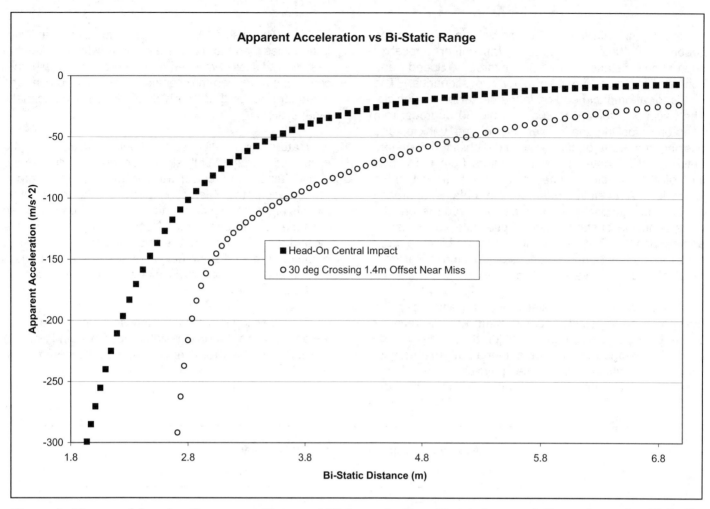

Figure 4 - Measured Acceleration versus Measured Distance for Two 40mph Approach Scenarios using Bi-Static Radar

VISION

When discussing vision systems we are really discussing three different types of systems: visible light, passive infrared vision (PIRV), and active infrared vision (AIRV). Vision - primarily visible light - systems have been presented in the literature, [3] and [1], as means of solving many of the problems related pre-crash sensing. This is in no small part due to the fact that vision systems are what humans use as our primary sensor for vehicle control. In fact vision sensors are the only sensor presented that would be able to cover the entire region of interest effectively while collecting control related information at the same time.

Vision sensors also collect data far beyond the telemetry data collected by most other sensors presented in this survey. Figure 5 shows an example of this. Based on this image we can tell that the road being traveled on is about to make a right turn, it is relatively flat, there are two lanes of travel, and most importantly there no objects in the probable path of travel that present a danger to the vehicle.The biggest drawback of vision systems of any type is not the sensors themselves but rather the computational overhead associated with extracting the information that most human drivers can gather from a

gather from a scene almost instinctively. Figure 6 shows lane makers detected using the Hough transform, a well understand algorithm for finding lines in images. This process alone is no trivial task [7], and this is just a small piece of the information that could be extracted from the scene. This simple task requires significant computer overhead to provide data at a reasonable rate.

Additionally, vision sensors themselves do not provide any telemetry data about objects in the scene. This data has be to gathered either with a separate sensor or inferred from a second vision sensor using stereovision triangulation.

Unlike the human eye, CMOS or CCD vision sensors do not automatically adjust to scene light and contrast. They must be tied to a system that correctly estimates the amount of light present and adjusts the gain of the sensor appropriately. This must happen rapidly and accurately as light conditions can change dramatically during either sunset or sunrise.

Figure 5 – Typical road scene

Figure 6 – Road scene with lane markers identified

The final drawback to vision based systems, especially visible light systems, is that their performance degrades rapidly in the presence of airborne particulates such as dust or precipitation. This problem is made worse by the fact that the conditions under which these sensors provide the least benefit are those in which the drivers need for such a system are the greatest.

NON-VISION PASSIVE INFRARED (PIR)

PIR sensors are commonly used in motion detectors for alarm systems as well as in automatic doors. These systems are robust and have low maintenance. However they are inappropriate for use in pre-crash sensing since

sensing since they rely on a known ambient null condition on which to base decisions regarding the presence of objects. For example at sunrise or sunset the sun produces far in excess of enough IR to trip such a system. While this condition can be accounted for in a fixed system it is likely to have problems on a platform that moves.

NON-VISION ACTIVE INFRARED (AIR)

AIR systems are used in many industrial and commercial applications to determine the presence of objects, including people. Direct AIR systems work with an emitter and detector that are placed at separate locations and aimed such that light from the emitter travels to the detector. Such systems detect objects when the detector no longer receives a signal from the emitter. Reflected systems work in much the same way except that they rely on a reflective surface to bounce the light from the emitter to the detector. Such a system relies on the emitter producing enough light to illuminate the target object so that it reflects more IR light than is present in the background.

AIR systems do not use time of flight measurements to collect distance information, but rather relay on the presence or absence of a return signal to determine if an object is present. Some systems exist that use the strength of the return signal to determine distance, however the distance measurements are only valid for objects whose reflectivity characteristics match the calibration object under carefully controlled conditions.

The primary limit to the range of an AIR system is the power and cost of the emitter. Focusing lenses on both the transmitter and receiver can be used to increase the range of the system as the cost of reducing the aperture.

As a result of their wide use and acceptance of AIR systems might seem to be a perfect fit for precrash sensing. The drawback of these types of sensors is that they only provide presence information. Considering that AIR systems are only slightly less expensive that ultrasonic systems and they only provide presence information rather than presence and distance they do not appear to be cost effective.

SUGGESTED SENSOR COMPONENTS

The primary goal of this system is to detect objects and gather telemetry information from them. A secondary goal is to collect classification data about the objects that have been identified as a threat. While there are any number of sensor combinations, sensors must be selected based not only on performance but also cost, availability, acceptance and maturity. To this end the following three sensors have been selected as good candidates for further investigation: Radar, Ultrasonic, and LIDAR. This means that the entire region of interest, except that right in front of the vehicle, is covered by two of the three sensors. Given that the cost and capabilities of each of the sensor technologies is constantly changing it is important to bear in mind that in the future there may be different and better combinations.

The radar was chosen because it gathers the most telemetry information under the widest range of environmental conditions at the highest frequency. Radars are one of the most expensive sensors available; however, since they are an integral part of ACC systems that are now becoming popular on commercial and luxury vehicles the incremental cost of adding this to the system is lowered. Figure 9 shows the relationship between the desired coverage region and the region covered by an

the region covered by an ACC radar.

The scanning laser or LIDAR is selected to be a complement to the radar. While current systems do not sample as fast as the radar they provide additional information. They provide some indication of object geometry is important and can be used to predict how the vehicle will deform during the impact. Second when used in combination with the radar the system should be able to determine when there is a fault and indicate this to the driver. Thirdly there is the possibility of object discrimination based on the type of signal from each sensor. Figure 7 shows one possible coverage region for a scanning LIDAR, note that a LIDAR could be designed to have an arbitrary aperature.

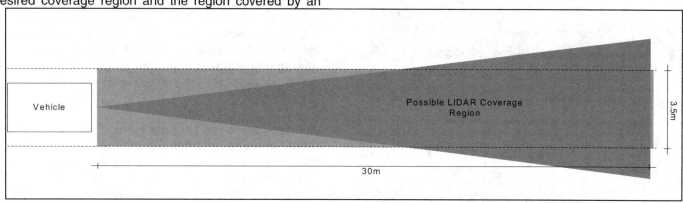

Figure 7 - LIDAR coverage region overlayed on suggested coverage region

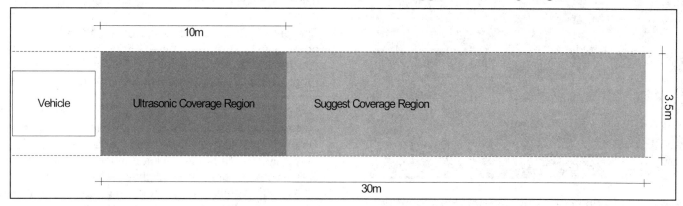

Figure 8 - Ultrasonic coverage region as compared to system coverage region

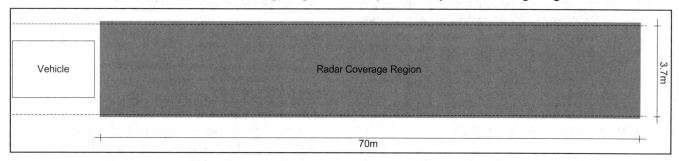

Figure 9 - Radar coverage region.

Ultrasonic sensor were selected as they are already common on vehicles today, they are inexpensive, and they cover an area directly in front of the vehicle that is not commonly covered by other available sensors. The region that they do cover is possibly the most important as it is where the data has the highest degree of validity and the prediction of an impact is most reliable. Figure 8 shows the portion of the region of interest covered by Ultrasonic sensors.

SYSTEM FUNCTIONALITY

The proposed system would work by first identifying the presence of an object using the RADAR. The object is then tracked using both the RADAR and the LIDAR so long as it is in the region of interest. At this point the system would begin to make predictions regarding the likelyhood that the object will collide with the vehicle and of closing velocity. As the object gets close the ultrasonic sensor tracks it and the system alerts the vehicles active safety systems of the likelihood of an imminent impact, along with information regarding the possible seriousness of the impact.

FUTURE WORK

While all of the above sensors are specified by their manufactures, for most of them there is sparse information regarding their performance relative to precrash sensing. A uniform test procedure needs to be established and the sensors need to be individually tested against a variety of objects under simulated environmental conditions. This is the next step of our research plan.

This paper has addressed the needs of precrash sensing as they relate to vehicles approaching objects in the path of travel. An important extension to this is expanding the coverage region so that a precrash system could potentially predict object approaching from the sides such as at an intersection.

Finally an algorithm must be developed that relates telemetric data of an object to the probability of impact. This involves not only generating significant simulations, but also real world testing to ensure that the simulations are representative.

CONCLUSION

A survey comparing the cost and capabilities of current state of the art sensor technologies has been present for use in intelligent vehicle design. A possible set of requirements for a precrash sensing system has been laid out. A framework for integrating several sensors together to meet these requirements has been presented.

ACKNOWLEDGMENTS

This research was funded by grant from the Office of Naval Research (ONR) through the California Central Coast Research Partnership (C3RP). We would like to thank Delphi for providing us with two radar systems for our use and evaluation.

REFERENCES

[1] Knoll, Peter, Schaefer, Bernd-Josef, Guettler, Hans, Bunse, Michael, & Kallenbach, Rainer, *Predictive Safety Systems – Steps Towards Collision Mitigation*, SAE 2004 World Congress & Exhibition, March 2004, Detroit, MI, USA, Session

[2] Breed, David, Sanders, W. Thomas, & Castelli, Vittorio, *A Complete Frontal Crash Sensor System – 1*, SAE S.A.E. transactions. 102, no. 6, (1993): 900

[3] Hamilton, Lisa, Humm, Lawrence, Daniels, & Yen Huan, *The Role of Vision Sensors in Future Intelligent Vehicles*, S.A.E. transactions. 110, Part 7 (2001): 613-617

[4] Cherry, James, *Applications of Ultrasonic Technology for Obstacle Detection*,

[5] Cartwright, Kert & Jindal, Gopi, *An Ultrasonic Proximity System for Automobile Collision Avoidance*,

[6] Beraldin, J.-A., Rioux, M., Livingstone, F.R., King, L., *Development of a Real-Time Laser Scanning System for Object Recognition, Inspection, and Robot Control*, SPIE Proceedings, Telemanipulator Technology and Space Telerobotics. Boston, Massachusetts, USA. September 7-10, 1993. Vol. 2057, pp. 451-461

[7] Chen, Ling, Chen, Hongjian, Pan, Yi, Chen, Yixin, *A Fast Efficient Parallel Hough Transform Algorithm on LARPBS*, The Journal of Supercomputing 29, no. 2 (2004), pp 185-195

[8] Statistics from the *Gesamtverband der Deutchen Versicherunswirtschaft e.V.* (Association of the German Insurance Indsutry) (2001)

[9] Crosby, Robert Radar detector for pre-impact airbag triggering. Patent 6,025,796. 9 December 1996

[10] Crosby, Robert Radar detector for pre-impact airbag triggering. Patent 6,097,332. 1 August 2000

New Algorithm for the Range Estimation by a Single Frame of a Single Camera

Shunji Miyahara
Visteon Japan, Ltd.

ABSTRACT

An algorithm, which determines the range of a vehicle on the road by a single image, is proposed. Since it uses a single camera, there is not a problem like the high cost in the stereo camera, the millimeter wave radar and the non-robust mechanical scanning in the laser radar. And it also gives the lateral position of the vehicle. So far it is difficult to get the range information from a single camera. The algorithm overcomes this drawback by introducing a new concept, "Range Window". It uses several portions of a captured image, namely windows. Each window is corresponding to a certain pre-determined range and has the fixed physical width and height. In each window, the size and position of an object in the image is estimated through the ratio between the widths of the object and the window and a score is given to the window. The window having the best score is determined as a best window. The range corresponding to the best window becomes an estimated range. The lateral position of the object is also calculated. This algorithm was applied to more than *4,500* real road images and showed *95%* detection rate for a motorcycle, sedan, minivan, truck and bus on rural, urban and city roads. The range accuracy was about +/- *1* [m]. The present maximum range is *50* [m]. This algorithm is effective for the short range application. This is expected to be applied to "Low speed follower".

INTRODUCTION

Since the middle of the 1990s, the vehicles with the Adaptive Cruise Control (ACC) system based on the laser radar have been introduced into the Japanese market. At the end of the 1990s, the vehicles with the system based on millimeter wave appeared in European and Japanese markets [1]. In 2004, vehicles with "Low speed follower (LSF)" have appeared in the Japanese market. This LSF system keeps the distance at low speed condition and stops the host vehicle automatically when the preceding vehicle stops. LSF is more useful and convenient than ACC, and contributes to the safety more than ACC. The present LSF is based on the laser radar. The radar is effective for the range measurement,

but has some difficulty to detect the lateral edge of a vehicle. In the case of the offset collision at low speed, the low speed follower might not work properly because of the lateral position accuracy and limited azimuth angle. Therefore the author considers a vision system, which is capable to detect the range as well as the lateral position of the preceding vehicle and low-cost, and here proposes it. The proposed system consists of only a single camera and calculates the range from a single frame of the camera image. Its algorithm is based on the Range-Window (RW). RW is a virtual window in the camera image and has a fixed physical size. The RW looks different, depending on its range. If there is a vehicle in the camera image, the ratio between the vehicle and the RW changes according to the range. The RW having the most adequate ratio is the most probable and its range becomes the estimated range.

So far there are several ways to measure the range and the lateral position of a preceding vehicle by a vision system. They are the stereo camera [2], a motion stereo [3] and a fusion system [4-6]. Unfortunately these approaches increase the cost and/or need some help of a radar. Recently a unique approach based on a single camera was proposed [7,8]. But it is based on the point of contact between the vehicle and the road. In a real situation, it seems difficult to identify the point regularly because of the shadow. Therefore the author proposes the new algorithm based on the RW.

Here the principle of the RW, the procedure to realize it, the application to real image, the discussion on some key points and the conclusion are described.

PRINCIPLE OF RANGE ESTIMATION

The principle of the algorithm is described. The basic assumption is that the host and preceding vehicles are on the same plane. Namely the road is assumed flat. The algorithm is based on the virtual windows at certain ranges. As seen in Fig.1, the windows standing vertically on the ground at different ranges are considered. The physical sizes of them are the same but they look different since the range is different.

Here the window of *4 × 2* [m] is chosen. Their sizes in the image of a camera, however, are different since the range is different as seen in Fig.2.

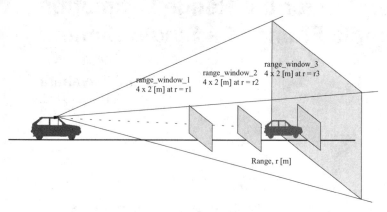

Figure 1: Principle of the Range-Window algorithm (RWA). The windows at three ranges, *r1*, *r2* and *r3*, are shown.

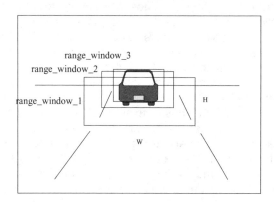

Figure 2: Appearance of the windows in a camera image. Size of the windows = W × H = 4 × 2 [m].

The position of the windows is calculated. For instance, the vertical position of the window in a picture, corresponding to the elevation angle, is considered. The lower and higher edges of the range_window_1 shown in Fig.3 (a) are calculated by Eqs.(1) and (2).

$$\theta_{1l} = \arctan(-r / hc) \qquad (1)$$

$$\theta_{1h} = \arctan(r /(hw - hc)) \qquad (2)$$

where θ_{1l}, θ_{1h}, *hc*, *hw* and *r* (=r1) are the elevation angles at lower and higher edges, the heights of the camera and the window, and the range, respectively. The module is [0, π]. The difference, $\Delta\theta_1 = \theta_{1l} - \theta_{1h}$, corresponds the height of the window in a picture. The horizontal position of the window in a picture, corresponding to the azimuth angle, is considered. The right and left edges of the range_window_1 shown in Fig.3 (b) are calculated by Eqs.(3) and (4).

$$\phi_{1l} = \arctan(-width\_w / 2r) + \pi / 2 \qquad (3)$$

$$\phi_{1h} = \arctan(width\_w / 2r) + \pi / 2 \qquad (4)$$

where ϕ_{1l}, ϕ_{1h} and *width_w* are the azimuth angles at right and left edges, and the width of the window, respectively. The module of "arctan" is [-π/2, π/2].

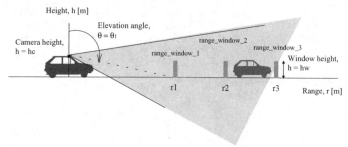

(a) Side view and the relation between a vehicle and the window.

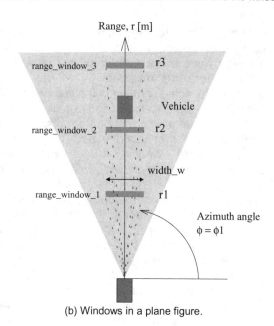

(b) Windows in a plane figure.

Figure 3: Position of the windows in a side view and a plane figure.

The object width is calculated through image processing for each window. In Fig.4 (a), the width of the object is about *30%* of the width of the window and then the estimated object width at *r1* is *1.2 [m]= 4 × 0.3*. For Fig.4 (b), the width of the object is about *45%* of the width of the window and the object width at *r2* is *1.8 [m]= 4 × 0.45*. For Fig.4 (c), the width of the object is about *80%* of the width of the window and the object width at *r3* is *3.2 [m]= 4 × 0.8*.

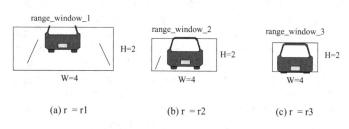

Figure 4: Object in each window.

44

Among the estimated object widths, *1.8* [m] for *r = r2* is the nearest to *1.7* [m], assumed vehicle width. Then the algorithm judges that the most probable window is *(b)* in Fig.4 and then the estimated range is the range of the window, *r2*.

In the judgment, other criteria can be added. For instance, the extracted object should have the height of *1.5* or *2* [m], or be on the bottom line since a vehicle cannot fly. The larger the height is, the more probable the object should be.

Since the proposed algorithm using a single frame of a single camera is based on the "Range-Window", this algorithm is called "Range-Window algorithm (RWA)" here.

PROCEDURE OF RANGE-WINDOW ALGORITHM

The procedure of RWA is described. The procedure is comprised of a pre-processing including an edge-enhancement, the process in a single window and the estimation of the range as shown in Fig.5. The process in the single window is comprised of line-segmentation, object-segmentation, scoring and best object-in-window determination.

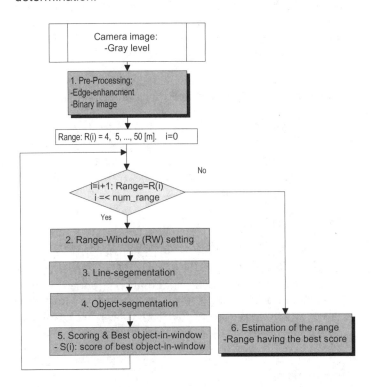

Figure 5: Flow chart of the range estimation based on the Range-Window algorithm (RWA). *R(i), I=1, 2, .., N*: pre-set range of the Range-Window. *S(i), I=1, 2, ..., N*: the score of best object-in-window.

PRE-PROCESSING

The original camera image (frame) is assumed monochromatic. The pre-processing makes the trinary image by edge-enhancement and binary image. In the edge-enhancement, the vertical one is adopted since the horizontal one tends to be influenced by a shadow.

Trinary image

Vertical edge-enhancement is applied to an original image and the cells in the image are classified into three groups as shown in Eq.(5). The trinary image is shown in Fig.6. Vertical Sobel filter is usually used.

$$i_{edge}(x,y) = \begin{cases} 1 .... \partial i(x,y)/\partial x > d_{threshold} \\ -1 .... \partial i(x,y)/\partial x < -d_{threshold} \quad (5) \\ 0 .... otherwise \end{cases}$$

where $i(x,y)$, $i_{edge}(x,y)$, x, y and $d_{threshold}$ are the intensity, trinary image, lateral cell number, vertical cell number and the threshold, respectively. The threshold depends on the average or distribution of the intensity in a camera image.

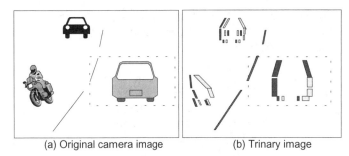

| (a) Original camera image | (b) Trinary image |

Figure 6: Trinary image based on the vertical edge-enhancement. The rectangle with dot line is the Range-Window. The trinary image in the window is used for the following image processing.

Binary image

Binary image is also calculated by a threshold, which is usually a median value. This is used to check the uniformity in the object area.

PROCESS IN A SINGLE WINDOW

The Range-Window is set by the pre-determined ranges. Refer to Fig.1 and Eq.(1) through Eq.(4). For each window, the "process in a single window" is implemented.

Line-segmentation

The feature points, cells having *+1* or *-1* in the trinary image as shown in Fig.6 (b), are connected and grouped. If the cells having the same polarity are adjacent vertically and have short lateral distance, they are connected and categorized into the same line-segment. The polarity can avoid inadequate connections.

Following the above process, some of the line-segments are merged into a single line-segment if their vertical spacing and lateral position discrepancy are small and their polarities are the same. This process is effective when the line-segments are discontinued.

Object-segmentation

This object-segmentation is explained by using the window in Fig.6 (b). An object is made from a pair of the line-segments having different polarity as shown in Fig.7 (b). The objects, o1, o2 and o3, are created. Other meaningless objects having the same polarity are automatically removed and this removal can reduce unnecessary calculation.

Scoring and best object-in-window determination

Scoring on the objects is made. The scoring is based on a "vehicle model" having the following criteria.

-Width of an object
-Height and vertical position of an object
-Length and aspect ratio of line-segments
-Uniformity in the object area calculated from the binary image

The score is given to each object as shown in Fig.7 (c). The object with the highest point is chosen as the "best object-in-window".

The score depends on the vehicle models, corresponding to a motorcycle, sedan, minivan and truck/bus. The selection of the vehicle models is based on the width and the height of the object.

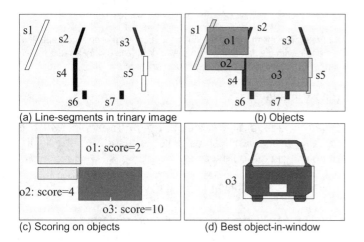

(a) Line-segments in trinary image   (b) Objects

(c) Scoring on objects   (d) Best object-in-window

Figure 7: Object-segmentation. Objects are created by a pair of segments having different polarity. Scoring on the objects is made. Then the best-object in window is determined.

ESTIMATION OF THE RANGE

After getting the best object-in-window and its score for each RW, the RW having the best score is picked up and identified as the most adequate window as shown in

Fig.8. Some kind of the threshold is necessary to eliminate the meaningless detection, which happens when there is no object. Namely if the best score is lower than a threshold, this RWA judges that there is no object.

The estimated range can be modified by the ratio between the width of the object with the best score and the pre-set width. This can mitigate the error coming from the discrete ranges. Here the pitch of the ranges is selected so that the ratio of adjacent ranges might be 1.2.

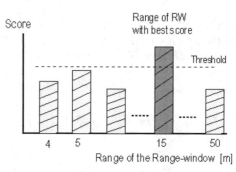

Figure 8: Estimation of the range. The range of the RW having the best score will be the estimated range.

APPLICATION TO REAL IMAGE

The proposed algorithm is evaluated through real data. A vehicle with a camera and a 77 GHz radar is run on the roads in Detroit in US. After the driving, the range estimation is made by the RWA in off-line and compared with the radar range. The lateral position is also calculated from the RWA.

CONDITION OF RANGE-WINDOW ALGORITHM

The ranges of the RWA are 4.0, 4.2, 5.0, 6.2, 7.4, 8.8, 10.4, 12.4, 14.8, 17.6, 21.0, 24.9, 29.7, 35.3, 42.0 and 50.0 [m]. The four vehicle models are used. The size of the windows is 4 × 2.2 [m]. At 4.0 [m], a half window, which is explained later, is used.

MEASURING SYSTEM

The CCD camera and the radar are installed on the rear view mirror and the grill as shown in Fig.9, respectively. The CCD image, radar information (range, relative velocity and azimuth angle) and vehicle information (velocity and yaw rate of the host vehicle) are sampled every 0.1 [sec]. The output of a camera, radar, velocity and yaw-rate are synchronously measured. The radar and vehicle information is not used in this RWA, but as comparison.

The CCD is monochromatic and has 320 x 240 cells. Azimuth and elevation angle widths of the camera are 48 and 35 [deg], respectively.

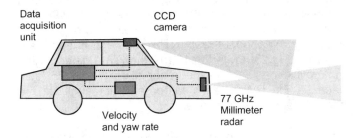

Figure 9: Measuring system by a CCD camera. Range obtained from the radar is used as a reference. Azimuth and elevation angle of the camera are 48 and 35 [deg], respectively.

MEASUREMENT AND RANGE ESTIMATION

The measurement was made on the highway, urban, local and city roads. The targets are a motorcycle, sedan, minivan, truck and bus. After the measurement, the algorithm is applied to each single frame of camera images in the off-line. It is applied to a variety of vehicles including 30 scenarios (scenario-A) having a vehicle and 14 scenarios (scenario-B) having no vehicle (refer to Appendix A). The scenario-A includes 3 motorcycles, 14 sedans & others, 6 minivan, and 7 trucks & buses. And then it is applied to vehicle following tests.

RESULTS

Variety of vehicles

The algorithm is applied to the camera images (pictures) of variety of vehicles. The results are shown in Fig.10. The discrimination between the vehicle existence and non-existence was correctly made. The non-existence corresponds to "zero distance" in Fig.10. The range estimation shows the reasonable performance except for the large target over the range of 50 [m].

For the 25 targets at shorter than 50 [m], the performance of this algorithm is described. The linear regression is applied to the range estimation as shown in Fig.10. The standard deviation was +/-1.2 [m] and the offset was 2.0 [m]. This offset expresses the distance between the position of the CCD camera and the radar adequately. The slope was lower than the unity by 0.14. The vehicle type identification was 80% probability. The minivan was sometimes identified as a sedan or truck. The lateral position was estimated precisely as shown in Fig.11.

Some of the camera images used in Fig.10 are shown in Fig.A1. The lateral positions in the images are exactly estimated as well as the range.

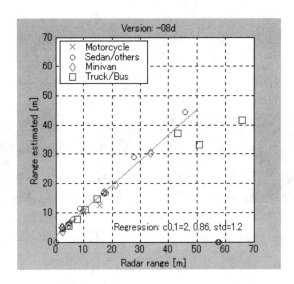

Figure 10: Comparison between the estimated and radar ranges. It is applied to 30 scenarios (scenario-A) having a motorcycle, sedan, minivan, truck and bus and 14 scenarios (scenario-B) having no vehicle. In scenario-B, the radar range is assumed zero. The vehicles in scenario-A are a motorcycle, sedan, minivan, bus and truck. The ranges of the RWA are 4.0, 4.2, 5.0, 6.2, 7.4, 8.8, 10.4, 12.4, 14.8, 17.6, 21.0, 24.9, 29.7, 35.3, 42.0 and 50.0 [m]. The solid line is the linear regression line (Estimated range [m]= 2.0 + 0.86 * Radar range [m]). Refer to Fig.A1.

Figure 11: Lateral position estimation: Small and large quadrilaterals are the estimated object area and the Range-Window, respectively.

For the 14 images having no vehicle, this algorithm recognizes the non-existence with 100% as shown in Fig. 10. In the figure the zero range corresponds to the non-existence. Some of the 14 images are shown in Fig. A2.

Process of RWA

The example of the process of the RWA is shown in Fig.12. The line-segments and objects are shown in Figs.12 (a) and (b), respectively. In the line-segmentation, the line-segments with the same polarity are connected if their vertical and lateral distances are small enough. The score at the several Range-Windows are shown in Fig.12 (c). The highest score is at the range of 7.4 [m]. The Range-Window and extracted object area are shown in Fig.12 (d). The object is precisely extracted.

 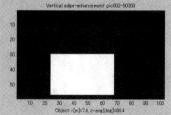

(a) Line-segments: Trinary image   (b) Object segmentation

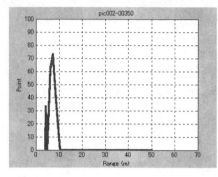

(c) Score at Range-Windows. At 7.4 [m], the score is the highest.

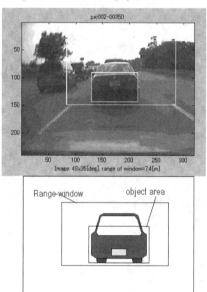

(d) Range-Window at 7.4 [m] and identified target

Figure 12: Process of the RWA: Range and lateral position estimation

Vehicle following test

The RWA is investigated through the application to vehicles running at low speed in rural area, urban area and a city road. The vehicles are a motorcycle, sedan and bus. The RWA is applied to a motorcycle cutting in and leaving for the adjacent lane. The result is shown in Fig.13. Although there were some estimation errors in Fig13 (a), the correlation between the estimated and radar ranges looks satisfactory. The quantitative

comparison is shown in Fig.13 (b). The deviation between the RWA and the radar was about 0.94 [m]. The detection rate, the rate providing a reasonable accuracy, was about 99%. The camera images are shown in Fig.13 (c). In Fig.13 (a), the RWA detected the motorcycle earlier than the radar at t= 3 and 26 [sec] because the RWA has a larger azimuth angle at the short distance than the radar.

The RWA is applied to a sedan in an urban area. The result is shown in Fig.14. The correlation looks satisfactory as seen in Fig.14 (a). The quantitative comparison is shown in Fig.14 (b). The deviation between the RWA and the radar was about 0.60 [m]. The detection rate, the rate providing a reasonable accuracy, was 97%. The camera images are shown in Fig.14 (c).

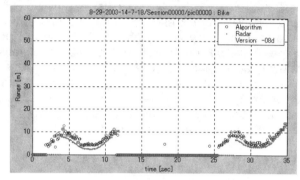

(a) Vehicle following test: motorcycle

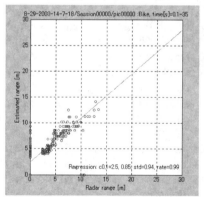

(b) Estimated and radar ranges: motorcycle

t= 10 [sec]   t= 30 [sec]
(c) Camera images

Figure 13: Application to the test to follow a low speed vehicle: motorcycle cutting in and out. The sampling of the images is every 0.1 [sec].

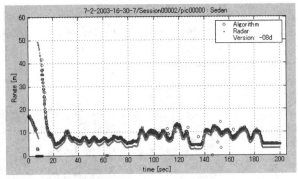

(a) Vehicle following test: sedan

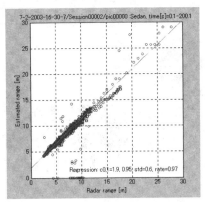

(a) Estimated and radar ranges: sedan

t= 60 [sec]   t= 120 [sec]
(c) Camera images

Figure 14: Application to the test to follow a low speed vehicle: sedan. The sampling of the images is every *0.1* [sec].

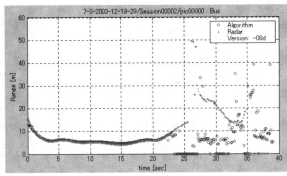

(a) Vehicle following test: Bus. The bus starts turning the right at the intersection at t= *24* [sec].

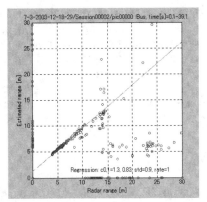

(b) Estimated and radar ranges: sedan

t= 0 [sec]   t= 30 [sec]: turning to the right
(c) Camera images

Figure 15: Application to the test to follow a low speed vehicle: Bus. The sampling is every *0.1* [sec].

The RWA is applied to a bus in a city road. The result is shown in Fig.15. The quantitative comparison is shown in Fig.15 (b). After t= *24* [sec], the bus is turning right and there is no preceding vehicle. Before the turning, the deviation between the RWA and the radar was about *0.90* [m]. The detection rate, the rate providing a reasonable accuracy, was about *100%*. The camera images are shown in Fig.15 (c). During the turning, the algorithm could not work properly because there were many vertical lines on the corner of the building as seen at t= *30* [sec] in Fig.15 (c).

This RWA was also applied to other motorcycles, sedans, minivans, trucks and buses running at low speed in highway, urban and rural roads although their results are not shown here. Overall accuracy and detection rate of about *4500* images were *1.04* [m] and *95%*, respectively.

The definition of the detection rate is: for the radar range of less than *50* [m], it is judged as "detected" if the range difference between the RWA and radar is less than *2.5* [m]. The definition of the accuracy is the standard deviation of the linear regression between the RWA and the radar for all the data in which the difference is less than *7* [m].

DISCUSSION

RANGE LIMITATION

The maximum range depends on the azimuth angle and number of cells of the camera. For the azimuth angle of *48* [deg] and the horizontal cells of *320*, the maximum range is approximately *50* [m]. This comes from the physical length corresponding to the cell at the range, about *14* [cm]. At the range of *50* [m], a vehicle can have only *13* cells by assuming the vehicle width of *1.8* [m].

Usually more than *15* cells in a vehicle are necessary to recognize a vehicle adequately in the RWA. Namely it is necessary to increase the number of cells or reduce the angle in order to expand the maximum range of the RWA. For the vertical cell, the same thing can be said and the vertical cell is sometimes more serious than the horizontal one. Another factor is the sharpness of the image. The object at farther range tends to have lower sharpness and the RWA results in the failure of the detection.

VEHICLE MODEL

The scoring on an object depends on the vehicle types because the motorcycle, sedan, minivan, truck and bus have much different characteristics in their rear view. For instance, the width of those vehicles ranges from *0.6* [m] to *2.5* [m] and the height of them ranges from *1.5* [m] to higher than *3.0* [m]. To evaluate those vehicles adequately, the present algorithm has four models corresponding to the motorcycle, sedan, minivan and truck/bus.

SIZE OF RANGE-WINDOW

The size of the RW is selected as *4 x 2.2* [m] here. The larger the size becomes, the higher the probability of the vehicle existence in the window will be. But unnecessary processing will be included and the calculation time will be increased. The author considers that an adequate lateral size might be a little larger than the lane width. The height of the RW may be from the ground to about *2* [m] since vehicle height is usually less than *2* [m].

OBJECT AT VERY SHORT RANGE

The developed algorithm sometimes does not work properly for the extremely short range of less than or equal to *4* [m]. The line-segments in the RW in Fig.16 (a) are shown in Fig.16 (a'). Unfortunately the edges of the lower left side of a preceding vehicle cannot be seen because the feature points in Eq.(5) are usually adaptively changed on the basis of the average or distribution of the intensity in an image. The upper area of the camera image is so bright that the edge-enhanced points are localized in the upper area. Usually the area higher than the horizon is usually much brighter than that lower than it.

Therefore the author proposes that the size of the RW be divided into a half vertically and the lower part be used for the processing. This half window process is shown in Fig.16 (b). The line-segments expressing a vehicle edge are extracted in Fig.16 (b') better than in Fig.16 (a').

APPLICABLE ROAD CONDITION

Through the application, it is judged that the RWA can apply to highway, rural and urban roads regarding the range, lateral positions and the vehicle existence judgment. But at the turning at the intersection in the city roads, the RWA had a problem as seen in Fig.15 (a) (t= *24~30* [sec]) as well as the radar.

 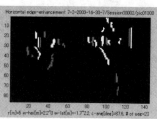

(a) Range-Window: *4 x 2.2* [m]   (a') Line-segments in RW

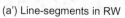

 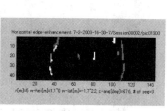

(b) Range-Window: *4 x 1.1* [m]   (b') Line-segments in RW

Figure 16: Half window at the very short range.

CONCLUSION

An algorithm, which determines the range of a vehicle on the road by a single frame of a single image as well as lateral position, has been developed. The algorithm overcomes the drawback of a conventional single camera, namely the low capability of the range estimation, and realizes the reliable range estimation by introducing a new concept, "Range-Window" algorithm. It uses several portions of a captured image, namely virtual windows. Each window is corresponding to a range and has the fixed physical width and height. In each window, the size and position of an object is estimated through the ratio between the widths of the object and the window, and the object is given a score. The window having the best score is determined as a best window. The range corresponding to the best window becomes an estimated range. This algorithm was applied to variety of vehicles running at low speed in rural and urban, and city roads. The range accuracy, which is calculated from the comparison with the *77GHz* radar, was about *+/-1* [m] and the detection rate was *95%*. The present range from the bumper is from *2* [m] to *50* [m]. It is found that this algorithm is promising for the short range application. Next step is the on-line test including the real time processing.

ACKNOWLEDGMENTS

The author thanks G. Sielagoski, F. Ibrahim, S. Fuks and A. Koulinichi, Visteon Co., for the technical discussion with them.

REFERENCES

1. W. D. Jones, "Keeping Cars from Crashing", IEEE Spectrum, pp.40/45, Sept., 2001
2. K. Hanawa and Y. Sogawa, "Development of Stereo Image Recognition System for ADA", Proceedings IEEE Intelligent Vehicles Symposium 2001, pp.177/182, 2001.
3. T. Kato, Y. Ninomiya and I. Masaki, "An Obstacle Detection Method by Fusion of Radar and Motion Stereo", IEEE Trans. Intelligent Transportation Systems, vol.3-3, pp.182/188, Sept. 2002.
4. M. Beauvais and S. Lakshmanan, "CLARK: a heterogeneous sensor fusion method for finding lanes and obstacles", Image and Vision Computing, 18, pp.397/413, 2000.
5. H. Higashida, R. Nakamura, M. Hitotsuya, K. F. Honda and N. Shima, "Fusion Sensor for an Assist System for Low Speed in Traffic Congestion Using Millimeter-Wave and an Image Recognition sensor", SAE2001, 2001-01-0800, 2001.
6. N. Shimomura, K. Fujimoto, T. Oki anf H. Muro, "An Algorithm for Distinguishing the Types of Objects on the Road Using Laser Radar and Vision", IEEE Trans. Intelligent Transportation Systems, vol.3-3, pp.189/195, Sept. 2002.
7. G. P. Stein, O. Mano and A. Shashua, "Vision-based ACC with a Single Camera: Bounds on the Range and Range Rate Accuracy", IEEE Intelligent Vehicles Symposium (IV2003), June 2003, Columbus, OH.
8. E. Dagan, O. Mano, G. P. Stein and A. Shashua, "Forward Collision Warning with a Single Camera", IEEE Intelligent Vehicle Symposium (IV2004), Parma, Italy, Jun. 2004.

APPENDIX A: SCENARIO-A AND -B

Some pictures used in Fig.10 are shown here. Some ones in Scenario-A are shown in Fig.A1. The range and lateral positions are estimated by the RWA. The objects are exactly identified. Namely the lateral position is estimated precisely. The range accuracy is already shown in Fig.10. Some pictures in Scenario-B are shown in Fig.A2. There is no vehicle in Scenario-B. The RWA judges this non-existence correctly.

Figure A1: Pictures in Scenario-A. Objects are a motorcycle, sedan, minivan, truck and bus.

Figure A2: Pictures in Scenario-B. There is no vehicle.

Vehicle Blind Spot Detection Using Anisotropic Magnetic Resistive Sensors

Aric Shaffer, Gurpreet Aulakh, Anya Getman and Ronald Miller

Manufacturing and Vehicle Design R & A, Ford Motor Company

ABSTRACT

This paper describes a method for detecting vehicles in the rear blind spot zones using magneto resistive sensors. The earth's magnetic field provides a uniform field over great distances and ferrous objects create a localized disturbance in the (X,Y,Z) component directions when placed near the sensor whether they are moving or standing still. Since this method does not provide an exact measurement of distance (passive vs. active sensing), filtering and extracting information from detailed analysis of variations in the magnetic field provide vehicle presence information and an estimated distance measurement. This system has been applied to achieve vehicle detection of a close proximity 6-10 ft from the side and rear of the vehicle.

INTRODUCTION

Each year millions of collisions occur in the US that cause a large number of fatalities and non fatalities with a high associated cost. The National Highway Traffic Safety Administration (NHTSA) estimates lane change and merging accidents account for 16% of all major collisions in the US mainly because of view obstructions, other drivers approaching too fast, or other visual issues.

Blind spot, lane change, and merging accidents do not cause many fatalities but are a statistically large contributor for property damage. These accidents cause traffic congestion problems that plague larger cities and are inclined to get worse as the world population increases. There are many blind spots around a vehicle but the most common is the zone located towards the rear of the vehicle on both sides. It is not easily visible using the rear view or side view mirrors leaving a known gap or zone as illustrated in Figure 1. The triangular zones are covered by existing side view mirrors.

Intelligent vehicles and sensing technologies have become a major area of research in recent years seeking to solve these problems with advances in electronics. Several systems in production already highlight new technologies such as adaptive cruise control using radar units that control the distance between two vehicles when traveling. Other new technologies such as scanning LASER's, cameras with image processing, near infrared distance measurement, and thermo-vision warning systems address the problem of blind spot accidents but at a large price to the consumer.

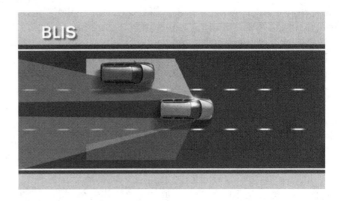

Figure 1 - Volvo Blind-Spot Lane Information System (BLIS) highlights zone not visible with standard issue mirrors.

This paper describes how blind spot problems might be solved using magnetic sensors at a much lower cost compared to the technologies listed above.

OUTLINE OF APPROACH

The earth provides a uniform magnetic field over a wide area. A ferrous object such as a car creates a local disturbance in this field whether it is moving or standing still. A Magneto-resistive sensor can detect the change in the earth's field when a vehicle is in close proximity of the sensor and by applying an electronic algorithm we can create a Blind-Spot Lane Information System (BLIS).

Magnetic sensors were mounted at several locations on the test vehicle to study the disturbance effects in the earth's magnetic field from different vehicles and noise sources. Experiments were designed to include several test schemes and map out sensitivity in three dimensional space. These include noise sources and vehicles of different size, weight, and material composition. By analyzing the variation of the X, Y, Z, components and magnitude, certain patterns are derived that can be used as a recognition tool.

The earth's field can typically be thought in terms of a dipole that provides a uniform field over a wide area of at least several kilometers (Figure 2).

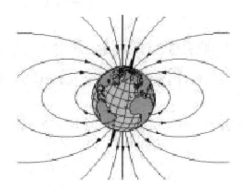

Figure 2: Earth's Magnetic Field

The presence of ferrous materials such as nickel, iron, steel near the magneto-resistive sensor creates disturbances in the earth's magnetic field that distort x, y, z field measurements. To accurately sense these distortions and apply them to blind spot detection, the presence of the earth's magnetic field must be taken into account when measuring other x, y, z fields. Also, the variance of the earth's magnetic field must be accounted for in different parts of the world. Differences in the earth's magnetic field are quite dramatic between North America, South America and Equator region. Another difficult challenge is accounting for the vehicle using these sensors and generating its magnetic distortion model for calibration.

As seen in the Figure 5 that the direction pointing to true north varies with location and changes over time, this is called magnetic declination. Typically these lines change over great distances (not within a few kilometers) and the time period that this field changes due to internal earth geology is very slow. Current magnetic north is in northern Canada but should reach Siberia by 2040. Given these challenges there are solutions to solve this problem by using GPS and/or special calibration algorithms to find magnetic north and compare to a vehicle magnetic calibration model.

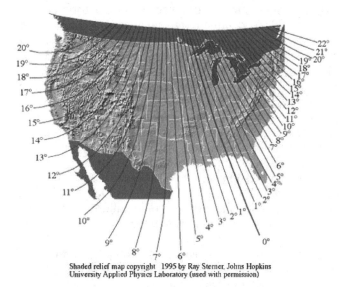

Shaded relief map copyright 1995 by Ray Sterner, Johns Hopkins University Applied Physics Laboratory (used with permission)

Figure 5: Magnetic Declination

When a vehicle or ferrous object is placed in a uniform field it creates a local disturbance as seen in Figure 3. This disturbance can be modeled as a composite of many dipole magnets resulting in a characteristic and repeatable distortion.

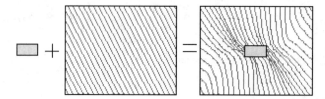

Ferrous Object + Uniform Magnetic Field = Field Disturbance

Figure 3: Magnetic Field Disturbance

Magnetic disturbances vary based on different types of vehicle such as cars, vans, trucks, buses, trailer trucks etc. When a vehicle passes close to the magnetic sensor it will detect its various parts and the field variation will reveal a detailed magnetic signature of it. The vehicle's presence can be detected using these variations with the use of pattern recognition and matching algorithms. Figure 4 is an example to highlight the test setup and how ferrous objects near the sensor disturb the magnetic lines.

The process is different from conventional sensing since measurement of a parameter such as distance cannot be taken directly from the sensor but is extracted from a variation in magnetic field. Typical or conventional sensors directly convert the desired parameter such as pressure or temperature into a proportional voltage or current output.

Figure 4: Magnetic Field Disturbance Near Sensor

The distortions are more obvious at the engine and wheel locations but can also vary depending on what ferrous items are in the interior, on the rooftop, or in the trunk locations. The net result is a characteristic distortion to the earth's magnetic field depends on the size of the ferrous material or vehicle size.

The mathematical approach is to measure the magnetic field components in the coordinate frame of the sensor device. Figure 6 illustrates the gravity and magnetic vectors contained in the earth's fields compared to a level plane.

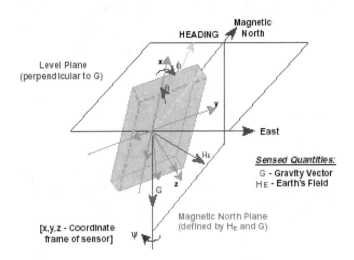

Figure 6: Gravity and Magnetic Vectors

The horizontal magnetic components (XH, YH) are used to determine the heading direction. These values can be found using for any roll and pitch orientation using the following formula:

$$X_H = X \cdot \cos(\phi) + Y \cdot \sin(\theta) \cdot \sin(\phi) - Z \cdot \cos(\theta) \cdot \sin(\phi)$$
$$Y_H = Y \cdot \cos(\theta) + Z \cdot \sin(\phi)$$

Θ and Φ are the rotational angles to transform compass to local level plane. Heading can be calculated using the formulas below:

$$
\begin{aligned}
\text{Heading} \quad &= 90 - \text{ArcTan}(X/Y) \cdot 180/Pi \quad (Y>0) \\
&= 270 - \text{ArcTan}(X/Y) \cdot 180/Pi \quad (Y<0) \\
&= 180 \qquad\qquad (Y=0 \text{ and } X<0) \\
&= 0 \qquad\qquad\;\; (Y=0 \text{ and } X>0)
\end{aligned}
$$

In addition to heading we are also interested in the way the magnetic field changes when an object is placed near the sensor, this will show the magnetic distortions in the X, Y, and Z directions.

TEST SETUP

In this next section we describe the vehicle test plan where we study static and dynamic testing conditions for several vehicles and noise sources. A few representative vehicles of different sizes, weights, and material construction categories are listed. In short we outfitted a test vehicle A with 4 mag sensors, GPS, ride height sensors, and some other inputs to confirm vehicle presence and distance as seen in Figure 7. Data was collected using a visual basic program run on a laptop to log digital and analog signals.

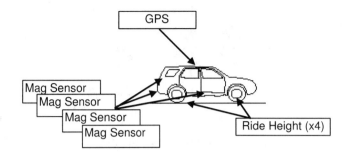

Figure 7: Test Setup

Figure 8 shows placement of several magnetic sensors and how the driver's view or blind-spot zone and magnetic sensing region are limited and overlap.

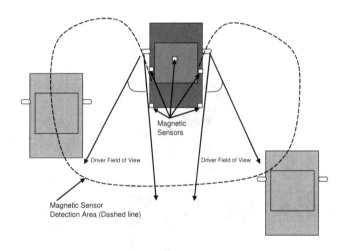

Figure 8: Sensor Placement

From the federal government regulations FMVSS111, each passenger car shall have an outside mirror of unit magnification. Field of view shall provide the driver a view of a level road surface extending to the horizon from a line, perpendicular to a longitudinal plane tangent to the driver's side of the vehicle at the widest point, extending 8 feet out from the tangent plane 35 feet behind the driver's eyes, with the seat in the rearmost position.

Figure 9 is an example of a blind-spot zone that is typical to driver side mirrors providing 13 degrees of viewing angle from normal viewing and up to 45 if the driver leans forward and to the side to extend the field of view.

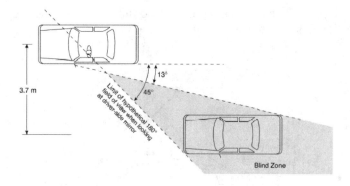

Figure 9: Blind Spot Zone

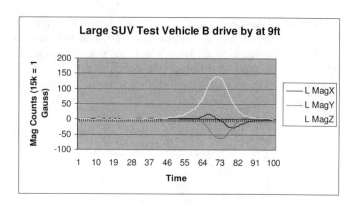

Figure 11: X, Y, Z Deflection as SUV Passes the Sensor

EXPERIMENTAL RESULTS

Using a large sized SUV (test vehicle B) we set up a matrix or test grid to map out sensitivity and proximity away from the vehicle and sensors. Looking at the (X,Y,Z) components we find that the sensitivity drops off quickly the further you are from the sensor, but we can still sense accurately out to 9-12ft on the SUV in Figure 10.

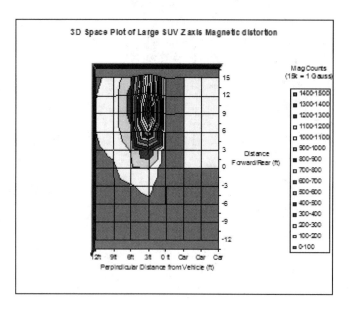

Figure 10: Large SUV Magnetic Deflection Space Plot of Z-Axis

In Figure 11 it can be seen that the Z axis is the prominent axis with the greatest change in magnetic variation. This plot shows the (X,Y,Z) components listed as L Mag(X,Y,Z) and zero offset to earth's magnetic field to highlight typical deflections in the component directions. This holds true for the matrix of several cars tested in variation studies.

Figure 12 shows the non linear nature of distance from the side of the vehicle and magnetic field distortion strength. This can be modeled as a nonlinear function similar to y = 1 / x.

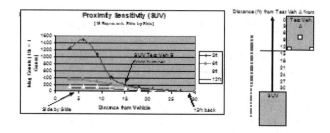

Figure 12: Proximity Sensing of Z-axis at Different Distances from the Sensor

Another vehicle choice at the opposite end of the spectrum for size, weight, and material composition is a composite body smaller passenger Test Vehicle C containing less ferrous material. Compared to the hefty SUV the sensitivity isn't as high considering the plastic body and smaller vehicle mass, but detection distances extend out 6-9 ft. In Red is what we would call the noise floor being 100 or less mag counts (15k mag counts = 1 Gauss). We conducted test from every angle and direction and this magnetic distortion plot is repeatable for all directions N,S,E,W. Please see figures 13 and 14 below.

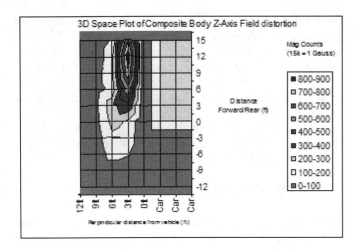

Figure 13: Composite Body Vehicle Magnetic Deflection Space Plot of Z-Axis

From our testing it looks like sensing vehicles seems a very promising using the magnetic sensor. If we take an already existing compass, modify it slightly, and move it to a different location it will provide additional benefits. Figure 14 below shows the magnetic distortions 3 ft out from the test vehicle side and started from 12ft behind the vehicle then moved in 3 ft increments up until the vehicles were side by side.

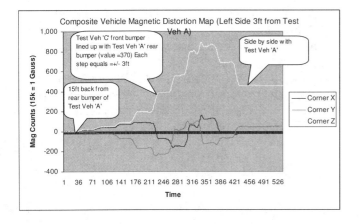

Figure 14: X,Y,Z Component Distortions at 3ft

NOISE SOURCES

Noise sources that generate/distort magnetic fields similar to those as in the vehicle testing include electrically powered components such as mirrors, windows, steering columns, powered seats, moon roofs, wipers, blinkers etc. Other noise sources listed below include:

Environment Dependent Noise Sources:

- Latitude/Longitude (Magnetic Declination and field variation around the globe)
- Electric Power Lines
- Bridges/Tunnels
- Sewer and piping
- Steel in concrete roads
- Geological magnetically attractive deposits such as iron
- Rail Road Tracks
- Solar Storms
- Buildings
- Manholes, light posts, stoplights (low speed consideration)

Vehicle Dependent:

- Vehicle Dynamics (acceleration, loading, turning, uphill/downhill, braking, etc.)
- Yaw
- Pitch
- Roll
- Vehicle Doors, (eg. Trunk open)
- Vehicle electrical loads (eg. mirrors, windows, moon roof, wipers, brake lights, seat motor, rear defrost, etc, use vehicle network and cal to adjust for)
- Vehicle Loading (eg 1-4 people influencing weight and ride height) maybe use tire pressure sensors
- Magnetic attractive material near sensor including (tool box, golf clubs, speaker box)
- Sensor Drift
- High power stereo systems
- Trailer towing

CALIBRATION METHOD

Similar to the strategy employed using electronic compasses the driver can drive in clockwise and counterclockwise directions to populate a look up table or generate an equation curve. This can be used in against another source such as a GPS that is not a magnetically-influenced reference for direction and magnetic compass heading (transmitted through satellite means) or by another magnetic sensor placed near the typical compass heading.

From Figure 15 it can be seen that direction and vehicle dynamics affect the magnetic field intensities for traveling in different directions, clockwise versus counter clockwise combining the effects of yaw, pitch, and roll.

Through a sensor fusion approach and neural networks these effects can be minimized. Strategic placement of the sensors can assist to cancel out noise effects. Sensor fusion using other sensors on the vehicle network such as accelerometers IVD, wheel speeds, and smart junction box info will greatly help in achieving a solution for calibrating the sensor to a vehicle magnetic model so that other vehicles in the blind-spot zone can be detected.

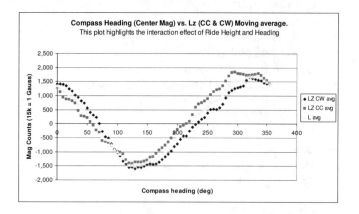

Figure 15: Z Component Driving in Circles for Calibration Curve Population

Deviation from the average value in Figure 15 is due from the changing near field nonlinear response characteristics of the body, suspension, and other magnetic materials in close proximity to the sensor. Vehicles or ferrous material to the side of the sensor always cause the Z-axis direction to increase, however where the material is located (above or below the sensor) can cause the Z-axis to increase or decrease. This is why we see a shifted sine or cosine function when driving in these circles and can be expressed by the equation below for the specific test vehicle equipped with the magnetic sensors.

$$Z\text{-axis} = 1570*COS(2*3.141*(\text{Sensor Data}+45)/360)+270$$

CONCLUSION

A Blind-spot warning system has been developed that is capable of detecting the presence of other vehicles in the defined zone 6-12 ft adjacent and rear of the vehicle and sensor placement. Smart algorithms using information from the vehicle bus provide a sensor fusion method to reject noise and use vehicle information for calibration strategy. Temporal and signal strength correlations applied to one or more sensors were able to detect vehicles at high speeds on the freeways and other road types and weather conditions.

One solution for quick implementation is to adapt the automotive commodity electronic compass for this application. This would yield cost savings as it is already an automotive grade component. With a huge market penetration of 30-50% in new vehicles, reusing this existing hardware will provide additional cost savings by relocating it to the rear of the vehicle with a slight change to software for enhanced operation to include vehicle detection and several other features compared to the typical compass heading information.

ACKNOWLEDGMENTS

We would like to thank Hanaan Elmessiri, Dan Miller, and David Dimeo for their research and project contributions.

REFERENCES

1. NHTSA (2000), "The Economic Impact of Motor Vehicle Crashes 2000," http://www.nhtsa.gov/people/ Economic/EconImpact2000/summary.htm
2. Michael J. Flannagan (2000), "Current Status and Future Prospects for Nonplanar Rearview Mirrors." SAE Technical Paper 2000-01-0324.
3. Michael J. Caruso and Lucky S. Withanawasam, "Vehicle Detection and Compass Applications using AMR Magnetic Sensors," www.ssec.honeywell.com
4. FEDERAL MOTOR VEHICLE SAFETY STANDARD (49 CFR PART 571) FMVSS 111 REARVIEW MIRRORS
5. Jim Riesterer, Topographical Maps and Magnetic Declination, http://geology.isu.edu/geostac/Field_Exercise/topom aps/ mag_dec.htm
6. Love, J. J. & Constable, C. G., 2003. Gaussian statistics for palaeomagnetic vectors, Geophys. J. Int., 152, 515-565, doi:10.1046/j.1365-246X.2003.01858.x

2004-01-1758

The Roles of Camera-Based Rear Vision Systems and Object-Detection Systems: Inferences from Crash Data

Michael J. Flannagan and Michael Sivak
The University of Michigan

ABSTRACT

Advances in electronic countermeasures for lane-change crashes, including both camera-based rear vision systems and object-detection systems, have provided more options for meeting driver needs than were previously available with rearview mirrors. To some extent, human factors principles can be used to determine what countermeasures would best meet driver needs. However, it is also important to examine sets of crash data as closely as possible for the information they may provide. We review previous analyses of crash data and attempt to reconcile the implications of these analyses with each other as well as with general human factors principles. We argue that the data seem to indicate that the contribution of blind zones to lane-change crashes is substantial. Consequently, as decisions are made about electronic countermeasures for lane-change crashes, the first goal should be to support drivers' own perceptual capabilities by eliminating blind zones, as could be done with camera-based displays. More elaborate systems that could bypass driver perception with artificial object-detection capabilities might also be useful, because they could address crashes attributable to driver weaknesses in areas beyond basic sensation, such as attention or decision making. However, design of such systems should take into account any remaining blind zone problems.

INTRODUCTION

The forward field of view is clearly of primary importance during the vast majority of driving. The maneuver of backing up is a major exception, but in that case the speeds involved are low enough that a driver's perceptual needs—for example, for distance information—are different in many ways from those involved in most driving. However, even at typical driving speeds there are some maneuvers—primarily lane changes and merges—in which a driver needs to be aware of traffic to the rear and sides. Rearview mirrors have been the primary means of providing drivers with that awareness; but limitations in their placement and fields of view have forced compromises between size of field of view, which is maximized by nonplanar mirrors, and quality of field of view, which is maximized by planar mirrors (Flannagan, 2000). Recent improvements in electronic sensors and displays have made new alternatives to mirrors more viable. Camera-based displays can now provide images to drivers with higher quality and less expense. Various other types of sensors, including radar and laser sensors, can provide drivers with awareness of nearby vehicles in more abstract forms, without full images of the traffic scene. The information from such sensors can be displayed to the driver in visual form (e.g., an indicator light might be displayed on a side rearview mirror to indicate the presence of a vehicle in a blind zone) or auditory form (e.g., an auditory warning might indicate to a driver that there is a vehicle in conflict with an intended or ongoing lane change).

The wider set of possibilities for rear vision afforded by new electronic aids may provide important benefits to drivers, but making the best choices among these possibilities requires that we understand drivers strengths and weaknesses with regard to rear vision. The purpose of this paper is to review recent analyses of crash data for the most relevant type of maneuver—lane changes—and to discuss what these analyses imply about driver needs for rear vision or detection aids.

Our review of the crash data is primarily intended to provide information about the likely benefits of the alternative countermeasures that we mentioned above—camera-based systems that provide drivers with images of surrounding traffic that are similar to those provided by rearview mirrors, and object-detection systems that provide drivers with information about the presence of other vehicles in more abstract ways. These two types of systems are naturally suited to address different deficits in drivers' awareness of the traffic situation: information deficits and attention deficits. The distinction we have in mind here is very similar to that made by Rumar (1990) when he argued that the basic driver error for crashes in general is late detection and that selection of countermeasures should be guided by a distinction between late detection problems caused by perceptual failures and those caused by cognitive failures. It is also closely related to the distinction that is often made in the human factors literature between displays and warnings.

Displays present information in a passive form; for the information to be used the driver has to actively attend to it. Warnings are designed to be relatively intrusive and actively acquire the driver's attention. In these terms, camera-based systems are usually designed as displays, and object-detection systems are usually—although not always—designed as warnings. Countermeasures that involve displays are appropriate when drivers lack access to certain specific information (such as visual information about vehicles in a blind zone). Countermeasures that involve warnings are appropriate when drivers are not attending to information that is already potentially available to them, either because they cannot sort through an overload of distracting information or because their overall arousal level is too low.

We will review two sets of analyses of lane-change crash data that were done in the 1990s—analyses that were done with different approaches and motivated by different issues. The first set of analyses was done on U.S. crash data and was intended to provide a comprehensive picture of the nature of lane-change and merge crashes (Wang & Knipling, 1994), and to provide guidance with regard to countermeasures for lane-change crashes (Chovan, Tijerina, Alexander, & Hendricks, 1994). The second set of analyses was motivated specifically by the issue of how planar and nonplanar rearview mirrors—and the blind zones that some mirror systems exhibit—affect overall safety (Luoma, Flannagan, & Sivak, 2000; Luoma, Sivak, & Flannagan, 1995; Schumann, Sivak, & Flannagan, 1998). These studies were done with data from Finland and the United Kingdom, which, unlike the U.S., use both planar and nonplanar driver-side rearview mirrors. The two sets of analyses came to somewhat different conclusions. In this paper we review both sets of analyses and attempt to determine what the combined results of the two imply about countermeasures for lane-change crashes.

COMPREHENSIVE CRASH-DATA STUDIES

Wang and Knipling (1994) used U.S. crash data, primarily from the General Estimates System (GES) and the Fatality Analysis Reporting System (FARS). They reported many results, which together provide an overview of the size and nature of the problem with lane-change and merge crashes. For example, in 1991 these crash types included about 244,000 police-reported crashes, 60,000 injuries, 6,000 serious injuries, and 224 fatalities. On average, lane-change and merge crashes tend to be less injurious to people than other types of crashes, accounting for 4% of all police-reported crashes but only 0.5% of fatalities. About 83% involve only property damage.

Chovan et al. (1994) cited some of the results from Wang and Knipling and reported additional results, some of which were based on an expert analyst's subjective assessments of police accident reports for samples of lane-change crashes. The work by Chovan et al. was especially intended to guide the development of countermeasures that fell under the domain that was then referred to as Intelligent Vehicle-Highway Systems (IVHS), and is now generally referred to as Intelligent Transportation Systems (ITS). Such countermeasures typically involve relatively elaborate electronic sensors and some form of data processing. In addition to their analysis of crash data, Chovan et al. discussed a range of possible ITS countermeasures and presented a model of lane changes that could be used in designing and evaluating such countermeasures. The aspect of their report that is most important for our current purposes is their conclusion with regard to the importance of blind zone issues within the general problem of lane-change crashes. Based on their analyses of crash data, they came to the following conclusion with regard to countermeasures:

"... these data suggest that crash avoidance concepts focused only on blind spot monitoring will not be adequate. Detection coverage over the full length of the SV [the subject vehicle, the one engaged in a lane change], on both sides, is needed" (p. 16).

They based their conclusion on two results from the crash data—a finding by Wang and Knipling (1994) that lane-change crashes are about equally likely to either side, and one of their own results from expert analysis of police accident reports. We will briefly review each of these results.

The finding from Wang and Knipling is shown in Figure 1, which shows the distribution of lane-change crashes for passenger vehicles by the location of the initial impact on the vehicle that changed lanes. Most of the impacts were on the sides, and they were about equally common on the right and left. Although they were not explicit about how they interpreted these data, Chovan et al. presumably assumed that blind zone problems will primarily result in initial impacts to the left side of the vehicle that changes lanes. That would be consistent with a belief that the worst blind zone problem is on the left side of passenger vehicles, where the exterior rearview mirrors are flat and therefore have limited fields of view. That belief is supported by data on the fields of view that are available to drivers of passenger cars and light trucks, taking into account how the mirrors are actually aimed in use (Reed, Lehto, & Flannagan, 2000; Reed, Ebert, & Flannagan, 2001).

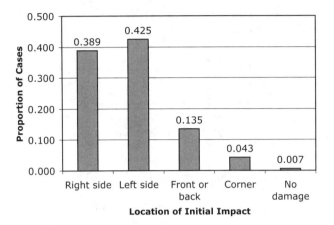

Figure 1. Distribution of lane-change crashes for passenger vehicles by location of initial impact. From Wang and Knipling (1994).

The analysis of police accident reports by Chovan et al. (1994) involved the categorization of crashes shown in Table 1. They identified five relationships that the two vehicles involved in a lane-change crash might be in prior to the lane-change maneuver. Three of these were variations of "proximity" crashes in which there was little or no longitudinal gap or difference in speed between the vehicles: forward overlap, side-by-side, and rearward overlap. "Forward" and "rearward" refer to the position of the vehicle that changed lanes in the maneuver relative to the other vehicle involved. Thus, a problem with the blind zone is most likely to be implicated in crashes coded as "forward overlap." In such cases, the driver who is intending to change lanes starts out just ahead of the vehicle that is struck, indicating that the struck vehicle is not in the driver's forward field of view and suggesting that it may be in a blind zone. The premaneuver relationships of the remaining crashes were designated as "fast approach" and were subdivided into "forward" and "rearward" (again based on the relative premaneuver position of the vehicle that changed lanes).

Using the categories in Table 1, Chovan et al. were able to classify 16 cases from the 1992 Crashworthiness Data System (CDS) data set and 66 cases from the 1991 GES data set. The results are shown in Figure 2. The great majority of the cases were assigned to the proximity subtype, indicating that in most lane-change crashes, just prior to the lateral movement by one of the vehicles,

the two vehicles are overlapping or nearly overlapping longitudinally and are traveling at about the same speed. In the CDS data set, which is biased toward more serious outcomes, most cases (70% of all the classifiable crashes) were assigned to the forward overlap variation, indicating that inability to see in a rearward blind zone may have played a role in causing them. In the GES data, the proximity crashes are more evenly distributed over the three variations. In the side-by-side and rearward overlap variations the vehicle that did not change lanes can be expected to be in the direct field of view of the lane-changing driver just prior to the lane change, suggesting that inability to see in a blind zone was not an issue in those crashes.

Table 1. Categories of premanuever relationships used by Choven et al. (1994).

Subtypes	Variations
Proximity	Forward overlap
	Side-by-side
	Rearward overlap
Fast approach	Forward
	Rearward

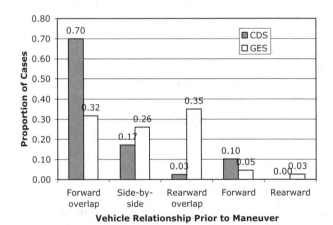

Figure 2. Distribution of lane-change crashes by vehicle relationship prior to maneuver for two crash data sets (CDS, GES). "Forward" and "rearward" refer to the position of the vehicle that made the lane change, relative to the other vehicle involved. From Chovan et al. (1994).

STUDIES FOCUSED ON BLIND ZONES

The set of analyses that were focused specifically on the roles of mirror fields of view and blind zones in lane-change crashes (Luoma, Flannagan, & Sivak, 2000; Luoma, Sivak, & Flannagan, 1995; Schumann, Sivak, &

Flannagan, 1998) all used the same general approach. They used crash data for passenger cars from countries in which both planar and nonplanear exterior rearview mirrors are reasonably common on the driver-side, and in which nonplanar mirrors are nearly universal on the passenger side. They determined how the frequency of lane-change crashes to the driver side was related to driver-side mirror type, using lane-change crashes to the passenger side as a control for exposure. The first study by Luoma et al. (1995) examined 407 Finnish crashes, and the study by Schumann et al. examined 3,038 British crashes. The second study by Loma et al. (2000) was a follow-up on their earlier report and included substantially more data. They examined lane-change crashes in Finland between 1987 and 1998. After eliminating cases in which the lane-changing driver was not considered fully at fault and cases in which drunk drivers were involved, the data set included 1,062 lane-change crashes. The results in all three studies were broadly consistent. Here we will review only the most recent of the three reports.

Luoma et al. (2000) used three categories to classify driver-side mirrors: multiradius (mirrors that are convex over their entire surface, but are more strongly curved toward the outboard edge), spherical convex (mirrors that have a single radius of curvature over their entire surface), and planar. Figure 3 shows the 1,062 cases analyzed by Luoma et al. classified by the driver-side mirror type and the direction of movement of the vehicle that changed lanes. Luoma et al. used the data in Figure 3 to calculate the relative effectiveness of different types of driver-side mirrors in reducing lane-change crashes to the driver side. Their measure of effectiveness used the following formula:

$$E = 100 \times \left(\frac{D_1/P_1}{D_2/P_2} - 1 \right)$$

where

D_1 is the number of crashes to the driver side for cars with mirror Type 1,

P_1 is the number of crashes to the passenger side for cars with mirror Type 1,

D_2 is the number of crashes to the driver side for cars with mirror Type 2, and

P_2 is the number of crashes to the passenger side for cars with mirror Type 2.

The effectiveness measure is thus an estimate of the percent reduction in driver-side lane-change crashes that would be expected for vehicles with mirror Type 1 relative to vehicles with mirror Type 2. If the measure is negative, mirror Type 1 is associated with fewer such crashes.

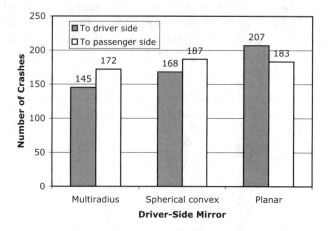

Figure 3. Crash data from Luoma et al. (2000).

Figure 4 shows values of the effectiveness measure, along with 95% confidence intervals, for two comparisons: nonplanar driver-side mirrors (combining multiradius and spherical convex mirrors) versus planar driver-side mirrors, and multiradius driver-side mirrors versus spherical convex driver-side mirrors. The relative effectiveness of multiradius and spherical convex mirrors is nearly neutral, and the corresponding confidence interval easily spans zero difference, indicating that the difference between these two types of convex mirrors is not statistically significant. Nonplanar mirrors on the driver side are associated with 23% fewer crashes to the driver side, and the confidence interval around that value excludes zero, indicating that the difference between nonplanar and planar mirrors is statistically significant. Because of the difference in fields of view associated with nonplanar and planar mirrors, this reduction may be a result of reducing or eliminating the blind zone on the driver side. These data presumably result from the combination of all aspects of how planar and nonplanar mirrors are used in the real world. If there are crashes resulting from misjudgments of speed or distance in nonplanar mirrors, those crashes should be included in this calculation, and the result should therefore be a *net* reduction in driver-side crashes resulting from the combined effects of larger fields of view and distortions of the fields of view.

The magnitiude of the benefit of nonplanar mirrors in the Luoma et al. (2000) data appears to be strongly influenced by driver age. Figure 5 shows the relative effectiveness of nonplanar and planar mirrors calculated separately for different age groups of the driver who changed lanes. These results suggest that the benefits of nonplanar driver-side mirrors are relatively high for the younger and older age groups but nearly neutral for middle-aged drivers. Luoma et al. discuss possible explanations for the age effect, including the possibility that younger drivers have trouble with the blind zone because they haven't had enough experience to be fully aware of it, and the possibility that older drivers have trouble with it because of decreased head and neck mobility.

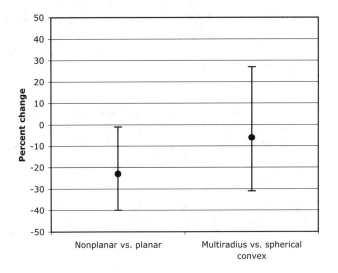

Figure 4. Effectiveness measures for comparisons between driver-side mirror types in terms of percent change in lane-change crashes to the driver side, with 95% confidence intervals (Luoma et al., 2000).

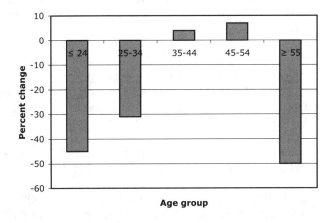

Figure 5. Effectiveness measures for nonplanar versus planar driver-side mirrors in terms of percent change in lane-change crashes to the driver side by driver age groups (Luoma et al., 2000).

COMPARISON ACROSS STUDIES

The conclusions of the two sets of analyses that we have reviewed have somewhat different emphases. However, the results are largely consistent, and we believe the conclusions are not as divergent as they might at first appear. Both sets of analyses suggest that blind zone crashes are a major component of lane-change crashes, and both are consistent with the conclusion that blind zone crashes are not the only significant component.

The most specific comparison that can be made across the two sets of analyses concerns the direction of movement of the vehicle that changes lanes in a crash. Figure 1 shows the distribution of lane-change crashes by point of initial impact for passenger vehicles in the U.S. data summarized by Wang and Knipling (1994). These data were part of the evidence used by Chovan et al. (1994) to argue that it is not adequate to address blind zone crashes. The relevant results from the Finnish data summarized by Luoma et al. (2000) are in Figure 3. The most comparable Finnish data are for cars with planar driver-side mirrors, since those vehicles had the same mirror configuration as the U.S. vehicles. The specific comparison is as follows: from Figure 1 (left .425, right .389) versus Figure 3 (driver side 207, passenger side 183). The ratio for Figure 1 is thus .425/.389 = 1.09 and the corresponding ratio for Figure 3 is 207/183 = 1.13. These results are reasonably close, suggesting that the role of blind zones in lane-change crashes may be similar in the U.S. and Finnish settings.

IMPLICATIONS FOR COUNTERMEASURES

It seems likely that any category of crashes—even a moderately specific one such as lane-change crashes—involves a mixture of causal mechanisms. The lane-change crash data suggest two major conclusions with regard to mechanisms: first, there appears to be a major subset of crashes that are related to visibility problems with blind zones, and, second, there nevertheless appears to be a substantial proportion of lane-change crashes that cannot be accounted for by blind zones or by any other simple deficit in driver vision that has been identified. Because crashes in this second set are not easy to assign to relatively specific causes, it is tempting to assign most or all of them to driver inattention.

To the extent that inattention is the problem, the most effective countermeasure would appear to be an active warning system—one that would not rely on the driver's monitoring of the roadway or of passive displays. Such a system would have to be able to detect potential conflicts and determine when to issue one or more levels of warning to the driver. Any system with the sensing and decision-making capabilities that would be required would be considerably more elaborate than a simple camera-based display, and by virtue of that level of "intelligence" would clearly fall into the domain of ITS solutions discussed by Chovan et al. (1994). It therefore seems likely that ITS solutions have an important role to play in any comprehensive approach to lane-change crashes.

However, because there appears to be a significant component of the overall lane-change problem that is attributable more to deficits in basic visibility (i.e., blind zones) than to deficits in attention, a comprehensive approach should also include consideration of how best to solve the visibility issues. In addition, if we assume that it is desirable to keep the driver aware of the traffic situation in as simple and direct a way as possible, then basic perceptual deficits, such as inability to see a vehicle in a blind zone, should be addressed (probably with optimized display systems) before addressing attention deficits (probably with warning systems). It would therefore be desirable to make sure that drivers are provided with the best possible fields of view, either from rearview mirrors or camera-based systems, before determining how best to supplement driver vision with intelligent aids.

Furthermore, even if intelligent aids for lane-change maneuvers are introduced without prior improvements in the blind zone situation, it would be important in designing such aids to recognize the effects of blind zones. For example, if a driver's lateral movement suggests that he or she is initiating a lane change, and an object detection system detects a vehicle in conflict with the apparent maneuver, the system should probably take into account whether or not the vehicle is in a blind zone as it decides when and how to warn the driver. The probability that the driver is actually aware of the conflicting vehicle, and the related probability that the lateral movement is not really indicative of an intention to change lanes, is probably higher for cases in which the vehicle is not in a blind zone. Also, the form of a warning should probably be stronger or more specific for cases in which a conflicting vehicle is in a blind zone. When a conflicting vehicle is in a driver's field of view even a mild, general warning might be sufficient to make the driver recognize the hazard.

CONCLUSIONS

Lane-change crashes appear to be caused by a mixture of problems with driver vision (blind zones) and problems with driver attention. Recent technical advances in "intelligent" aids for drivers, especially in object-detection systems that can warn drivers of vehicles that may be in conflict with lane changes, have the potential to address these problems and may be particularly valuable for addressing problems with driver attention. However, there have also been recent advances in camera-based rear vision systems that offer more options for addressing blind zone problems than have been available with rearview mirrors. A comprehensive approach to the problem of lane-change crashes should consider the best solutions to the basic problems of driver vision that can be provided by rearview mirrors and camera-based imaging systems before determining the best design for intelligent driver aids for changing lanes.

ACKNOWLEDGMENTS

We wish to thank Ichikoh Industries, Ltd. for their generous support of this research.

REFERENCES

1. Chovan, J. D., Tijerina, L., Alexander, G., & Hendricks, D. L. (1994). *Examination of lane change crashes and potential IVHS countermeasures* (No. Report No. DOT HS 808 071). Washington, D.C.: Department of Transportation, National Highway Traffic Safety Administration.

2. Flannagan, M. J. (2000). *Current status and future prospects for nonplanar rearview mirrors* (SAE Technical Paper Series No. 2000-01-0324). Warrendale, Pennsylvania: Society of Automotive Engineers.

3. Luoma, J., Flannagan, M. J., & Sivak, M. (2000). Effects of nonplanar driver-side mirrors on lane change crashes. *Transportation Human Factors, 2,* 279-289.

4. Luoma, J., Sivak, M., & Flannagan, M. J. (1995). Effects of driver-side mirror type on lane-change accidents. *Ergonomics, 38*(10), 1973-1978.

5. Reed, M. P., Ebert, S. M., & Flannagan, M. J. (2001). *Mirror field of view in light trucks, minivans, and sport utility vehicles* (No. Report No. UMTRI-2001-1). Ann Arbor: The University of Michigan Transportation Research Institute.

6. Reed, M. P., Lehto, M. M., & Flannagan, M. J. (2000). *Field of view in passenger car mirrors* (Report No. UMTRI-2000-23). Ann Arbor: The University of Michigan Transportation Research Institute.

7. Rumar, K. (1990). The basic driver error: late detection. *Ergonomics, 33*(10/11), 1281-1290.

8. Schumann, J., Sivak, M., & Flannagan, M. J. (1998). Are driver-side convex mirrors helpful or harmful? *International Journal of Vehicle Design, 19,* 29-40.

9. Tijerina, L. (1999). Operational and behavioral issues in the comprehensive evaluation of lane change crash avoidance systems. *Transportation Human Factors, 1*(2), 159-175.

10. Wang, J.-S., & Knipling, R. R. (1994). *Lane change/merge crashes: Problem size assessment and statistical description* (No. DOT HS 808 075). Washington, D.C.: National Highway Traffic Safety Administration.

Obstacle State Estimation For Imminent Crash Prediction & Countermeasure Deployment Decision-Making

Kwaku O. Prakah-Asante, Mike K. Rao and Gary S. Strumolo
Scientific Research Laboratory, Ford Motor Company

ABSTRACT

Predictive crash sensing and deployment control of safety systems require reliable and accurate kinematic information about potential obstacles in the host vehicle environment. The projected trajectories of obstacles in the path of the vehicle assist in activation of safety systems either before, or just after collision for improved occupant protection. This paper presents an analysis of filtering and estimation techniques applied to imminent crash conditions. Optimization of design criteria to achieve required response performance, and noise minimization, are evaluated based on the safety system to be activated. The predicted target information is applied in the coordinated deployment of injury mitigation safety systems.

INTRODUCTION

Methods to intelligently deploy safety systems with advanced sensor systems are areas under development in the automotive industry. Added information about the crash environment and vehicle interior from radar/vision systems has provided opportunities for development of new sensing and decision-making systems[1,2,3,4]. Intelligent algorithms in conjunction with advanced sensor information decide on what systems to deploy under various crash scenarios. Figure 1 shows statistics from the National Highway Traffic Safety Administration (NHTSA) [5], which indicate that about 47% of passenger car crashes were caused by frontal impacts, side impacts accounted for 31%, and 21% of all passenger crashes were caused by rear impacts. The highest occurring crash type, frontal imminent crash, is mainly referred to in this paper, although the techniques discussed are applicable to other crash types.

Estimation of the relative kinematic information between the host vehicle and the target vehicle allows for restraints activation before collision. Predictive sensing algorithms make imminent-crash decisions based on outputs from radar and vision sensors. When an accident is unavoidable, discrimination of the threat potential of obstacles in the vehicle path requires high confidence. Consequently, potential hazards in the sensor field-of-view require effective detection and tracking. Object tracking is performed to minimize variability in range and range-rate measurements, by applying filtering and estimation techniques to predict future target positions.

Various filtering and estimation techniques including Kalman filtering, alpha-beta filters, fixed gain filters, and adaptive filters have been employed in the literature as methods for smoothing and target state estimation [6,7,8,9]. For imminent crash decisions, time-to-impact values are in the order of milliseconds requiring fast processing for meeting impact estimation and deployment control performance requirements. This paper presents an analysis of filtering and estimation techniques with application to imminent crash conditions for a radar based system. The design and selection of the estimation method used for imminent collision prediction is critical, as well as the ability to be robust in range and range-rate prediction in the presence of missed detections. In addition the type of estimation method must take into account the countermeasure to be deployed to minimize the potential of false alarms. The impact of estimator design criteria on performance and noise minimization, are evaluated for specific safety system activation. Subsequent decision-making for activation of the safety systems is based on the predicted obstacle range, and velocity obtained from the estimators.

Section two of the paper presents analysis of filtering and target state estimation for imminent crash prediction. Section three presents collision countermeasure decision-making based on predicted kinematic information. Conclusions are presented in section four.

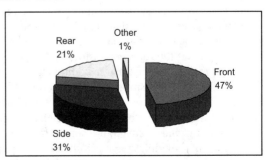

Figure 1. Vehicle crashes by point of impact

FILTERING & TARGET STATE ESTIMATION
FOR IMMINENT CRASH PREDICTION

Presented in this section is an analysis of filtering and target state estimation for imminent crash prediction using a radar based system. The design and synthesis of methods to account for effective noise minimization, tracking, and the type of countermeasure to be deployed is discussed.

Figure 2 shows a block diagram of the architecture of an integrated system for target state estimation and imminent crash prediction. Obstacle kinematic data from anticipatory radar sensors after object detection and signal processing are sent to the data association and correlation system. The assigned track outputs are input to the filtering and estimation algorithms for disturbance rejection and target state prediction. The kinematic information obtained is then sent to the decision-making system. The crash prediction and restraints deployment decision-making system determines the potential for a collision to occur, and intelligently activates or pre-arms restraints. The decision-making algorithms are based on the predicted obstacle range, and velocity information obtained from the estimators.

Example injury mitigation systems include airbags, reversible pretensioners, and pre-impact brake-assist systems. Activation of the above mentioned safety systems require reliable range and closing velocity information. The filter and estimator design criteria should therefore be tailored to the safety application to quantify the relative degree of robustness for the safety system operating range. Figure 3 shows time-to-impact values for relative velocities from 6-14 m/s for a vehicle near-zone range of 0-2m. Time-to-impact for the near-zone are in the order of 25 ms to 350 ms. Fast computation is required for decision-making and has to be accounted for in estimation techniques. Filtering and estimation techniques in addition have to account for head on collisions, offset and oblique crashes, and provide information for maneuvering situations especially for deploying countermeasure before crash occurs. Safety systems activated before collision require effective estimation systems to account for a maneuvering vehicle in near miss situations to avoid potential false alarms. Stringent measures are required to counteract false alarms especially if the safety system is not actively reversible. In the case of pre-arming of restraint systems the final activation decision is made after confirmation of impact. Noise minimization and accurate closing velocity measurement are more critical.

Noise in radar sensor data is typically from sensor calibration errors, and measurement statistical noise. Filtering and estimation methods are required to reduce the error in the measured position, and to estimate the target velocity or acceleration to predict future target positions. Challenges for estimator design include the fact that in addition to noisy radar measurements, the target's position is observed only at discrete points in time. The intent of the target is not known, and the motion model of the target is not known. Consider design and synthesis of filtering and estimation schemes for a 24 GHz pulse radar systems providing range information. The update rate of the sensor is about 20 ms. Correlation and data association is accounted for and a measurement assigned to a

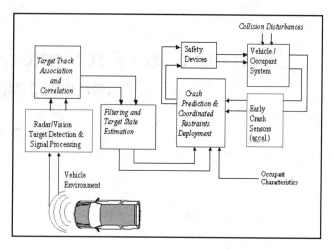

Figure 2. Imminent Crash Prediction and Safety System Deployment Decision-Making

track is used to update the track.

Low pass filtering minimizes sensor noise, and the velocity can be obtained from finite difference approximation. The low pass filter is of the digital form

$$y_n = \gamma y_{n-1} + (1-\gamma)r_n \qquad (1)$$

where y is the filtered output, r is the radar range input at sample time n, and γ is the characteristic coefficient of the filter. The low pass filtering minimizes sensitivity to high frequency noise. The rate-of-change of the range can be computed from the backward-difference approach with the discrete form of the differential operator s given by,

$$s = \frac{\left(1 - z^{-1}\right)}{T} \qquad (2)$$

where T is the sample time, and z the discrete transform operator, to provide velocity. This method considers motion along a single coordinate only, or decoupled coordinates, and a uniform sampling rate. Noise entering the different coordinates is assumed to be uncorrelated, and contain constant process and measurement noise. In the case of target obstacles very close to the vehicle, if the radar sensor

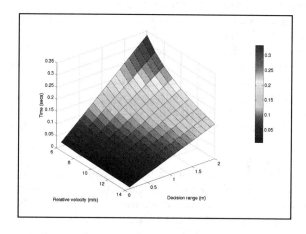

Figure 3. Time for near-zone scenario predictive decision-making

provides contacts for the target overtime at a periodic rate with no missing contacts or false contacts, such a simple filter could meet performance specifications. Faster computation is obtained for the filter, however, it is not optimal if noise statistics change with time, and the estimated velocity is more susceptible to noise. If reliable Doppler velocity is available filtering of the velocity would only be required. For imminent crash conditions such as pre-arming restraints, such a systems could suffice if the performance specification for range and closing velocity accuracy are met. Countermeasures activated before impact would pose more challenges for the filters in the presence of a maneuvering target.

The Alpha-Beta filter is another example of fixed gain filters, which is considered for filtering and estimation for imminent crash conditions. A prediction model for range and velocity is given by

$$x_n^{'} = \hat{x}_{n-1} + \hat{v}_{n-1}T \quad (3)$$

$$v_n^{'} = \hat{v}_{n-1} \quad (4)$$

where x is the range and v is the velocity predicted values, and T is the sample time. Update equations are represented by

$$\hat{x}_n = x_n^{'} + \alpha\left(x_n^{sens} - x_n^{'}\right) \quad (5)$$

$$\hat{v}_n = v_n^{'} + \frac{\beta}{T}\left(x_n^{sens} - x_n^{'}\right) \quad (6)$$

where α and β are the steady state gains for updating the radar measurements. The steady state gains are expressed in terms of the tracking index

$$\Lambda = \frac{qT^2}{\sigma} \quad (7)$$

The tracking index is proportional to the ratio of the motion uncertainty and the measurement uncertainty. The value q is chosen to reflect the maximum acceleration to track, T is the sensor sample period, and σ is based on the statistics of the measurement noise. Based on the computed tracking index, the α and β are obtained as,

$$\alpha = -\frac{1}{8}\left[\Lambda^2 + 8\Lambda - (\Lambda + 4)\sqrt{\Lambda^2 + 8\Lambda}\right] \quad (8)$$

$$\beta = 2(2-\alpha) - 4\sqrt{1-\alpha} \quad (9)$$

The filter requires a constant process and measurement noise, and a uniform sampling rate. The filter architecture facilitates relatively straightforward real-time implementation and could be a viable cost effective option if fast computational resources are not available. When collision is unavoidable the predicted range and closing velocities can effectively be used if they are within performance specifications. However, measures are required to validate the estimation accuracy for decision-making to avoid erroneous command generation of safety devices. Performance of the alpha-beta filer is diminished for smaller values of the tracking index. If the tracking index is small, the alpha-beta filter constant gains will be small. An evasive action by the host vehicle in the presence of a target would be more of a challenge for tracking

with this filter. A detection system to determine if a maneuvering action is taking place can be used to tailor activation of countermeasures within the performance limitations of the estimated obstacle kinematic information.

Consider the Kalman filter for filtering and state estimation of the radar range and Doppler velocity. The vehicle dynamical sensor system affected by external disturbances can be represented in state space as

$$\underline{x}_{k+1} = \Phi_k \underline{x}_k + B_k \underline{w}_k \qquad w_k \approx N(0,Q) \quad (10)$$

$$\underline{z}_k = C_k \underline{x}_k + \underline{v}_k \qquad v_k \approx N(0,R) \quad (11)$$

where $\underline{x}_{k+1}$ is the state vector at time $t = k+1$, Φ is the state-transition matrix which determines successive states, w is the system process noise which is normally distributed with zero mean and Q variance. Q is the process noise matrix, B is the matrix relating process input disturbances to the state. C is the output measurement matrix, $\underline{z}$ is the vector of measurements, $\underline{x}$ the state vector, and v the radar measurement noise. The following update equations are used to determine the state variables of range and velocity. The predicted values at time $t=k+1$ is given by

$$\underline{x}^{'}_{k+1} = \Phi_k \hat{\underline{x}}_k \quad (12)$$

and the prediction error covariance matrix given by

$$P^{'}_{k+1} = \Phi_k P_k \Phi_k^T + B_k Q_k B_k^T \quad (13)$$

The update gain is given by

$$K_{k+1} = P^{'}_{k+1} C_{k+1}^T \left[C_{k+1} P^{'}_{k+1} C_{k+1}^T + R_{k+1}\right]^{-1} \quad (14)$$

and the estimates of the range and velocity states given by

$$\hat{\underline{x}}_{k+1} = \underline{x}^{'}_{k+1} + K_{k+1}\left[\underline{z}_{k+1} - C_{k+1}\underline{x}^{'}_{k+1}\right] \quad (15)$$

The estimate error covariance is given by

$$P_{k+1} = \left[I - K_{k+1}C_{k+1}\right]P^{'}_{k+1} \quad (16)$$

where P is the covariance matrix. The Kalman filter minimizes the variance of the estimation error in the radar measurements.

The Kalman filter can be used in varying imminent crash environments by changing a few parameters. Noise statistics can change in time with the position. The filter automatically accounts for missed detection and non-uniform sampling rates, which improves prediction and countermeasure decision-making. The provided measure of the estimated target state accuracy through the covariance matrix facilitates decision-making should erroneous data become available and minimizes potential false alarms. The motion model can be augmented and fused with track-level data from multiple sensors, including the vehicle dynamics for path estimation and driver in the loop models, or in choosing the best track estimate reported by several sensors. Incorporation of the

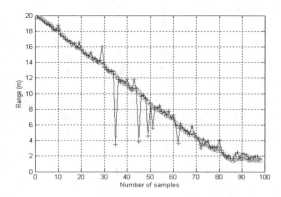

Figure 4. Filtering and state-estimation of range data

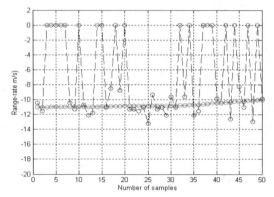

Figure 5. State estimation in the presence of missed detections

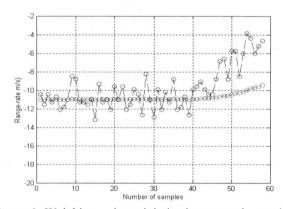

Figure 6. Weighing noise minimization more than trajectory change

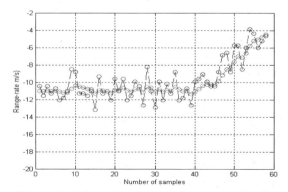

Figure 7. Noise and trajectory change optimization

Kalman filter in imminent crash systems improves robustness to handle deviations from the assumed motion model.

Application of the Kalman filter or Extended Kalman filter for non-linear systems is computationally intensive compared to fixed gain filters; implementation of the steady-state gains reduces that requirement. Effective motion models and measurement models are necessary, in addition to knowledge of the statistics of the radar measurement noise. If computational resources are available for real-time implementation, and prediction can be made within 40-150 ms before collision without detrimental delays the Kalman filter is a viable option. Effective modeling of target maneuvers requires augmenting the architecture of the filter.

Adaptive estimation methods can be applied to account for vehicle trajectory changes. Variable gain filters have the ability to adapt to changes in obstacle trajectory. Essentially, determine when a target maneuver starts or ends and adjust the estimator coefficients accordingly. For example, for the Kalman filter design, changing the process covariance Q when maneuver is detected, or adjusting the filter model when a maneuver is detected is applicable. The maneuver can also be modeled as a deterministic unknown input. Another consideration is the application of multiple parallel filters based on different maneuver levels. Each filter is based on a different motion model. Different target models are formed and the information combined in an optimum way in the form of the interactive multiple model (IMM). The overall target kinematic state estimate is then a weighted combination of the estimates from the multiple individual filter estimates. There are many trade-offs in using the IMM algorithm, in particular with respect to how many filters to use and which and how many different motion models need to be considered, and if real-time computational resources are cost-effective.

Figure 4 shows data obtained from an instrumented vehicle with a 24 GHz radar sensor, which provided range and velocity from Doppler measurements of a target. The update rate of the sensor was 20 ms. Figure 4 also shows Kalman filtering of the reported range in conjunction with a pre-filter to remove outliers from the data based on the data characteristics. Noise in the range data is filtered, and the relative range effectively tracked. The impact of missed detection on filtering and estimation for imminent crash decision-making was investigated by considering Doppler range-rate information from data with considerable variability in a time span of 50 samples. Figure 5 shows filtering and estimation of the range-rate data with a Kalman filter weighted for noise minimization. Noise disturbance rejection was considered priority. Potential error in the measured range-rate is minimized. If detections are not reported it is important the estimation algorithms effectively track the target dynamics. If satisfactory estimation is not obtained form a particular sensor, information is sent to the safety decision-making systems to employ other supplemental measures from other sensors.

Figures 6 and 7 show performance of designed estimators for data obtained from the instrumented vehicle. The data duration encompasses a scenario where there is a change in the trajectory of the reported Doppler range-rate. In Figure 6 the Kalman filter design was targeted mainly for noise minimization. Effective noise minimization was achieved for up to about 40 samples. However when there was a trajectory change the filter was not as robust in estimating the change. In Figure 7, the filter accounts for the trajectory change. The filter design is more robust to unexpected changes. The design targeted both noise suppression and trajectory track dynamics. The process noise covariance was large enough to cover the trajectory change. An adaptive filter scheme, which detects when maneuvering starts and adjusts filter structure and parameters accordingly effectively, improves the performance of filtering and state estimation techniques. The design and synthesis of filtering and estimation techniques for imminent crash prediction are implemented in the context that they improve decision-making of the safety systems, they are computationally manageable, and performance robustness is maintained. Subsequent, decision-making for activation of respective safety systems are based on the predicted obstacle range, and velocity information obtained from the estimators.

COLLISION COUNTERMEASURE DECISION-MAKING

Advanced occupant protection systems require added information about crash conditions, and adaptive systems to facilitate tailoring of safety systems for improved injury mitigation. This section presents collision countermeasure decision-making based on the predicted target information obtained from remote sensor information after filtering and state estimation. Example collision countermeasures include restraints pre-arming, safety systems activated before impact including motorized pretensioners, and pre-impact brake-assist systems.

A typical frontal impact takes about 100-120 ms during which time the occupant's longitudinal movement is restrained by the seat belt and cushioned by the airbag system. Anticipatory crash sensing may provide means for earlier air bag deployment initiation in crashes, and thereby allow more time for moderate bag inflation than is now reasonable with current crash sensing technology. The seat belt pretensioner removes slack in the seat belt webbing. This is instrumental in keeping the occupant in the seat prior to the deployment of airbags. Reversible pretensioners have been suggested as a way to establish the position of an occupant before a collision, from information from predictive sensors that provide information about the vehicle external environment. The time-constants for the pretensioner range between about 140-200 ms. In order to achieve optimum functionality, the microprocessor controlled restraint systems are required to deploy at appropriate times, and be well coordinated to meet related safety objectives including minimization of head and chest decelerations, upper neck forces and moments, and chest deflection.

Figure 8 shows a schematic of the imminent crash prediction and the deployment decision-making systems integrated with airbags, a reversible pretensioner, and a brake system. Potential obstacle state dynamics in the vehicle environment, and host vehicle dynamics are sent to the crash prediction block through a control area network. The output from the decision-making block is sent to the local restraint systems algorithms to augment the decision-making process for improved system functionality and robustness. After determining that a crash is imminent the system decides what restraint devices are to be activated and recommends deployment of the safety device before/after the collision. A signal is sent to the restraint Restraints Control Module (RCM) to assist in decisions for multistage air-bag activation, should collision occur. The decision-making system systematically decides what safety actuators are to be deployed and at what time they should be deployed before or after impact.

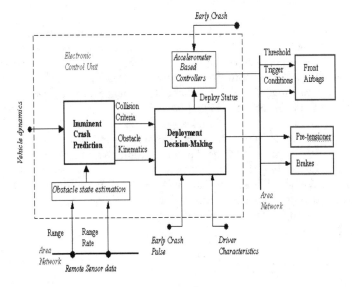

Figure 8. Advanced safety sub-systems for coordinated countermeasure decision-making

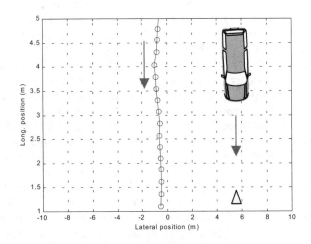

Figure 9. Collision scenario with host vehicle and target obstacle

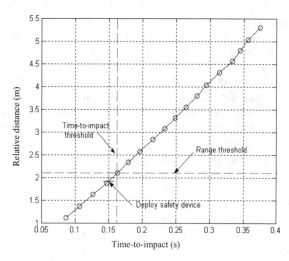

Figure 10. Predicted collision and deployment decision-
making status

Consider real-time decision-making for activation of reversible pretensioner based on obstacle state information. The hardware consists of a 24 GHz radar sensor equipped vehicle prototype incorporated with electronics and microprocessors to facilitate real-time embedded computation. The vehicle speed is measured with an independent optical sensor, and a radial potentiometer measures the steering wheel angle position. A Linear Variable Differential Transformer (LVDT) is used to estimate the accelerator-pedal, and brake-pedal position. The brake system pressure is measured with a silicon strain gage, and the angular velocity of the vehicle is measured with a yaw rate sensor. Figure 9 shows a test scenario for frontal deployment activation of the pretensioner. A target obstacle is placed in the path of the host vehicle for non-destructive testing. The obstacle is run over by the sensor-equipped vehicle with a speed of about 13.2 m/s.

Figure 10 is a plot indicating the predicted collision time, and decision-making for activation of the motorized pretensioner. Control command generation to activate the safety device is based on the predicted time-to-impact, once system safeguards are satisfied. A range threshold of about 2.1 m, and a time-to-impact of about 0.16 s were determined as thresholds for activation based on the pretensioner time constant, and the relative velocity of 13.2 m/s. As depicted in Figure 9, the signal to deploy the pretensioner is activated about 0.15 s before impact. Activation of the resettable pretensioner mechanism before collision assists in establishing the occupant position before a potential collision occurs for improved injury mitigation. Subsequently, a confirmation signal is sent to the airbag RCM with the status of pretensioner deployment to prepare for multistage air-bag activation after impact is confirmed. The pretensioner is only activated if sensing, estimation, and occupant characteristic requirements are satisfied. If satisfactory estimation is not obtained diagnostic messages are sent to the safety decision-making systems to employ other supplemental safety measures based on other sensor information. New and advanced safety systems are subject to broad testing before introduction in production vehicles to account for performance robustness and reliability under various vehicle collision, and near-miss scenarios.

CONCLUSIONS

An analysis of filtering and estimation techniques applied to imminent crash conditions was presented. The design, and synthesis of methods to account for effective noise minimization, tracking, and the type of countermeasure to be deployed was presented. From an integrated system perspective, it is critical to account for the countermeasure dynamics, crash scenarios, applicable potential maneuvers, and computational resources in design. Filter and estimator designs targeted both noise suppression and trajectory change dynamics. Adaptive filter schemes, which detect maneuvering, states and adjust filter structure and parameters accordingly, improves the performance of filtering and state estimation. Predicted target information was applied in the activation of a reversible collision countermeasure device. The integrated Imminent Crash Prediction and Safety System Deployment Decision-Making system demonstrated the methodology for augmenting the functionality of safety systems for improved safety.

REFERENCES

1. Becker, J.C., Simon, A.,"Sensor and Navigation Data Fusion for an Autonomous Vehicle," Proceedings of the IEEE Intelligent Vehicles Symposium," pp.156-161, 2000.
2. Swihart, W.R.,and A. F. Lawrence, "Investigation of sensor requirements and expected benefits of predictive crash sensing," SAE Paper No. 950347.
3. Prakah-Asante, K.O., Rao, M.K., Morman,K.N., Strumolo, G.S., "Supervisory Vehicle Impact Anticipation & Control of Safety Systems," IEEE CCA/ISIC Publication, pages 326-330, 2001.
4. Woll,J.D.,"Radar Based Vehicle Collision Warning Systems," SAE 94C036.
5. NHTSA, *Traffic Safety Facts*. U.S. Department of Transportation., 2000.
6. Castella, F.R., "An Adaptive Two-Dimensional Kalman Tracking Filter," IEEE Trans. Aerospace and Electronic Systems, AES-16: pp.822-829, 1976.
7. Stiller et al., "Multi-sensor Obstacle Detection and Tracking", IEEE International Conference on Intelligent Vehicles, pp. 451-457, 1998.
8. Kalata, P.R., "The Tracking Index: A Generalized Parameter for Alpha-Beta-Gamma Target Trackers". IEEE Trans. Aerospace and Electronic Systems, AES-20: pp.174-182, 1984.
9. Wellstead, P.E., and Zarrop, M.B., Self-tuning Systems: Control and Signal Processing, John Wiley & Sons. West Sussex, England, 1991.

2001-01-0317

A 360x226 Pixel CMOS Imager Chip Optimized for Automotive Vision Applications

Frank J. Schauerte, John R. Troxell and Randy A. Rusch
Delphi Delco Electronics Systems

Larry M. Oberdier and Marie I. Harrington
Delphi Research Labs

ABSTRACT

Multiple automotive systems are now being developed which require an imager or vision chip to provide information regarding vehicle surroundings, vehicle performance, and vehicle passenger compartment status. Applications include lane departure, lane tracking, collision avoidance, as well as occupant position, impaired driver, and occupant identification. These applications share many requirements, including robust design, tolerance for the automotive environment, built in self-test, wide dynamic range, and low cost. In addition, each application has unique requirements for resolution, sensitivity, imager aspect ratio, and output format. In many cases, output will go directly to vehicle systems for processing, without ever being displayed to the driver. Commercial imager chips do not address this wide spectrum of requirements. A CMOS imager chip has been designed to address these unique automotive requirements. An automotive grade, 1.2 micrometer minimum feature size process, which offers proven automotive environmental durability, has been used to fabricate two prototype imager chips. Results for both 128x128 pixel array chips and 360x226 pixel array chips are documented. Both chips incorporate capabilities for enhanced dynamic range, near infrared light sensitivity, and support for on-chip vision processing functions. Chip images are documented using LabVIEW®.

INTRODUCTION

An optimal control is a function which minimizes some specific cost functional, or, performance measure [1]. The performance measure constrains the optimization. Two prototype imager chips have been designed, both constrained for automotive vision applications. This optimization is described by comparing results for the initial 128x128 pixel chip and the new design, a 360x226 pixel chip, within the automotive vision constraint set.

Again borrowing terminology from optimal control theory, progress between these two prototype imagers may be viewed as a single jump in a steepest descent algorithm for the optimization of CMOS imagers for automotive applications.

Automotive vision applications may be differentiated from mainstream vision applications in a variety of ways. Fundamentally, mainstream applications are designed to be visually stimulating, whereas automotive vision systems must be designed to minimize driver distraction. Mainstream vision demands full color, accurate image reproduction, and photographic or video aspect ratios. Such applications as these are realized with CCD or CMOS sensors [2], operating over a visible light spectrum of 0.45 to 0.7 micrometers. Mainstream visual cameras require reasonable dynamic range, but operate in a benign indoor or outdoor environment.

Automotive vision systems are designed as sensor systems to give the driver information either directly, using some in-vehicle display system, or, indirectly, providing visual or audio cues to the driver in other ways. Thus an imager system could directly provide a video signal for safe passing (blind spot detection) or near obstacle avoidance, or, indirectly, as in occupant position sensing or highway lane tracking. These applications are monochrome, may involve variable aspect ratios, pixel size and pitch, require very wide dynamic range and automotive durability. Although some automotive applications utilize sensors made of exotic materials to provide night vision at light wavelengths greater than 5 micrometers, many automotive applications can be addressed by more conventional silicon sensors over the visible to near infrared region of the spectrum, from 0.45 to 1.1 micrometers.

The last decade witnessed a dramatic increase in CMOS imager technology. Not all of this technology is particularly useful in the automotive environment. Automotive vision applications have their own constraint set. Automotive vision may require a dynamic range as high as six orders of magnitude. Automotive imagers must collect light at very low levels, and yet still resolve objects in direct sunlight. The fundamental constraint for any automotive vision system is that it must have a high enough dynamic range to resolve objects under all driving conditions.

Another constraint concerns practical resolution requirements. Automotive applications may benefit from a variety of imager aspect ratios, but each imager will benefit from being comprised of as small a pixel as possible. What is required is a small pixel to maximize imager resolution while simultaneously minimizing the size of the chip. This constraint tends to produce prototypes of the largest array of the smallest pixels that can be practically packaged.

A third constraint of the automotive vision environment is unique in that any application must minimize driver distraction. Some applications may require a quick glance, such as providing the driver a means for viewing his blind spot while passing, but most will not. The automotive vision system will be used as sensor input for computational vision. Computational vision performs calculations from data extracted from the imager, such as the generation of edge locations in a highway lane tracking system. This information probably will not be displayed to the driver as an image. The system will use the image sensor data to calculate when the vehicle is following another vehicle too closely, or is moving too close to the edge of the lane or the edge of the road. The automotive vision system will then warn the driver or take corrective action in some other fashion.

The last constraint is generic for any automotive system: provide features a customer will pay for at an affordable cost with customary automotive durability. The prototype imager chips described here exhibit properties that uniquely address these automotive requirements. The integrated image sensor/pixel output electronics have been fabricated in an established automotive process. The single crystal silicon sensor is sensitive not only over the visible light spectrum, from 0.45 to 0.7 micrometer wavelength, but also into the near infrared portion of the spectrum, out to about 1.1 micrometers. Rather than utilizing three pixels to yield visible light data from the red, green, and blue bands, pixels are light sensitive over the visible and near infrared regions. The standard CMOS process allows future up-integration of circuitry to achieve the computational aspects of the vision system, in order to increase system speed by minimizing post-processing.

Both prototype imager chips were designed with these constraints in mind. Imager results from the 128x128 pixel chip, design considerations which led to the current 360x226 imager chip, and some imager results from that chip are presented here. The goals were to maximize dynamic range, minimize pixel size in an automotive

quality CMOS process, and look toward products to provide the driver information without distractions.

CHIP DESIGN

The imager chips are designed to take advantage of a mature, automotive qualified 1.2 micron CMOS process. This robust fabrication process results in imager chips with good performance and excellent durability for automotive applications.

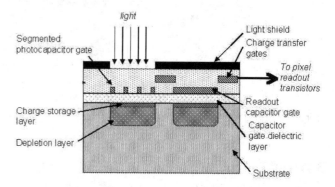

Figure 1: Photocapacitor Based Pixel Layout

The photosensitive pixel structure of both imager arrays is summarized schematically in Figure 1. Incident light passes through a segmented photocapacitor gate structure and stimulates the generation of electron-hole pairs in the underlying silicon. These charges are attracted by the electric field of the capacitor depletion region and collected in the inversion layer directly beneath the gate electrode. Charge is then transferred to the inversion layer of a readout capacitor structure through the action of a charge transfer gate, from where it is detected and converted to a voltage by the readout transistors. With the exception of the photocapacitor at each pixel of the array, the entire imager chip structure is covered by an opaque aluminum light shield layer.

Photocapacitor segmentation is done at charge diffusion length increments in order to capture photons from the blue end of the spectrum. Such photons do not penetrate through the thickness of a continuous polysilicon photocapacitor gate to contribute to charge generation. Voltage applied to a polysilicon gate segmented at narrow charge diffusion length intervals will still create a continuous depletion layer below the segmented gate. This structure will be more sensitive to blue light as blue photons that pass between the photocapacitor segments will strike the underlying silicon directly and thus contribute to the charge collected, increasing overall signal strength.

Both prototype imager chips essentially create a light sensitive capacitor from the gate of a CMOS transistor. A CMOS fabrication process was initially considered, in part, in order to enable the design of a non-destructive pixel readout structure. Many conventional imager chips [3] employ a photodiode structure which is not conducive to multiple readouts of the data from a given pixel. The capability of data storage over some extended period of time, which is provided by the use of capacitor structures, enables multiple use of pixel data for the on-chip calculations required in a windowing operation to detect scene edges in future on-chip computational vision processing systems.

Figure 2 shows a comparison of the sizes of the two prototype imager arrays. The relatively large size of the pixel (72 by 71.2 microns) used in the 128 x 128 pixel array imager chip caused consideration of a modified pixel design for higher density pixel arrays. This modified design is still functionally equivalent to that shown in Figure 1. Consequently the second generation chip, with a 360 by 226 pixel array, employed a different design. This design incorporated a shrink by a factor of 4 in the photocapacitor structure and a shrink by a factor of 8 in the charge readout capacitors. The net result was that the overall pixel dimensions were reduced to 25.6 by 62.5 micron.

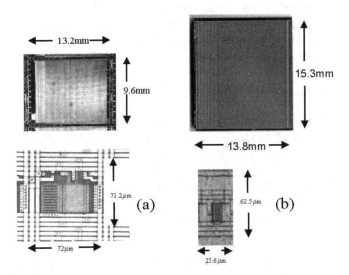

Figure 2: Chip and Pixel sizes for two prototypes: (a) 128x128 array. (b) 360x226 array.

Automotive applications require a higher dynamic range than can be provided by the basic photodetection mechanism. The intrinsic dynamic range for silicon based imaging devices is of the order of 60 dB. This corresponds to the measurement of the largest signal at about 1 V, and the smallest measurable signal to be about 1 mV. The two prototypes employed different methods to extend their dynamic range beyond this 60 dB intrinsic level. The 128x128 array design incorporated the ability to improve dynamic range by varying the pixel integration time; that is, the time during which photogenerated charge is collected for each pixel of the array. The 360x226 pixel chip design incorporated a different capability by providing a variable image data compression function [4], which can serve to enhance the dynamic range of the data supplied by the imager. Circuitry was incorporated into the imager chip to

allow for the introduction of a logarithmic variation in the amount of charge stored at each pixel. This capability was achieved by providing for a step-wise variable voltage to be applied, defining the photocollection well depth with respect to a charge sink. Thus, if charge is collected too rapidly (corresponding to a very bright object at a given location within the image), some of that photogenerated charge is dumped to the charge sink, preventing the pixel from saturating prematurely. This function was implemented using multiple flip-flops on each array row, and defining 8 time steps, corresponding to half of the frame time, a quarter of the frame time, an eighth of the frame time and so on down to one 256th of the frame time.

Operation of each imager chip is based upon a rolling integration interval. This means that with the exception of the row currently being read out, all other rows of pixels may be collecting light, subject to the biases supplied to the pixel and row compression latches. This technique results in improved imager sensitivity. For reasonable frame rates (30 frames per second or greater) this is a practical approach for automotive applications. Following an integration time, during the single row read operation, collected charge is transferred to the readout capacitor, and the charge induced on the gate of this readout transistor modulates the output voltage on the column readout line.

For the 128x128 imager, selection was provided via three input pins as to how many rows ahead of the row currently being read charge was being collected. Thus charge could be collected 1,2,4,8,16,32,64, or 127 rows ahead of the row currently being read. Given the frame rate, these eight choices directly provide eight different exposure times for gathering information from a given scene. This directly extends the dynamic range of the imager by over 40 dB.

Although the 360x226 chip was designed to enable the logarithmic compression of image information, since the dump gate is controlled by 8 independent voltage sources off chip, this same compression circuitry can be used to collect charge for whatever combination of 8 time divisions of the frame are desired. This is functionally similar to the integration interval charge collection designed with the 128x128 pixel chip.

Electronically active circuit elements, and the lines used to route the signal and power lines to those elements, combine to limit the light sensitive area of the pixel array. This active area is typically less than 25% for most CMOS pixel imager designs. Overall chip dimensions for the 360x226 array are 13.8 mm by 15.3 mm. Although this represents a relatively large chip, these dimensions must be evaluated in the context of the use of a 1.2μm design rule IC fabrication process. Transferring this design to a 0.35 μm process would result in a significant reduction in chip size. This shrink can be accomplished without the reduction in imager sensitivity which sometimes occurs with smaller feature fabrication processes, due to the use of optically absorbing silicide gate structures[5]. This effect is mitigated in the present design by the use of the segmented photocapacitor structure. In this case, the pixel area would be approximately 12μm by 32μm, assuming NO change in the area of the photosensitive region of the pixel. Corresponding overall chip dimensions would be approximately, 5mm by 8mm, representing a reduction in area by nearly 80%. Although this represents an extreme extrapolation from the present chip, it suggests that such a reduction of process minimum feature size would allow doubling of the resolution of such a chip (to a 720 x 500 pixel array) while still maintaining overall chip dimensions comparable to those of the present chip.

IMAGER SYSTEM DESIGN

Both imager chips were connected to custom analog/digital interface boards, which provided the necessary power supply, signal conditioning and analog-to-digital converter functions. Ten bit A/D chips have been used, but only the 8 most significant bits are utilized for the images presented in this paper.

Figure 3 shows a block diagram of the entire 360x226 imager chip, as well as of the addressing electronics built on the driver interface board. The imager array is comprised of 360 columns and 226 rows. Chip timing is based upon 256 rows. The difference resulted from a need to reduce the overall dimensions of the imager chip in order to make use of the chosen 121 pin ceramic PGA (pin grid array) package. A second innovation related to the chip size was the decision to incorporate 4 simultaneous row readouts. Although this resulted in the need for 4 output amplifiers and 4 A/D converters, it enabled a more efficient pitch match between the dimensions of the pixel and the dimensions of the column readout circuits. In addition, it reduced the data rates required for each of the A/D converters. Pixel biasing and clock signals are provided from the driver board, as are up to 8 distinct voltage levels to set the biases supplied to the array by the row compression latches.

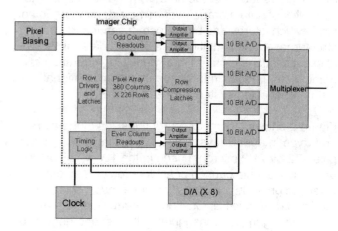

Figure 3: 360x226 Imager Chip and Readout Block Diagram

The output signals from this board were fed into a PC via direct memory access, and LabVIEW® was used to form the images from this data that are presented herein. LabVIEW® has enabled demonstration of the multiple capabilities of the imager chip without the need to build new hardware test boards, and has been essential in the progress of this program.

IMAGER SYSTEM PERFORMANCE

Figure 4 shows a model road scene in which a 600 W lamp was used to illuminate the scene from the forward direction in order to collect of images using both imager chips. A variety of lighting conditions were used in order to demonstrate the performance of the basic photodetection mechanism, without augmentation by dynamic range enhancement techniques.

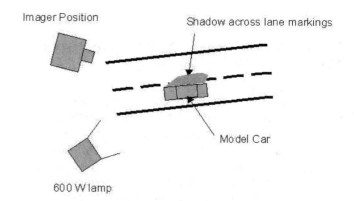

Figure 4: Configuration of Camera and 600W Illumination Lamp for Model Car Images

Figure 5 details the improvement in output image information content that can be achieved by the implementation of dynamic range enhancing circuitry. Images from the model road scene with relatively short charge integration time (Fig. 5a) results in good image detail for brightly illuminated portions of the scene, but essentially no information regarding regions that are in the shadow of the vehicle model.

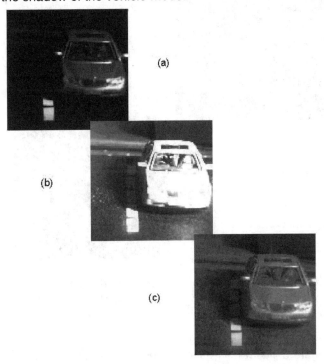

Figure 5: Images of model scene taken with 128x128 pixel imager. (a) Image collected with 2.1 ms charge collection time constant (b) 33.3 ms (c) composite

Alternatively, when a longer charge integration time is used (Fig. 5b), features in the poorly illuminated regions can be detected, but details of the car, subjected to high illumination conditions, become washed out. Making use of the ability to vary integration times in this imager chip, a composite image can be generated. This capability, built in LabVIEW® software, combines the feature content of the short time constant frame (Fig. 5a) and that of the long time constant frame (Fig. 5b) to form a single composite frame (Fig. 5c) with enhanced dynamic range. Thus, at the cost of a reduction in imager frame rate, a high dynamic range imaging capability can be achieved.

Figure 6: High dynamic range image under 600W illumination (b) High dynamic range image under room light conditions. In both images, lane markings are visible

Two composite output images from the 128x128 array are shown in Figure 6. The first output image was taken with both normal room light and a 600W halogen lamp (70cm away) illuminating the screen. The second output image was taken with the 600W lamp switched off, and only room light illuminating the scene. The ability to resolve both the vehicle and the lane markings, under this wide range of illumination conditions, represents a visual confirmation of the high dynamic range capability of this imager.

It is important to note that this enhanced dynamic range image is achieved without the use of logarithmic compression techniques, either in hardware, or in software. Logarithmic compression techniques operate on the detected signal voltage amplitude, compressing more information content into a given range of signal level. An example of this technique is given by two images generated from the 360x226 pixel array in Figure 7.

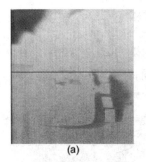

(a)   (b)

Figure 7: Images from 360x226 array, (a) without compression, (b) with compression, both in the presence of both room light and the 600W halogen lamp

IMAGE PREPROCESSING AND EDGE DETECTION

Drivers must spend most of their time operating the vehicle, not looking at images. However, image data will be used to extract information from road scenes which will enable the driver to track his position within the lane, avoid collisions, detect front seat passenger occupation characteristics, and so on, in future automotive products. Information will be extracted from imager data electronically.

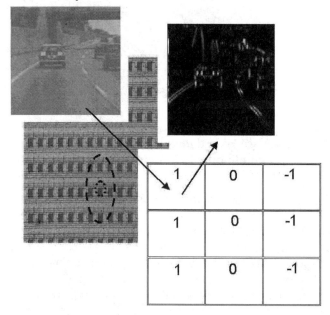

1	0	-1
1	0	-1
1	0	-1

Figure 8: Edge detection algorithm example

One example of such information extraction is illustrated in Figure 8, where a 35mm photograph of a road scene has been viewed by the 128 x 128 pixel imager. Once the scene data is read into a computer, an edge detection algorithm is applied. This algorithm uses information at each pixel from its eight surrounding neighbor pixels to generate edge data across the entire pixel array.

Figure 9: Model car image before and after application of a 3x3 edge detection algorithm, accentuating both vertical and horizontal edges

Figure 9 illustrates image processing with the previously used model car scene by a similar 3x3 edge detection algorithm. Again, the initial scene is captured by the imager chip, and then processed to generate data which highlights the location of scene edges.

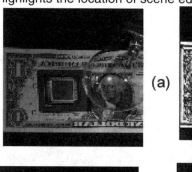

(a)

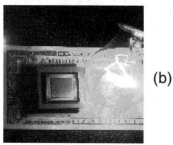

(b)

Figure 10(a) Direct image and edge image with 40W lamp off, (b) Direct image and edge image with 40W lamp on. Details of the dollar bill and bulb filament (above Washington) are discernable in both cases

Figure 10 reveals how edge information can be captured from the same scene under two very different lighting conditions. The top figures show the imager chip and dollar bill with the 40 W light bulb not illuminated. The application of the 3x3 edge detection algorithm to this scene provides great image detail, including in the area beneath the transparent light bulb envelope. The lower set of figures shows the same scene with the 40W light bulb illuminated. In this case, direct image data can be seen beneath the light bulb, although the contrast ratio is reduced. This is in part due to the limitations of the monitor and/or printer displaying the output gray-scale image. The edge detected image is generally not limited by the dynamic range of the output device, and consequently, the edge detected detail directly beneath the light bulb is still evident.

CONCLUSION

Automotive vision applications are uniquely constrained. Products designed to meet these constraints will require high dynamic range, affordable resolution, desirable functionality, automotive durability, and still must not distract the driver.

ACKNOWLEDGMENTS

The authors acknowledge the contributions of former team member Charles N. Stevenson to all phases of this work. All facets of this chip were directly aided by the many talented people of the Delphi Microelectronics Center in Kokomo. Dave LaRosa, Dan Pear, Ray Loar, Frank Szorc, Don Ford and Bob Grossman were particularly essential. Ron Lesperance has contributed to possible edge detection algorithms. The photographic experience and skill of Randy Shafer are also gratefully acknowledged.

REFERENCES

1. Donald E. Kirk, *Optimal Control Theory: An Introduction,* EnglewoodCliffs:Prentice-Hall, Inc,1970

2. E. M. Lockyer, "Applications Hold the Key to Imager Choice," Photonics Spectra, March, 1979, pp. 80-90.

3. E. R. Fossum, "CMOS Image Sensors: Electronic Camera-On-A-Chip," IEEE Trans. Electron Dev. <u>44</u>, pp. 1689-1698, 1997.

4. S. Decker, D. McGrath, K. Brehmer, and C. Sodini, "A 256 x256 CMOS Imaging Array with Wide Dynamic Range Pixels and Column-Parallel Digital Output," IEEE Journal of Solid-State Circuits, vol 33, 2081-2091, 1998.

5. H.- S. Wong, "Technology and Device Scaling Considerations for CMOS Imagers," IEEE Trans. Electron Dev. <u>43</u>, pp. 2131-2142, 1996.

1999-01-1302

A Simple CCD Based Lane Tracking System

Frank S. Barickman
National Highway Traffic Safety Administration

Duane L. Stoltzfus
Transportation Research Center Inc.

ABSTRACT

A low cost system has been developed to measure a vehicle's lateral position relative to the lane markings on a roadway. The system is capable of tracking white or orange lines, solid or dashed edge lines, while operating in daylight or at night. The tracking system is comprised of two "off-the-shelf" black and white charge coupled device (CCD) video cameras along with commonly available electronic components. The lane tracking system is capable of outputting real time data at 30Hz through an analog output. Using the data from this sensor system it is possible to detect lane changes, determine the magnitude and duration of lane exceedances, and other metrics commonly used by researchers in the transportation community. This paper will discuss the design and performance of the system, processing of the raw lane tracker data, and the benefits and limitations of the technology.

INTRODUCTION

Lateral position of a vehicle relative to the roadway's lane boundaries has been an important measure for many of the National Highway Traffic Safety Administration's (NHTSA) crash avoidance research programs. Drowsy driver research (Weirwille, 1996), run-off-road research (Pomerleau, 1995), lane change merge research (Chovan, et al, 1994), and automated highway systems research (Hatipoglu, et al, 1997) are several examples of programs in which lateral position measurement or lane tracking has been an important measure. This metric has been used to address issues related to driver attention as well as the performance of vehicle control and collision avoidance systems.

Historically, lateral position has been measured by calibrating a video image to the roadway using a mapped field of view allowing the data to be reduced manually by an observer. This method is very time consuming and labor intensive when dealing with large amounts of data. Over the last several years, many advances have been made in automatic sensing of vehicle lateral lane position. Generally, there have been two methods of sensing

which have become popular. Vision based systems such as the Rapidly Adapting Lateral Position Handler (Pomerleau, 1995), Ohio State University's Demo '97 vehicle (Ozgüner, et al, 1997), and others have become quite popular. Most of these systems use a forward looking video camera. Some are capable of outputting data other than lateral lane position such as radius of curvature and can even continue to provide lane tracking data when there are no painted edge lines present. A major drawback of a vision based system is cost. These systems are estimated to cost greater than $10K per unit (small quantities 1-20).

A second type of technology being used for measuring lateral lane position includes sensors which work with an installed infrastructure. These methods rely on magnetic nails, RADAR striping (Ozgüner, et al, 1997), or other devices installed on the roadway and work in conjunction with sensors on the vehicle. One advantage of these technologies is that, in theory, they offer better performance on roadways obscured by snow or debris. Drawbacks to these technologies are that an investment needs to be made into the infrastructure since they will not work on existing roadways.

Given the importance of this measure and lack of availability of commercial systems, NHTSA's Vehicle Research and Test Center (VRTC) has developed an approach that meets the need for a low cost lane position measurement system to be used in it's research programs. The latest version of the system uses a low cost and small (1.25" x 1.25") black and white video camera as the sensor. The sensors can be mounted in the side view mirrors or on the rear end of a passenger vehicle. The video signal from the camera is fed to the analog and digital processing board developed by engineers at VRTC. The cameras used in the VRTC system are sensitive to near infrared light and thus the road can be illuminated at night with infrared light that is not visible to humans. A complete system, both left and right side trackers, costs approximately $1000 in small quantities (10 systems) which includes parts and labor. It is anticipated that this cost can be greatly reduced if the system is produced in larger quantity.

GENERAL SYSTEM OVERVIEW

The lane tracking system is a simple system which provides accurate lateral lane position measurements. For the system to function properly, the roadway must have visible painted lane edge markings. The system is capable of outputting raw data in real time representing how far the painted edge line is from the center of the vehicle at an update rate of 30 Hz. The system is built from commonly available electronic components which are available off the shelf from a variety of electronics stores. The lane tracker is powered directly from vehicle power, requiring 12VDC unregulated input at 200mA.

DESIGN OF THE SYSTEM

The lane tracking system uses any black and white camera capable of outputting a standard National Television System Committee (NTSC) video signal. The system performs analog signal processing, looking for a section of the video that has the signature of a painted edge line. By using an off the shelf camera the flexibility of mounting the camera is greatly increased and all the circuitry for the iris and scan control that would be associated with using a one dimensional line scan camera are eliminated.

A major part of the "processing" circuitry is an LM1881 video synchronization (sync) separator integrated circuit (IC). This IC accepts an NTSC video signal input and provides vertical and composite (combined vertical and horizontal) sync output pulses. These pulses allow using relatively simple circuitry to "process" the video signal, by providing information concerning the time location of the horizontal and vertical sync pulses. The vertical sync pulse from the LM1881 is used to find the middle of the video image.

At the start of each frame, an 8.3 ms delay is triggered by a one shot timer. After this delay, a scan line timer is started 9μs after the next horizontal sync pulse. The 9μs timer allows the "back porch" of the NTSC signal to be ignored so that only the scan line intensity information is analyzed. Figure 1a displays the raw scan line intensity information which would be analyzed by the system. Figure 1a also shows an offset in the video signal. This is very common due to non-uniform background intensities, such as shadows, surface variations and glare. The raw video signal is high pass filtered to remove low frequency intensity changes. With only edges remaining, the scan line is analyzed for a step increase in intensity followed by a step decrease in intensity (i.e. edge detection, see Figure 1b.). The tracker will only trigger if the video has a step increase in intensity and then a subsequent step decrease in intensity within a certain window of time. Therefore this requires that for a painted edge line to be found it needs to be a certain width. This allows the system to work better if there is a shadow cast by the vehicle on a sunny day.

For the system to track a roadway line, two conditions must be met. First the step increase and decrease must exceed a threshold set during construction of the system. Second, the time between the step increase and step decrease must fall within a set period.

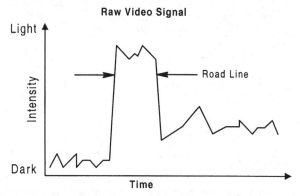

Figure 1a. Raw video signal intensity from one video scan line.

The thresholds are programmable and are optimized to provide reliable line detection while minimizing the tracking of false targets. The upper and lower thresholds are shown in Figure 1b. During construction a test pattern is placed into the field of view of the tracker sensor which provides a contrast ratio of approximately 1.3:1. The contrast ratio is determined by measuring the luminance of the simulated line divided by the luminance of the background image. The thresholds are adjusted so that the system triggers just when the signal exceeds the background intensity level by 1.3.

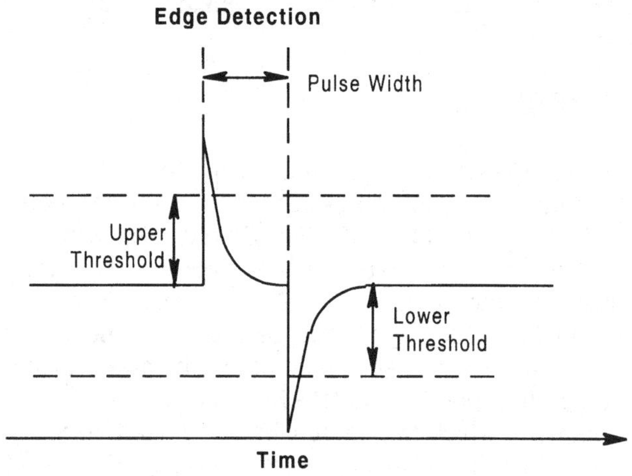

Figure 1b. Edge detection after the raw video signal is high pass filtered.

The time between the step increase and step decrease is used to filter out false triggers that the system may track by solely relying on the thresholds for triggering. By setting a minimum and maximum time width for the pulse, a basic line width detection scheme is implemented. The pulse width must be greater than 1 s and less than 4.5 s wide. This approximately correlates to a minimum line width of 5 cm and a maximum width of 20 cm which compensates for geometrical distortions in the 2D image.

If the system is triggered, the value of the scan line timer is outputted to a digital to analog converter (DAC). The value of the scan line timer indicates the location of the edge line within the horizontal field of view of the camera. The analog output from the 8 bit DAC can then be recorded by a data acquisition system or other device. The output value of the DAC is held until a new value is sent from the scan line timer. The analog output is a 0 to 2.5 volt signal.

A status bit was added in the latest version which provides an output that is logic high when the system is triggered, and a logic low if not triggered during the scan line. This output can be recorded along with the analog output signal providing an indication of when the DAC output is not being changed, thus indicating that the tracker is not tracking an edge line.

INSTALLATION

The lane tracking system is installed in one of two ways depending on the requirements for the data being collected. Both methods require the sensor's field of view to be aimed directly towards the ground as seen in Figure 2. The cameras are setup so that the left edge of the image, when viewed on a television monitor, is parallel and near to the vehicle's longitudinal centerline. It has been found that a downward looking approach provides a better contrast ratio under a variety of weather and lighting conditions than a forward looking approach, thus providing a better signal- to- noise ratio for detecting the edge lines.

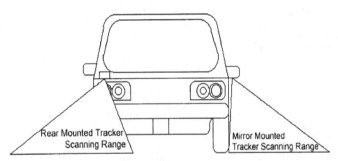

Figure 2. Lane tracker mountings with their associated fields of view.

The first method installs the sensors on the rear of the vehicle. Using this method, the sensors are installed in a short amount of time. The configuration allows for lateral position measurement, lane bust detection, and measures the extent of a lane exceedance. This method of installation has been successfully used for VRTC testing in the areas of drowsy driver research, driver workload research, and car following research. There are several drawbacks to this type of installation. First, the sensor is not easily integrated into the vehicle. Figure 3. displays how the sensors are integrated on the rear of the vehicle. Second, over long term naturalistic data collection periods, the potential exists for the cameras to be "bumped" by the subject entering the vehicles trunk or vandalized in a public parking lot. "Bumping" or vandalizing the cameras would significantly change the systems calibration.

Finally, field experience has uncovered problems with glare off of the rear bumper of some vehicles caused by ambient sunlight or following vehicle's head lamps.

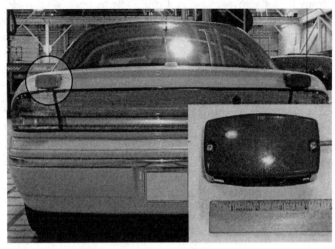

Figure 3. Tracker sensors mounted on the rear of a vehicle.

The second method of installation integrates the sensors into the side-view mirrors of the vehicle. This method is very unobtrusive and robust. System calibration is generally set once with this method since the cameras are not easily disturbed and they are not as sensitive to glare like the rear method. Mirror adjustments made by the driver do not effect the calibration. This configuration allows for lateral position measurement, however lane bust detection and lane exeedance measurements are limited because the edge becomes blocked by the vehicle. Another drawback of this method is that it takes more time to install the sensors. Some side-view mirrors do not have enough room to internally house the small cameras. Figure 4. displays the bottom of a side-view mirror with a tracking camera installed.

Figure 4. Side-view mirror with lane tracker camera integrated.

CALIBRATION

Calibration is completed on the system after the CCD cameras have been mounted. Statically, a line is swept from the minimum to the maximum range of the system.

The output voltage at various distances measured from the center line of the vehicle are recorded. This data is then fit with an exponential equation to compensate for the geometry of the tracking system. Although other methods such as interpolation with a calibration look up table could be used, the exponential fit provides sufficiently accurate results as can be seen in Figure 5. This calibration should be completed before and after the system is used for testing. It is important that the sensors do not move during testing to maintain accuracy.

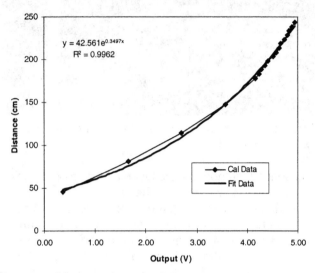

Figure 5. Measured verses fit lane tracker data.

PERFORMANCE

Using the exponential fit data, the system accurately measures the distance to a painted edge line within approximately 6 cm over the systems operating range. The accuracy and full scale operating range will vary based on the installation height and the optics used with the sensors. Table 1 displays the calibration data used for Figure 5. Table 1 also displays the calculated absolute value of the error from applying the exponential fit. For this installation, the maximum operating range for the tracking system was found to be approximately 2.4 m. This distance is typically seen in most installations with about a +/- 0.2 m difference. Accuracy will vary accordingly with full scale operating range. As the operating range becomes larger, the accuracy will degrade. Typical accuracy is within 6 cm with a resolution of approximately 2 cm over an operating range of 2.5 m.

DATA PROCESSING

The analog outputs are processed with several functions to clean up the raw data and derive other measures. Some of the measures that are typically derived from this data include lane width, deviation of the subject vehicle from the center of the roadway, lane exceedance extent/duration, and lane changes.

Raw data is initially processed using a spike removal routine. The routine looks for large spikes of a programmable width which deviate from the average value. Figure 6 displays the raw data against the spike removed data. The noise spikes in the raw data are seen because the tracker is detecting objects which meet the criteria of a valid line before the edge line. However, by looking at the data, the general structure of the system tracking a road edge line can be seen. Figure 6 displays a very noisy segment of the tracking data. In many instances the raw data is cleaner than illustrated. The spike removal function tries to recreate the structure that the human observer can see.

Table 1. Calibration data used in Figure 5.

Volts (V)	Distance from Center Line (cm)	Exp. Fit (cm)	ABS Error (cm)
0.36	45.7	48.3	2.6
1.66	81.3	76.1	5.2
2.70	114.3	109.4	4.9
3.56	147.3	147.8	0.5
4.16	177.8	182.3	4.5
4.24	182.9	187.5	4.6
4.28	188.0	190.1	2.2
4.36	193.0	195.5	2.5
4.42	198.1	199.7	1.5
4.51	203.2	206.0	2.8
4.57	208.3	210.4	2.1
4.63	213.4	214.9	1.5
4.67	218.4	217.9	0.5
4.73	223.5	222.5	1.0
4.80	228.6	228.0	0.6
4.83	233.7	230.4	3.2
4.88	238.8	234.5	4.3
4.94	243.8	239.5	4.4

The spike removed data is then low pass filtered and used to calculate a lane width. The lane width is calculated by summing the left and right tracker values. The lane width value is only updated when both trackers are valid, otherwise the previous value is held. Since gross changes in lane width happen infrequently, this method has proven to work reliably.

Combining the left and right tracker data with the current lane width information, a channel is created indicating the vehicle's deviation from the center of the roadway lane. In the event that one side of the system stops tracking, the lane deviation can still be calculated using data from the other tracker and the lane width data.

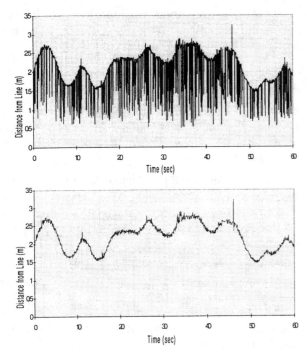

Figure 6. Raw tracker data compared to processed data with spikes removed.

Using the measures described above, various maneuvers can be detected such as lane changes. Lane changes can be detected provided that one of the lane tracker sides is functioning correctly throughout the maneuver. A routine has been implemented in an attempt to detect them automatically. Figure 7a displays a vehicle's distance from the center of a roadway throughout a lane change maneuver over time. To detect a lane change, the vehicle's distance from the center of the roadway is differentiated to obtain a lateral velocity. The lateral velocity is then processed with the spike removal routine so that the resultant data can be integrated to obtain a continuous lateral position as seen in Figure 7b. The lateral position is then analyzed by the routine for changes in direction (ie. lateral velocity approaching zero) as can be seen in Figure 7c. The amplitude of the continuous lateral position is measured between zero crossings of the lateral velocity. If the delta of the amplitude is greater than 1.5 meters, the maneuver is considered to be a lane change.

LIMITATIONS

Although the system performs well on roadways with good quality, painted edge lines, in the real world not all roadways have them. In a recent test program conducted at the VRTC to study the effectiveness of detecting driver drowsiness, validity of the tracker was tested. The test program involved eight subjects. The subjects' personally owned vehicles were instrumented with a VRTC lane tracking system. The participants selected were traveling on long distance trips that they had already planned on making, i.e. college student driving home for semester break. Data was recorded, during daytime and nighttime,

throughout the duration of their trip only when they were traveling at speeds above 70 kph. Subjects were not instructed to follow any pre-described route nor were they asked to limit their travel to a particular type of roadway. It is believed that this data set accurately characterizes the realistic performance of the lane tracker system on typical public roadways.

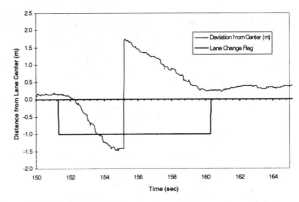

Figure 7a. Vehicle's distance from the center of the roadway through a lane change maneuver.

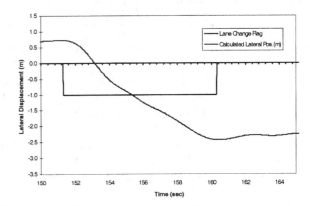

Figure 7b. Calculated lateral position through a lane change maneuver.

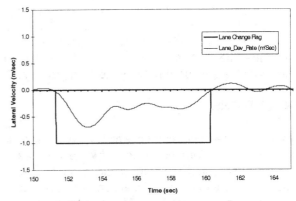

Figure 7c. Vehicles calculated lateral velocity throughout a lane change maneuver.

A channel representing the validity of the tracker data is generated as part of the post-hoc processing. The validity flag indicated that throughout the 89 hours of data collected, the system provided valid lane position approximately 62% of the time as can be seen in Table 2.

An observer confirmed the accuracy of the validity flag by manually reviewing time history traces of the raw tracker data. No significant discrepancies were found by the manual observer.

Table 2. Performance of the lane tracking system operating in the open road environment.

Subject	Amount of data (Hours)	Tracker Valid (%)	Amount of Valid Tracker Data (Hours)
1	8.2	71%	5.8
2	12.1	70%	8.5
3	14.3	70%	10.0
4	13.8	37%	5.1
5	2.8	78%	2.2
6	14.5	63%	9.1
7	13.8	47%	6.5
8	9.8	87%	8.5
Overall	89.3	62% ($\bar{x}$)	55.7

Table 2 also breaks out the performance of the lane tracker on a subject by subject basis. It should be noted that the worst performance of the lane tracker system was 37%. This low performance was attributed to a hardware problem and not the roadways traveled. The sensors were mounted on the rear of this subject's vehicle. Reviewing recorded video from this subject, a noticeable glare caused from sunlight reflecting off of the vehicle's bumper was observed. The glare caused a flare in the optics of the CCD sensor, creating a vertical stripe through the video image. This resulted in the system falsely tracking the flare instead the painted edge line.

The condition was replicated in the lab using a bright light source. It was observed that this condition could happen on any vehicle with a protruding bumper. Vehicles with glossy/waxed plastic and/or chrome bumpers seem to be affected the most. A shield has been fabricated and installed into the sensor to remedy the flare problem.

CONCLUSIONS

A system has been developed which measures vehicle lateral position in reference to the painted edge lines of a roadway. The system can be installed on almost any vehicle and can resolve lateral position within 6 cm over a range of approximately 2.5 m. The lane tracking system is capable of operating under a variety of light and weather conditions.

The system was found to report 62% valid data while operating under real world conditions. Although the tracker does have performance limitations, significant benefits are attainable when using the system. Improvements in efficiency of data analysis are realized through the elimination of the need for intensive manual data extraction and reduction. As a result, the time and cost of data analysis are greatly reduced.

REFERENCES

1. Weirwille, Walter. (1996). "Driver Status / Performance Monitoring." *Proceedings from the Technical Conference on Enhancing Commercial Motor Vehicle Driver Vigilance.* December, 1996.

2. Pomerleau, Dean. (1995). "Program Overview: Run-Off-Road Collision Avoidance Using IVHS Countermeasures." *Peer Review of the National Highway Traffic Safety Administration Program.* 43-51. Intelligent Transportation Society of America.

3. Chovan, J. D., Tijerina, L., Alexander, G., & Hendricks, D. L. (1994, March). *Examination of lane change crashes and potential IVHS countermeasures* (Report No. DOT HS 808 701/DOT-VNTSC-NHTSA-93-2). Cambridge, MA: U.S. Department of Transportation Volpe National Transportation Systems Center.

4. Hatipoglu, C., Redmill, K., Ozgüner, Ü. (1997). Steering and Lane Change: A Working System. *Proceedings from IEEE Conference on Intelligent Transportation Systems.* Boston, Massachusetts, November, 1997.

5. Ozgüner, Ü., Baertlein, C, Cavello, D., Farkas, C., Hatipoglu, C., Lytle, S., Martin, J., Paynter, F., Redmill, K., Schneider, S., Walton, E., Young, J. (1997). The OSU Demo '97 Vehicle. *Proceedings from IEEE Conference on Intelligent Transportation Systems.* Boston, Massachusetts, November, 1997.

1999-01-1235

Performance Assessment of a Side Object Detection System

R. Steven Hackney
A.L.I.R.T. Advanced Technology Products

ABSTRACT

A great deal of research and development has been invested in collision avoidance systems. A few such systems are now available commercially, including backup warning systems, backup vision systems for heavy vehicles, forward impact warning systems for heavy vehicles, and side collision avoidance systems for heavy vehicles.

Relatively little data are published regarding basic sensor performance. A limited amount of performance data have been presented in the form of fleet accident avoidance, predictive reliability models, and individual driver accuracy ("Right clear?" questioning samples). However, few instances of actual sensor accuracy measurements are presented in the literature.

This paper discusses parameters which can be used to characterize sensor performance. Methods of data collection are also considered, and two measures are recommended: Detection Accuracy and Mean Time Between False Alarms. Real-world data collection on public roadways is recommended.

Finally, performance data are presented for a sample sensor. This performance measurement is based on approximately 75 hours of data collected on public highways using prototype and early production samples of a side collision avoidance system.

Overall detector performance showed Accuracy of Detection of 99.32%, with a Mean Time Between False Alarms of approximately 3 minutes. Performance showed a correlation with time of day which is reflective of decision factors built into the algorithm used by the system.

INTRODUCTION

Collision avoidance ("CA") systems have been the focus of considerable research and development effort in recent years. Much of this effort has been dedicated to the development of cost effective, reliable sensor systems.

Despite these efforts, and despite the relatively large quantity of published work in the area, there have been few explicit discussions of actual performance of sensors. Detailed studies have measured the performance of the overall system, made up of the sensor, the enunciator and the driver: however, these have made only passing reference to the performance of the sensor itself.

There have also been few discussions of performance characterization parameters ("PCP") to be measured, and there is no general agreement on the form that these should take. System specifications, for example, are vague or silent on the issues of detection reliability and false alarm rates.

This paper presents a discussion of the characteristics which might be considered in selecting performance parameters for sensor measurement. Two specific PCPs are suggested as suitable. Finally, efforts to measure an existing sensor system against these parameters are described, and an overall performance measurement of the system is presented.

PERFORMANCE MEASUREMENTS PUBLISHED TO DATE

Data which have been published to date include a number of approaches to performance measurement. Each of these presents difficulties.

Extensive testing of a number of different Heavy Truck CA systems has been conducted [1,2,3,6]. Testing has been based on a number of techniques, including staged tests using known targets (e.g. a pedestrian or a target vehicle following a specified path), and measurement of improvement in driver accuracy with respect to "right clear" questioning while driving on public roadways. These tests focus on either overall system performance (i.e. sensor plus display unit plus driver) or on measurement of design parameters for the sensory system (e.g. size and location of detection zones). Only a very limited amount of information is presented regarding actual sensor performance.

Other published performance data include reports of overall accident rate reductions, and samples of data collected by recording systems [5]. These reports do not characterize reliability or accuracy of the sensor, and neglect confounding factors such as regression toward the mean in fleet accident tracking.

ASPECTS OF SENSOR PERFORMANCE TO BE MEASURED

There are two important characteristics of sensor performance which must be summarized in PCPs. These are reflected in the two types of errors that a sensor can have: false negatives (i.e. missed detections) and false positives (i.e. alarms indicated when no dangerous condition is present).

The measures should be calculated in a way that makes them independent of environmental variables, such as driving style or characteristics of surrounding traffic. For example, a parameter such as mean time between false negatives would be greatly affected by the volume of surrounding traffic (i.e. mean time between *opportunities* for false detection) during data gathering. For this reason, it would be a poor choice of PCP.

RECOMMENDED MEASURES

Overall performance of a sensor in a CA system can be characterized by two parameters.

QUANTIFYING FALSE NEGATIVES: ACCURACY OF DETECTION

- **Accuracy of Detection (AD)** – number of accurate alarms divided by total number of vehicles which *should* have been detected.

This parameter offers some independence from traffic density, which is a key factor likely to vary in real-world testing.

However, some additional measurement may be required to "qualify" vehicles which should have been detected – variances in relative velocity or lane positioning may mean that some apparently "false" negatives are actually "true" negatives.

QUANTIFYING FALSE POSITIVES – Two options are available for quantifying false positives (false alarm rates). These are Mean Time Between False Alarms, and False Alarm Rate.

- **Mean Time Between False Alarms (MTBFA)** – total number of test minutes divided by the number of alarms signaled when no vehicle was in the detection zone.

This measure is suitable for characterizing sensors which provide continuous data (i.e. sensors which are always on). The sensor described in this paper is of this type.

This parameter is somewhat problematic, since false alarm rates may be influenced by road characteristics, environmental conditions, or other factors which may cause a false alarm with a given sensor. These influences can be mitigated by testing across a wide range of conditions.

- **False Alarm Rate (FAR)** – total number of incorrect alarms given, divided by the total number of sensor "inquiries" when no hazard was present.

This measure is suitable for characterizing sensors which provide discrete data in response to a query by a system controller or by the driver. An example of this would be a sensor which is interlocked with the vehicle turn signal: the sensor would provide information about the presence of a hazard only when the driver "interrogated" the sensor by means of the turn signal.

CORRESPONDENCE BETWEEN MTBFA AND FAR – Note that correspondence between MTBFA and FAR can be calculated. To do this, it is necessary to know the mean duration of false alarms (MDFA). With this information, the relation between the two measures is according to the following formula:

FAR = MDFA / MTBFA

This equation describes the percentage of time that a false alarm condition will exist. This corresponds to the probability that a false positive "answer" will be given in response to interrogation of the sensor.

MEASURES NOT RECOMMENDED – Measures which are less appealing, and should not be used include:

- False to good alarm ratio. While initially appealing, this can be greatly affected by outside variables: if there are few (or no) opportunities to signal an accurate alarm this ratio will necessarily be high. For example, testing showed variation in this value from 0.33 to 1.20 with changes in only traffic volumes (Figures 1 and 2).
- Overall fleet accident improvement rates. This value may be affected by a number of other concurrent factors, such as driver training, regression towards the mean, etc. This can be particularly true in situations where CA systems are introduced as a reaction to a period of time when fleet accident performance is particularly poor.

SENSOR DESIGN PARAMETERS NOT INCLUDED IN RECOMMENDED PCPS

There are a large number of other measures which can be taken of a sensor's performance. Examples include:

- range;
- detection response speed;
- definition of "dangerous condition" which warrants an alarm;
- severity, type, and other characteristics of an alarm; etc.

These characteristics (and others) are all critical to the function of any CA sensor system, and values for each of these characteristics are implicit in the design of each sensor.

However, these characteristics are fundamentally different than the PCPs presented. The PCPs attempt to describe how well a sensor system performs, while the design parameters are independent variables specified by the design engineers in developing the system.

For example, engineers writing a specification for system performance may describe the geometry of a lateral blind spot zone as a certain size and shape relative to the host vehicle, as part of the definition of "dangerous condition" requiring an alarm. This definition may or may not correspond to the size and shape chosen by the design engineers at the potential sensor supplier.

However, both sets of engineers should be able to agree on the performance of the sensor system, as measured against one specification, and as characterized by the PCPs.

DATA COLLECTION METHODOLOGY – In addition to choosing the correct PCPs, it is important that any sensor performance data be collected under conditions that reflect actual use of the sensor. This requires that a considerable body of data be collected across a range of conditions in order to produce a valid estimate of performance.

The large quantity of data required gives rise to a temptation to simplify measurement by limiting data collection to a small number of "worst case" scenarios. This approach assumes that an easier "real world" situation will give rise to better performance than a minimum acceptable performance in these "worst case." Unfortunately this assumption may not be valid, for several reasons:

1. The designer of the "worst case" will have implicit assumptions about the underlying sensor technology. A "worst case" scenario designed for this assumed technology may be far from a "worst case" for some other technology, producing inaccurate evaluations for some technologies. Detection of a small motorcycle, for example, may be a worst case scenario for an active radar system. Conversely, exposed high temperature components on the motorcycle may make this a relatively easy scenario for a passive infrared sensor.

2. Conversely, there may be elements of the scenario that render a technology completely ineffective, in an artificial and inaccurate way. A passive infrared technology, for example, depends on the energy emitted by a warm target vehicle: a scenario presuming detection of a stationary inanimate target by means of radar would result in inaccurate evaluation.

3. Engineers designing a sensor to meet a few specific performance scenarios will inevitably optimize sensor performance for these situations, and the design scenario may no longer be "worst case." Worse, sensor optimization for these limited scenarios may actually degrade performance in some real-world situations.

ALIRT BSD Test Results Record

| Date | 27 AUG | Tester | BSH | Temp. | 28 |

Location: 401 WEST FROM MISSISSAUGA

BSD Setup: S   1=Mod100 proto; 2=Mod200 no conf.; 3=conf + bin; 4=pre-prodn heavy shutter; 5=pre-prodn light shutter; 6=Mod200 prodn; 10=computer

Weather: 1   1=sunny; 2=light cloud; 3=moderate cloud; 4=overcast; 5=light/moderate rain; 6=downpour

Driving Style: 3   1=aggressive; 2=moderate; 3=mild

Comment: HEAVY TRAFFIC. COMPARE WITH 27 AUG 17.15.

Start Time	17.10							
Stop Time	17.17							

True Positive	HHT HHT II							
False Positive, Correlation	IIII							
False Positive, No Correlation								
False Negative								

Test006.xls - Blank Collection Sheet

Figure 1. Data collection sheet with light traffic volume

ALIRT BSD Test Results Record

Date _27 AUG_   Tester _RSH_   Temp. _25_

Location _401 WEST FROM DUTTON_

BSD Setup _5_   1=Mod100 proto; 2=Mod200 no conf.; 3=conf + bin; 4=pre-prodn heavy shutter; 5=pre-prodn light shutter; 6=Mod200 prodn; 10=computer

Weather _1_   1=sunny; 2=light cloud; 3=moderate cloud; 4=overcast; 5=light/moderate rain; 6=downpour

Driving Style _2_   1=aggressive; 2=moderate; 3=mild

Comment: _LIGHT TRAFFIC   COMPARE WITH 27 AUG 17:10_

Start Time	_19.15_									
Stop Time	_20.14_									

True Positive	~~HH~~ ~~HH~~									
False Positive, Correlation	~~HH~~ ~~HH~~ //									
False Positive, No Correlation										
False Negative										

Print Date 8/15/98   Test006.xls - Blank Collection Sheet

Figure 2. Data collection sheet with heavy traffic volume

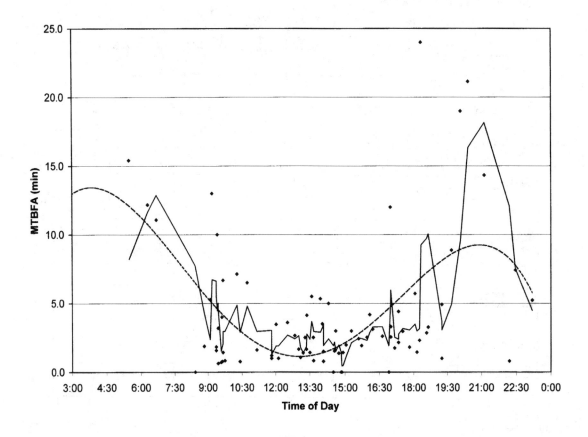

Figure 3. MTBFA vs. Time of Day

As a consequence, data should be collected as much as possible in real world situations (i.e. on public roadways under normal everyday driving conditions). As much as practical, testing should include targets that are randomly selected vehicles driven by the general public, not particular vehicles chosen by the testers.

Care should be taken to collect data across the entire range of conditions for any parameter which might affect sensor performance. For example, performance should be measured across a range of weather and traffic conditions, driving styles, times of day, etc.

Finally, data should be gathered in specific tests designed as "worst case" scenarios based on individual sensor technology design. For example, a worst case test for a microwave radar or active infrared system may involve a second system-equipped vehicle operated adjacent to the test system to determine performance under an interference scenario. For a passive infrared system, it may be testing at noon on a clear day, when sun loads produce the greatest background noise for the signal of interest.

The key consideration here is that the "worst case" scenario is tailored to the specific weaknesses of each technology, and thus may be different for each system. Further, worst case testing is conducted as a supplement to real-world performance measurement, and not as a replacement.

DESCRIPTION OF THE SENSOR TESTED

The tested sensor is a passive infrared blind spot sensor (side collision avoidance sensor) designed for use on light vehicles on multilane highways [8].

The sensor detection zone extends approximately 10 meters rearward from the location of the sensor on the driver's side exterior mirror, and laterally approximately 3 meters from the side of the vehicle. Actual size and location of the zone can vary depending on "aiming" of the sensor during user install. All data in this evaluation were gathered with the sensor installed according to the manufacturer's instructions.

The device is battery powered and is not wired to the vehicle. Control of the device is achieved solely through an on/off switch, and no turn signal interlock is provided. Characterization of false positives is therefore best achieved through the MTBFA measure described above.

No user adjustment (e.g. sensitivity) is provided, other than variation in aiming, as noted above.

This device has been developed to implement a proprietary sensor technology in a low-cost, simple device in the retail aftermarket.

DATA GATHERING

Data were gathered for internal technical purposes during development of the sensor. Methodology evolved slightly over the extended test period to a "tally sheet" approach (Figures 1 and 2).

Data gathering consists of:

1. Completing background and environmental data. This includes information such as date, sensor setup information (e.g. engineering revision level) weather, etc. In addition, several less obvious parameters such as driving style, location, and tester are recorded for later correlation analysis.

2. Space is provided for recording a number of start time and stop times. This was provided to allow proper recording of brief stops (e.g. gasoline, rest stops) which are common during normal highway driving.

3. The majority of the sheet consists of a tally section broken into areas for True Positive; False Positive with Correlation; False Positive with No Correlation; and False Negative. (A "correlated" false positive is an alarm for which the tester could determine the cause.)

4. When a second person was available to record data, the driver would inform the recorder of the tally for each incident, while the recorder would act as a check on the driver's accuracy. In cases where no second person was available to act as a recorder, the driver would mentally keep count of the four "bins" (e.g. 3-2-0-0) and record the data by "hash marks" as driving allowed.

Tests were of varying duration, according to the travel requirements of the testers. In this respect, the distribution of samples (e.g. time of day, weather, etc.) were a random sample of real world conditions.

Each test, corresponding to a tally sheet, was recorded as a separate entry on a simple computer data base. Since calculations were weighted on the basis of time and not number of tests, this had no effect on calculations of PCPs.

A new tally sheet or (test entry) would be started according to the tester's best judgement. This would typically be at the start of a travel period, or when there was a significant change in the environmental conditions (e.g. change from clear skies to rain).

ANALYSIS OF DATA

A total of 90 tests were recorded, averaging 54 minutes per test. These were reasonably evenly distributed across times of day and urban/rural highway.

Data were tagged according to the reliability of the test, as noted on the tester's record. For example, early prototypes of the sensor were determined to be sensitive to moisture leakage. Certain tests showed evidence of this, and these tests were flagged to allow removal during analysis.

RESULTS OBTAINED

Overall accuracy of detection of 99.32%, after exclusion of 6 tests totaling 7.8 hours, during which design malfunctions or other errors were known to have occurred. AD values ranged between 100% and 95.1%.

MTBFA was 3.23 minutes across all valid tests. A high MTBFA of 20 minutes was recorded for one 146 minute test, and approximately 25% of tests showed MTBFA greater than 10 minutes.

MTBFA showed a correlation with time of day. This was expected, since the infrared-based technology can be susceptible to noise from solar heating of the background road surface in the field of view.

This correlation is shown in Figure 3. This plot shows for MTBFA for each test, plotted against time of day. The plot also shows moving average across 3 data points (jagged line), and 4th order polynomial curve fit (smooth line) .

MDFA for the sensor was not measured during data gathering. However, engineering estimates based on the signal processing algorithm place MDFA at approximately 2.2 seconds: limited data gathering suggests that this estimate is accurate. Using this value, FAR can be estimated at:

FAR = MDFA / MTBFA

= 2.2 sec / (3.23 minutes x 60 sec.)

= .01135

= 1.14%

In other words, this sensor would be expected to provide false alarms in response to approximately 1.1% of all interrogations of the sensor.

POTENTIAL FOR SENSOR PERFORMANCE IMPROVEMENT

During the design process and algorithm development, the ALIRT sensor has been adjusted to maximize AD, at the expense of a lower-than-desirable MTBFA. This represented a conscious choice on the part of the designers that the device should be fail safe.

Design constraints for this sensor included: (i) energy budget of less than 5 mA at 3.0 V; (ii) manufacturing cost to allow retail selling price under $100US; (iii) user install with minimal hand tools in less than 5 minutes. These constraints forced circuit simplicity and prevented turn signal interlock wiring.

These factors suggest two potential sources of improvement in the performance of this sensor:

1. Additional development of sensor circuitry and/or signal processing algorithm. Fusion with other low cost sensing elements may be desirable, although this may require increase in energy budget and/or unit cost.

2. Turn signal interlock would provide an immediate order of magnitude increase in apparent false alarm rates. This would of course force an increase in complexity of installation, which is undesirable in the retail aftermarket.

REFERENCES

1. Garrott, W.R., Flick, M.A. and Mazzae, E.N.: Hardware Evaluation of Heavy Truck Side and Rear Object Detection Systems. SAE Technical Paper 951010, 1995, Society of Automotive Engineers

2. Mazae, E.N. and Garrott, W.R.: Human Performance Evaluation of Heavy Truck Side Object Detection Systems. SAE Technical Paper 951011, 1995, Society of Automotive Engineers.

3. Mazzae, E.N., Garrott, W.R. and Flick, M.A.: Human Factors Evaluation of Existing Side Collision Avoidance System Driver Interfaces. SAE Technical Paper 952659, 1995, Society of Automotive Engineers.

4. Rossow, G., von Mayenburg, M. and Patterson, C.: Truck Safety Technology for the 21st Century. SAE Technical Paper 952260, 1995, Society of Automotive Engineers.

5. Woll, J.D.: Vehicle Collision Warning System with Data Recording Capability. SAE Technical Paper 952619, 1995, Society of Automotive Engineers.

6. Mazzae, E.N., Garrott, W.R. and Cacioppo, A.J.: Utility Assessment of Side Object Detection Systems for Heavy Trucks. Proceedings of the Human Factors and Ergonomics Society 38th Annual Meeting, 1994, Human Factors and Ergonomics Society.

7. Tijerina. L. and Garrott, W.R.: A Reliability Theory Approach to Estimate the Potential Effectiveness of a Crash Avoidance System to Support Lane Change Decisions. SAE Technical Paper 970454, 1997, Society of Automotive Engineers.

8. Patchell, J.W., and Hackney, R.S.: New Thermal Infrared Sensor Techniques for Vehicle Blind Spot Detection. SAE Technical Paper 970176, 1997, Society of Automotive Engineers.

CONTACT INFORMATION

The author can be reached at:

ALIRT Advanced Technology Products
33 Grovepark Street
Richmond Hill, Ontario
Canada L4E 3L5

Phone: (905) 773-2257
Fax: (905) 773-3437
Email: eng@alirt.com
Internet: www.alirt.com

Vehicle Environment Sensing by Video Sensors

**Bernd Huertgen, Werner Poechmueller, Christoph Stiller,
Andreas Heiner, Christoph Roessig and Jens Goldbeck**
Robert Bosch GmbH, Advanced Research and Development, Hildesheim, Germany

ABSTRACT

This paper is concerned with the description of a vision system designed for use within future driver assistance systems. The system comprises a high dynamic range CMOS camera, dedicated evaluation hardware based on commercial available PC technology as well as algorithms for the interpretation of video scenes in video real time. The sensor is employed for vehicle environment sensing. Tasks include lane boundary detection, object detection and road sign recognition.

INTRODUCTION

Video sensors have become an interesting research topic during the 80ies when vehicle researchers turned to investigate automated highway systems and automated driver support systems [3], [7]. The intriguing aspect of video information is caused by the fact that man controls vehicles mainly by visual information. No wonder that due to the large amount of information perceived by the human eye and processed by the brain much information in the current transportation infrastructure is passed visually. Examples are lane markings and road signs which, in their current form, can hardly be recognized and interpreted by other than optical sensors.

Another advantageous aspect of video sensors is their passive nature since no emission of signals from the vehicle itself is necessary. Hence no problems appear with federal admission, disturbing interaction with other traffic participants or public concern about electromagnetic emissions.

Because of the efforts in Europe, Japan and the United States regarding automated highway research, video sensing in vehicle applications is often associated with automated hands-off driving, i.e. lateral and longitudinal steering is entirely under computer control. However, automated driving comes only at the end of a long chain of possible applications for video sensors in vehicle environment sensing. Warning functions and functions increasing the comfort of driving without overtaking active control of vehicle movement are less critical with respect to safety demands. Moreover, from the technical point of view these are realizable within the next years. Some examples are parking aids, lane departure warning, intelligent headlight control systems, road sign assistance and distance warning systems. More sophisticated approaches perform intervention into lateral and longitudinal control of vehicle movement up to a certain degree. Examples are intelligent cruise control for longitudinal and heading control for lateral steering.

Due to the fact that current system designs do not perform well under all conditions it is a crucial point that the limits of these systems are well defined. This demands for self and mutual supervision and the capability for informing the user. This leads together with a human like behavior to a transparent system which is absolutely necessary for a broader acceptance by the customer.

The potential of video sensors for vehicle applications is being investigated. Current CCD technology and video components from the consumer market do not fulfill automotive requirements, i.e. thermal stability, shock resistance, environmental illumination range, reliability, functional robustness and more. To meet these challenges new paths in system design have to be pursued. Only by considering the whole signal flow from optics through sensor, sensor electronics, digital signal processing electronics, algorithms up to the final actuator and/or driver information system a powerful video sensor system can be realized.

This paper presents an appropriate system design mainly consisting of a high luminance dynamic CMOS video camera, a video signal processing and evaluation system together with interface circuitry connecting the sensor to the vehicle's CAN infrastructure. The system is used for the development of new comfort, assistance, and control functions. Associated algorithms are designed to perceive vehicle environment data. Basically the vehicle position within the lane, vehicle tilt and pan angle, lane curvature, and the relative position of other road users or objects as well as information provided by road signs are of interest.

The remainder of the paper is organized as follows: The overall system design is described in the next chapter. Special emphasis is put on image acquisition, image processing as well as data interface, data visualization and integration into the test vehicle. The following chapter deals with basic image sequence evaluation software, e.g. detection of lane boundaries, detection and classification of road signs, object and obstacle detection and tracking. The paper concludes with some results and a brief summary.

OVERALL SYSTEM DESIGN

The entire system comprises an image acquisition unit, an image processing unit, a data interface, and a data visualization unit.

The image acquisition system is mounted below the interior rear mirror and observes the area in front of the car (see Fig. 1). Additionally a rear camera may be mounted for some applications such as parking aid. Image data is digitized within the camera and transferred via optical fiber to the preprocessing unit. The preprocessing unit allows simple operations on the image data without burdening the main processor, e.g. convolution for low and high pass filter, data subsampling, and a construction of resolution pyramids.

Figure 1. Interior of the test vehicle

The preprocessed image data are then fed via the PCI-bus into the main memory of the image evaluation unit where it can be processed by various image interpretation algorithms. Examples are lane detection, object detection, and road sign recognition. The aim of these algorithms is to extract parameters which describe the physical state of the vehicle itself regarding the street, the position of other objects in the surrounding of the vehicle, the meaning of road signs or the like. These parameters are written on the vehicle's CAN bus to which other control or supervisor units have access.

IMAGE ACQUISITION – In most video based approaches for vehicle environment sensing CCD cameras are used. Against the advantages of high contrast, low price, and compactness stands the disability of CCD cameras to cope with high luminance dynamics. Espe-

cially real world outdoor scenario demands for an extremely high luminance dynamic range from frame to frame and even more important, within one frame. Typical situations include rising or setting sun, headlights or reflections thereof on wet street surfaces during night. Plenty of similar situations may be observed in a normal drive and with most of them the human driver can cope easily. But many situations overtax standard CCD cameras due to their linear mapping function from light intensity to output voltage. This results in extremely degraded image quality or even complete blindness of the camera for a couple of frames. It is obvious that in these cases image processing software and therewith the entire system fails. Hence it is essential to enhance the cameras brightness dynamic capabilities in order to cope with those situations which frequently occur in normal traffic.

For the purpose of enhanced luminance dynamic standard CCD cameras employ highly sophisticated circuitry for electronic shutter and/or iris control as well as advanced design of the single CCD cells. Despite this effort, the luminance dynamic range still remains unsatisfactory and too low for the envisaged applications particularly when the dynamic within one frame is considered. A possible solution to this problem is to realize a monotonic non-linear mapping function whose sensitivity decreases with increasing light intensity. A natural choice are modern CMOS imagers which exploit the logarithmic behavior of MOS transistors operating in the subthreshold region [9]. Sensors of this type are able to cope with most of the above described critical situations and therefore have been chosen for the vision system described in this paper.

Fig. 2 shows a typical traffic situation in which a CCD based camera fails to provide acceptable image quality, both, in the background where certain bright details overmodulate the CCD sensor and in the foreground where the illumination is too low to modulate the sensor. Hence information from both areas are lost. In contrast the CMOS sensor supplies image data in the dark as well as in the bright parts of the scene.

Another reason in favor of an entirely new camera design is that most CCD based cameras are adopted to the CCIR or NTSC television standard. This means that they use an interlaced image format with a 4:3 aspect ratio and a frame rate of 25 or 30 frames per second, respectively adopted to the countries alternate current frequency. These characteristics have been established under the constraints of television broadcasting and surely do not cover the needs of a vehicle environment sensing system for which a broader image format like 2:1, progressive scan and a higher frame rate are desirable.

Figure 2. Exit from tunnel into bright daylight.
Top: standard CCD camera
Bottom: high dynamic range CMOS camera

CMOS camera – In order to take full advantage of the CMOS imager and ideally suit the image acquisition unit to the needs of a vision based vehicle environment sensing system, several prototypes of a high dynamic range camera based on a CMOS imaging chip were developed. The main features of the cameras are a high brightness dynamic range of 1:100.000, high resolution of 512x256 pixels, 2:1 aspect ratio, progressive scan, partial image read out capability, 10 bit resolution, logarithmic compression, digital image data output, and on board CAN interface for camera control. Due to the partial image read out capability the frame rate is not fixed to 25 or 30 frames per second but may vary in a wide range depending on the size of the image area to be read out in one frame. With the current design 20 Mega pixel per second can be read out leading to a maximum full frame rate of 160 frames per second. If only parts of the image are of interest the possible frame rate may increase roughly linearly with decreasing image size. Fig. 3 shows the first prototype of the CMOS camera.

The data flow within the camera is as follows: At first an individual but time invariant offset value which can be determined in the production process of the imaging chip

is subtracted from the output of each sensor element. This compensates for the CMOS specific fixed pattern noise. The corrected analog image data from each pixel is then fed into two parallel 10 bit A/D converter stages. The resulting digital data are multiplexed into a single bit stream and converted for optical transmission. The optical link between the camera and the evaluation unit has been chosen to avoid disturbances caused by the noisy car environment. Cheap plastic fibers have been found to be sufficient for transmission distances up to 10m. Further versions will employ the IEEE 1394 FireWire standard mainly for its higher physical robustness compared to the very sensitive optical fiber.

Figure 3. First prototype of high dynamic range CMOS camera with sensor board (left), digitizer board (middle) and interface board with power supply (right)

Preprocessing – In computer vision high level algorithms rely on certain low level image preprocessing such as lowpass filter, e.g. for anti aliasing or highpass filter for edge detection. In order to remove the computational burden of those structurally simple preprocessing algorithms from the main processor, an additional preprocessing unit is employed. It mainly consists of an optical receiver, a demultiplexer, a dedicated filter processor and a subsampling device, all realized in a field programmable gate array. The data flow is organized such that simultaneously multiple images can be provided for further processing depending on the actual programmable adjustment of the module. For example, the module may provide the original camera image, two gradient images in x- and y-direction, respectively, and additionally subsampled versions of the original image. Due to these filter and subsampling capabilities the image evaluation software may rely on resolution pyramids increasing processing speed significantly in most applications.

PCI-Interface and Image Processing Unit – The preprocessed image data have to be transferred from the preprocessing module to the main memory in order to give the main processor access to the image data. This is the task of the PCI-Interface capable of transferring approximately 100 Mbyte/s. The image processing unit itself consists of an ordinary single processor Pentium PC either MMX 200 MHz or PII, 333 MHz depending on the configuration. Optionally the PCI board may be equipped with an additional PowerPC processor which replaces the PC host and then serves as stand-alone system for rapid prototype development.

Figure 4.   PCI-Interface with optionally installed PowerPC processor

IMAGE PROCESSING – The main task of the image processing algorithms is to extract data about the car's ego position and movement and its environment including course of the lane, position and movement of other objects, information about road signs and more [1], [6]. In this sense the video system is used as measurement unit for vehicle environment data. Knowledge of this data enables the envisaged applications like lane departure warning, heading control, autonomous cruise control, obstacle detection, speed warning etc. The following paragraphs describe three main basic algorithms which are detection of lane boundaries, detection of objects and classification of road signs.

Common to all algorithms is the need for robustness, both, against environmental conditions like bad weather, e.g. rain, dust, fog or dazzling light and also against mismatches between the real world and the used models. Another requirement is that all algorithms should be able to perform processing in real time on standard PC hardware which demands for extremely high computational efficiency.

Detection of lane boundaries – A crucial point of vehicle environment sensing is to determine the ego position of the vehicle regarding the actual lane and additionally the

course of the lane. Our approach is based on the detection of lane boundaries under the assumption of their parallelism in each point [4]. For this purpose a model based approach has been chosen in which the vehicle is modeled as a dynamic system and the lane is modeled as piecewise clothoidal [3], [7], [8]. The vehicle/lane model is completely described by eight parameters, three describe the state of the vehicle, the remaining five describe the course of the lane:

- lateral offset of the vehicle regarding the middle of the lane
- tilt angle of the vehicle regarding the street plane
- pan angle of the vehicle regarding the tangent of the lane at the actual position of the vehicle
- width of lane
- horizontal curvature and horizontal change of curvature
- vertical curvature and vertical change of curvature

The algorithm works as follows: Emerging from the knowledge of the actual state which is the vehicle's position and the course of the lane the system dynamic is employed to predict a new state. The new state consists of the assumed position of the vehicle and the estimated curvature of the lane. This leads to the predicted position of the lane boundaries within the new image frame. Around these hypothetical lane boundaries the algorithm searches for lane markings indicated by white lines or simple luminance or texture edges. If those so called measurement points are of satisfactory quality which is determined by their preference direction and their markedness, they are used to estimate a new parameter set. For the estimation and prediction process a robust extended Kalman filter is applied.

In order to minimize the expenditure for special car equipment, e.g. steer angle sensor or yaw rate sensor, the system has been designed to provide full functionality even without any external data. However, if this data is available it may be used to enhance robustness, accuracy and reliability of the entire system.

Due to the fact that prediction accuracy decreases with increasing prediction time between two images and additionally is affected by the renunciation of external sensors it is a crucial point that the time between two processed frames be diminished to the absolute minimum. Our experience showed that a critical upper limit is video real time which is 40 ms (33.3 ms) between two successive frames. The latest version of the lane detection algorithm processes one frame in about 10 ms using a 200 MHz Pentium enabling a frame rate of approx. 100 Hz.

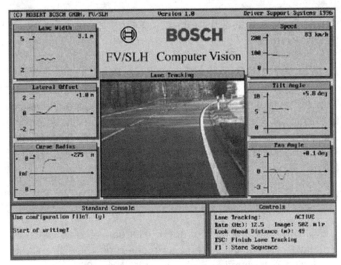

Figure 5. Screen dump from in-car real time test system running the lane detection algorithm

Object detection – The second main function of the video system is the detection of those objects which are located above the street plane, e.g. other traffic participants like vehicles, pedestrians, cyclist or other objects like trees or crash-barriers. Although a monocular object detection system is theoretically possible, a binocular camera is employed in the presented applications [1], [5]. This is mainly for the sake of detection reliability in critical situations in which a monocular system is likely to fail, e.g. in the case of moving shadows or when the vehicle itself does not move.

For this purpose two identical and genlocked cameras are mounted in the area of the interior rear mirror such that their optical axes are almost parallel. The baseline of the camera rig is approximately 30 cm allowing the cameras to be put into the rear mirror in future versions. In a semiautomatic calibration process the exact relative orientation of both cameras to each other and between the camera rig and the vehicle is determined which is necessary for all successive operations.

The basic task within the object detection algorithm is to find points in the left and right camera image which correspond to the same 3D real world point. By evaluating these so called disparities one can distinguish between points from the street plane and those from objects outside the plane. Simultaneously the actual orientation of the vehicle regarding the street plane is updated by evaluating only those disparities belonging to points of the plane. The update is carried out by a Kalman filter approach which additionally enables the orientation of the vehicle to be predicted for the time when the next image is processed.

The 3D coordinates of object points are determined by triangulation. Then the resulting clusters of 3D measure points are grouped to single objects for which their length, width and height can be calculated. By additionally exploiting the temporal consistency of the physical world the object detection is stabilized and its robustness against measurement errors is increased significantly.

The actual implementation achieves a frame rate of 25 Hz on a 333 MHz P II processor at a resolution of 256 by 256 pixel.

Figure 6. Typical traffic scene on German autobahn. Distance of objects are measured by stereo object detection

Road sign detection and classification – Another interesting application for vehicle environment sensing is detection and interpretation of road signs. This is mainly due to the fact that most traffic information, e.g. road signs or traffic lights is provided visually. A challenging task is to detect road signs in complex environment solely on the basis of monochrome image data as provided by the CMOS camera system introduced above. Since color information is not available the detection algorithm has to extract the information about possible road signs from edge information only [2]. For the real time tests we restricted ourselves to circular signs, e.g. speed limits or other prohibiting signs including their counterparts but the hierarchical structure of the algorithm allows also to include other types of road signs for the expenditure of an increased computational effort.

The algorithm is threefold. In the first step the image is searched for circular structures. This is done by generating an edge orientation image from two orthogonal gradient images which are provided by the preprocessing unit. A matching process follows in which the edge orientation image is correlated with templates describing circular structures. Points with high correlation coefficient indicate positions of possible sign candidates. These candidates are then tracked from frame to frame in order to compensate for outliers and measurement errors, respectively. The sign candidate is then fed into a classification unit which employs a statistical method. The result of the classification process is the meaning of the pictograms inside the circular structure.

The computational burden of the algorithm mainly is caused by the detection part. This is due to the fact that color information which would simplify detection of road signs significantly is not available. However the 333 MHz P II processor is capable of processing 25 frames per second at a resolution of 512 by 256 pixels.

Figure 7. Detected road sign

PERFORMANCE RESULTS – The presented algorithms have been tested under various weather conditions ranging from bright sunlight, rising and setting sun to rainy and foggy night time operation on rural roads, state roads and highways. Under "good" weather conditions all algorithms perform very well. However, problems have been encountered for road sign detection during night operation because of a low temporal dynamic of the actual CMOS imager at low illumination intensities.

Another remaining task is the visibility of objects and lane boundaries in the case of snow, fog or heavy rain. Under these severe conditions the availability of the entire system decreases. This point must be improved in future designs as well as the temporal sensor dynamic.

SUMMARY

In the preceding chapters a video based vehicle environment sensing system has been presented. The system consists of a high dynamic range CMOS camera system, an image preprocessing unit and a PC based image evaluation unit together with some interface circuitry. Three algorithms - lane detection, object detection and road sign recognition - have been presented which perform the basic tasks for a vehicle environment sensing system. All algorithms work in real-time on standard PC hardware

which means that the processing time for one frame is below 40 ms.

ACKNOWLEDGEMENT

The authors wish to thank the German federal ministry for education and research (BMBF) for partial support within the programs "Elektronisches Auge" (Electronic Eye) and "Mobility and Transport in Intermodal Traffic (Motiv).

REFERENCES

1. C. Stiller, W. Pöchmüller and B. Hürtgen: "Stereo vision in driver assistance systems", Proceedings IEEE Int. Conf. Intell. Transportation Systems, 3701, Nov. 1997

2. P. Seitz and G.K. Lang: " Using local orientation and hierarchical feature matching for the robust recognition of objects", Proceedings SPIE Int. Conf. Visual Com. and Image Processing '91, pp. 252-259, 1991

3. E. D. Dickmanns and B. Mysliwetz: "Recursive 3D Road and Relative Ego-State Recognition", IEEE Trans. on Pattern Analysis and Machine Intelligence PAMI-14, pp. 1999-214, 1992

4. J. Goldbeck, G. Draeger, B. Hürtgen, S. Ernst and F. Wilms: "Lane Following Combining Vision and DGPS", Proceedings IEEE Int. Conf. on Intelligent Vehicles '98, pp. 445-450, 1998

5. C. Rössig, R. Herrmann and M. Hoetter: "Continuous Estimation of the Road Plane for Measuring Obstacles Using a Stereo Rig", Proceedings Image Processing and Applications IPA '97, pp. 394-397, 1997

6. Q.-T. Luong, J. Weber, D. Koller and J. Malik: "An integrated approach to automatic vehicle guidance", Proceedings ICCV'95, pp. 52-57, 1995

7. U. Franke: "Real time 3D-road modeling for autonomous vehicle guidance", Proceedings Scandinavian Conference on Image Analysis, pp. 316-323, 1991

8. R. Risack, P. Klausmann, W. Krüger and W. Enkelmann: "Robust Lane Recognition Embedded in a Real-Time Driver Assistance System", Proceedings IEEE Int. Conf. on Intelligent Vehicles '98, pp. 35-40, 1998

9. U. Seger, H.-G. Graf and M.E. Landgraf: "Vision assistance in scenes with extreme contrast", IEEE Micro, pp. 50-56, 1993

COLLISON WARNING SYSTEMS

ITS Technology in Automotive -- Who's on First?

Gerald D. Conover
PRC Associates

ABSTRACT

This paper examines the status of important ITS technology developments at major automakers and key Tier One suppliers around the world.

Technology topics include navigation, telematics, vehicle dynamics control, parking aids, adaptive cruise control, lane departure warning, active safety systems, predictive safety systems (e.g., braking assist, collision warning, and collision avoidance), and vehicle-to-vehicle and vehicle-to-roadside communications.

Each company's ITS technical acumen is assessed and compared to its competitors. Leaders in each technical area are suggested.

The paper concludes with postulation of important future trends in automotive ITS.

INTRODUCTION

The whole notion of which organizations or firms are ahead or behind in the development and deployment of ITS products and services is almost ludicrous. After all, ITS is supposed to bring on a host of social and economic benefits, not the least of which are reductions in traffic incidents and deaths.

But the world is a competitive place. Each weekend, thousands of young men and women go onto the fields of sport to compete against one another, rack up the highest score and be declared the winner – or in some cases, the champion.

And so it is with ITS. Like any human pursuit, some people are compelled to measure their accomplishments against those of others – in similar agencies or competing companies – to see who is biggest and best.

The auto industry has been highly competitive for over a hundred years, so it is logical – almost predestined – that automakers and suppliers should peer out of their headquarters at one another to try to figure out who has done the most with ITS. Some popular ways to count success around the world's auto industry are (1) the number of new applications of technology or (2) sales volumes in either units sold or dollars (or yen or euros) of revenue.

We prefer to look at the number of different applications of the various ITS technologies to measure the mettle of automakers and suppliers.

THE FRAMEWORK

To bring some order to the world of automotive ITS applications, we propose the following analytical framework:

	Autonomous	Connected

Driver Assistance	• GPS Navigation
	• Adaptive Cruise Control
	• Park Assist
	• Night Vision (Vision Enhancement)
		• Telematics
		• Emergency Calling (Collision Notification)
		• Vehicle Infrastructure Integration
		  – With Other Drivers
		  – With Infrastructure

Vehicle and Traffic Safety	• Lane/Road Departure Warning • Lane Keeping • Sensor Fusion (enabler) • Blind Spot Monitor (Lane Change Assistance) • Pre-Collision Protection (Forward Collision Warning/Assistance) • Rear Impact Warning • Rear View Television/Obstacle Detection • Peripheral Monitoring • Drowsy Driver Warning	• Intelligent Speed Adaption • Intersection Collision Warning

Figure 1. Analytical Framework

These technology applications and their allocation around the "picture frame" are certainly open to debate. Others will have other frameworks that better suit their respective arguments.

For this analysis the areas of discrimination are (1) use – to aid the driver or to promote safety – and (2) whether the applications stand alone or need to be connected to some infrastructure off the vehicle in order to deliver the expected benefits.

There are several technology applications that have not been included here – electronic stability control and rollover warning or protection, for example. While these applications certainly provide new levels of safety enhancement they represent improvements in vehicle dynamics control rather than in ITS[1] functionality.

GPS NAVIGATION

Over the last fifteen years, on-board vehicle navigation using GPS as the main positioning engine has become essentially ubiquitous in the major auto consuming countries. As China becomes an automotive power, it too is reaching out to GPS navigation as an essential capability for commercial, government and private drivers.

Offering on-board navigation is more a marketing choice for an automaker than a matter of technical capability. Numerous navigation systems – including very capable GPS engines and highly developed routing algorithms – are available from Tier One and Tier Two automotive suppliers (mainly Japanese and European). Excellent maps can be had from American[2], European and Japanese digital geographic information firms.

Navigation systems are universally available in Japan in both premium and everyday automotive brands. Virtually all European premium and many mainstream cars also have navigation systems available. Many are optional extras although there are others installed as standard. Navigation systems are just starting to catch on with American automakers – Chrysler Group, Ford and General Motors all offer navigation system options on some of their cars and light trucks.

ADAPTIVE CRUISE CONTROL

Adaptive Cruise Control[3], or ACC, is the "killer app" for ITS in this decade not only for its own capabilities but also for the other ITS applications that its radar is spawning – pre-collision warning and blind-spot monitoring, to name only two.

ACC uses radar – either microwave or laser (each has its own operational strengths and weaknesses) – to (1) locate a vehicle in the lane ahead, (2) determine the "target" vehicle's distance from our "own" vehicle, and (3) calculate the relative speed between one's own and the target vehicle. These last two dimensions are sometimes referred to as "range" and range-rate."

There are those who would argue that ACC is a safety system and thus belongs in that "pane" of the analytical framework "window." That notion certainly has merit. We chose the driver assist window because of the way ACC is marketed by automakers. They prefer to position ACC as a driver aid in order to avoid potential tort liability from offering a forward-collision-warning system – the promise might be greater than the deliverable.

Although Mercedes made the first offer of ACC to the retail automotive market in Europe in 1999, in fact Jaguar made the first retail delivery of ACC about a month before Mercedes.

In only a few short years ACC has expanded worldwide with offerings on numerous cars and some SUVs in Europe, America and Japan. Virtually all European auto brands offer ACC, mostly on their premium models. Japanese brands offering ACC include Daihatsu, Nissan (including Infiniti), and Toyota (including Lexus). Cadillac is the only Big Three brand to offer ACC – it's a Delphi design based on a Raytheon sensor.

Numerous Tier One suppliers offer turnkey ACC systems including Continental Automotive Systems, Delphi Automotive, Denso, Hitachi, Siemens VDO Automotive, TRW and Valeo (this is only a partial list of ACC suppliers).

Second-generation ACC systems have just come to market. These new embodiments are termed "full-authority" – they operate all the way from highway speeds down to a complete stop. Gen I ACC systems cut out at around 20-30 km/h. The first of these is offered by Mercedes on its 2005 S Class – Continental is the supplier. Several other Tier One suppliers have Gen II ACC systems developed and ready for sale to automakers.

PARK ASSIST

Ultrasonic park assist is a relatively old technology application. Japanese suppliers were "shopping" ultrasonic park assist systems in America in the mid-1970s. First applied to vehicle rear bumpers to mitigate the occurrence of low-speed contact with objects and other vehicles, park assist has quickly moved to front bumpers for added protection in highly styled vehicles where the front edge may not be easily apparent to the driver.

Park assist can be found on almost all vehicle brands around the world, in all major markets. This is another example of marketing decisions setting the automotive product application limits. The technology is in place and well-proven.

There have been some applications of radar to park assist. This is a higher cost approach than ultrasonics for close-distance protection, but some applications use radar to reach an additional several meters back of the vehicle to give earlier identification of potential obstacles. Visteon and some others have fielded this.

Park assist has not been promoted as a safety device by automakers. It has only been implied that pedestrians or small children would benefit from early detection by a slowly backing vehicle. Suppliers have been more aggressive in this regard than have automakers – tort concerns again.

Denso and Toyota were among the first automotive companies to work on "total parking space definition" using several sensors to determine the length and depth of a parking space, then offering guidance to help the driver park in the space safely. Other suppliers, notably

Aisin AW and Continental, have developed systems that will use the parking space data to steer a vehicle into a parking space – the driver still controls accelerator and brakes.

Aisin's Intelligent Parking Assist was jointly developed with Toyota, and helps the driver in all reversing parking operations. By establishing the target parking position, the vehicle's parking movements are accomplished with automatic steering.

The driver need only control the speed of the vehicle. In the Toyota Prius adaptation, this means operating only the brake.

Volkswagen showed a fully automated parking concept vehicle several years ago, but it used an impossibly expensive 360-degree array of radar and ultrasonic sensors as well as a large computer to work the algorithms.

NIGHT VISION

Night vision and vision enhancement technologies have been under development since the late 1980s – several research applications were demonstrated in 1994 as part of Europe's PROMETHEUS program.

Night Vision was #2 on JD Powers' consumer research list of emerging technologies – until price was mentioned. At $1,500, it dropped to #17.

There are two technical approaches to night vision, one using the far infrared spectrum (FIR) and another using near infrared frequencies (NIR).

Numerous suppliers have developed night vision systems. Among them are Siemens VDO, ITT, Delphi (using Raytheon technology), Autoliv (for BMW) and Robert Bosch.

General Motors was the first automaker in the world to offer night vision (FIR) as a regular production option (on some 2000 model year Cadillacs). Although the first-year take rate approached 20% for the $2,250 option, the rate quickly dropped to 2% when it stayed stubbornly stuck. GM dropped night vision from the Cadillac DTS concluding that a business case could not be realized. The system Raytheon developed for Cadillac used an Uncooled Focal Point Array that works well at 72°F. Raytheon used a $20 part to keep the temperature of the array stable, heating or cooling it as necessary.

Toyota has had similar experience with night vision on Lexus, where the percentage of high-end LX470 models shipped with the $1,750 NIR option has dropped from a peak of 26% a few months ago to just 5% recently[4].

Honda has developed what it calls the "world's first Intelligent Night Vision System" which is targeted to reduce night-time pedestrian casualties. The system is available on the Japanese home market Honda Legend.

Honda uses FIR cameras to detect pedestrians in or approaching the vehicle's path and provides the driver with visual and audible warnings. The Intelligent Night Vision System uses images obtained from two far infrared cameras positioned in the lower section of the front bumper to detect the position and movement of heat-emitting objects and determine whether they are in or approaching the vehicle's path. Based on object's size and shape, the algorithms determine if it is a pedestrian.

The latest automakers to jump into the night vision pool are BMW (NIR) and Mercedes (FIR). Shown on the 7 Series and S Class, respectively, the systems are expected to retail in Europe for more than €2,000.

While it is arguable that night vision could be the "next big thing" in automotive features, the price will have to come down by at least half to get to the mainstream kinds of volumes that suppliers and automakers need to fulfill a sustaining business case.

LANE AND ROAD DEPARTURE WARNING

The US National Highway Traffic Safety Administration says single vehicle road departure crashes represent the most serious type of crash on national highways. There were 927,000 single vehicle off-roadway crashes in 2002 and 12,360 of them resulted in fatalities. There are many different causes of these types of crashes, including weather and vision problems, driver impairment, inattention and improper driving behaviors. Roadway departure systems alert drivers when they are drifting out of their lanes and off the roadway – and when they are approaching an upcoming curve at excessive speed[5].

The first, and so far only, commercially available lane departure warning system has been offered by Iteris for commercial vehicles, and via a licensing agreement with Valeo for light vehicles.

The Valeo system installed on Infiniti products (FX, M and Q45 models) recognizes lane markings through the use of a small camera mounted behind the rearview mirror. The camera's signal and the vehicle's speed are sent to the system's microprocessor which combines the information to calculate both the distance between the vehicle and the lane marking and the lateral velocity to the lane marking. If it is determined that the vehicle is leaving the lane, a combination of visual and audible warnings are given. In some embodiments the car's braking systems are activated to help return the vehicle to its original lane.

The Infiniti LDW system is temporarily disabled by driver activation of the vehicle's turn signals, which informs the system's control unit of intended lane changes. The system also features a manual cancel switch, which allows the driver to turn the system off when desired. The system automatically resets when the vehicle is restarted.

On the 2005 Infiniti FX 35, lane departure warning was offered as part of the DVD Mobile Entertainment System with Lane Departure Warning option package priced at $1,750. LDW likely represents only a small fraction of the total package price.

Aisin's Lane Departure Warning System, jointly developed with Toyota, also measures the distance between a car and the white or yellow lane line but with a rear-mounted camera (which is also used by the parking assist system), and alerts the driver with an audible beep if the vehicle is about to cross a highway lane.

LANE KEEPING

Lane keeping is the next big step beyond lane departure warning, Here an electronic "big brother" applies a corrective torque to the vehicle's steering system to bring the car or truck back to the correct driving line automatically.

While no automaker has put a lane keeping system into regular production, there are numerous development programs underway at automaker and supplier laboratories.

One typical lane keeping system is being developed by Continental Automotive Systems. Like lane departure warning, lane keeping determines the vehicle position relative to the lane markings, but offers active support in keeping the vehicle centered on the lane. The driver always retains control – he or she can feel the recommended steering reaction as a gentle movement of the steering wheel, but his or her decision to act or not takes priority.

Mazda's Lane Drive Aid System automatically determines the vehicle's lateral position using a camera and provides steering assistance that helps the driver keep the vehicle in the middle of the lane. When the system judges that the vehicle is in danger of drifting out of the lane, it issues a warning.

Further, the system uses radar sensors on either side of the vehicle to monitor adjacent lanes alongside and aft of the vehicle. If the driver starts to move into a lane without realizing that a vehicle is there, the system issues a Blind Spot Warning.

Mitsubishi's Lane Trace Assist uses an on-board camera to monitor the lane dividing lines and applies a corrective torque to the electric power steering when it senses the vehicle is starting to depart from its lane.

The Lane Keeping Assistance on Toyota's Estima HEV uses electric power steering guided by a video camera that monitors lane markers to assist steering to help keep the vehicle in the intended lane. The system constantly applies the small amount of steering effort required to keep the vehicle near the center of its lane.

SENSOR FUSION

We must take a short detour – a digression – here to talk about sensor fusion. The foregoing discussion of various driver assistance systems has covered applications based on single sensor inputs. While this is the current state of the art, we would be remiss if we didn't say something about the next generation of ITS vehicle sensors with will feature combinations of sensors working like partners to deliver richer arrays of data that drivers and their support systems need.

Denso, for one, has developed additional sensors that detect road conditions and objects ahead of a vehicle. Denso's lidar sensor (laser radar) or millimeter-wave radar sensor is used with Adaptive Cruise Control systems. The millimeter-wave radar sensor is also used in the Pre-Crash Safety Systems, which can help lessen damage in an imminent collision. A vision sensor is used in the Lane Keeping Assist system that helps the driver stay in the center of a lane.

By combining multiple sensors into a single system, more highly accurate safety systems can be created. Combining a millimeter-wave radar sensor and a vision sensor, a sensing system that recognizes the shape, size and type of object can be developed. The radar sensor detects the position and relative velocity of the object, while the vision sensor detects the type and attributes of the object.

As a result, it is possible to accurately identify the positional relationship between the object and the vehicle, as well as the object's relative movement, leading to a safety system that is better tailored to each situation.

As part of the effort to realize even more advanced safety systems, Denso is developing vision processing technology that allows the detection of pedestrians from among multiple objects.

BLIND SPOT MONITORING

Also known as lane change assistance, the goal of the blind spot monitor is to keep a driver from changing lanes when there is another vehicle aft of and to the side of one's own vehicle. This is another PROMETHEUS development that is finally "coming home."

Modern blind spot monitors use radar (some microwave, although laser seems to be increasing in popularity with developers) to look in an area in about a 90 degree arc at each rear corner of a vehicle. If the directional signal is activated, the corresponding radar looks into its assigned area and returns a warning if there is a vehicle in the "blind spot."

Some designs put the sensor in the outside rearview mirrors. This allows for a narrower arc, but increases range requirements. This mounting alternative does give a view into more of the real "blind spot" – the area behind the driver's peripheral vision but ahead of the outside mirror's field of view.

Hella and Siemens VDO, for example, have designed a corner-mounted sensor system, while Mobileye, an Israeli company, has developed a system with the outside mirror mounting position.

Mercedes is evaluating another route with its exterior monitoring system. Cameras integrated into the housings of the outside rearview mirrors monitor traffic to the side of and behind the vehicle when it is parked. If another car or a bicycle is approaching from behind, the system will automatically disable the doors for a brief period to avoid the risk of a collision when they are opened. A warning signal sounds at the same time accompanied by a red danger symbol that illuminates in the mirror glass.

PRE-COLLISION PROTECTION

Current pre-collision warning systems are tied into adaptive cruise control, using the ACC's forward-looking radar to detect potential collision "targets" ahead of one's own vehicle. Typically these systems take several actions to prepare for the potential collision:

- Adjust the seat back angles of the front seats (and in some cases the rear seats) to an angle best suited for "taking" the potential collision,

- Tighten the seat belts,

- Alert the air bags to be prepared to be quickly deployed in the event of a crash, and

- Pre-load the Brake Assist system, if present, so that the brakes can be applied by the driver with maximum force.

Toyota claims to be the first automaker to have developed a radar-based collision safety system – it came out with its Pre-crash Safety System in February 2003. The system uses millimeter-wave radar and a camera to detect other vehicles and obstacles on the road ahead and warns the driver when a collision is imminent. In August 2003, Toyota added Pre-crash Brakes to the system that applies the brakes when the driver fails to react in time. In July 2004, it adopted an "image fusion" method, which adds camera-gained image information to information obtained by millimeter-wave radar. Toyota has further advanced its Pre-crash Safety System by adding a feature that monitors the

driver's alertness to determine if he or she is looking straight ahead.

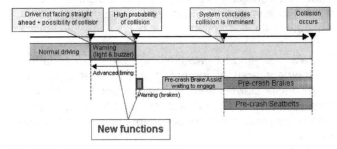

Figure 2. Toyota Pre-Crash Safety System Operation

Honda announced in May 2003 that it had developed the world's first Collision Mitigation Brake System (CMS), which predicts longitudinal collisions and assists brake operation to reduce impact on occupants and vehicle damage. This system determines the likelihood of a collision based on driving conditions, distance to the vehicle ahead, and relative speeds, and uses visual and audio warnings to prompt the driver to take preventive action. It can also initiate braking to reduce the vehicle's speed.

Mercedes' PRE-SAFE system was first announced in Europe in 2001 and came to market in 2003. Initially only adjusting the seat angles and closing the sunroof, the system was quickly expanded to include all the other functions now considered to be part of any pre-crash system.

The evolution of many of these sensor-based detection systems is toward fully automatic collision avoidance and mitigation systems.

REAR IMPACT WARNING AND PROTECTION

As has been demonstrated in the foregoing discussion, most of the attention in longitudinal car-to-car collisions has been on warning the following driver and activating protection systems in his or her vehicle.

While a number of companies have been designing sensor and activator packages to help the occupants of the leading vehicle, Mazda has been the most communicative about its developments via its SensorCar development concept.

Two optical sensors (charge coupled devices or CCDs – really small TV cameras) mounted 2.5 inches apart in the SensorCar's rear bumper monitor surrounding vehicle traffic. The CCD units send their data to a dedicated computer that compares distance, approach angles and closing rates of the other vehicles to determine whether any of them is about to collide with the SensorCar.

If the system determines that a significant rear impact is probable – contact at a closing rate of 6 mph or greater – it sounds an audible alarm from the rear audio speakers and illuminates a warning icon to call attention to the danger.

If the approaching vehicle speed indicates that a collision is imminent, a motorized retractor tightens the front seat belts.

Mazda's rear-impact warning system needs further development and refinement to assure that it screens out all false alarms and detects all potential impacts. Once refined, it could be tied into multiple active safety systems to protect vehicle occupants or avoid accidents entirely.

REAR VIEW TELEVISION

Although closely allied with ultrasonic park assist, rear view TV provides sufficiently more information and lets the driver actually see what objects might pose a danger, or be in danger, when a vehicle is backing. Thus it has been classified as a safety system in this paper.

Rear view TV had its genesis in Japan almost a decade ago – a logical combination of Japanese expertise in making small, rugged television systems with the shared display (for navigation and convenience systems) that had previously taken up residence in the upper center part of the instrument panel. The TV camera is activated when the vehicle is shifted into reverse gear. Nissan, Toyota and Honda quickly jumped on the bandwagon.

Nissan (through its Infiniti brand) was the first Japanese company to bring rear view TV to Western markets. Toyota (through its Lexus channel) and Honda also offer the systems in North America.

The cameras are usually part of a package with other options putting total package prices in the $2,000 to $4,000 range. The Big Three are starting to catch up. A rear view camera system will show up as a regular production option on GM's new 2007 full-size SUVs.

Many Japanese consumer electronics firms offer rear view TV systems in the aftermarket, a trend that is spreading to North America as well. Aftermarket systems are priced between $300 and $700 plus installation.

PERIPHERAL MONITORING

Many automakers are looking at ways to combine the outputs of sensors looking forward, backward and to the sides into a comprehensive system that monitor all around the periphery of the vehicle – hence, the term "peripheral monitoring."

Nissan's Around View Monitor substantially reduces blind spots by displaying a 360-degree view of the area around the vehicle on an instrument panel monitor. The

easy-to-understand images displayed in real time are helpful in supporting the driver's maneuvers when parking, by assisting the driver in steering the vehicle easily and accurately into the intended parking space. The system is a further advancement of Nissan's Rear View Monitor, Blind Spot Monitor and Lane Departure Warning applications.

Toyota and Mazda are also experimenting with peripheral monitoring. Image-processing technology enables their systems to detect and warn the driver about children and obstacles within the vehicle's blind spots and to detect and warn the driver about vehicles that are approaching from either side.

Subaru has been researching for at least a decade the use of binocular vision systems for hazard detection and monitoring. Subaru's "eyes" are its frontal recognition sensors, which are adapted from Subaru's Sensor Fusion technology in its Active Driving Assist system. It includes a pair of stereo cameras, as well as a millimeter-wave radar unit from the Legacy Touring Wagon 3.0R.

The stereo camera has been improved as have its recognition algorithms. Processing speed has been accelerated in the new stereo cameras and recognition has been enhanced. In addition, the camera unit and its controller are now packaged together in one module to be more compact and lower cost.

The new stereo camera device can not only detect multiple objects (cars, pedestrians, bicycles), road traffic lines, and conditions (climate changes, light changes, and other conditions), but it also calculates the distance between the car and an object (even at night) as well as the inter-vehicle speed.

DROWSY DRIVER WARNING

Considering the number of traffic incidents and injuries that can be ascribed to driver inattention and drowsiness, it is amazing to see how long it has taken technological solutions to come to fore.

Drowsy driver detection is another of those PROMETHEUS technologies that was demonstrated in vehicles in the mid-1990s. Early systems tried to use changes in the steering wheel reversal rate to predict when a driver was tired (tired drivers correct less often), but these were unable to avoid a high rate of false positive indications (drowsiness was indicated even though the driver was actually alert).

Later experiments tried video detection, using small TV cameras to watch the driver's face the check the location of certain facial features, such as the outboard edges of the eyes and mouth, or eye blink duration and frequency. This technology has continued in development. In the US, Carnegie Mellon University has done extensive work in this area.

The Johns Hopkins University has done work using a 24-GHz transceiver both to illuminate the head and upper torso of the driver. The technique monitors and quantitatively measures several indices, such as the general activity level, the speed, frequency and duration of eyelid closure, the rate of heartbeat and respiration, by analyzing the Doppler components present in the reflected signal.

A drowsy index is calculated based on general activity. As fatigue sets in, this activity will drop below a certain level. A warning can be given at this time notifying the person that he or she is becoming sleepy.

A supplemental index monitors the eye activity for changes associated with increased levels of drowsiness and for eye closure. This second index allows for warnings of increasing intensity as the onset of sleep occurs[6].

Toyota's Pre-Crash Safety System, discussed earlier, comes closest to using sensors to check driver condition, even if Toyota's system determines only head position, not sleepiness or other inattention.

Watching the driver – to detect drowsiness or operation under the influence – is an increasingly popular topic with law enforcement, regulators and automakers because the inattentive or impaired driver is the cause of a large fraction if injury and fatal auto crashes in all developed countries.

Denso is among the leaders among Japanese suppliers for its work on technology that will detect the driver's condition, such as facial recognition, blink and closed eyes detection, and electrocardiograph measurement. Facial recognition identifies an individual by taking a picture with a camera and processing the image.

Blink and closed eyes detection applies different processing to images taken of the driver in order to measure eyelid aperture and detect any lack of attention from sleepiness.

Electrocardiograph measurement is not as far along as the driver monitoring technique, but is envisioned to measure the driver's heart rate through electrodes installed in the steering wheel, then display results on the car navigation monitor or transmit them to an external location such as a physician or family member.

Let's change focus now from autonomous systems to the "connected systems", those that need some electronic link to the roadside or other infrastructure to operate.

TELEMATICS

The telematics industry in Japan in dominated by Toyota and its G-Book, just as General Motors's OnStar dominates in the US.

Toyota recently initiated G-BOOK ALPHA, a more advanced telematics service that consolidates "safety and security", "driving intelligence" and "amusement".

In the event of a traffic accident or medical emergency, the HELPNET service (standard feature) connects users to a dedicated operator and enables summoning of ambulance or other assistance with an onboard microphone. Some new Toyota models come with a system that automatically connects to the operator if the air bags are activated. If there is no response from inside the vehicle, the operator will call for an ambulance.

Optional (e.g., Lexus) G-Security detects breaking into vehicles and tracks vehicle position. For "driving intelligence", the G-BOOK ALPHA service is equipped with a new "G-Route Search" function that combines traffic information from the Vehicle Information and Communication System (VICS) with past statistical data to forecast road traffic conditions and suggest routes that avoid traffic congestion.

The "amusement" area includes Toyota's on-demand car audio system. When vehicles are shipped, encrypted data for more than 10,000 songs are made available on the navigation system's hard drive with a digital rights management system. The customer can then download and purchase a decryption key from a network for the musical pieces that he or she wants to enjoy

Mazda recently adopted Toyota's G-BOOK ALPHA after its own attempt at offering proprietary telematics failed.

Nissan continues to offer its own Telematics system. CARWINGS provides a host of helpful services such as fastest route calculations based on statistical traffic data, real-time traffic updates, weather reports and news. The system supports e-mail notification of the driver's present location, live operator assistance for destination searches and navigation system questions and emergency roadside assistance services, among other services. The current CARWINGS system also provides seamless communication through the use of a Bluetooth-enabled mobile phone.

EMERGENCY CALLING

Also known as Automatic Collision Notification, this is probably one of the most sought-after features from modern telematics systems – the ability to have an on-board mobile phone automatically call emergency services in cases when an air bag deploys.

General Motors added this feature to its OnStar system, receiving wide acclaim from traffic safety professionals and the car and light truck-buying public. GM has, of course, already installed its call centers in North America to handle "ACN" and other calls (but has closed the one center it had previously opened in Europe),

The European Commission has made "eCall" a priority application in its vast new "eSafety" program. In the European vision, eCall is an emergency call either generated manually by vehicle occupants or automatically via activation of in-vehicle sensors when an accident occurs. When activated, the in-vehicle eCall system establishes a voice connection directly with the relevant Public Service Answering Point which is a public authority or a private eCall center. At the same time, a minimum set of data – including key information about the accident such as time, location and vehicle description – is sent to the Public Service Answering Point operator receiving the voice call. The minimum set of data may also contain the link to a potential Service Provider by including its IP address and phone number[7].

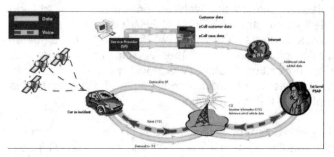

Figure 3. European eCall Network Operation

Mercedes launched its own call-center-based telematics system – TELEAID – which is capable of receiving eCall alerts from vehicles and relaying emergency requests to the appropriate public agencies. Fiat also operates a small telematics call center for its customers in Italy.

Japan has had a more unified approach to emergency calling. The Japanese police have been promoting the Universal Traffic Management Systems (UTMS). The Help System for Emergency Life Saving and Public Safety (HELP) is one of subsystems of UTMS.

Emergency calls from mobile phones have been increasing very rapidly in Japan with the expanding use of mobile phones, and account for more than half the total number of emergency calls. It is difficult to identify the location of a caller and this can delay an emergency vehicle's arrival.

In the HELP system, when an incident occurs while driving, an in-vehicle unit and a cellular phone automatically transmit caller location information obtained from GPS to the operation center, and then automatically activates voice communication.

HELP reduces the time lag from occurrence to reporting of an accident by 40% (from 15 minutes to 9 minutes). All command and control centers of prefectural police headquarters in Japan are equipped to receive location and other information obtained from the HELPNET center. HELP is available everywhere in Japan.

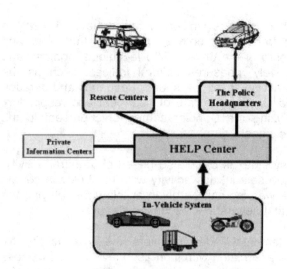

Figure 4. Japanese HELP Emergency Calling System

In order to further improve HELP, automakers in Japan are being encouraged to provide in-vehicle units equipped with the HELP system in all new cars[8].

VEHICLE INFRASTRUCTURE INTEGRATION

VII is a brave (and bold) comprehensive initiative to connect all the cars in the United States with each other and with the roadside (and thus a variety of incident management and emergency services).

The notion behind VII is fairly simple. If vehicles (or their drivers) can be made aware of the intentions of other vehicles (and their drivers) it may be possible to keep these vehicles from running into each other and thus reduce (or eliminate) traffic incidents and the resultant injuries and deaths. Similarly, if vehicles can communicate in real-time with the roadside infrastructure a variety of safety, emergency and convenience features can be made available to the driver and vehicle passengers.

The VII initiative enjoys support from US federal government, a number of trade and research groups, and the principal automakers doing business in North America (both traditional and "new" domestic manufacturers).

Current efforts are focused on determining if the investment necessary to equip new vehicles and the roadway infrastructure with communications is warranted and can be synchronized between the public and private sectors. Studies are underway to achieve at least proof-of-concept for the technology and to develop business cases sufficient to convince public and private sector leaders to invest.

A project to make VII ubiquitous across the United States is probably of greater magnitude than the construction of the Eisenhower Interstate Highway System. No one knows for sure what the total investment would be in a comprehensive network of roadside units that would communicate with and relay messages to vehicles moving on the road. The computer network to process in real time the millions of asynchronous communications between vehicles and infrastructure has yet to be defined.

Nor has anyone figured out how the more than 3,000 road agencies in the US will be convinced to install and operate the roadside units. Perhaps there is a private sector role here.

A similar idea is being pursued in Europe by the Car2Car Consortium among others. The Car2Car Communication Consortium was organized by European vehicle manufacturers and is open to suppliers, researchers and other partners. The Car2Car Communication Consortium wants to improve traffic safety and efficiency by means of inter-vehicle communications.

The Car2Car Consortium hopes to establish an open European industry standard for Car2Car communication systems based on wireless LAN components and to guarantee European-wide inter-vehicle operability. The development of active safety applications is being promoted through the specification, prototyping and demonstration of the Car2Car system. The Consortium also plans to develop realistic deployment strategies and business models to speed-up market penetration.

The radio system for Car2Car is derived from the standard IEEE 802.11 (Wireless LAN). As soon as two or more vehicles are in radio communication range, they connect automatically and establish an ad hoc network. The range of a single Wireless LAN link is limited to a few hundred meters, so every vehicle is also a router that allows sending messages to farther vehicles via multi-hop. The routing algorithm is based on the position of the vehicles and is able to handle fast changes of the ad hoc network topology[9].

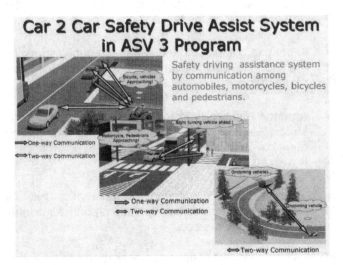

Figure 5. Vision of Car-to-Car Communication in Japan

Japan has integrated the notion of vehicle-to-vehicle and vehicle-to-roadside communication into its Advanced Safety Vehicle program. ASV-3 (the third ASV program)

is reaching its conclusion. Demonstrations of the various architectures and technologies are on-going in 2006[10].

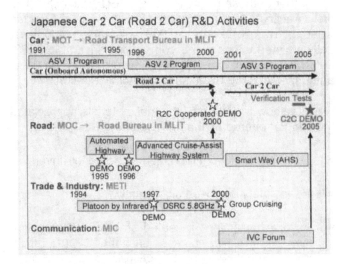

Figure 6. Japanese Car-to-Car R&D Activities

General Motors (Cadillac) and BMW recently demonstrated vehicle-to-vehicle (V2V) communications without benefit of any external communications infrastructure. These respective manufacturers' research cars talked to one-another via an 802.11p WLAN operating in the 5.9 GHz band. Vital information about vehicle location, status (e.g., stopped) and the nearby surroundings (e.g., its raining) were transmitted back to following vehicles so their drivers could take appropriate action. Vehicle location was by GPS, but relative (car-to-car) rather than absolute (on the surface of the planet), so relative positioning error was about one foot.

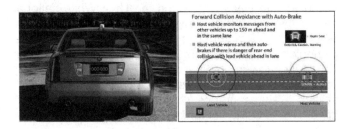

GM equipped one Cadillac with collision avoidance systems that advised the driver when another vehicle was in its blind spot or applied the car's brakes with full authority in case a collision with another V2V-equipped vehicle was eminent.

INTELLIGENT SPEED ADAPTION

Viewed by some as the ultimate automotive invasion by "Big Brother" and by others as a logical and necessary move to improve traffic safety, Intelligent Speed Adaption (or ISA) provides on-board advice to drivers to slow down or in extreme embodiments takes control of the vehicle throttle and actually slows the car or truck down to the posted legal speed limit.

The technology of "smart road signs", capable of transmitting instructions to passing vehicles via some form of short-range communication architecture was part of another of those pesky PROMETHEUS projects. The first practical implementation of ISA was in Sweden where it was tested by 10,000 voluntary test drivers, both private motorists and commercial drivers, in 1999-2002. The trial was a joint venture between local authorities and the Swedish Road Administration.

Results show that most of the trial motorists appreciated the support offered by the ISA system.

- Motorists felt ISA helped them to be better drivers.

- Road safety improved significantly.

- As a result of less repeated braking and stopping, travel time remained unchanged (a slight improvement was noted) despite lower maximum speeds.

- A more even driving speed helped reduce fuel consumption and emissions.

- There was a very high level of acceptance for ISA in urban areas.

- Drivers believed that ISA was the best available measure, along with police surveillance, to solve road safety problems on streets in built-up areas compared with physical measures such as road humps[11].

Other European countries are experimenting with ISA including The Netherlands and the UK. While national auto clubs generally support continued experimentation, they and the automakers are concerned that the technology is still nascent and additional technological and legal impediments need to be overcome before ISA is ready for application as a mandatory traffic control system. These groups do support a rollout in which drivers could opt to volunteer to use.

INTERSECTION COLLISION WARNING

Our final ITS technology application for examination is intersection collision warning. While traffic incidents at intersections are of concern to every traffic agency, they are of particular concern in Japan where many urban intersections can involve small, narrow streets crossing much wider and busier main roadways. In addition, there are many "blind" intersections that are particularly dangerous for crossing vehicular traffic as well as pedestrians.

In the typical embodiment, information on intersections that are without traffic signals is gathered from car navigation system data. Images from the vehicle's cameras are analyzed to detect stop lines and stop signs. Based on the vehicle's speed and distance to the

stop line the system determines whether the vehicle is traveling at a speed that will enable it to stop by the time it reaches the stop line.

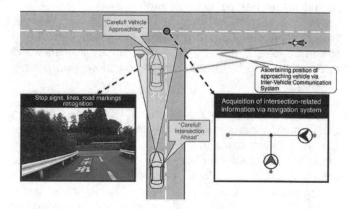

Figure 7. "Smart" intersection with "smart" vehicles

If the vehicle is going too fast, the system delivers an audio warning while signaling the driver to slow down with the application of gentle, intermittent braking.

Once the automobile has come to a stop, the Inter-Vehicle Communication System detects the position of any approaching vehicles and assists the driver in determining whether it is safe to proceed through the intersection[12].

CONCLUSION – WHO IS REALLY "ON FIRST?"

The foregoing discussion paints a very broad picture of which ITS developments that impact cars and trucks are being pursued by whom and where.

There are several general conclusions:

- European automakers, notably those in Germany, are developing and fielding a plethora of automotive safety systems that make use of modern computer and communications technology.

- Japanese companies and Japanese public agencies are working very closely on cooperative systems, those that involve both vehicles and roadways.

- America has the lead in telematics due to General Motors' dedication to its OnStar business unit. The contribution of independent telematics provider ATX is also important.

- Satellite radio is entering the telematics fray by offering real-time traffic situation reports for twenty US metropolitan areas.

This is not to say that each country or region is operating with tunnel vision. On the contrary, there is research (mostly) and development work (some) on every technology application in every region.

There is also a popular notion that some premium vehicles built by European and Japanese companies are "de-contented" when they are exported to North America because "Americans don't understand technology – they are afraid of it." This is not entirely true – for example, Nissan maintains its ITS applications when it sends Infiniti-branded products to America.

When it comes to vehicular applications, there are several conclusions evident:

- Premium vehicles "have all the luck" when it comes to applying ITS technologies. The principal exception to this rule is Ultrasonic Park Assist which is widely applied around the world – because it's cheap.

- One can find similar pricing and application practices with GPS navigation, but in this case its feature content that drives the price. Higher priced, premium vehicles have more features on their navigation systems – color displays, hard-disc drive data storage and route planning algorithms. A small car's navigation system, on the other hand might have a monochrome display, flash card or CD memory and position location.

- European and Japanese premium marques have most of the applications because their domestic buyers are technologically aware and want (and in some cases demand) new safety and convenience enhancing technologies.

- Nissan brings to market a variety of technological wonders – cars and light trucks absolutely laden with multiple, high-capability, high-value ITS applications.

On the supplier side, virtually every Tier One supplier who trades in automotive electronics is developing some ITS systems. In Europe its Continental AG, Siemens VDO Automotive and Valeo. In Japan, the lead is with Denso, Aisin and Hitachi, but many others are also active. The US effort is led by Delphi with the foregoing European and Japanese firms also very active.

If you want to see the future face of new automotive technology in general and ITS in particular, watch Mercedes, Toyota and Honda. These companies have little fear of technological risk – in fact, they seem to revel in it – because at the end of the day it brings them image and, in the long term, image brings customers, revenue and profits.

REFERENCES

[1] ITS is defined in this paper as the application of modern computer and communications technology to road transportation.
[2] In this paper the terms "America" and "American" refer to all of North America – the USA, Canada and Mexico.
[3] Adaptive Cruise Control has also been called "active cruise control" or "intelligent cruise control" in other literature and automotive publicity and sales materials.
[4] *Business Week Online*; September 14, 2005
[5] NHTSA Vehicle Safety Rulemaking and Supporting Research Priorities: Calendar Years 2005-2009 – Update January 2005
[6] The Johns Hopkins University, Applied Physics Laboratory
[7] eCall Toolbox; SCOPE eSafety Observatory; ERTICO
[8] Morita, Masatoshi; "Current Status and Future Plans for HELP"; International Telecommunications Union; March 2005
[9] The Car2Car Consortium
[10] Furukawa, Yoshimi; "Overview of R&D on Active Safety in Japan"; undated (2004 or 2005)
[11] "ISA for increased social responsibility and competitiveness"; Swedish Road Association; undated
[12] "Honda Completes Development of ASV-3 Advanced Safety Vehicles"; Honda press release; September 5, 2005

CONTACT

Gerald Conover is managing director of PRC Associates, an advisor on matters of product, technology and corporate management to firms and agencies in the intelligent transportation systems and services community.

Before founding PRC Associates, Mr. Conover enjoyed 32 years at Ford with assignments in Car Product Development, Manufacturing Operations, and the Technical Staffs. From 1991, his focus was on development and implementation of intelligent transport systems product and supply strategies for Ford and its worldwide affiliates.

Mr. Conover is active in numerous professional organizations including the Society of Automotive Engineers, the Intelligent Transportation Society of America (having served on its Board and Executive Committee), and the Board and Organizing Committee for the ITS World Congress, as well as local and national charitable and community organizations.

Mr. Conover is the 2005 recipient of the SAE Delco Electronics Intelligent Transportation Systems Award.

2005-01-1472

A Secure Wireless Protocol for Intersection Collision Warning Systems

Srinivas R. Mosra and Syed Masud Mahmud
Department of Electrical and Computer Engineering, Wayne State University

ABSTRACT

Radar and infrared technologies can detect impending rear-end and lane-change collisions. However, these technologies cannot detect impending intersection collisions because they require line-of-sight communications. Wireless communication technology will be a viable technology for detecting intersection collisions. In this paper, we assumed that every vehicle is equipped with a wireless communication unit and every intersection has a wireless unit called the Intersection Traffic Controller (ITC). All vehicles near an intersection communicate with the corresponding ITC to send their dynamic information such as speed, acceleration, lane number, road number, and distance from the intersection. Though the wireless technology will be a viable technology for developing intersection collision warning systems, it is subject to various types of security attacks unless the system is properly designed. This paper presents a secure wireless protocol for intersection collision warning systems with a detailed bit level description of the protocol. The security is maintained using a chain of digital certificates issued by various federal and state organizations. The vehicles validate the certificate of the ITC using the public key issued by a federal organization, such as the US Department of Transportation. This paper also presents the performance analysis of the intersection collision warning system.

Keywords: Security, Digital certificates, Intersection collisions, collision avoidance.

1. INTRODUCTION

Intersections induce more attention for safety analysis than other roadway elements due to the fact that many intersections are found to be relatively crash-prone spots from a safety point of view. In 2002, approximately 3.2 million intersections related crashes occurred, representing 43 percent of all reported crashes. There were 9,612 fatalities (22 percent of total fatalities) due to crashes that occurred at or within an intersection environment. The cost to society for intersection crashes is approximately $96 billion a year [1]. In order to reduce the number of traffic accidents and to improve the safety and efficiency of traffic, the research on Intelligent Transportation System (ITS) has been developed for many years in many countries [2]. New driving assistance systems such as night vision and collision warning systems (CWS) have been designed, tested, and deployed using the advanced technologies like sensing, computing, and communication technologies [3, 4, 5, 6, 7]. Despite the fact that intersection collision accounts for almost 43% of all crashes, intersection collision avoidance systems received less attention than the forward collision avoidance systems [5, 8]. The reason, besides the fact that the intersection collision problem is more complicated than rear-end crash, is the limitation of the radar technology, the most widely used object sensing method in vehicle collision avoidance systems. Most radar systems require line-of-sight for object detection. This renders ineffective collision warning/avoidance system that requires line-of-sight for threat detection.

Differential global positioning systems (DGPS), electronic compasses, roadside sensors, etc are the current technologies that are under investigation to avoid intersection collisions. There are several disadvantages of these technologies. For example, the GPS signals determine the location of the vehicle with some errors and in some areas, especially in downtown areas with very tall buildings; the signals may not be detected. If there are multiple lanes on the road, then the roadside sensors aren't effective, as they cannot detect all the vehicles. In our earlier work we have developed an intelligent architecture for issuing intersection collision warnings using wireless communications, as they do not require line-of-sight. Using wireless communication technologies, all the vehicles approaching the intersection communicate with the Intersection Traffic Controller (ITC), which is installed at the intersection. The function of the ITC is to broadcast the status of the intersection. Vehicles cooperatively share the critical information such as location, velocity, acceleration, etc with the ITC for collision anticipation and threat detections [9].

Though the wireless technology will be a viable technology for developing intersection collision warning systems, the risks are inherent in wireless technology unless the system is properly designed. The loss of confidentiality and integrity and the threat of denial of service (DoS) attacks are risks typically associated with wireless communications. Many current communications protocols and commercial products provide inadequate protection and thus present unacceptable risks for any operations [10]. We must actively address such risks to protect their ability to support essential operations, before deployment of wireless technologies.

In this paper, we propose a secure wireless protocol for intersection collision warning systems by using the concept of digital signatures, certificates and certificate chains. Digital certificates are issued by various federal and state organizations. The rest of the paper is organized as follows. Background material along with the previous work done on the intersection collision avoidance systems is presented in Section 2. A secure protocol using digital certificates is presented in Section 3. The procedure for the validation of the ITC is presented in Section 4. Performance analysis of the protocol is presented in Section 5. Suggestions and comments are presented in Section 6 and conclusion is presented in Section 7.

2. BACK GROUND MATERIAL RELATED TO INTERSECTION COLLISION WARNING SYSTEMS AND WIRELESS COMMUNICATIONS

The major reasons for intersection crashes are driver inattention, faulty perception, and impaired or obstructed vision. Thus, a timely warning of upcoming intersection should be helpful. General human factor issues considered in an Intersection Collision Avoidance System (ICAS) include what information to present to the driver (warning content), when to present it (timing of warning), and how to present it (type of warning modality) [11]. Lloyd et al. [11] tried to use brake pulsing as haptic warning for an intersection collision avoidance countermeasure. This Intersection Collision Avoidance System (ICAS) involves technologies such as multiple beam radar system, geographical information system (GIS) and a Global Positioning System (GPS). The GIS-GPS uses data derived from an in-board map database to detect an upcoming intersection, identify the traffic control requirement at the intersection, and determine the vehicle's distance to the intersection. This information combine with data regarding vehicle's dynamic state and driver's vehicle operation is used to calculate whether the vehicle can stop before entering the intersection. A warning is provided if the system detects that the vehicle will not stop before the intersection zone [11]. This haptic warning system (HWS) focuses on unsignalized (stop-sign-controlled) intersections. An ICAS for signalized intersection needs integration with the traffic control infrastructure therefore is more complex than an ICAS for unsignalized intersections.

Ferlis [12] outlined an infrastructure collision avoidance concept for straight crossing path crashes at signalized intersection. The infrastructure-based intersection collision-avoidance systems use roadside sensors, processors, warning devices, and traffic signals to provide crossing assistance to motorists and pedestrians. The sensors identify potential violators by determining the speed and/or deceleration rate of each vehicle approaching towards the intersection. The processing system would identify vehicles at an upstream control point that are unlikely to stop at the intersection. Once a violator is identified, warnings would be conveyed to the violator and other drivers on adjacent approaches to the intersection.

White and Eccles [13] designed an infrastructure-based intersection collision-avoidance system (ICAS) to prevent left-turn crashes with opposite-direction traffic. The major cause of left turn across the path of opposite-direction traffic crashes is the inability of the left-turning motorist to adequately perceive the required gap for a left turn. The proposed ICAS compares turning times and the approach times of the opposite-direction vehicle. Using this comparison, ICAS provides guidance to the left-turning motorist in the form of a dynamic sign to assist in the left-turn decision. This system relies solely on presence detectors, which have good reliability in varying weather or traffic conditions. The major limitation of this system is that it does not actually detect speeds and relies on assumed speeds. The performance of the ICAS will be susceptible for vehicles that deviate greatly from those assumed values.

In our earlier work we developed Intelligent Architecture for Issuing Intersection Collision Warnings (IAIICW) [9]. This architecture requires that every vehicle must be equipped with a wireless device to communicate with the Intersection Traffic Controller (ITC). We assumed that all intersections are installed with an ITC, and all roads of the intersection are equipped with a mechanism to determine the road number, lane number and distance from the intersection. We also assumed that all vehicles are capable of detecting the road number, lane number and distance from the intersection. The ITC broadcasts the condition of the intersection using messages and also receives the information from the vehicles approaching the intersection. As a vehicle approaches an intersection, using the infrastructure system present at the intersection, the vehicle's onboard computer knows on which road number, lane number it is present and the distance from the intersection. The vehicle receives its speed and acceleration from the in-vehicle network. This information is sent to the ITC through the wireless communication using the specific message format. The vehicle's onboard computer receives the broadcasted messages from the ITC and checks the information in the message that corresponds to the road number and lane number of the vehicle. From the message, the vehicle's on-board computer determines whether the signal at the intersection is green, yellow or red and the time left for the signal to change from its current state to the next state. The on-board computer calculates the time left for the vehicle to reach the intersection and checks whether the vehicle can pass the intersection without violating the traffic signal. From the message, the vehicle's on-board computer knows if any other vehicle has violated the traffic signal. If any vehicle violates the traffic signal, the on-board computer issues pre-warning to the driver indicating that a vehicle has violated the traffic signal

Intersection collision avoidance systems that use wireless communication technology must be protected from various types of security attacks. Digital identities and security technologies enable the security services like authentication, authorization, digital signature and encryption to exchange messages over the wireless network. A brief background of security techniques is presented in the following subsections of this paper.

2.1. DIGITAL CERTIFICATES

There is a broad range of applications for digital certificates: electronic banking, electronic payment systems, e-mail communication, identification in communication with public authorities (e.g. transportation, tax declaration, court documents, electronic passports, public health service, etc.), electronic contracts, selective web access, selective database access, etc. Public-key cryptography is a key-factor for the solution of the transaction security problems arising with the commercial use of the Internet: *authenticity*, *integrity*, *confidentiality* and *non-repudiation* [14, 15].

Public key cryptography is based on the use of key pairs. When using a key pair, one of the keys, referred to as the private key is kept secret and under the control of owner. The other key, referred to as the public key, can be disseminated freely for use by any person who wishes to participate in security services with the person holding the private key. The private key and public key are mathematically related but it remains computationally infeasible to derive the private key from the knowledge of public key. In theory, any individual can send the holder of a private key a message encrypted using the corresponding public key and only the holder of the private key can decrypt the secure message. Similarly, the holder of the private key can establish the integrity and origin of the data he sends to another party by digitally signing the data using his private key. Any one who receives the data can use the associated public key to validate that it came from the holder of the private key and verify the integrity of the data has been maintained.

2.2. KEY AND CERTIFICATE MANAGEMENT

The distribution and management of the public key is the crucial point in the procedures described above. It must be guaranteed that the key really belongs to the respective person (or e-mail address or authorization role). A means to guarantee this is the use of digital certificates. They are digital documents containing the public key, the name of the possessor, the digital signature of the *certification authority* (CA) that issued the certificate and the certificate validity period. In this way the problem of key management is reduced to the public key of the CA. Once in possession of the trustworthy public key, the end user is able to verify all certificates issued by the certification authority. The function of a CA is therefore the verification of the identity of the certificate holder. The certification of identity is only the simplest form of a certificate. Similar extensions are provided with role-based systems as the Simple Public Key Infrastructure (SPKI) [16]. With the use of application-specific extensions, the function of the certification authority is extended to the verification of the respective attributes of the certificate holders.

2.3. TRUSTED THIRD PARTIES AND CROSS CERTIFICATION

The Nation-wide use of certificates causes the emerging of a large number of certificate issuers. One cause for this is that a certificate issuer needs a certain regional presence in order to verify the identity of a person. From this point of view, an organization issuing certificates consists of a large number of locally operating entities, independent from each other. For the end users, the management of different trustworthy public keys is not applicable, because each of these would have to be transmitted in a secure way. This problem can be solved by the use of *cross certificates*. These are certificates issued by a CA certifying another CA. In this way, an end user is able to verify a certificate issued by a CA whose public key was not directly transmitted to the end user. For the verification, there must only be a link via cross certificates to the CA whose trustworthy public key is with the end user. This link is also called *certification path* or *chain of trust*. The CA whose trustworthy public key is provided is called *trusted third party* [17, 18]. Using these mechanisms, a system can be built that consists of several certification authorities issuing certificates for individuals but also building links between each other using cross certificates. In an ideal situation, each end user is able to verify the certificates of any other person using only one trusted third party in this system. The combination of certification authorities linked to each other via cross certificates and the end users is called *public-key infrastructure* [19].

3. SECURED ITC AND WIRELESS PROTOCOL

In this section of the paper we briefly explain the protocol for the ITC that has to be broadcasted and then explain the protocol for the digital certificates.

3.1. ITC PROTOCOL

The role of ITC in the Intersection Collision Avoidance System (ICAS) is to broadcast the condition of the intersection for the traffic crossing the intersection [9]. The broadcasted messages have a specific format so that all the vehicles can understand easily. The length of the message varies depending upon the number of roads, lanes and signal lights present at an intersection. The messages are broadcasted in the form of packets. Figure 1a shows an overview of the message format that is broadcasted by an ITC.

All messages begin with a Start of Message (SOM) field. This serves to identify the beginning of a message. The second field in the message is always the ITC field, which contains 16-bit ITC number field. This field is used to identify the intersection. The third field in the message is the vehicle violation field. This field contains one bit, which is used to determine whether any vehicle has violated the traffic signal.

Field number four is called the Intersection status field. This is a variable length field. The first four bits are used to specify the number of roads present at the intersection. Following these four bits are road information fields. The length of road information fields depends upon the number of roads present at the intersection.

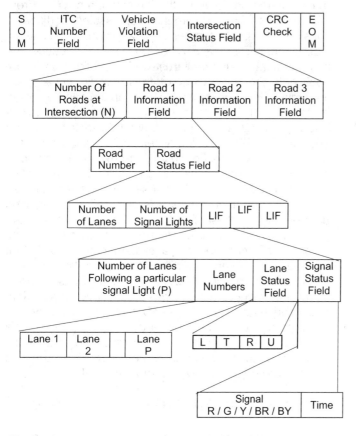

Figure 1a: An overview of the ITC message format.

The road information field is again divided into a 4-bit road number field and a road status field. The length of the road status field depends upon the number of signal lights and lanes present on a particular road. The first part of the road status field is the number of lanes field, which is four-bit wide. This field specifies the number of lanes present on a particular road. The second field is the number of signal lights field, which contains two bits. This field specifies the number of signal lights present for a particular road. Following the number of signal lights field is the lane information field. The lane information field is of variable length. This field is repeated depending upon the number of signal lights present on a road.

The lane information field is divided in to four sub fields. The first sub field contains four bits. This field specifies the number of lanes following a particular signal light. The second sub field specifies the lane numbers of the road following the respective signal light. The third sub field is lane status field, which contains four bits. These four bits are L bit, T bit, R bit and U bit, which specify whether the vehicles on a particular lane can go left, through, right or make a U turn, respectively or not. The last sub field is the signal status field, which has two fields. The First field is the signal field which is of three bits and specifies the signal status i.e. whether the signal is red, green, yellow, blinking yellow or blinking red and the second field is the time field which contains 32 bits, specifying the time left for the signal change.

Following the Intersection status field is the CRC field. It consists of 16-bit Cyclic Redundancy Check code, which allows the receivers to verify the correctness of the received message. The final field of the message is the End Of Message (EOM) field, which servers as the end of the message.

S O M	Vehicle Road number	Velocity	Acceleration	Distance From the intersection	C R C	E O M

Figure 1b: Format of the vehicle to ITC message

Figure 3b shows the format of the message that is to be received by the ITC from the vehicles approaching towards the intersection. The first field is the Start Of Message (SOM) field, which serves to identify the beginning of a message. The second field is of four bits specifying the road number of the vehicle. The third field contains 12 bits, and it specifies the velocity of the vehicle. The fourth field contains 16 bits, and it specifies the acceleration of the vehicle. Following the acceleration field is the Distance field, which has 16 bits to specify the distance from the intersection. The next field is the CRC field. It consists of a 16-bit Cyclic Redundancy Check code, which allows the ITC to verify the correctness of the received message. The last field is the End of Message (EOM) field. The total length of the message that has to be broadcasted by the vehicle is 96 bits.

3.2. DIGITAL CERTIFICATE FORMAT:

Our secured system is based on digital signatures. A digital signature is a kind of cryptographic check sum where the vehicle's On Board Unit (OBU) can verify the checksum and thus convince itself that the messages it received is the one that was sent by an authorized ITC.

Digital certificates not only carry the public key but also some information about what it's authorized to do. In our system, OBUs (the units in cars) contain only public key and a range of valid dates. The public key is issued to the OBU by USDOT at the time of manufacturing the vehicle. ITC's, on the other hand contain their public key, unique serial number, valid dates and digital signature. Fig 1c shows the format of the digital certificate.

S O M	Hierarchy level	Issuer ID	Validity	Subject ID	Public Key	Digital Signature	C R C	E O M

Figure 1c: Format of the Digital certificate

The digital certificate contains nine fields. The first field is Start Of Message (SOM) field, which contains 16bits and is used to identify the beginning of the certificate. The second field is the hierarchy level field, which is one byte long. The lower 3 bits specify the hierarchy level and the remaining bits are reserved for the future use. The third field is the Issuer ID field. This field serves to identify which Certificate Authority (CA) is signing the certificate. The length of this field is three bytes. Field number four is validity field. This field specifies how long the certificate is valid and is 4 bytes long. The fifth field is Subject ID field, which is 3 bytes long. This field specifies the ID of the certificate recipient. Field number six is called Public Key field. This field is 16 bytes long and is used to determine the public key of the certificate recipient. Seventh field is the Digital signature field with 16 bytes. This field is used to verify the signature signed by the certificate issuing authority. Next field is the CRC field, which is two bytes long and is used to verify the CRC check. The last field is the End Of Message (EOM) field and is served to specify the end of message with 2 bytes length. The total length of the digital certificate is 49 bytes.

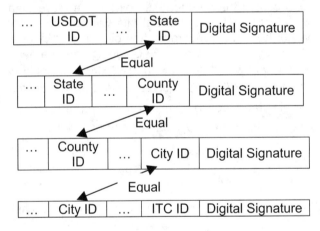

Figure 2a: Certificate Chain.

4. ITC VALIDATION

In our architecture we used unidirectional chain of certificates for validating the ITC messages in which USDOT acts as a certifying authority and is trusted by everyone. During the manufacturing of the vehicle the USDOT's public key is embedded in the vehicles OBU. USDOT issues and signs a certificate for each state.

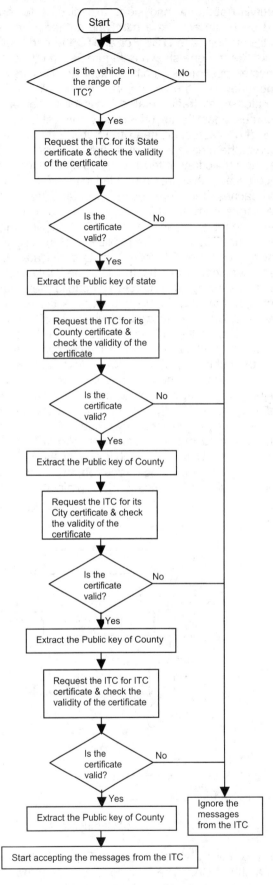

Figure 2b: Process of obtaining the public key and verification of ITC

Each state acts as a certifying authority and signs a certificate for each county. Each county acts as a certifying authority and signs a certificate for each city and each city acts as a certifying authority and signs a certificate for each ITC. For an ITC the certificate chain looks like the one shown in Figure 2a.An ITC is issued a (private, public) key pair by its city. This key pair is embedded in the hardware of the ITC. The ITC broadcasts all traffic related messages after encrypting the messages by its private key. All vehicles that are near the intersection need the public key of the ITC to decrypt the messages sent by the ITC. The vehicles can get the public key of the ITC by going through the chain of certificates that are embedded in the ITC. As a vehicle approaches the intersection it starts receiving the messages from the ITC. The vehicle's OBU ignores the messages it received until the OBU receives the public key of the ITC. To get the public key of the ITC the OBU first requests the ITC for the state certificate. From the state certificate sent by the ITC the OBU verifies the signature using the USDOT's public key, which is embedded in its software and extracts the public key of the state. After extracting the public key of the state, the OBU requests the ITC for the county certificate. Using the state's public key it verifies the county certificate and gets the public key of the county. Next the OBU asks for the city certificate and verifies the certificate using the count's public key and extracts the city's public key from the certificate. Finally the OBU requests for the ITC certificate and using the city's public key it verifies the ITC certificate and extracts the public key of the ITC. Once the OBU gets the public key of the ITC, it starts accepting the messages received from the ITC. The whole process of verification is explained in brief using Figure 2b.

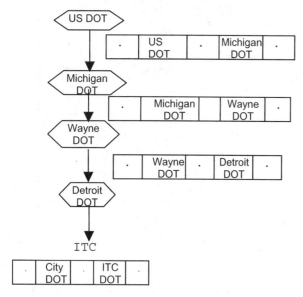

Figure 2c: Hierarchy of Certificate Authorities

Let's examine a simple example. Figure 2c illustrates the hierarchy of CAs. The hexagons represent the CAs, the arrows represent certificate issuance, and the sectional rectangles represent certificates. Consider an ITC, which is at the intersection of Warren and Woodward roads, is broadcasting the status of the intersection and a vehicle V1 is approaching the intersection and attempting to

verify the ITC certificate. We construct a certification path between the ITC certificate and the US-DOT certificate. The certification starts from the US-DOT certificate and works its way to the ITC. The path is US-DOT→Michigan-DOT→Wayne-DOT→Detroit-DOT→ITC. The vehicle's OBU verifies the Michigan-DOT certificate and confirms that US-DOT is the issuer of Michigan-DOT certificate and hence trusts the Michigan-DOT. Similarly the vehicle's OBU verifies the certificate of Wayne-DOT and confirms that Michigan-DOT is the issuer of Wayne-DOT certificate and so on until the ITC certificate is verified.

5. PERFORMANCE ANALYSIS

An ITC will broadcast (in real time) the information of the approaching vehicles and change of traffic signals to all the vehicles approaching towards the intersection [9]. The ITC broadcasts the message once in every 10 milliseconds so that the vehicles receive the updated signal information and keep track of the situation at the intersection. The total length of the message that is broadcasted by the ITC with N number of roads, L_n lanes per road and S signal lights per road is $N(42S + 4L_n + 10) + 69$ bits [9]. For wireless communications, there is a huge overhead for sending raw data. The actual amount of overhead depends on the specific coding technique used for the wireless communication. For example, for the Rate 1/3 FEC (Forward Error Correction) coding, three copies of every raw data bit are sent through the air. Thus, for the Rate 1/3 FEC coding the overhead is going to be more than 200%, because some additional bits will be necessary for packet headers, synchronization bits, end of frames, etc. Similarly, for the Rate 2/3 FEC coding, the overhead is going to be more than 50%. In this paper, we assume an overhead of 100% for our analysis. Let B_b be the bandwidth required by an ITC to broadcast its messages. The value of B_b can be expressed as:

$$B_b = \frac{N(42S + 4L_n + 10) + 69}{5000} \text{ Mbps} \qquad (1)$$

Table I: Bandwidth, B_b, required for an ITC to broadcast its messages

Number of Roads (N)	Number of Lanes per Road (L_n)	Number of Signal States per Road (S)	Length of message in Bytes	Bandwidth required in Mbps (B_b)
3	2	1	31.125	0.049
3	3	2	48.375	0.077
4	3	2	61.625	0.098
4	4	3	84.625	0.135
8	8	4	428.625	0.349

Table I shows the bandwidth required by an ITC to broadcast its messages. From Table I it is seen that the maximum bandwidth required for broadcasting the messages is 0.349 Mbps. As explained in Section 3.1, the length of the message that has to be broadcasted by the vehicle is 96 bits. Let N_v be the total number of vehicles at an intersection. Consider that the vehicles are communicating with the ITC every t ms. Let B_v be the bandwidth required by an ITC to accept messages from all the vehicles. If we consider an overhead of 100% in converting the raw bits into wireless packets, then the bandwidth B_v can be expressed as:

$$B_v = \frac{2*96*N_v}{1000t} = \frac{N_v}{5.2t} \text{ Mbps} \tag{2}$$

Table II shows the bandwidth, B_v, required by an ITC to accept messages from N_v vehicles. Table III shows the distance traveled by a vehicle in t ms.

Table II: Bandwidth, B_v, required by an ITC to accept messages from N_v vehicles.

Number of vehicles at intersection (N_v)	Bandwidth, B_v, for different values of t (in Mbps)		
	T=10 ms	t=20 ms	t=30 ms
50	0.962	0.481	0.321
100	1.923	0.962	0.641
150	2.885	1.442	0.962
200	3.846	1.923	1.282
250	4.808	2.404	1.603
300	5.769	2.885	1.923
400	7.692	3.846	2.564

Table III: Distance traveled by the vehicles in t ms.

Velocity of the vehicle (V) mph	Distance traveled by the vehicle in t ms (feet)		
	T=10 ms	T=20 ms	T=30 ms
30	0.44	0.88	1.32
40	0.58	1.17	1.76
50	0.73	1.46	2.2
60	0.88	1.76	2.64
70	1.02	2.05	3.08

Generally it is recommended that a driver should maintain at least a 2-second distance between his vehicle and the vehicle at the front. Let us consider the situation where a driver is maintaining only a 1-second distance instead of a 2-second distance. In this situation the distance maintained between the vehicles is $1.47V$ ft, where V is the velocity of the vehicle in mph. Let L_v be the average length of a vehicle in feet and R be the

range of the ITC in feet. So there are maximum $R/(L_v + 1.47V)$ vehicles within the range of the ITC in each lane. The total number of vehicles at a given intersection with N number of roads, L_n lanes per road is

$$N_v = \frac{NL_nR}{L_v + 1.47V} \tag{3}$$

Let B_w be the total bandwidth available from the wireless communication system, and B_a be the bandwidth available for authentication. The value of B_a can be expressed as:

$$B_a = B_w - (B_b + B_v) \text{ Mbps} \tag{4}$$

Let L_d be the length of the digital certificate in bytes. Assuming 100% overhead in converting raw bits into wireless packets, we can say that the time required by each vehicle for authenticating the ITC is

$$T = \frac{16*L_d}{B_a} \text{ µSec.} \tag{5}$$

Total time required for N_v vehicles to authenticate the ITC is

$$T_{nv} = N_vT \text{ µSec.} \tag{6}$$

From Equation 6, the maximum distance a vehicle can travel during the authentication process is given by

$$D = \frac{1.47T_{nv}V}{10^6} \text{ ft} \tag{7}$$

As explained in Section 3.2, the length of the digital certificate is 49 bytes. Table IV shows the distance traveled by the vehicle during the authentication process for different ranges of ITC. Here the intersection is assumed to have four roads with four lanes per road. From Table I, the bandwidth required for four roads with four lanes per road is 0.135 Mbps. The average length of the vehicle is assumed to be 17 feet and the total bandwidth available from the wireless communication system is assumed to be 9 Mbps, which is same as that of the DSRC systems.

Figure 3a shows the time required for N_v vehicles to authenticate the ITC as a function of the vehicle speed for different communication ranges of the ITC. Figure 3a shows that for higher communication range of the ITC, more time is needed by the vehicles to authenticate the ITC. The reason is that as the range of an ITC increases, more vehicles are available with its range. Thus, more time is needed to authenticate the ITC by all the vehicles. Figure 3a also shows that when all vehicles are moving slow, more time is needed for ITC authentications. This is due to the fact that when vehicles move slower, the gaps between consecutive vehicles decrease. As a result, there are more vehicles within the range of an ITC. Therefore, more time is needed to authenticate the ITC by all the vehicles within its range.

Figure 3b shows the distance that a vehicle will go through while it is authenticating an ITC. This figure shows that for an intersection with four roads and four

lanes per road, the vehicles will move only a fraction of a foot during the ITC authentication process. Figure 3c shows similar results for an intersection with 8 roads and 8 lanes per road with and an ITC range of up to 500 feet. Figure 3c shows that even for a very large intersection with 8 roads and 8 lanes per road in each direction, the vehicles will move only a few feet during the ITC authentication process. Thus, we can say that the time required for ITC authentication by all the vehicles is not too much, and it is acceptable for all practical purposes.

Table IV: Maximum distance traveled by a vehicle during authentication process (N=4, Ln=4, S=3)

Range of ITC (R)	Velocity of the vehicle (V) mph	Number of vehicles at intersection (N_v)	Bandwidth required by N_v vehicles (B_v) mbps	Bandwidth available for authentication (B_a) mbps	Time required for authentication of one vehicle (T) msec	Time required for authentication of Nv vehicles (T_{nv}) msec	Maximum distance traveled by a vehicle during authentication (D in ft)
400	10	202	1.29	7.56	0.10	20.92	0.31
400	20	138	0.88	7.97	0.10	13.56	0.40
400	30	105	0.67	8.19	0.10	10.05	0.44
400	40	84	0.53	8.32	0.09	7.91	0.47
400	50	71	0.45	8.40	0.09	6.62	0.49
400	60	61	0.39	8.47	0.09	5.64	0.50
300	10	151	0.96	7.89	0.10	14.99	0.22
300	30	79	0.50	8.35	0.09	7.41	0.33
300	50	53	0.33	8.52	0.09	4.87	0.36
300	60	46	0.29	8.56	0.09	4.21	0.37
200	30	52	0.33	8.53	0.09	4.78	0.21
200	40	42	0.26	8.59	0.09	3.83	0.23

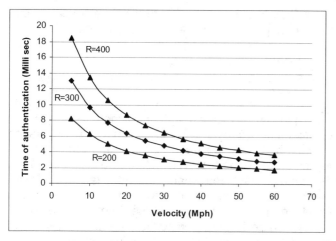

Figure 3a: Time required for authenticating an ITC by all the vehicles within its range (for N=4 and Ln=4).

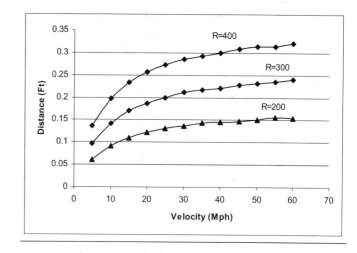

Figure 3b: The distance a vehicle can travel during an ITC authentication process for different ranges of the ITC (for N=4 and Ln=4).

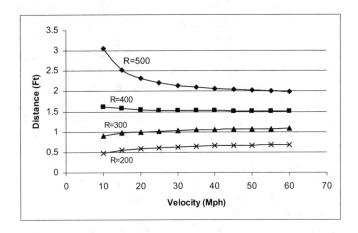

Figure 3c: The distance a vehicle can travel during an ITC authentication process for different ranges of the ITC (for N=8 and L_n=8).

Minimum Required Range of an ITC

A person needs about a second time to react to the warning and apply brakes to stop the vehicle. Let the speed of a vehicle at an intersection be V_o feet per second (fps). Let us assume that an average driver can decelerate a vehicle at the rate of about a feet/sec^2. From the basic laws of kinematics we have $S = V_o^2/2a$ feet, where S is the distance traveled by the vehicle from the time the brake is pressed until the vehicle comes to the stop position. Tables V and VI show the total distance traveled, with a reaction time of one second, by a vehicle before it stops after the warning has been issued.

Table V: Distance traveled by a vehicle before it stops with a deceleration of 20 ft/sec^2

Velocity (V) mph	Distance (S) feet	1 Second Reaction Distance (RD) feet	Total Distance (TD) feet
30	48	44	92
40	86	59	145
50	134	73	208
60	194	88	282

Table VI: Distance traveled by a vehicle before it stops with a deceleration of 15 ft/sec^2

Velocity (V) mph	Distance (S) feet	1 Second Reaction Distance (RD) feet	Total distance (TD) feet
30	65	44	109
40	115	59	173
50	179	73	253
60	258	88	346

Normally it takes some time for two wireless devices to establish a link after they come within their range. Let us assume that a vehicle needs one-second time to establish a wireless link with an ITC. On highways, the intersections are far away from each other and the speed limit is 55 mph. Let us consider a situation where a driver is going at 60 mph. Since an average vehicle can decelerate at the rate of 15 to 20 ft/sec^2, from Table VI we see that for a vehicle speed of 60 mph, the maximum distance traveled by a vehicle from the instance the warning is given until the vehicle is stopped is 346 feet. The distance traveled in one second during the link establishment time is 88 ft, and the distance traveled during the ITC authentication process is under couple of feet at a vehicle speed of 60 mph (see Figures 3b and 3c). Thus, the total distance traveled by a vehicle, from the time the vehicle starts communicating with an ITC until it stops when a warning is issued, is 346+88+2 = 436 feet. From the above analysis it is clear that on highways the range of an ITC should be greater than 436 feet.

In downtown areas, the intersections are closer to each other and the speed limit is 35 mph. Let us consider a situation where a driver is going at 40 mph. From Table VI, we see that for a vehicle speed of 40 mph, the distance traveled by a vehicle from the instance the warning is issued is 173 feet. The distance traveled during the link establishment time is 59 feet, and the distance traveled during the ITC authentication process is under 2 feet. Thus, the total distance traveled by a vehicle, from the time the vehicle starts communicating with an ITC until it stops when a warning is issued, is 173+59+2 = 234 feet. Hence, for downtown areas the range of an ITC must be greater than 234 feet.

Table VII: List of parameters used in the performance analysis.

Parameter	Description
N	Number of roads at intersection
S	Number of signal states per road
L_n	Number of lanes per road
V	Velocity of the vehicle in Mph
R	Range of ITC in feet's
N_v	Number of vehicles at intersection
L_v	Average length of the vehicle in feet's
B_w	Total bandwidth available
B_a	Bandwidth available for authentication
B_b	Bandwidth required for broadcasting messages
B_v	Bandwidth required by N_v vehicles
L_d	Length of the digital certificate in bytes
T	Time required for authentication of one vehicle
T_{nv}	Time required for authentication of N_v vehicles
D	Distance traveled by the vehicle during the authentication
S	Distance traveled by the vehicle during the application of brakes

6. SUGGESTIONS AND COMMENTS

When the intersections are very close to each other, say within 300 feet from each other, a vehicle that moves from the region of one ITC to another may not get enough time to establish a link with the new ITC, authenticate the new ITC and stop when a warning is issued unless the vehicle moves very slow, say under 30 mph. In this case, a vehicle may need to establish a link with the next ITC before it leaves the intersection of the previous ITC. In other words, the vehicle will communicate with two ITCs: the current one and the next one. However, before the vehicle leaves the intersection of one ITC, it will have limited communications with the ITC of the next intersection. We would like to investigate this case in our future work.

7. CONCLUSION

In this paper, we have presented a secure wireless protocol for intersection collision warning systems. We provided the detailed description of the digital certificate format along with the authentication process. We also investigated the feasibility of implementing the protocol using the available technology. Suggestions are made to improve the functionality of the protocol for downtown areas where the intersections are very close to each other.

REFERENCES

1. A Compilation of Motor Vehicle Crash Data from the Fatality Analysis Reporting System and the General Estimates System, Traffic Safety Facts 2002 Overview, DOT HS 809 612, National Highway Traffic Safety Administration, U.S. Department of Transportation.
2. R. French, Y. Noguchi, and K. Sakamoto, "International competitiveness in IVHS: Europe, Japan, and the United States," in *Proc. 1994 Vehicle Navigation and Information Systems Conf.*, Yokohama, Japan, July 1994, pp. 525–530
3. P. Zador, S. Krawchuk, and R. Vocas. Final Report–Automotive collision Avoidance (ACAS) Program. Technical Report DOT HS 809 080, NHTSA, U.S. DOT, August 2000. 2
4. R. Kiefer, D. LeBlanc, M. Palmer, J. Salinger, Z. Deering, and M. Shulman. Development and Validation of Funtional Definitions and Evaluation Procedures for Collision Warning/Avoidance Systems. Technical Report DOT HS 808 964, National Highway Traffic Safety Administration, U.S. Department of Transportation, August 1999. Final Report.
5. J. Pierowicz, E. Jocoy, M. Lloyd, A. Bittner, and B. Pirson. Intersection Collision Avoidance Using ITS Countermeasures. Technical Report DOT HS 809 171, NHTSA, U.S. DOT, September 2000. Final report.
6. L. Tijerina, S. Johnston, E. Parmer, H. Pham, M. Winterbottom, and F. Barickman. Preliminary Studies in Haptic Display for Rear-end Collision Avoidance System and Adaptive Cruise Control System Applications. Technical Report DOT HS 808 (TBD), NHTSA, U.S. DOT, September 2000.
7. G. M. Corporation and D. -D. E. Systems. First annual report: Automotive collision avoidance system field operational test. Technical Report DOT HS 809 196, NHTSA, U.S. DOT, December 2000.
8. Z. Yang, T. Kobayashi, and T. Katayama. Development of an Intersection Collision Warning System Using DGPS. In *Intelligent Vehicle Systems (SP-1538)*. SAE World Congress, Detroit, Michigan, March 2000. 2000-01-1301.
9. Srinivas Reddy Mosra, Shobhit Shanker and Syed Masud Mahmud "An Intelligent Architecture for Issuing Intersection Collision Warnings", Proceedings of the 3 Annual Intelligent Vehicle systems Symposium of NDIA, National Automotive Center and Vectronics Technology, June 21 –24, 2004, Traverse City, Michigan
10. Tom Karygiannis, Lese Owens "Wireless Network Security, 802.11, Blue tooth and Handheld Devices", National Institute of Standards and technology (NIST) Technology Administration, U,S Department of Commerce, Special Publication 800-48
11. Lloyd, M.M., Wilson, G.D., Nowak, C.J., and Bittner, A.C., Brake Pulsing as Haptic Warning for an Intersection Collision Avoidance Countermeasure, in *Transportation Research Record 1694*, TRB, National Research Council, Washington, D.C., 1999.
12. Ferlis, R.A., Infrastructure Collision-Avoidance Concept for Straight-Crossing-Path Crashes at Signalized Intersection, in *Transportation Research Record 1800*, TRB, National Research Council, Washington, D.C., 2002.
13. White, B., and Eccles, K.A., Inexpensive, Infrastructure-Based Intersection Collision-Avoidance System to Prevent Left-Turn Crashes with Opposite-Direction Traffic, in *Transportation Research Record 1800*, TRB, National Research Council, Washington, D.C., 2002.
14. Bhimani, A. "Securing the Commercial Internet". *Communications of the ACM,* 39(6), pp. 29-35 1996.
15. R.Rivest, A.Shamir and L.Adleman, "A Method for Obtaining Digital Signatures and Public-Key Cryptosystems", Communications of the ACM 21 (Pages 120-126), 1978.
16. Carl M. Ellison, Bill Franz, Butler Lampson, Ron Rivest, Brian M. Thomas, and Tatu Ylönen. SPKI certificate theory, Simple public key certificate, SPKI examples. Internet draft, IETF SPKI Working Group, November 1997.
17. ITU-T *ITU-T Recommendation X.509: Information Technology - Open Systems Interconnection - The Directory: Authentication Framework*. ITU, 1993 and ISO/IEC 9594-8 1993.

18. Rüppel, R. A. and Wildhaber, B. "Public Key Infrastructure - Survey and Issues". In *Trust Center - Grundlagen, rechtliche Aspekte, Standardisierung und Realisierung*(Ed, Horster, P.) Vieweg Verlag, Wiesbaden, pp. 197-212 1995.
19. Schneier, B. *Applied Cryptography - Protocols, Algorithms, and Source Code in C.* John Wiley & Sons, Inc., New York 1996

CONTACT

Srinivas R Mosra
Graduate Student
Wayne State University
630 Merrick St Apt #307
Detroit, MI-48202
Phone: 313-550-3658
Email: srinivasreddy@wayne.edu

Dr. Syed Masud Mahmud
Associate Professor
Department of Electrical and Computer Engineering
Wayne State University, Detroit, MI-48202
Phone: 313-577-3855
Fax: 313-577-1101
Email: smahmud@eng.wayne.edu

A Monocular Vision Advance Warning System for the Automotive Aftermarket

Itay Gat and Meny Benady
Mobileye Vision Technologies Ltd.

Amnon Shashua
Hebrew University

Abstract

Driver inattention and poor judgment are the major causes of motor vehicle accidents (MVA). Extensive research has shown that intelligent driver assistance systems can significantly reduce the number and severity of these accidents. The driver's visual perception abilities are a key factor in the design of the driving environment. This makes image processing a natural candidate in any effort to impact MVAs. The vision system described here encompasses 3 major capabilities: (i) Lane Departure Warning (ii) Headway Monitoring and Warning (iii) Forward Collision Warning. This paper describes in detail the different warning features, the HMI (visual and acoustic) application design rules, and results of a study in which the system was installed in a commercial fleet and passenger vehicles.

1 Introduction

While the number of cars manufactured each year continues to grow - so do the figures of motor vehicle accidents. The alarming data show that around 10 million people around the world are injured in MVA each year. 20-30% of them are severally injured and around 400,000 are fatal injuries, resulting in death [1]. Research has shown that driver inattention and poor judgment are the major causes of MVAs. Extensive research has shown that intelligent driver assistance systems can significantly reduce the number and severity of these accidents. [2].

Controlled experiments have shown that when vehicles are equipped with crash warning systems, accidents are reduced by 78%. Providing the driver with 0.5 second alert to a rear-end collision can prevent as much as 60% of this type of accidents. The figures are even more impressive if the alarm is given 1 second in advance: a reduction of up to 90% is achieved (cited in [3]).

According to the National Highway Traffic Safety Administration (NHTSA), more than 43% of all fatal MVAs reported in 2001 involved a lane or road departure. This statistic increases every year, making it the single largest cause of automotive highway fatalities in the United States alone [4, 5].

NHTSA estimates that more than 1.5 million police-reported MVAs involve some form of driver inattention: the driver is distracted, asleep or fatigued, or otherwise 'lost in thought'. Driver distraction is one form of inattention and is a factor in more than half of these MVAs. The presence of a triggering event distinguishes a distracted driver from one who is simply inattentive or 'lost in thought.'

In most cases, failure to maintain safe headway can be attributed to driver inattention and/or misjudgment of distance. It has been shown that drivers tend to overestimate their headway and consequently drive with short and potentially dangerous headway [6]. An intelligent system may serve a double purpose in this case, both as an alert, and as an tool for 'educating" the drivers. It was further shown that even imperfect systems are quite helpful in positively impacting drivers' habits [7].

Current solutions and alternative technologies

Most popular solutions today are based on Radar and Lidar technologies. The radar technology measures reflections from metal objects, and takes into account the Doppler effect in order to provide relative speed information. The Lidar systems use laser-beams to measure distance.

The high cost of radar systems limits their usefulness and the are on the whole restricted to high-end vehicles. Although Radar is unaffected by weather and lighting conditions, sensor data from the radar is extremely limited in the context of trying to interpret an extremely complex and dynamic environment. In most cases, the combination of smart processing with radar data works well for the constrained application of a distance control in highway driving, but there are situations where no matter how much processing is performed on the radar data, the data itself does not reflect the environment with a high enough fidelity to completely interpret the situation. Spatial resolution is relatively coarse for the detected field of view, such that detections can be improperly localized in the scene and object size is impossible to determine. The effects of this are that small objects can appear large, radically different objects appear similar, and position localization is only grossly possible. This leaves room for improvement, which becomes important as the sensing technologies are applied toward safety features.

Given that the the driving environment is designed around the human driver's ability for visual perception it may look natural to search for vision solutions. Therefore, another family of solutions is based on image systems with two sensors that can provide depth information in the image. Such systems are still rather expensive, require accurate calibration among the cameras. Moreover, the ability to provide depth information is only for the short range (up to 20m), whereas most of the vehicles on the road are much farther away from us.

Monocular vision systems are starting to emerge but they are usually focusing only on one aspect of the problem - e.g. lane departure warning [4]. It turns out that in many situations providing warning based on one modality may be too limited. For example lane departure system would gain a lot from insertion of information about vehicles on the road (blocking the view on the lanes). Furthermore, higher level of information about lanes can be of aid - for example unstable driving within a lane (indicated by lateral velocity) may be an important indication of intelligent systems.

This paper describes Mobileye's Advance Warning System (AWS) product which is based on technology enabling detection and accurate measurement of lanes, road geometry, surrounding vehicles and other information using monocular camera. The AWS description with its underlying. technology is given in the second section, whereas the third section provides thorough analysis of system performance. Finally the conclusion summaries the system capabilities.

2 System description

The technology of Mobileye enable the detection of lane marks and vehicles in complex environment together with various measurements on these objects. The analysis is carried out using a monocular vision system thus creating a reliable and cheap solution.

2.1 Building a detection scheme

Mobileye's detection system architecture loops through the following modules:

1. **Generate candidate regions of interest:** a systematic scan of the image for rectangular shaped regions at all positions and all sizes would be computationally unwieldy. An attention mechanism filters out windows based on lack of distinctive texture properties and in-compliance with perspective constraints on range and size of the candidate vehicle. On average, the attention mechanism generates 75 windows (out of the many thousands of candidates which could be generated otherwise) per frame which are fed to the classifier.

2. **Single frame classification:**

   The core of the detection process lies in the classificationstage, where each region of interest is given a score that represent the likelihood of that region to be a vehicle. We are using several classification schemes throughout the system and they can be rather degenerate such as the nearest neighbor approach which employs relatively sophisticated local features such as those used by [12], or integration via a cascaded classifier such as the hierarchical SVM approach used by [11]. A particulary powerful scheme we employ borrows from the idea of the recognition-by-components using a 2-stage classifier algorithm. Namely, we breakdown the region of interest into sub-regions, create a local vector representation per sub-region, feed each of the local feature vectors to a discriminant function and integrate the local discriminant results by a second-stage classifier . The crucial difference from the conventional paradigm is the way we handle the training set. Since the number of local sub-regions are small we generate multiple local discriminants (one per local sub-region) by dividing the training set into mutually exclusive *training clusters*. The idea behind the subset division of the training set is to breakdown the overall variability of the class into manageable pieces which can be captured by relatively simple component classifiers. In other words, rather than seeking sophisticated component classifiers which cover the entire variability space (of the subregions) we apply prior knowledge in the form of clustering (manually) the training set. Each component classifier is trained multiple times —

once per training cluster — while the multiple discriminant values per subregion and across subregions are combined together via Adaboost [13].

3. **Multi-frame Approval Process:** candidates which survive the single frame classification thresholds are likely to correspond to vehicles. However, due to the high variability of the object class and the high levels of background clutter it is conceivable that coincidental arrangements of image texture may have a high detection score — an ambiguous situation which is likely to be unavoidable. Additional information collected over a number of frames are used in the system for further corroboration.

4. **Range measurement:** a more detailed description of our range and range-rate measurement are given in the next section.

The four basic steps above are also coupled with supporting functions such as host vehicle ego-motion (of Yaw and Pitch) [10], robust tracking — and of primary importance the classification scores of background sub-classes which include licensed vehicles, poles, guard-rails, repetitive texture, lane mark interpretation, bridges and other man-made horizontal structures. The sub-class scores play an important role in the final decision-tree multi-frame approval process.

2.2 Providing range and range-rate using monocular vision

Range to vehicles and range-rate are two important values required for any vision-based system. In this section only the essence of the algorithm is described where a full description can be found in [8].

As the data is collected from a single camera range must be estimated by using perspective. There are two cues which can be used: size of the vehicle in the image and position of the bottom of the vehicle in the image. Since the width of a vehicle of unknown type (car, van, truck etc) can vary anywhere between $1.5m$ and $3m$ a range estimate based on width will have only about 30% accuracy.

A much better estimate can be achieved using the road geometry and the point of contact of the vehicle with the road. We assume a planar road surface and a camera mounted so that the optical axis is parallel to the road surface. A point on the road at a distance Z in front of the camera will project to the image at a height y, where y is given by the equation:

$$y = \frac{fH}{Z} \qquad (1)$$

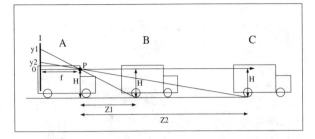

Figure 1: *Schematic diagram of the imaging geometry (see text).*

where H is the camera height, and f is the focal length of the camera (both given in meters).

Figure 1 shows a diagram of a schematic pinhole camera comprised of a pinhole (P) and an imaging plane (I) placed at a focal distance (f) from the pinhole. The camera is mounted on vehicle (A) at a height (H). The rear of vehicle (B) is at a distance (Z_1) from the camera. The point of contact between the vehicle and the road projects onto the image plane at a position (y_1). The focal distance (f) and the image coordinates (y) are typically in *mm* and are drawn here not to scale.

Equation 1 can be derived directly from the similarity of triangles: $\frac{y}{f} = \frac{H}{Z}$. The point of contact between a more distant vehicle (C) and the road projects onto the image plane at a position (y_2) which is smaller than (y_1).

To determine the distance to a vehicle we must first detect the point of contact between the vehicle and the road (i.e. the wheels). It is than possible to compute the distance to the vehicle:

$$Z = \frac{fH}{y}. \qquad (2)$$

Figure 2 shows an example sequence of a truck at various distances. The distance from the horizon line to the bottom of the truck is smaller when the truck is more distant (a) than when it is close (b and c).

This outcome fits in with our daily experience, that objects that are closer to us are perceived as lower then objects that are further away. The relationship demonstrated here may be further used to deal with situations in which the assumption of planner road does not hold (e.g. starting to climb a hill or bumps on the road).

Beyond the basic range estimation, it is also important to provide information about the range-rate or the time it would take to cross the distance to current in path vehicle - *time to contact*. The human visual system is able to make very accurate *time to contact* assessments based on the retinal divergence (scale change of the target) and, therefore can be used in our monocular system. It can be shown that

(a)

(b)

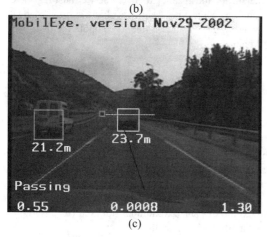

(c)

Figure 2: *A typical sequence where the host vehicle decelerates so as to keep a safe headway distance from the detected vehicle. The detected target vehicle (the truck) is marked by a white rectangle. As the distance to the target vehicle decreases the size of the target vehicle in the image increases.*

if one assumes constant velocity, it is possible to obtain a simple relationship between the scale change and the time-to-contact:

$$TTC = \frac{t}{S-1}. \qquad (3)$$

where TTC is the time to contact, S is the scale change between two consecutive images and t is the time difference between them. It is possible to add acceleration and deceleration into the computation [9].

2.3 Application

The Advance Warning System (AWS) provides a set of warnings for the driver based on the vision technology described above:

- **Lane Departure Warning (LDW)** The LDW module detects lane boundaries, finds the road curvature, measures position of the vehicle relative to the lanes, and provides indications of unintentional deviation from the roadway in the form of an audible rumble strip sound. The system can detect the various types of lane markings: solid, dashed, boxed and cat-eyes, and also make extensive use of vehicle detection in order to provide better lane detection. In the absence of lane markings the system can utilize road edges and curbs. It measures lateral vehicle motion to predict the time to lane crossing providing an early warning signal before the vehicle actually crosses the lane. Lane departure warnings are suppressed in cases of intentional lane departures (indicated by activation of turn signal), braking, no lane markings (e.g. within junctions) and inconsistent lane markings (e.g. road construction areas).

- **Headway indication and warning** The headway monitoring module provides constant measurement of the distance in time to the current position of the vehicles driving ahead in the same lane. The ability to indicate current in-path vehicle is dependent upon the information from the lanes detection module. While insufficient distance keeping is a major cause of MVAs, it is difficult for many drivers to judge this distance correctly while considering the traveling speed of the vehicle. The AWS headway display provides a visual indication when insufficient distance is being kept to the vehicle ahead, as well as a clear numeric display (in seconds) which provides an accurate cue for driving habits improvement for the driver.

- **Forward Collision Warning (FCW)** The FCW module continuously computes time-to-contact to the vehicle ahead, based on range and relative velocity measurements. An advanced image processing algorithm

126

determines whether the vehicle ahead is in a collision path (even in the absence of lane markings) and provides audio warnings to the driver at predetermined time intervals prior to collision (e.g. 2.5, 1.6 and 0.7 seconds) alerting the driver to the danger and allowing appropriate action such as braking or steering away from the obstacle ahead [9]. The system uses information about driver actions (e.g. braking) to suppress warnings in situations that are under the driver's control.

2.4 AWS applicative overview

The AWS system consists of:

- **SeeQ** Real-time image processing unit running at 30 FPS on the *EyeQ* vision system on chip that include a Compact High Dynamic Range CMOS (HDRC) camera. The units' size is 3 over 5 cm and it is located on the windscreen (see figure3).

- Display and interface unit located in front of the driver

- A pair of loudspeakers for providing directional warnings

The interface unit is also connected to signals from the vehicle (vehicle speed signal - VSS, indicators and brake). The system is turned on shortly after ignition. An example of the video display is provided in figure 4. In this situation the system is active and has detected a vehicle in current path. The headway value for this vehicle is 1.2 seconds which is enough headway according to these settings. The driver sensitivity mode is indicated in the example as *NORMAL*.

There is a small set of commands that the driver can pass to the AWS. Among them are a volume control and sensitivity control. Making the system more sensitive raise the frequency of alerts - for example, alerts when the driver is first getting close to crossing lanes are provided.

Apart from the warnings, the system also provides a visual indication of its availability. The system may be unavailable under two conditions:

1. **Failsafe** the system has low visibility and cannot provide alerts at this period. The problem may in some cases be a temporary one (e.g. low sun causing the image to be unusable, or dirty windscreen)

2. **Maintenance** The system is not receiving the required signals for computation (either the signal from the camera or speed signal from the vehicle).

(a)

(b)

Figure 3: *View of the SeeQ. (a) a road facing view of the camera. (b) Processing unit at the rear side of SeeQ*

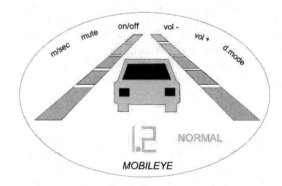

Figure 4: *Example of AWS display.*

(a)   (b)

Figure 5: *Simultaneous capture of frontal and road images. (a.) Image of Road near the left wheel with 10cm markings from the wheel (shown in red). The distance from lane mark obtained by AWS (1.43m) is shown in green. (b). Frontal image showing vehicle detection (blue and green) and lanes (green). Note that the lane mark segment appearing in (a) is also visible at the lower part of (b).*

3 Performance and results

In order to evaluate the performance of the AWS system, two types of evaluations were carried out:

- Quantitative analysis: measured performance of each feature of the system.

- Qualitative analysis: effectiveness of the system as evaluated by professional and non-professional drivers.

3.1 Performance envelope

In order to facilitate the examination of accuracy of the AWS the following infrastructure was used:

Lane departure warning

A dual camera system was used in order to measure the accuracy of lane departure warning. The usual view of the driving scenario was accompanied with a camera placed above the wheel and looking downward on the road. The images coming from both cameras were synchronized by using an interchangeable sampling scheme. A view on the outcome of such a system is shown in figure 5.

headway control

The two issues that need to be examined in this situation are the availability of vehicle detection and the accuracy of distance measurement. The availability of the AWS is measured by manually inspecting video clips recorded while driving. The accuracy is measured by comparing the distance estimation of AWS with that obtained by Radar sys-

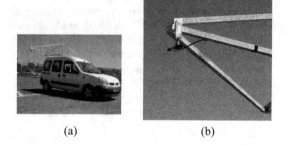

(a)   (b)

Figure 6: *The remote mounting structure.*

tems recorded simultaneously. A matching algorithm was used to synchronize the radar measurements and the ones provided by AWS.

Forward collision warning

The performance of this capability is the most difficult to measure, as situations in which our vehicle is within less then 2 seconds from collision with the vehicle in path are rare and dangerous. The only solution is to simulate such conditions, either using balloon cars, or by placing the camera at a safe distance from the actual vehicle. We chose the second solution, and used a camera placed on a large Rod positioned above the vehicle. This camera displayed images of what a vehicle in crash situation could "see". The infrastructure used for this test is shown in figure 6.

3.2 Qualitative assessment

In field tests (clinic) the system was installed in several vehicles, drivers were instructed on the use of the system and were asked to drive normally and report the performance of the AWS. The most important question arising from this clinic was whether the system contributed to the safety feeling of the driver.

In order to asses this information we asked the drivers to answer several questionnaires:

- Pre driving questionnaire, in which we addressed the characteristics of the drivers (e.g. age, gender) and expectations from such system.

- After each drive: to report specific problems.

- After the whole process: we addressed more high level issues.

An offline kit that enables the recording (both video and log data) of alerts and other related events was installed in each vehicle. Throughout the drive, the following general information was also collected: as:

- Distance traveled
- Speed
- Time of day
- Eye gaze towards the AWS display

The information is currently being processed, and final results will be available soon.

3.3 Results

The performance envelope of the AWS was examined in various scenarios (e.g. highway, urban, day, night). The following statistics were based on a database of 30 hours of driving data randomly selected from over 300 hours of data from Europe, Japan, USA and Israel.

- **Lane Departure Warning** Availability values obtained were:
  - Well marked highway (day/night): 99.5%
  - Poorly marked highway (day/night): 97.5%
  - Country road (day/night): 98.6%
  - Bott's dots (day/dusk): 89.8%

The average absolute error in measuring position of the lanes was 5cm and the false warning produced were less then 1 per 2 hours of average driving.

- **Headway control** The availability of vehicles detection was 99.5% The accuracy as measured in average absolute error in range is:
  - Vehicles up to 40m: 5.6%
  - Vehicles up to 60m: 6.0%
  - Vehicles up to 80m: 6.4%
  - All vehicles: 7.6%

A typical performance is shown in Figure 7.

- **Forward Collision Warning** In all of the testing carried out using the collision simulation the warning was produced. The average time in advance was 1.5 sec. The false rate of the system was less then 1 false per 5 hours of driving and even those falses were short in nature (lasted less then 0.5 seconds). A typical performance is shown in Figure 8.

The subjective assessment in the clinic process is currently being evaluated.

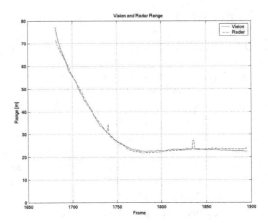

Figure 7: *Range for a typical sequence where the host vehicle decelerates so as to keep a safe headway distance of 24m from the detected vehicle. The lead vehicle was traveling at 80KPH.*

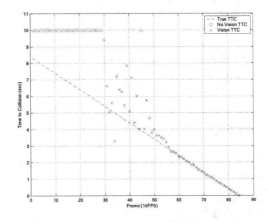

Figure 8: *Time to contact computed using vision divergence. Using the remote mounting infrastructure it is possible to compute the exact time to contact and to compare the computed results with it. The results show that reliable estimate is given up to 4 seconds before collision.*

4 Conclusion

The AWS is an advance system for the automotive aftermarket that offers a suite of active safety applications for accident reduction. The main purpose of the AWS system is to alert the driver and increase the awareness of drivers to dangerous situations that may be caused by weariness or by other distractions during driving.

Based on a single camera located on the front windscreen, the AWS detects and tracks vehicles on the road ahead providing range, relative speed, and lane position data. In addition, the system detects lane markings and measures and monitors distance to road boundaries.

A small display unit and a pair of left and right speakers inside the car provide timely audio and visual warnings, allowing the driver to react to various types of dangerous situations and to reduce the risk of accidents. It was demonstrated that a single low-cost camera can provide this information with satisfactory accuracy. The application built around the technology is such that it keeps a balance between supplying sufficient information and avoiding a surpass of alerts, that the driver would find annoying. Furthermore, the driver has the option of influencing the system's sensitivity and controlling the alert level provided. The performance of the system shows that both from the performance and the applicative points of view the system provides extra-value for the driver, and can reduce accidents and help in education of drivers for safer drive.

A second generation product that enhances the system capabilities and includes pedestrian protection is now being developed using advanced implementations of the principles used in the current AWS.

References

[1] Organization for Economic Cooperation & Development - Paris.

[2] The National Safe Driving Test & initiative Partners (CNN: http://edition.cnn.com)

[3] National Transportation Safety Board, Special Investigation Report - Highway Vehicle- and Infrastructure-based Technology For the Prevention of Rear-end Collisions. NTSB Number SIR-01/01, May 2001

[4] Iteris (http://www.iteris.com)

[5] P. Zador, S. Krawchuck and R. Voas Automotive Collision Avoidance System (ACAS) Program/First Annual Report. NHTSA - National Highway Traffic Safety Administration (http://www.nhtsa.dot.gov/) DOT HS 809 080, August 2000

[6] National Safety Council, Defensive driving course [Course Guide] 1992.

[7] A. Ben-Yaacov, M. Maltz and D. Shinaar , Effects of an in-vehicle collision avoidance warning system on short- and long-term driving performance. In *Hum Factors.*, pages 335 -342, 2002

[8] G. Stein and O. Mano and A. Shashua Vision-based ACC with a Single Camera: Bounds on Range and Range Rate Accuracy In *IEEE Intelligent Vehicles Symposium (IV2003)*, June 2003

[9] O. Mano, G. Stein, E. Dagan and A. Shashua. Forward Collision Warning with a Single Camera., In *IEEE Intelligent Vehicles Symposium (IV2004)*, June. 2004, Parma, Italy.

[10] G. Stein, O. Mano and A. Shashua. A Robust Method for Computing Vehicle Ego-motion In *IEEE Intelligent Vehicles Symposium (IV2000)*, Oct. 2000, Dearborn, MI.

[11] A. Mohan, C. Papageorgiou, and T. Poggio. Example-based object detection in images by components. In IEEE Transactions on Pattern Analysis and Machine Intelligence (PAMI), 23:349-361, April 2001.

[12] D. G. Lowe. Distinctive image features from scale-invariant keypoints. International Journal of Computer Vision, 2004.

[13] Y. Freund and R. E. Schapire. Experiments with a new boosting algorithm. In Proceedings of International Conference on Machine Learning (ICML), pp. 148-156, 1996.

Wireless Multicasting for Remote Software Upload in Vehicles With Realistic Vehicle Movements

Radovan Miucic and Syed Masud Mahmud
Electrical and Computer Engineering Department, Wayne State University

ABSTRACT

Future vehicles will have many features that include, but are not limited to, drive-by-wire, telematics, pre-crash warning, highway guidance and traffic alert systems. From time to time the vehicles will need to have their software modules updated for various reasons, such as to introduce new features in vehicles, the need to change the navigation map, the need to fine tune various features of the vehicles, etc. A remote software update has a number of advantages, such as it does not require consumers to take their vehicles to the dealers, and the dealers do not need to spend time on vehicles on an individual basis. Thus, remote software updates can save consumers' valuable time, as well as cost savings for the vehicle manufacturers.

Since wireless links have limited bandwidth, uploading software in thousands of vehicles in a cost-effective and timely manner is a challenge. Another major issue related to the remote software update is the security of the update process. In another paper, we addressed the security issue of the update process. In this paper, we present a wireless multicasting technique for uploading software in vehicles. Since the servers that will be broadcasting the software are located in some permanent places, and the vehicles are located all over the country, the software upload process has to depend on an infrastructure such as the cellular infrastructure. We have developed simulation models to determine the performance of our proposed wireless multicasting technique for remote software uploads. We simulated hundreds of vehicles distributed around a city area. In the simulation model, we assumed realistic speeds for the vehicles depending upon where the vehicles were located. The vehicles transfer the software from a buffer to an electronic control unit (ECU) when their ignition is off. Our simulation results show that if the multicast technique is used instead of the unicast technique (one vehicle at a time), then the software can be updated in a very cost-effective manner. The paper will give a detailed description of our technique and provide numerous results of the simulations collected for various distributions of the systems and vehicles.

INTRODUCTION

Today, most of the electronic controller units (ECUs) have in-vehicle programming (IVP) capabilities. Some of the ECUs in the modern vehicle are IC (instrument cluster), ABS (antilock brake system), ECM (engine control module), PCM (power control module), BCM (body control module), HVAC, infotainment systems, radio, MP3 player, DVD, On Star, door module, seat module, airbag module, navigation module, etc.

Sometimes software needs to be updated for various reasons, such as: calibration update, a problem with the existing software in modules, a recall of the vehicle due to software error, a new implementation, or the addition of new features. Some component suppliers only have a limited time to develop their products, and sometimes products go into production with very little testing. Such components are at high risk of unacceptable behavior later on, and it is very useful to have a cost-effective reprogramming system in place. Nowadays, if there is a need for a software update, a customer usually goes to the dealer, and the technician reprograms the module by physically plugging a programmer into the connector that has access to the intervehicele bus or busses such as CAN, LIN, J1850, UBP, SCP, etc. The programmer then uploads software into the vehicle's modules. Technicians at a dealership upload new software into vehicles one vehicle at a time. If there is a vehicle recall due to software error, dealers would have to reprogram hundreds of thousands of vehicles. The process of uploading software into each vehicle is an inconvenience for both the dealers and the customers. The car manufacturers lose money and the customers lose confidence because of this process. For each module to be programmed, it will take time for setup and actual servicing, including taking the customer's time to make the appointment, physically bringing the vehicle to the dealer, and spending time while the vehicle is being serviced. In addition, it takes months to update the software in all the vehicles, and there is no guarantee that all customers will bring in their vehicles for service.

Our motivation is to bring down the cost of updating modules by broadcasting the software that will be updated in particular modules in vehicles. Cellular towers are currently located in cities, towns and near major roads where the most vehicles are located. Car manufacturers can jointly work with major cellular companies to have some channels reserved for their needs, or lease the channels when needed. Software can be updated in vehicles by using either a unicast or a multicast process. In a unicast process, the vehicles will receive software on an individual basis, but in a multicast process, all vehicles within the range of a tower will receive software in parallel. However, updating software using the multicast process will need significantly less bandwidth than that needed by the unicast process.

Security of the transferred software is an important aspect of multicast programming. A vehicle network must have a cost-effective security system. The desired level of security needs to have a good balance of cost and extensiveness [3]. We will mainly focus on the wireless update process and leave security issues for another paper. This paper presents simulation models for performing remote software updates in vehicles. The simulation results show that updating software using the multicast process is very cost-effective compared to using the unicast process.

RELATED WORK

Rooftop networks, proposed in [1], provide an alternative to wired networks. The rooftop network is not mobile. In this network, antennas are mounted on building roofs, and several can be installed in metropolitan areas. The antennas are placed in line-of-sight orientation to their neighbors. The routing algorithm for rooftop networks proposes self-configuration for hundreds of thousands of such nodes in a metropolitan area. However, the routing algorithm faces significant scalability and security challenges.

GeoMote [2] developed a routing algorithm for TinyOS (Operating System for Tiny Networked Sensors). Low-cost, densely placed networked sensors with limited available memory, power and processing resources are called "motes". Every mote must know its own location in order to participate in the network. Three types of motes: GeoRouters, GeoGateways and GeoHosts are used in the networks. GeoRouters run routing algorithms to decide how the messages are to be routed based on the geographical locations.

OnStar, Vetronix and Networkcar have remote diagnostics programs. If remote diagnostics programs prove to be successful, OEMs will be sure to invest more to exploit remote diagnostics [4].

SOFTWARE SIMULATION DESCRIPTION

Our software consists of the simulation of three major elements: a map, the towers and the vehicles. These elements interact with each other. The vehicle movement and the tower broadcast are represented on a map, as shown in Figure 1.

The map, shown in Figure 1, is an array representing city blocks. City blocks consist of empty spaces and roads. In our simulation model, a vehicle cannot exist in an empty space - it can only be on a road. Roads intersect with each other. Software simulation considers situations in which a vehicle can travel east or west, north or south, and north, west, south, and east at intersections (see Figure 2).

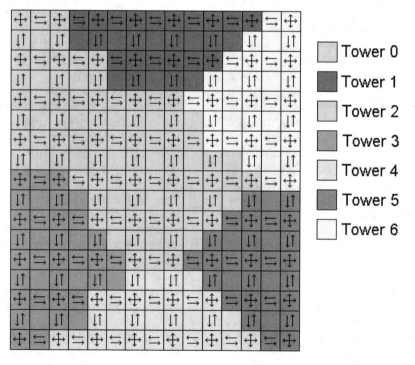

Figure 1: Map of the city blocks and towers' coverage areas

The towers are represented on the map. Each city block is under some tower coverage. Only the tower coverage of the roads is of interest to us. The towers broadcast software to their coverage area. For simplicity, we have assumed that there are no overlaps of the coverage areas of the towers. Assuming there is no overlap of tower coverage areas is reasonable, because a vehicle receiver can have a signal strength detector and the detector may be set to only accept the strongest signal.

Vehicles are simulated to be traveling on these roads. The vehicle position and direction is determined in the initialization phase of the simulation software. A vehicle may change direction when it reaches an intersection. When a vehicle enters an intersection, it randomly decides to change or not to change its current direction. For simplicity, if a vehicle reaches a boundary of the map, it reverses its direction. The vehicle receives software packets from the tower while it is moving. The vehicle may miss receiving a packet due to a weak signal and noise, or due to a change in the coverage area.

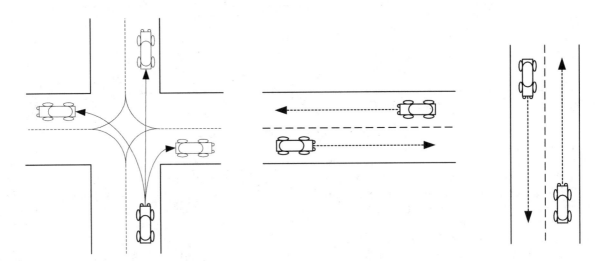

Figure 2: Possible vehicle movement in the simulation model

The original equipment manufacturer (OEM) or the supplier decides that a certain ECU needs an update. The OEM sends the software for a target ECU to the central server (CS). The CS uploads the software to the nearest tower, and the tower starts broadcasting the software to the vehicles within its coverage area. The tower also sends the software to its neighboring towers. The software must contain information about the vehicle model and year and about the ECU. The vehicle must recognize if the software that is being received is targeting one of its ECUs. If the software is targeting an ECU on the vehicle, the vehicle will buffer the received software. If the software is not targeting an ECU of the vehicle, the vehicle will ignore that software. The received software must contain version control information. The vehicle will compare the ECU's current software version with the received software version. The vehicle accepts received software only if the version of the broadcast software is newer than the version of the software currently available in the ECU. The strategy of matching the new software with the current software in the target ECU will be the topic of a different paper. In our simulation, we will only consider vehicles that have an ECU in need of an update.

Software Packets: The software is divided into smaller pieces of code, called packets. A tower sends packets sequentially, from the first packet to the last. Vehicles accept software packets that are being wirelessly sent from the tower. A vehicle may fail to receive a packet for various reasons, such as the presence of noise, transition from one tower to the next tower, etc. If a vehicle starts receiving a packet and does not receive the next packet for some time, or if it receives the last packet sent but some prior packets are missing, then it will start sending periodic "complain" messages indicating that the vehicle's target ECU has not been programmed. After a tower has sent all of the software packets it waits for some time to determine whether any vehicle sends a message indicating that the target ECU has not been programmed. We use the symbol VNP for any message that indicates that a vehicle has not been programmed, and the symbol WP to indicate the time during which the vehicle waits for the VNP message. A VNP message will identify what software is missing from a vehicle. If a tower receives a VNP message during the period WP, it starts transmitting the entire software all over again. If a vehicle exits and stays out of the coverage area after the WP period it will not be updated: the software will remain in the tower's buffer during the WP period. If a tower does not receive the VNP message, the tower flushes its buffer to make it free for other activities. If the tower receives a VNP message after the WP period, it will request the software from the central server, and the entire process of broadcasting the software is then repeated. Figure 3 shows the broadcasting timeline. The entire process includes the following steps, as shown in Figure 3:

1) The tower starts broadcasting the software.
2) The tower finishes broadcasting the software.
3) The tower waits in the WP state.
4) The tower receives a VNP message.
5) The tower starts broadcasting the software.

6) The tower waits in the WP state and does not receive any VNP message.
7) The tower deletes the software from its buffer.

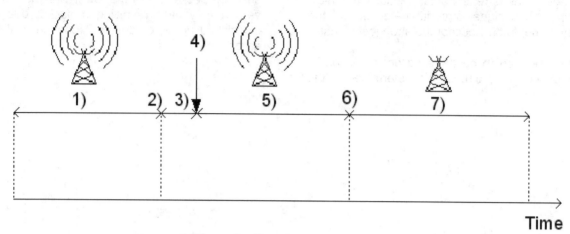

Figure 3 – Broadcasting timeline

There must be a dedicated device in the electrical architecture of the vehicle, perhaps a new ECU (WECU – Wireless ECU) which is capable of wireless communication. The vehicle will have a buffer large enough to store memory of the largest ECU in the vehicle's network architecture. The vehicle will receive the software via the WECU and save it in its buffer. The target ECU is reprogrammed from the buffer when it becomes convenient, i.e. safety critical ECUs may be updated when the vehicle stops and the ignition is off. Other ECUs may be updated on the fly. The WECU will determine if it is safe to update the target ECU. The designated ECU will check the battery condition and other target-specific conditions. Updating a vehicle without the customer's consent can pose problems. For example, the vehicle owner has memorized his/her favorite seat position. Their saved seat information will be lost during the software update, and the customer will notice the difference.

Simulation Process – Our software simulation executes in loops. Each loop takes the smallest time interval, here called time instance TI. In our simulation, the value of TI is 12.5 ms. The entire packet has been broken down in the simulation program into many small packets, and the size of each packet is 1KB (kilobytes). We assumed that a vehicle needs 8 time instances (TI) to receive a packet. In other words, the speed of the wireless communication is 10 KB/sec.

Broadcast Delay - The delay between a pair of neighboring towers is one hop. We consider one hop delay to be equal to one packet delay, or 8-TI delay. We assumed that tower T0 is the closest tower to the server. Since different towers are located at different hop distances away from Tower T0, different towers start broadcasting the software after a different length of delay, as shown in Figure 4. Tower T0, which is closest to the server, has no delay in broadcasting the software.

Signal Strength – The wireless signal is the strongest in the immediate vicinity of a tower. The signal strength of the broadcast in a city block depends on the distance from the city block to the broadcasting tower. We have defined four levels of signal strength. The city blocks closest to the tower have the strongest signal. The strongest signal is represented by using four fully colored bars, as shown in Figure 5. The city block that is farthest from the broadcasting tower has the weakest signal and it is represented with one fully colored bar, as shown in Figure 5. The vehicle in the city block with the strongest signal has the highest probability of receiving broadcasted packets. The vehicle in the city block with the weakest signal has the least probability of receiving broadcasted packets.

RESULTS

We ran our simulation with various input parameters. Figure 6 shows the outcome of the simulation. The cost, which is the number of times the towers had to rebroadcast the entire software, is shown on the Y-axis. The size of the software in packets is shown on the X-axis. The average cost to rebroadcast the software increases with the number of vehicles on the roads. As we mentioned earlier, packets may be lost due to low signal strength or due to movements of the vehicles from one tower region to another.

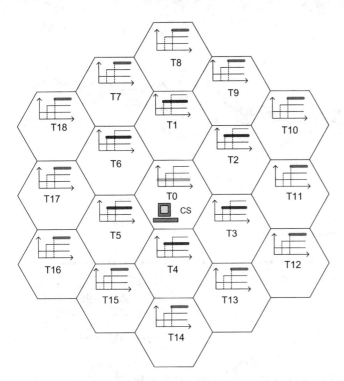

Figure 4 – Towers' broadcast delays.

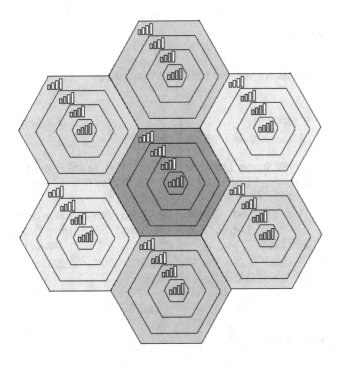

Figure 5 - Towers' signal strength

We assumed east-west and north-south roads connecting two intersections to be 800 meters long. Also, we assumed the length of the intersection to be 50 meters. Let's say that the bandwidth of the wireless link is 10 KB/sec. If a vehicle is moving at 100 km/hour it will travel 0.347 meter per time instance (12.5 ms). For simplicity, the simulation program updates the vehicle's position on the map after every 10 meters.

(Number of packets) x (number of instances per packet) x (time instance) = time for 1 software broadcast

For example, software containing 1500 packets will need $1500 \times 8 \times 12.5 \times 10^{-3}$ s = 150 sec to be uploaded in vehicles.

The signal strength at a vehicle depends on its distance from the nearest tower. In our simulation, a vehicle has to be in the coverage area during a software broadcast for 8 consecutive time instances in order for it to receive a complete packet. There is a possibility that the vehicle may receive corrupted data in any of the 8 time instances while a packet is being received, resulting in the loss of the packet. Figure 6 shows the results of the simulation runs for the following parameters.

- The probability of losing a packet is 10% in a region where the signal strength is the weakest (signal where only one bar is green, as shown in Figure 5)
- The probability of losing a packet is 5% in a region where the signal strength is nearly the weakest (signal where two bars are green, as shown in Figure 5)
- The probability of losing a packet is 2.5% in a region where the signal strength is nearly the strongest (signal where three bars are green, as shown in Figure 5)
- The probability of losing a packet is 0.01% in a region where the signal strength is the strongest (signal where all four bars are green, as shown in Figure 5)

Figure 6 is divided into two parts. The bottom part of Figure 6 shows the result for the case which assumes that the vehicles only lost packets because they were moving from one tower region to another. More specifically, this means that while a vehicle was receiving a packet it did not stay within the coverage of the tower from which it was receiving the packet for eight time instances (TI). For this case, the towers only broadcast the software a couple of times. The top part of Figure 6 shows the result for the case which took into consideration the vehicle movements and signal strengths. Figure 6 shows that the vehicles lost some packets mostly due to weak signals rather than due to vehicle movement. Thus, when signal strengths were taken into consideration, the towers had to broadcast the software significantly more times than when we were just considering vehicle movement. Figure 6 shows that the towers had to broadcast the software containing 1500 packets 18 times to load it into 500 vehicles. However, had the vehicles been uploaded one at a time instead of all 500 vehicles in parallel, then the towers would have been broadcasting the software 500 times. Thus, there is significant bandwidth saving when a group of vehicles is uploaded in parallel. This means that when compared to the unicast technique, a software upload using the multicast technique will result in significant cost savings.

135

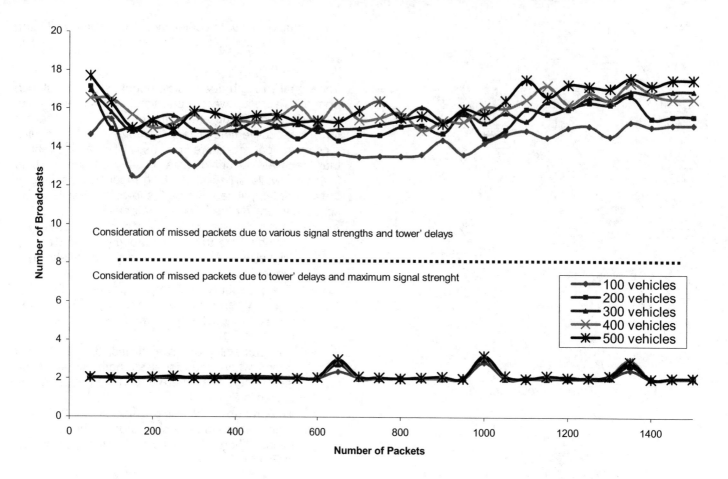

Figure 6 – Number of Broadcasts versus Number of Packets for various numbers of vehicles.

CONCLUSION

A study has been done to evaluate the effectiveness of software updates in vehicles using the multicast technique. We did a simulation assuming vehicles would move as predicted on city block maps. As the vehicles move from one city block to the next, they also move from one tower region to another tower region. According to the simulation results, updating software in vehicles using the multicast process is very cost-effective compared to that using the unicast process. Since the software can be updated using existing cell phone networks, no additional infrastructure is necessary for updating software in vehicles.

REFERENCES

1. Thanos Stathopoulos, John Heidemann, and Deborah Estrin. A Remote Code Update Mechanism for Wireless Sensor Networks. Technical Report CENS-TR-30, University of California, Los Angeles, Center for Embedded Networked Computing, November, 2003. http://www.isi.edu/~johnh/PAPERS/index.html

2. Peter Broadwell, Joseph Polastre, and Rachel Rubin, "GeoMote: Geographic Multicast for Networked Sensors." Class Project Paper 2001. 21

March 2004. Available: http://www.cs.berkeley.edu/~pbwell/

3. Mark Krage and Laci Jalics. "The Many Faces of Security," SAE, 2004-01-0236, 2004.

4. The Hansen Report on Automotive Electronics. "Remote Diagnostics— the Next OEM Frontier" dec. 2003/jan. 2004 Available: http://www.vetronix.com/pdf/press/vtx_2004-01-xx_hansen.pdf

CONTACT

Radovan Miucic received a B.S. degree in Electrical Engineering from Wayne State University, Detroit, in 2001 and his M.S. degree in Computer Engineering from Wayne State University, Detroit, in 2002. Currently he is working on his PhD degree, and as a software engineer in the automotive industry.

Email: radovanmiucic@hotmail.com

Syed Masud Mahmud, Ph.D.
Associate Professor
Department of Electrical and Computer Engineering
Wayne State University, Detroit, MI 48202
Phone: (313) 577-3855
Fax: (313) 577-1101
Email: smahmud@eng.wayne.edu

DEFINITIONS, ACRONYMS, ABBREVIATIONS

ECU – Electronic control unit.

CS – Central server. A computer repository for software to be broadcasted.

OEM – Original equipment manufacturer.

SFTE – Software for targeted ECU

VNP – The vehicle's targeted ECU is not programmed. A message from a vehicle to the tower indicating that some packets are missing in the vehicle's buffer.

TI –Time instance. This is the smallest time unit in the software simulation of loop time.

WP – Wait for VNP. The period of time the tower waits for a VNP message after a broadcast.

WECU – Wireless ECU.

2004-01-0204

Secure Inter-Vehicle Communications

Syed Masud Mahmud, Shobhit Shanker and Srinivas Reddy Mosra

Department of Electrical and Computer Engineering, Wayne State University

ABSTRACT

The study done by the U.S. National Highway Traffic Safety Administration (NHTSA) shows that developing automotive collision warning and avoidance systems will be very effective for reducing fatalities, injuries and associated costs. In order to develop an automotive collision warning and avoidance system, it will be necessary that the vehicles should be able to exchange (in real time) their dynamic information such as speed, acceleration, direction, relative position, etc. The only way to exchange the vehicles' dynamic information will be through wireless communications. The communication links among vehicles must be secured. Otherwise, hackers may inject some misleading data into the inter-vehicle messages to make the vehicle systems malfunction. In this paper, we have presented a technique for exchanging vehicles' dynamic information in a secure mode. We also investigated the feasibility of implementing secure inter-vehicle communication links using today's technology. Our study shows that current technology will allow us to build such a system.

INTRODUCTION

The aim of this paper is to provide an architecture to enable information exchange between vehicles in a secure way. The federal government incurs losses of millions of dollars every year in terms of vehicle accidents, and 90% of these losses are due to collisions [1].

During the last several decades, the improvements in seat belts, air bags, crash zones, and lighting have significantly reduced the rate of crashes, injuries and fatalities. In spite of these improvements, each year in the United States, motor vehicle crashes still account for about 40,000 deaths, and over 130 billion in financial losses [1]. A significant focus and interest has risen to develop active safety technology and collision avoidance systems. In order to develop an automotive collision avoidance system, it becomes imperative that the vehicles exchange their dynamic information such as speed, acceleration, position and direction in real time. With the advent of wireless ad hoc networking technologies, this can be done in a fairly accurate and feasible manner. The goal of the Automotive Collision

Avoidance Systems (ACAS) is to detect and warn the driver of potential hazard conditions. Therefore, it is of extreme importance that the data integrity should be taken care of, which means that the system should be immune to security attacks. It should be ensured that all communication between various vehicles is done in a secure fashion. Otherwise, a person with malicious intent might inject incorrect information into the inter-vehicle wireless links leading to a disaster, not only for the driver but also for the hundreds of others on the road.

For two devices (in our case, two vehicles) to communicate securely, it is necessary that they share some kind of secret key(s). In a client-server based architecture, this is done by the server using protocols such as SSL, VPN tunneling [2], etc. However in ad hoc networks there is no client-server architecture. The topology of ad hoc networks is dynamic and is constantly changing [3].

Ronald Miller and Qingfeng Huang have done some work in the area of ACAS using wireless communication technologies [4], but their systems are not secure, and can easily be attacked by hackers. In our architecture, security aspects such as authentication, authorization, and data integrity have been implemented. In our proposed architecture, we have suggested a device known as an NDM (Network Device Monitor). The concept of using an NDM has been proposed earlier in our previous work [5]. For an in-vehicle (intravehicle) network, the NDM acts as a key distribution center located inside the vehicle. The NDM acts as a gateway device, which all the devices use to communicate with each other. The NDM is also responsible for maintaining the sessions of the in-vehicle wireless network. The detailed operations of the NDM have been explained in the subsequent sections of this paper.

In our earlier work, we proposed having an NDM in a vehicle for taking care of the security of intravehicle networks. In this paper, we have proposed to have another level of Network Device Monitors to take care of the security of intervehicle networks. This new level of NDMs should be embedded in towers that will be alongside the roads. These towers, known as the *Intelligent Transportation Towers*, are responsible for providing seamless secure communication between the vehicles on the roads. An ITT (Intelligent Transportation Tower) will function in the same way as a cell phone tower. Since cell phone towers cover almost all road systems, the contracts for maintaining ITTs can be given to the companies that maintain the cell phone towers. In that case, the same set of towers can serve both purposes: maintaining cell phone networks as well as providing secure communication mechanisms among intelligent vehicles.

SECURE VEHICULAR COMMUNICATIONS

In this section, we first present background material including our prior work on building secure in-vehicle network, and then we present our proposed technique for building secure inter-vehicle networks. After that we present system requirements for our proposed technique.

BACKGROUND MATERIALS

During the last several years, interest in using wireless communication technologies have grown significantly. Bluetooth features are becoming very common in cell phones, PDAs, laptops, etc. More and more homes are getting broadband connections and using Wi-Fi technology for in-home wireless networking. The automotive industry has also started introducing Bluetooth technologies to build in-vehicle wireless networks. The Bluetooth-enabled cell phone fitted in the 2003 Saab 9-3 car can access any other Bluetooth-enabled devices in the car, such as a PDA [6]. We hope, in the future, other automobile companies may also introduce wireless networking technologies in their vehicles. If intra and inter-vehicle wireless networks can be implemented at reasonable costs, then the government may also require all auto manufacturers to have wireless networking features in every vehicle, so that the vehicles can exchange messages among themselves (in real time) to issue pre-crash warnings. The vehicles must communicate among themselves through secure wireless links to avoid any disasters from hackers.

Securing any type of communication links involves three key requirements. First, the links must be protected from *eavesdropping*, so that unauthorized persons can't access private information. Second, the end users must be authenticated before anything is sent to or received from them. Third, the communication links must be protected from tampering by hackers.

If all information is transmitted in encrypted form, then that should protect the communication links from eavesdropping. If standard techniques, such as Cyclic Redundancy Check (CRC) [7], [8] Check Sum [8], Hamming Code [9], etc. are used with the information before it is encrypted, then that should protect the links from tampering. Yet the most vital task in establishing a secure communication link is authenticating the parties

on both sides of the communication link and exchanging their encryption keys.

In our earlier work, we developed a technique for securing an in-vehicle network [5]. A brief description of our earlier work is presented in the following subsection to make the readers familiar with the technique for developing a secure in-vehicle network. After that we present the technique for building inter-vehicle secure wireless networks.

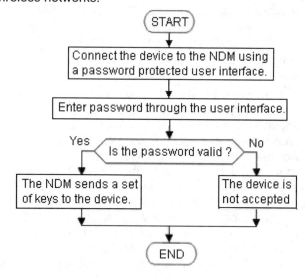

Figure 1: Registration process of a device to the NDM

A Secure In-Vehicle Network:

For in-vehicle devices to securely communicate with each other, they should use some secret keys. The length of the keys should be sufficiently large so that the keys are not susceptible to attacks such as brute force attacks. In order to build a secure in-vehicle network, we used a device called the Network Device Monitor (NDM). The NDM is responsible for distributing secret keys to all the in-vehicle devices. All devices to be used in the vehicle are initially registered to the NDM through a password protected user interface. The user interface could be a keypad, an infrared link or a very short-range (say, a few inches) wireless link. For every wireless device to be used in the vehicle, the password-protected human interface is used only once during the lifetime of the device. During the registration process of a device, the NDM sends a set of authentication keys to the device, and also keeps a copy of the keys in its own memory. Figure 1 shows the registration process of a device to the NDM. When a device wants to join an in-vehicle communication session, the NDM authenticates this device using one of its authentication keys. Different in-vehicle devices are given different sets of authentication keys, so that the keys of other devices can't be obtained by using a lost or stolen device. A lost or stolen device can also be deregistered from the NDM using the password-protected user interface. This deregistration process is necessary to protect the in-vehicle wireless network from unauthorized users who may have the lost or stolen device in their possession.

Assume that the authentication keys of a device are k_1, k_2, .. k_n, where, $n \geq 1$. Later on, when the device will try to join a wireless communication session, the NDM will use one of the keys of the device to authenticate the device. Different authentication keys will be used for different communication sessions of the device. Key k_i will be used during the ith session of the device after the device has been registered. During the nth session of the device, the NDM will send another set of authentication keys to the device, for the device to join future sessions.

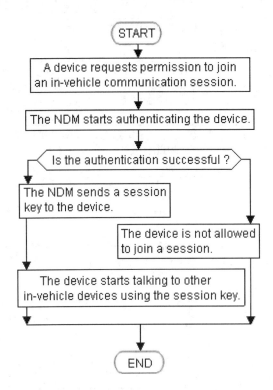

Figure 2: Authentication of a device by the NDM

The NDM sends a session key to a device, in encrypted form, after the device has been authenticated by the NDM. The device then uses this session key in order to communicate with other in-vehicle devices. Figure 2 shows the authentication process of a device by the NDM. During a particular communication session, all devices of the vehicle use the same session key in order to encrypt and decrypt the messages to be exchanged. From time to time the NDM changes the session key, so that the wireless devices of a vehicle don't communicate too long using the same session key. Changing the session key from time to time is necessary to protect the communication session from brute force attacks by hackers.

PROPOSED TECHNIQUES FOR BUILDING SECURE INTER-VEHICLE WIRELESS NETWORKS

In this section of the paper we have presented a technique for building secure inter-vehicle wireless networks. This technique requires that every vehicle must be equipped with a wireless device to communicate with the neighboring vehicles as well as

with the Intelligent Transportation Towers (ITTs). An ITT will be responsible for authenticating the vehicles when the vehicles come within the range of the ITT. If the contract for maintaining the ITTs is given to companies that maintain the cell phone towers, then the same tower can be used for providing service to the cell phone networks as well as to the intelligent transportation networks. An intelligent vehicle will have three different types of wireless links: 1) a link for the in-vehicle wireless network, 2) a link for inter-vehicle wireless networks, and 3) another link for the vehicle to ITT communications. Our goal is not to advocate a particular wireless technology to be used in the vehicles. Any wireless technology, such as Bluetooth, Wi-Fi, etc. can be used for communications. Bluetooth or any similar technology that allows devices to form ad hoc networks can be used for in-vehicle as well as for intervehicle wireless networks. A short-range (say 10-meter) technology can be used for in-vehicle networks, and a long-range (say a couple of hundred meters) technology can be used for intervehicle networks. The links for vehicle to ITT communications must have a longer range so that a vehicle will be able to communicate with at least its nearest ITT. Wi-Fi or Cell Phone technology can be used for the vehicle to ITT communications. The protocol for vehicle to ITT communications can be a client-server protocol, where the vehicles are the clients and the ITTs are the servers.

When a vehicle is manufactured a wireless device will be installed in it. After the wireless device has been installed, a set of keys will be given to it. A copy of these keys will also be kept in a secure central server. The secure central server can be maintained by any one of the following organizations: a) the manufacturer of the vehicle, b) the supplier which makes the wireless device, c) the company which is responsible for maintaining the intelligent transportation towers, or d) any other trusted organizations. Different vehicles will be given different sets of keys. These keys will be securely kept in the central server. If the secure central server of a particular vehicle is going to be maintained by the manufacturer of that vehicle, then every auto company will have its own central server. Authorized organizations will be able to access the keys of a vehicle by using the vehicle's ID (VID) number. If the central servers are maintained by the auto companies, then the VID of a vehicle will indicate which server to go to for accessing the keys.

Authentication Process:

Every ITT will be equipped with a device called the Network Device Monitor (NDM). When a vehicle enters a road, it will send its VID to the nearest ITT. The NDM of the ITT will then check its own memory to determine whether or not the keys of the vehicle are locally available. If the keys are not locally available, then the ITT will check to see whether or not the keys are available in the adjacent ITTs. If the keys are not available in the adjacent ITTs either, then the ITT will access the central server of the vehicle to access the keys. The ITT will then try to authenticate the vehicle using one of the keys. Based on a built-in protocol, both the vehicle and the central server know which key to use for the current authentication. If the keys were locally available in the NDM of the ITT, then the ITT also knows which key to use for authentication. After a vehicle is authenticated by an ITT, both the vehicle and the ITT will generate a common encryption key. After that, all communications between the vehicle and the ITT will take place in encrypted form. The ITT will use different encryption keys for different vehicles, so that the communication between a vehicle and an ITT remains private.

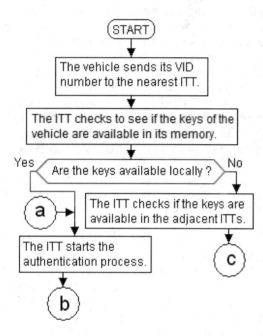

Figure 3a. Authentication of a vehicle and distribution of the session key.

After all authentication keys of a vehicle are used, the central server of the vehicle will issue a new set of authentication keys to the vehicle via the network of ITTs. This new set of keys will be used for future authentications.

A vehicle needs a session key in order to communicate with its neighboring vehicles. The session key is given to the vehicle by the ITT after the vehicle has been authenticated. The ITT sends the session key to the vehicle in encrypted form. The vehicle then talks to other neighboring vehicles using the session key.

If the authentication of the vehicle by the ITT fails, then the ITT will not send any session key to the vehicle. The ITT will warn all vehicles within its range about the presence of an unauthorized vehicle on the road. The ITT will also send a warning message to the highway patrol officers informing them about the presence of an unauthorized vehicle. Figures 3a, 3b and 3c show the entire authentication process of a vehicle by an ITT.

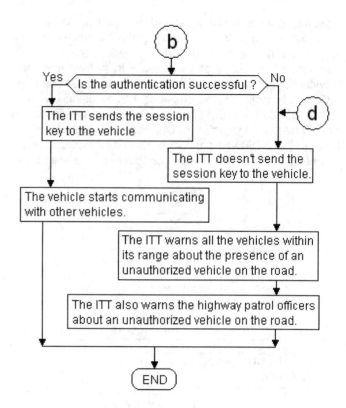

Figure 3b. Authentication of a vehicle and distribution of the session key (contd.).

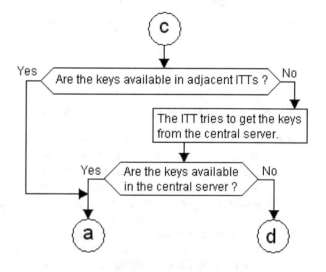

Figure 3c. Authentication of a vehicle and distribution of the session key (contd.).

Communication among vehicles:

Figure 4 shows a snapshot of a road with thirteen vehicles (V1 through V13) and two ITTs (ITT-1 and ITT-2). All thirteen vehicles are moving from the left to the right. Vehicles V1 through V7 are within the range of ITT-2, and vehicles V7 through V13 are within the range of ITT-1. The thirteen vehicles form three clusters. Vehicles V1 through V4 are in Cluster 1, vehicles V5 through V9 are in Cluster 2, and vehicles V10 through V13 are in Cluster 3. It is assumed that the clusters are far away from each other, so that the vehicles of one cluster don't need to form a network with the vehicles of the other clusters. The vehicles of Cluster-1 will use the session key of ITT-2 to communicate among themselves. The vehicles of Cluster-3 will use the session key of ITT-1 for communicating among themselves. The vehicles of Cluster-2 are in the range of both ITT-1 and ITT-2. Since vehicles V5, V6 and V7 are in the range of ITT-2, they will communicate using the session key of ITT-2. Similarly, since vehicles V7, V8

and V9 are in the range of ITT-1, they will communicate using the session key of ITT-1. Since vehicles V5 and V6 moved into the range of ITT-2 from the range of ITT-1 just a while ago, they still know the session key of ITT-1. Thus, V5 and V6 will communicate with V8 and V9 using the session key of ITT-1. From time to time, every ITT will change its session key in order to protect the inter-vehicle communications from Brute Force attacks.

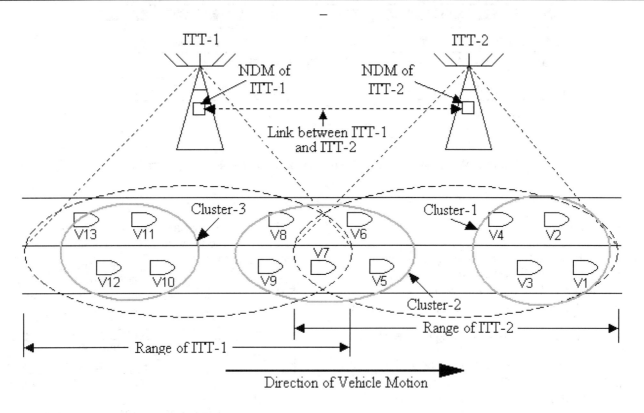

Figure 4. A road with two ITTs and three clusters of vehicles.

The communication between a pair (*vehicle, ITT*) is private. Hence, when a vehicle talks to an ITT, no other vehicles can understand that conversation. Thus, when a session key is sent from an ITT to a vehicle, it is sent in a secure mode. If the session key is long enough, and also if the ITT changes the session key from time to time, then it will be difficult for someone to crack the key using some kind of external hacking device (not a part of the vehicle electronics) . Unauthorized vehicles, such as stolen vehicles, unidentified vehicles, vehicles without valid registration, suspended vehicles (vehicles that have lost the privilege of being on the road due to prior violations), etc. will not be issued any session keys by the ITTs. In fact, an ITT can issue warning messages to all authorized vehicles and highway patrol officers within its range, informing them about the presence of an unauthorized vehicle within the neighborhood. The drivers of the authorized vehicles can then take extra precaution in driving their vehicles. If the authorized vehicles are equipped with *Automatic Collision Avoidance System* (ACAS), then these vehicles can automatically disable their ACASs and go back to

manual mode after receiving warnings about the presence of unauthorized vehicles on the road.

SYSTEM REQUIREMENTS FOR BUILDING SECURE INTER-VEHICLE WIRELESS NETWORKS.

In this section we present an estimate of memory and bandwidth requirements for an ITT. The memory and bandwidth requirements have been determined for the *worst-case scenarios*. In order to determine the memory requirement of an ITT, we assumed that the road system is completely jammed. This means that traffic is bumper-to-bumper. In order to determine the bandwidth requirement of an ITT, we assumed that a vehicle maintains a distance of *one second* (distance traveled in one second) from the vehicle in front of it. Note that, for safe driving, a distance of at least two seconds is recommended behind the vehicle in front of you, but we used a distance of one second to determine the *worst-case requirement*.

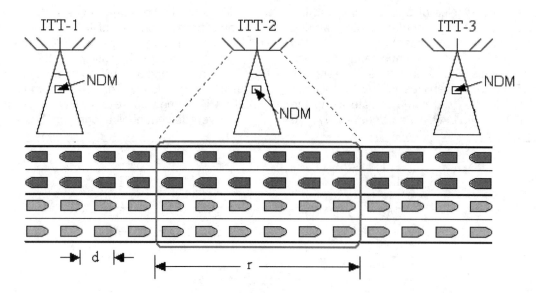

Figure 5. A 4-lane road with three ITTs and bumper-to-bumper traffic.

Memory Requirement:

Let's assume that each tower covers a distance of r miles of the road, each vehicle takes up a space of d feet, and there are k lanes on the road. Then the total number of vehicles within the range of an ITT on a road is $\dfrac{5280rk}{d}$. Let's assume that the length of each key is x bytes and an ITT must keep y number of keys per vehicle within its range. Then the size of memory required to keep the keys of the vehicles of a road is $\dfrac{5280rkxy}{d}$ bytes. If there are n such roads within the range of an ITT, then the total size of memory required to keep the keys of all the vehicles within the range of an ITT is

$$M = \frac{5280nrkxy}{d} \text{ bytes.} \qquad (1)$$

We have determined the size of memory assuming that the traffic on the road is bumper-to-bumper (*worst-case scenario*), as shown in Figure 5. For such a case, we assumed that the average space needed by a vehicle is 25 feet (d = 25). We also assumed that each key is 128 bits (16 bytes) long and at any given time four keys need to be kept for every vehicle within the range of the ITT. Table I shows a list of all the parameters used to determine the memory and bandwidth requirements.

For **rural driving,** an ITT may have to take care of only one or two highways. Also, the highway may not have more than four lanes. Table II shows the memory required by an ITT in rural areas. This table shows that the size of memory required by an ITT to keep the keys of the vehicles is in the order of a few hundred kilobytes, which is not a significant size.

Table I: A list of parameters used to determine the memory requirement of an ITT.

Parameter	Description
r	Range of an ITT, in miles, along a road.
d	Average space, in feet, needed by a car for bumper-to-bumper traffic.
k	Number of lanes per road.
x	Length of a key in bytes.
y	Number of keys to be kept per vehicle.
n	Number of roads within the range of an ITT.
m	Number of roads in every mile of a downtown area.
M	Size of memory needed by an ITT to keep the keys of all the vehicles within its range.
B	Bandwidth required from each ITT to maintain the secure inter-vehicle networks.

For **city driving**, especially in the downtown area of a big city, an ITT may have to take care of many roads, and each road may have many lanes. Assume that there are m roads per mile of a downtown area. If the range of an ITT is r miles along the East-West and North-South directions, then the ITT will have to take care of vehicles for $2mr$ number of roads as shown in Figure 6. Replacing the parameter n, used in Equation (1), by the term $2mr$ we get the following expression for the size of memory required by an ITT in a downtown area.

$$M = \frac{10560mr^{2}kxy}{d} \text{ bytes} \qquad (2)$$

For downtown areas like New York, Los Angeles, Chicago, etc, we can assume that there are approximately eight roads per mile (m =8) along both East-West and North-South directions. Table III shows the memory required by an ITT in the downtown area of a big city. This table shows that the size of memory required by an ITT to keep the keys of the vehicles is in the order of a few megabytes, which is not that significant either (by today's standard).

Table II: Memory required by an ITT in rural areas.
(For $d = 25$, $x = 16$ and $y = 4$)

Number of Roads (n)	Range of an ITT in miles (r)	Number of Lanes in the Road (k)	Memory Required in Kbytes (M)
1	2	2	52.8
1	2	4	105.6
1	3	2	79.2
1	3	4	158.4
2	2	2	105.6
2	2	4	211.2
2	3	2	158.4
2	3	4	316.8

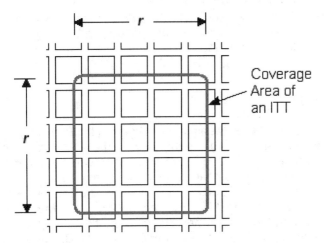

Figure 6. Coverage area of an ITT in a Downtown.

Table III: Memory required by an ITT in a big downtown area.
(For $d = 25$, $m = 8$, $x = 16$ and $y = 4$)

Range of an ITT in miles (r)	Number of Lanes in each Road (k)	Memory Required in Mega Bytes (M)
2	2	1.65
2	4	3.30
3	2	3.71
3	4	7.43

Note that in addition to the memory size shown in Tables II and III, some more memory is needed by each ITT to keep and run the program for maintaining secure intervehicle networks. The size of this additional memory is fixed and doesn't depend on the number of vehicles within the range of the ITT.

Bandwidth Requirement:

An ITT will have to communicate (in real time) with all the vehicles entering its range. The ITT will have to exchange several messages with every vehicle to authenticate the vehicle and then issue the session key. Let's assume that every message will contain about 256 bytes including the header, data and checksum, which is a reasonable assumption. Let's also assume that a total of six messages need to be exchanged between a vehicle and the ITT to go through the process of authentication and issuance of the session key. Thus, the ITT needs to exchange about 1.5 Kbytes of data with every vehicle entering its range. If the vehicles are keeping a one-second distance between them and the vehicles in front of them, then the total number of vehicles entering into the range of the ITT in every second will be equal to nk. Hence, the bandwidth necessary from an ITT to communicate with all the vehicles in its range is $1.5\,nk$ KB/sec (Kilobytes/sec). For a downtown area $n = 2mr$. Hence, the required bandwidth for a downtown area is $3mrk$ KB/sec. Some additional bandwidth is necessary from each ITT to receive the keys of the vehicles from the central server. The additional bandwidth needed from an ITT is equal to $nkxy$ bytes/sec and $2mrkxy$ bytes/sec for rural and downtown areas, respectively. Hence, the total bandwidth needed from an ITT can be expressed as

$$B = \left(1.5 + \frac{xy}{1024}\right)nk \text{ KB/sec} \quad \text{(for rural areas)} \quad (3)$$

and

$$B = \left(1.5 + \frac{xy}{1024}\right)2mrk \text{ KB/sec} \quad \text{(for a downtown)} \quad (4)$$

Table IV shows the bandwidth requirement from an ITT in a rural area. It is seen from Table IV that for two roads and four lanes per road, the bandwidth needed from an ITT is 12.5 Kbytes/sec, i.e. 100 Kbits/sec, which is not a significant requirement.

Table V shows the bandwidth requirement from an ITT in a large downtown area. It is seen from Table V that if the range of an ITT is 3 miles and if every road has four lanes, then the bandwidth needed from an ITT is 300 Kbytes/sec, i.e. 2.4 Mbits/sec, which can be obtained using Wi-Fi or similar technology.

Table IV: Bandwidth required from an ITT in a rural area.
(For $x = 16$ and $y = 4$)

Number of Roads (n)	Number of Lanes in the Road (k)	Bandwidth Required in Kbytes/sec
1	2	3.125
1	4	6.250
2	2	6.250
2	4	12.500

Table V: Bandwidth required from an ITT in a big downtown area.
(For $m = 8$, $x = 16$ and $y = 4$)

Range of an ITT in miles (r)	Number of Lanes in each Road (k)	Bandwidth Required in Kbytes/sec
2	2	100
2	4	200
3	2	150
3	4	300

From our above analysis, it becomes clear that using today's technology we can build ITTs with sufficient memory to keep the keys of all the vehicles within the range of the ITT. Today's technology will also allow us to build ITTs with sufficient bandwidth capabilities to support secure communications among intelligent vehicles.

CONCLUSION

Exchanging vehicles' dynamic information such as speed, acceleration, position, direction, etc. is necessary to build collision warning and avoidance systems. Secure intervehicle communication links are necessary to protect the intervehicle messages from tampering by hackers. In this paper, we presented a technique for maintaining secure communication links among intelligent vehicles on the road. We provided detailed descriptions of the key exchange mechanism for maintaining security. Intelligent Transportation Towers, like cell phone towers, are necessary to authenticate vehicles on the road. We investigated the feasibility of implementing Intelligent Transportation Towers (ITTs) using today's technology. From our analysis we found that by using today's technology it is possible to build ITTs with enough memory and bandwidth capabilities.

REFERENCES

1. P.L. Zador; S.A. Krawchuk; R.B. Voas, "Final Report -- Automotive Collision Avoidance System (ACAS) Program", Performed by Delphi-Delco Electronic Systems, Contract No: DTNH22-95-H-07162, Washington, DC, August 2000, DOT HS 809 080.

2. How Virtual Private Networks Work. http://computer.howstuffworks.com/vpn.htm.
3. Bluetooth Specification Version 1.0B, Vol 1, 1999.
4. Ronald miller and Q. Huang "An Adaptive Peer to Peer Collision Warning System, " Proceedings of IEEE Vehicular Technology Conference (Spring), Birmingham, Alabama, May 2002.
5. Syed Masud Mahmud and Shobhit Shanker, "Security of Wireless Networks in Intelligent Vehicle Systems," Proceedings of the 3rd Annual Intelligent Vehicle Systems Symp. of NDIA, National Automotive Center and Vectronics Tech., June 9 – 12, 2003, Traverse City, Michigan, pp 83 – 86.
6. "Top 10-Techno Cool Cars". IEEE Spectrum, February 2003, pp. 30-35.
7. P. L. Higginson, P. T. Kirstein: On the Computation of Cyclic Redundancy Checks by Program. The Computer Journal 16(1): 19-22 (1973).
8. Nirmal R. Saxena, Edward J. McCluskey: Analysis of Checksums, Extended-Precision Checksums, and Cyclic Redundancy Checks. IEEE Transactions on Computers 39(7): 969-975 (1990).
9. Kevin T. Phelps, Mike LeVan: Kernels of Nonlinear Hamming Codes. Designs, Codes and Cryptography 6(3): 247-257 (1995).

CONTACT

Dr. Syed Masud Mahmud has received his Ph.D. degree in Electrical Engineering from the University of Washington, Seattle, in 1984. He received his B.S. degree in Electrical Engineering from Bangladesh University of Engineering and Technology (BUET), Bangladesh, in 1978. He received a Gold Medal for securing the highest percentage of marks among all the recipients of B.S. degree from BUET in 1978. During his doctoral study at the University of Washington, he was one of the four recipients of Physio Control Fellowship. Currently he is an Associate Professor in the Department of Electrical and Computer Engineering of Wayne State University. He published over 60 technical papers in refereed journals and conference proceedings in the area of Vehicle Multiplexing, Hierarchical Multiprocessors, Cache Coherence Protocols, Interconnection Networks, Wireless Security, and Simulation Techniques. In 2002 he received the President's Teaching Excellence Award of Wayne State University. He has been listed in a number of Who's Who. (Email: smahmud@eng.wayne.edu, Phone: 313-577-3855)

2001-01-2534

Advanced Electronic Chassis Control Systems

Debbie Sallee and Ross Bannatyne
Motorola Transportation Systems Group

ABSTRACT

The objective of this paper is to examine the state-of-the-art technologies driving the development of advanced electronic chassis control systems. In addition, the paper provides an overview of how the future systems will function and reveals the benefits to the consumer by these advanced solutions.

Advanced electronic chassis control systems naturally segment into several categories including advanced electronic braking systems, advanced electronic steering systems, electronic suspension systems, and collision warning / avoidance systems. Together, these intelligent, automated, vehicle control systems offer the consumer a safer driving experience with a higher performing, more fuel efficient, and environmentally friendly automobile.

Key enabling technologies to support these systems include low-cost, robust 'failsafe' solutions, high performance processors capable of signal processing, a whole array of new sensor technology and a new high speed serial communications protocol which will be highly dependable. All of these new technologies will be discussed from a perspective of how they enable the electronic chassis control system.

INTRODUCTION

There are several electronically controlled chassis systems that enhance safety by optimizing the interface between the tires and the road surface, either in the longitudinal, lateral or vertical directions. Figure 1 illustrates the popular chassis control systems and the associated directional dynamics on which they act.

To optimize dynamic stability in the longitudinal direction, there are three popular systems: Antilock Braking Systems (ABS), Four wheel drive (4WD) and Traction control systems. Note that conventional 4WD systems typically use a transfer box with viscous coupling that operates when a difference in the speed of rotation between front and rear wheels occurs. The electronically controlled system is more efficient, as considerable slip is

not required before the 4WD operates and a better optimization of drive-line torsion, traction and braking capacity may be achieved.

In the vertical direction, Roll stabilization and Active-suspension systems may be implemented, although these are still in their infancy in terms of the actual number of vehicles that include such systems. A sensor to detect the Roll motion of the vehicle could also be employed in order to implement a rollover protection system, enabling hidden rollover bars in a convertible in the event of an accident. The same sensor could be used with the airbag system in order to determine when a roof airbag should be fired as well as the new inflator curtain type airbags which protect occupants from glass splinters and intrusion during a rollover crash.

FIGURE 1 – CHASSIS CONTROL SYSTEMS

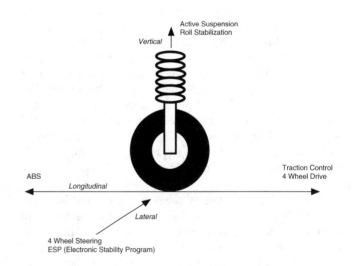

Lateral stability is the third directional dynamic factor in overall vehicle chassis control and safety. As well as Four wheel steering (4WS) that increases stability whilst cornering at high speed, new systems are being introduced to compensate for understeer or oversteer by the driver. The industry has settled on the name 'Electronic Stability Program' (ESP) for these systems after being referred to as many names such as

Integrated Vehicle Dynamics, Automatic Stability Management Systems, and others. ESP will be discussed later in this paper.

Taking the concepts of ESP slightly further, a fully integrated chassis control system would control the functions of suspension, steering and braking seamlessly and would require real-time information on all six degrees of freedom of the vehicle as well as information on the status of each systems control variables and a real-time communication link with other relevant systems such as Powertrain. It is normal today for the Traction Control System to communicate to the Powertrain system in order to adjust throttle angle whilst applying braking forces to achieve optimum traction. It is expected that one system will control the inter-operability of all of these related subsystems in the near future.

Antilock Braking Systems (ABS), Traction Control Systems (TCS) and Electronic Stability Programs (ESP) are all very closely related and can often be standardized on a single electronic control platform.

If a chipset solution is available, the development time associated with a product can be significantly reduced as the evaluation of many products that are required to interact can be avoided. A total chipset need not necessarily come from one semiconductor supplier, although it often does.

BRAKING SYSTEMS

The real value of a chipset is illustrated when it can be applied across a number of products that are closely related, as in the case of Anti-lock braking systems and vehicle control systems. Basic system requirements can be met with a chipset while interchangeability with pin-for-pin compatible variants allows upgradability to higher performance systems. The chipset is the basic platform on which to build. Braking and chassis control applications are a good example. Vehicle chassis control systems are often based on a single platform, but vary in features and functionality depending on the vehicle model in which it is implemented.

As the only real differences between the basic ABS, TCS and ESP systems is a relatively small amount of incremental hardware and software, a chipset approach to these solutions works very well. The software is written in a modular style (i.e. a software module for 'Wheel speed', 'Traction control', etc.) and the electronic components performance is determined with worst case system requirements in mind.

A block diagram representing a chipset solution for ABS / TCS / ESP is illustrated in Figure 2. There are four basic elements in every automotive electronic control unit - conditioning system inputs, conditioning system outputs, processing and 'housekeeping' functions such as the

maintenance of a stable power supply. Three basic semiconductor technologies are applied, HCMOS for the processing portion, Analog ASICs for the input / output conditioning and housekeeping and some PowerFETs for driving power stages, in this case switching the hydraulic pump motor which can be peak rated at over 100A. With these three basic technologies, infinite partitioning options are possible.

FIGURE 2 – BRAKING / CHASSIS ELECTRONIC CONTROL SYSTEMS

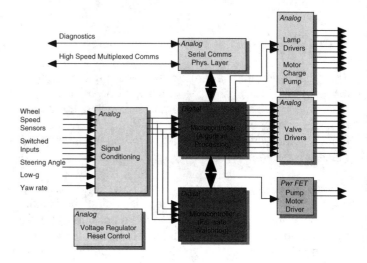

The Input portion is provided to translate all the analog input signals to clean digital waveforms that can be applied directly to the microcontroller i/o. The diagram illustrates all of these sensor inputs grouped together in a single device. Although a single input conditioning device is possible, it is seldom implemented as such. This device, because of the interfacing for steering angle, low-g and yaw rate sensors would only be required in the ESP system and would probably not be cost effective in the basic ABS system (this would of course depend on ultimately how many units of ABS v ESP systems were planned). For this reason, usually at least two interface devices are specified, the second of which is added to the basic chipset when an ESP system is being developed.

There are usually two processing elements in the processing portion of the circuit. These are required as a 'failsafe' system is required. The purpose of the failsafe system is to ensure that any faults in the electrical / electronic system are self diagnosed and result in the system switching itself off safely, leaving the conventional hydraulic brakes fully functional ('pressure increase mode'), with the absence only of the ABS control. Theoretically, a single microcontroller can observe and check each part of the system with the exception of its own sanity. Hence a second, failsafe, microcontroller is used to observe the operation of the master microcontroller.

The output conditioning portion of the electronic control system is, like the input portion, implemented in analog based technology. Analog ASIC technology allows basic logic to be incorporated onto these devices to enhance the performance. This 'smart' functionality is used for diagnostics and to enhance failsafe operation. For example if a short or open circuit is detected at an output driver, the ASIC device could communicate such a status to the microcontroller in order to ensure that the system remains in a safe operating mode.

ADVANCED TECHNOLOGIES FOR BRAKING SYSTEMS

The M68HC12 microcontroller architecture has been used widely in electronic braking systems. The M68HC12 was also developed specifically for real-time embedded control applications and has custom specific features that have been developed especially for ABS applications. One such feature is the Enhanced Capture Timer (ECT) that has been implemented on the M68HC12BE32, M68HC12D60 and M68HC12DG128.

The Enhanced Capture Timer consists of a 16-bit, software programmable counter driven by a prescaler. The timer can be used for many purposes including input waveform measurements while simultaneously generating an output waveform. There are 8 input capture / output compare channels, four of which include a buffer called a holding register. This allows two different timer values to be 'memorized' without the generation of an interrupt. Four pulse accumulators are associated with the four buffered channels in order to count pulses during a time specified by a 16-bit modulus counter.

The purpose of using the ECT in a braking application is to offload timer interrupts that are associated with the wheel speed sensors. The input frequency of the pulses is from around 6Hz per Km/h to a maximum of 6Khz for 250Km/h. This would provide an interrupt, on average, around every 160uS. The ECT will count pulses from each wheel (using four timer channels) in order to generate the information required to calculate wheel speed every control cycle loop - after the necessary information has been saved in the input capture register, holding register and pulse accumulator, only one interrupt will be generated per cycle.

Figure 3 illustrates the block diagram of the ECT Timer. Data is latched into the Input Capture Holding Register and latched into the Pulse Accumulator Holding Register. At the end of a control loop cycle, all the relevant information required to calculate wheel speed during that cycle is available to be read directly from the CPU in these registers. The wheel speed can be calculated if the time of the first and last pulse is known along with the number of pulses acquired during the cycle.

As braking and steering systems become more complex and future systems become networked together to provide additional safety features, the performance requirements of the central algorithmic processor will see a significant increase in required capabilities.

FIGURE 3 – ECT BLOCK DIAGRAM

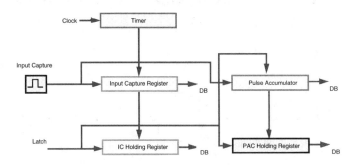

STEERING SYSTEMS

Although it could be argued that the steering system is not, strictly speaking, a safety system, there is no argument that it is a safety-critical system, and as such requires carefully implemented electronic controls. The steering system will, in the future, be closely integrated with other chassis control functions such as braking and suspension controls to form an overall chassis control system.

There is currently a trend towards implementing direct-assist electric motor steering systems from the more conventional electro-hydraulic power steering systems. In the event of a system fault, the direct-assist system requires additional safety considerations to ensure that the driver must always overcome any motor torque required to retain control of the steering. In this respect, unlike the ABS system, the direct-assist system must be fault-tolerant - in the event of a failure, steerability must be maintained. For this reason, additional protective elements are designed into the controller (typically smart diagnostics). Both systems are shown in Figure 4.

Both systems have similar controller architectures that include MCUs and power stages, although the requirements of theses components will differ depending on the motor type and its associated control strategy. A simple PM (permanent magnet) DC motor would typically by controlled by an average performance 8-bit MCU such as a M68HC11.

ADVANCED TECHNOLOGIES FOR STEERING SYSTEMS

As chassis control for steering systems becomes more complex, additional fast math capability may be required in addition to the micro-controller functionality. The algorithmic controller must now provide capabilities such as control oriented instructions, higher code density, and easy programming capability utilizing high level languages. Some designers have considered DSP's to provide this capability but are finding that more integrated devices which provide both digital signal processing capabilities as well as micro-controller capabilities may be a better fit.

FIGURE 4 – ELECTRONIC POWER STEERING SYSTEMS

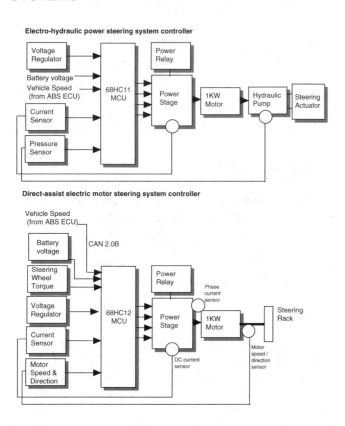

One such device is the automotive version of Motorola's Hawk processor that was specifically developed to address the needs of future steering applications as well as a variety of automotive motor control applications. Figure 5 is a block diagram of the Hawk processor which combines the microcontroller unit (MCU) ease of use with the speed of a DSP to deliver the best of both worlds. A Harvard style architecture has been implemented which uses both a load / store bus (for data) and an instruction bus (for instructions) providing three 16-bit data addresses buses and three 16 bit address busses. Most microcontrollers use a Von Neumann style architecture implementing a single bus for both data and instructions. The Harvard architecture is more powerful as it supports parallelism in fetching data and instructions.

SUSPENSION SYSTEMS

Although electronically controlled vehicle suspension systems can optimize road holding, handling / stability and ride quality, there has been a disappointing amount of progress in bringing such systems to the market at a price which is considered affordable for the average consumer. One of the main reasons for this, besides cost, is that suspension characteristics must be modified dynamically. The electronic control unit could certainly execute such a complex algorithm in real-time without issue, but in order to actuate the suspension elements (hydraulic, pneumatic or electromechanical), a significant amount of power is required. It is expected that the trend toward higher voltage capability in the vehicle (42 volts) will help facilitate mainstream active suspension systems.

FIGURE 5 – 'HAWK' MCU / DSP HYBRID PROCESSOR

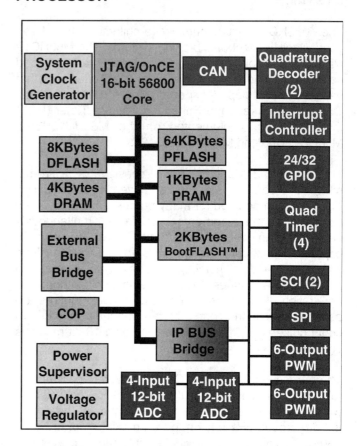

Figure 6 illustrates an example of a block diagram of a basic active hydraulic suspension system. Although the control loop cycle time will depend on the actuation time of the hydraulic pressure valves, there are many inputs which must be evaluated during the control cycle. For this reason, a very high performance microcontroller at the heart of the system will be required. Each wheel unit will require a g-sensor input, vehicle height input and information will also be required on the braking, acceleration and steering behavior of the vehicle. Several proposals have been made to simplify the control algorithm by using a fuzzy logic based approach. Fuzzy logic is a useful technique when a very complex and difficult to define algorithm needs to be implemented. By defining the control operation using fuzzy logic, it is possible to solve the problem much more intuitively and the exploit the tolerance for imprecision in the result.

ADVANCED TECHNOLOGIES FOR SUSPENSION SYSTEMS

With more demanding, high performance requirements for the next generation active suspension control, devices such as the MPC555 microcontroller are being selected by automotive designers. A block diagram of the MPC555 is shown in Figure 7. This diagram illustrates the CPU, various peripherals attached to the IMB3 bus, the flash EEPROM and RAM memory arrays and the integration module which contains all control / arbitration functions as well as the interface to the outside world. This has been integrated with a number of very complex peripherals and a significant amount of memory, to form a high performance automotive microcontroller. Although the throughput of the CPU (that includes a floating point unit) is very high, most of the peripherals are intelligent, performing many operations with minimal or zero CPU intervention.

FIGURE 6 – BASIC HYDRAULIC ACTIVE SUSPENSION SYSTEM

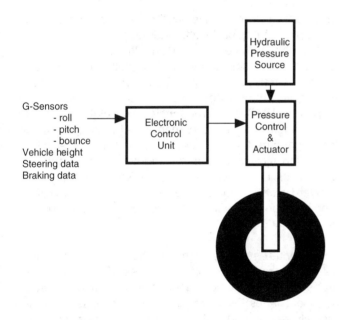

In order to address the suspension control application, the microcontroller has certain characteristics to ensure that it is robust and will operate in harsh electrically noisy environments, such as under the hood of a vehicle. One such characteristic is the operating temperature range of up to 125 degrees centigrade. Another is the dual power supply configuration. Although the processing feature sizes dictate that the CPU must operate with a nominal 3.3v supply voltage, a 5v i/o system is provided to ensure that the chip can interface easily with its neighboring system devices.

FIGURE 7 – MPC555 BLOCK DIAGRAM

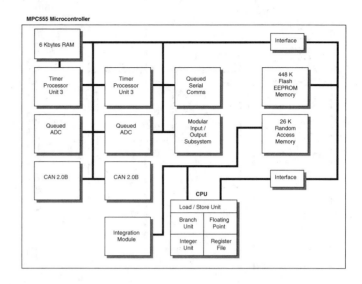

The MPC555 implements a 32-bit effective address, integer data types of 8, 16 and 32-bits, as well as floating point data types of 32 and 64-bits. The CPU also includes four execution units; an integer unit, a load / store unit, a branch processing unit and the floating point unit. A Harvard style architecture has been implemented which uses both a load / store bus (for data) and an instruction bus (for instructions). Most microcontrollers use a Von Neumann style architecture which implements a single bus for both data and instructions. The Harvard architecture is physically larger and more complex, but is more powerful as it supports parallelism in fetching data and instructions.

The MPC555 includes some very high performance peripherals. One such peripheral is the Timer Processor Unit 3 (TPU3). The TPU3 is an on-board co-processor that has been developed for timing control functions. Operating simultaneously with the main CPU of the MPC555, the TPU3 processes instructions, schedules / processes real-time hardware events, performs input / output and accesses shared data without CPU intervention. This results in a higher CPU throughput as setup and service times for timer events are minimized or eliminated. A block diagram of the TPU3 is given in Figure 8.

There are two TPU3 peripheral modules on the MPC555. Each TPU3 provides 16 independent programmable channels and pins. Any of these channels can be programmed to perform any timing function and each channel can be synchronized to one of two timer count registers with programmable prescalers. Many timing functions have already been written for the TPU and these functions are available from the supplier.

FIGURE 8 – TIMER PROCESOR UNIT 3

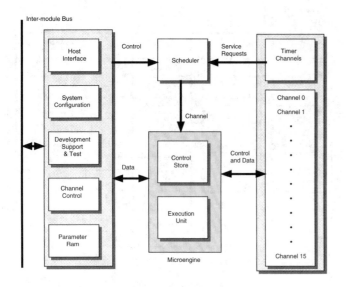

Another approach to adding 'intelligence' to the peripherals in order to reduce the amount of interaction they require from the CPU is to add a 'Queue'. The queue is an array of RAM that is used as an automated buffer by storing data that is used by the peripheral module. The queue is used with the Analog-to-Digital Converter modules (QADC) and the Serial Multi-Channel Module (QSMCM). There are two QADC modules, each with 16 independent channels. The resolution is 10-bit and the typical conversion time is 10uS. The queue can be used to automatically store results until the buffer is full. This offloads the number of interrupts which the main microcontroller CPU is subject to. In the case of the QSMCM, the queue can be used to preprogram transfers, again having the effect of reducing overhead. The QSMCM includes two asynchronous serial communications channels as well as one synchronous channel.

Flash EEPROM has been provided on the MPC555 as the program memory. There is a total of 448 Kbytes as well as another 64 Kbytes associated with the peripheral modules and control registers. In addition, there are 26 Kbytes of RAM. Manufacturers of various products (including automobiles) often wish to revise software in the field. Unless there is a method of reprogramming the memory array remotely, typically this problem requires the Electronic Control Unit to be removed and replaced. This is a time-consuming and expensive process for the Original Equipment Manufacturer, not to mention a greater inconvenience to the owner. Flash EEPROM provides the technology that allows such field revisions.

COLLISION WARNING / AVOIDANCE SYSTEMS

Although collision warning and avoidance systems may still be regarded as being in their infancy in terms of vehicle penetration, their perceived value in enhancing safety and reducing accidents is high. As the driving public ages, they expect to continue to be able to drive safely, whilst their reaction time is increasing and other senses like sight and hearing are diminishing. Work on radar systems for vehicles has been taking place since the 1950s and has been well documented - however most of this work did not lead to practical and economically viable products. This was mainly on account of the limited existing electronic controls on vehicles that could be re-used, and the high cost of enabling technologies such as semiconductors (mainly analog based at the time) and radar systems.

There are two categories of systems available or under development; passive collision warning systems and active collision avoidance systems. The passive system will detect a hazard and alert the driver to risks, whereas the active system will detect the hazard and take preventative action to avoid collision if possible. Both types of systems require object detection, the main difference in them being how a collision diverting event is actuated following object detection - by the driver, or automatically.

Both active and passive systems operate on the same principles of object detection although the active system will control throttle, braking and in the future, steering systems in order to avoid front collisions. There are several different techniques used for obstacle detection, the main approaches being a scanning laser radar sensor, FMCW (Frequency Modulated Constant Wavelength), or else a camera used in conjunction with an algorithm that will detect hazardous objects. This detection system is usually mounted at the front of the host vehicle in order to detect objects in the vehicles forward path. Other techniques may involve a combination of different sensors, including 'back-up' sensors.

For frontal systems, long range and large azimuth resolution radar is required because of the high forward speed of vehicle and the need to determine objects in adjacent lanes. The forward range of these systems is usually about 100 - 200 m ; this distance would give around 3 - 6 seconds warning of a stationary hazard when the host vehicle is traveling at 100 km/h. It is important that frontal systems can distinguish when there is more than vehicle in front positioned very closely but in different lanes. Frontal radar require a higher frequency of operation (and thus shorter wavelength) than rear systems as better azimuth resolution is obtained at higher frequencies.

Note that some manufacturers prefer to call the frontal active system 'Autonomous', 'Intelligent', 'Active' or 'Adaptive' cruise control. These systems are a subset of a collision avoidance system and unlike conventional cruise control, will adapt to the speed of slower vehicles ahead automatically. Only a frontal radar system is required for these systems, although it will become more common to implement object detection sensors at the sides and rear of the vehicle in the near future.

A key difference between the object detection system used in active and passive systems is that the active system will require more accurate object recognition, so as to discriminate against collision avoidance maneuvers against objects such as road signs.

Basic object detection is relatively straightforward. The most challenging problem is determining if an object is potentially hazardous, whilst traveling at speed and where many objects are present. It may be possible to detect all obstacles, but if a warning is given to the driver under these circumstances, there will be false alarms that will be is irritating to the driver and a 'cry wolf' phenomenon may result, that will defeat the purpose of having a warning. In the case of a collision avoidance system, automatic braking on account of a false alarm is likely to be dangerous.

The most popular technology to implement the frontal collision warning system today is the scanning pulse-based radar. The principle of operation of the scanning pulse-based radar is very straightforward; the time of flight of a pulse is measured which is proportional to range.The scanning radar transmits an pulse of light in a horizontal line, back and forth (hence 'scanning'). Distance is calculated easily as the microcontroller timer can determine the time interval from the transmitted pulse and the received pulse. After each pulse transmission, the receiver looks for an echo pulse, hence the transmission is not continuous. The pulse radar is often referred to as Laser Radar as a pulse laser diode is used as the emitting device.

As the transmission frequency is phase coherent from pulse to pulse, it is also possible to measure Doppler shift of the target; the Doppler shift can yield motion, speed and direction. Doppler shift occurs as the frequency of the waves are shifted relative to the receiver as the source of the waves gets closer or moves further away. Consider a police car traveling with its siren on. As it approaches, the sirens pitch becomes higher, then it gets lower as the vehicle moves away. This change in pitch occurs because the there is a shift in the frequency of the sound waves. This phenomenon allows us to determine motion - whether an object is traveling towards a receiver, or further away. By measuring the rate of change of the frequency shift, speed can also be determined.

A simplified control circuit for this type of system is shown in Figure 9. The circuit consists of a microcontroller which executes the control algorithm and generates output signals to control the laser diode. The laser diode signal is reflected via a system of mirrors and lenses (indicated by the 'optics' block) which is controlled by a stepper motor. The motor is used to step through different

positions and allows the beam to be deflected horizontally in a scanning motion.

In each position, the beam will be reflected back to a complementary positioned mirror, through a laser diode and back to the microcontroller. The microcontroller will then calculate distance. The time value is measured using a counter (integrated with the microcontroller) which is enabled when the pulse is transmitted and read when the input is received from the signal amplifier. Because the speed of light is 3×10^8 m/s, the microcontroller clock speed must be reasonably high in order to measure distance with acceptable resolution.

A photo diode is also used in the system to determine if that the optical port of the system is clean and free from debris. If the port glass is dirty, the laser beam pulse may be scattered and performance can be affected. Bad weather can also affect performance although this may be overcome by increasing the output power of the laser pulse. The addition of the photo diode to detect clarity of the optical port is necessary, as this type of frontal scanning pulse radar system is likely to be mounted at the front of a vehicle where debris and dirt from the road is commonplace. One solution is a 'wiper' system for the optical port.

FIGURE 9 – SIMPLIFIED SCANNING RADAR CONTROL CIRCUIT

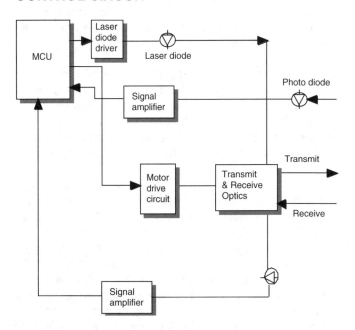

FUTURE SYSTEM DEVELOPMENTS : BRAKE-BY-WIRE

There is a high expectation that today's standard braking system which uses hydraulic fluid will be replaced, in the near future, by fully electrical systems. Although there are still some challenges to be overcome to realize this vision, the expected advantages of such a system over a conventional hydraulic system are such that the motivation to develop 'Brake-by-Wire' is strong. Some of the factors that have been identified as advantageous are as follows;

-   No brake fluid. Ecologically friendly and reduced maintenance.
-   Lighter weight
-   Fuel economy (pad clearance control)
-   Increased performance (brakes respond more quickly)
-   Minimized brake wear (more control of friction material application)
-   More simplistic / faster assembly and testing (modular structure)
-   More robust electrical interfacing.
-   No mechanical linkages through bulkhead (enhanced safety)
-   Electronic architecture is more easily upgradable
-   Consistent characteristics of pedal, constant travel
-   Significantly fewer parts than a hydraulic-based system

There are several issues which are being addressed in order to realize 'Brake-by-Wire'. One issue is the actuation energy required for braking. A disc brake requires approximately 1000 watts of actuation energy and a drum brake requires around 100 watts. Although the 12v-based vehicular electrical system does not easily support the high power requirements required for electrical brake actuation, future developments to higher voltage supplies on the vehicle will facilitate this.

Another major challenge facing the brake-by-wire system is the requirement for fault-tolerance. In systems where the hydraulics have been completely removed, no independent back-up actuation system exists - this means that rather than employ systems which fail-safely, a fault-tolerant (or 'fail-operational') system is required.

Although many clever techniques can be employed to enhance the safety of fault-tolerant systems, the underlying approach is to provide redundancy. If nodes or ECU's fail, back-ups must exist that come on-line without destroying existing system integrity. The degree of fault-tolerance and where it is employed is likely to differ from application to application, but it could reasonably be expected that important sensors and controllers are replicated and that certain components need not be. In addition to the obvious requirement for redundancy, the serial communications between each of the nodes in the system must also support fault-tolerance.

Each wheel would likely have a motor controlled actuator with an associated control circuit. If an individual wheel

unit was to fail, the vehicle could still be braked to a stop using the remaining three wheel units (assuming that the control algorithm was intelligent enough to apply to correct brake forces in the correct sequence to maintain lateral stability). As a failure of this type would not be catastrophic, the wheel units would not be required to be replicated. However if the brake pedal position sensor was to fail, this could be catastrophic, so it is likely that a redundant sensor would be added to this part of the system.

The facilitating technologies required before brake-by-wire becomes a mainstream solution are thus an enhanced vehicular power management system and a cost-effective fault-tolerant serial communications architecture. It will also be a major market discontinuity for the industry to migrate to an entirely electrical solution from a mainly mechanical / hydraulic solution.

From a semiconductor standpoint, the enabling technologies for conventional electronically controlled braking systems and brake-by-wire systems are very similar. The major differences are the employment of fault-tolerant technology and motor control technologies. As fault-tolerance is a semiconductor process independent concept, this implementation will present no significant challenge. The implementation of motor controlled actuators is likely to drive cost-effective semiconductor technology to a higher temperature capability in order to withstand the temperatures generated in the vicinity of the braking actuator. Figure 10 illustrates a possible configuration of a brake-by-wire system.

FIGURE 10 – BRAKE-BY-WIRE CONFIGURATION

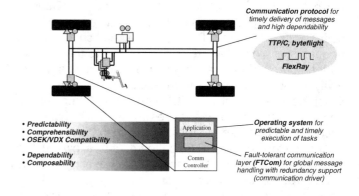

Figure 11 illustrates a photograph of a demonstration system that provides the viewer with an interactive experience to convey the system concepts and benefits of a brake-by-wire system.

FIGURE 11 – BRAKE-BY-WIRE DEMONSTRATION SYSTEM

CONCLUSION

As the automotive industry looks to the future for more advanced, more integrated, and safer chassis systems for the vehicle, a variety of new systems requirements and semiconductor technologies will be combined to meet the need. Systems designers of tomorrow's systems must ensure not just fail-safe operation but fully fault-tolerant operation. Semiconductors will meet this need with a variety of algorithmic processors, offering various performance levels, both MCU, DSP, and in some cases both processing capabilities on one device as well as specific automotive functionality to enable new features. Chassis systems of the future will bring braking and steering systems together with advanced communications technologies. The primary motivation will most certainly be to improve driver safety. However, the benefits will extend well beyond safety to include a myriad of new functionality enabling a cleaner environment, increased fuel economy, and greater reliability.

2000-01-1354

Estimation of Crash Injury Severity Reduction for Intelligent Vehicle Safety Systems

Wassim G. Najm, Marco P. daSilva and Christopher J. Wiacek
Volpe National Transportation Systems Center

ABSTRACT

A novel methodology is presented to estimate the safety benefits of intelligent vehicle safety systems in terms of reductions in the number of collisions and the number and severity of crash-related injuries. In addition, mathematical models and statistics are provided to support the estimation of the crash injury reduction factor in rear-end, lane change, and single vehicle roadway departure collisions. Simple models based on Newtonian mechanics are proposed to derive Δv, the change in speed that a vehicle undergoes as a consequence of crashing. Statistics on the distribution of vehicle types and weights in the United States are supplied, which are needed for Δv estimation. Moreover, mathematical equations are derived to estimate the average harm per collision. Finally, statistics on the average harm per occupant are obtained from the 1994 and 1995 Crashworthiness Data System crash databases.

INTRODUCTION

Intelligent vehicle safety systems (IVSSs), such as rear-end collision warning systems, have the potential to increase the safety of motor vehicles. The safety benefits of IVSSs can be measured by the number of collisions that might be avoided and by the number and severity of crash-related injuries that might be alleviated if drivers were assisted by such systems. An engineering-based methodology was recently developed to estimate system effectiveness and predict potential safety benefits in terms of the number of collisions that might be avoided with the assistance of an IVSS [1]. This methodology utilizes as input non-crash data collected from driving simulator experiments, test track studies, and field operational tests. Crash data are not available from field operational tests since they are usually limited to few instrumented vehicles operating for a relatively short period of time. The reduction in the number of collisions is projected based on driver and vehicle performance in safety-critical driving conflicts *with* and *without* the assistance of an IVSS. A safety-critical driving conflict leads to a collision if the driver does not attempt any avoidance action.

This paper describes a novel methodology to assess the impact of IVSSs on the number and severity of crash-related injuries. This methodology estimates an injury reduction factor in both the number of injured persons due to collisions that might be avoided and the severity of injuries due to lower-impact collisions that might not be avoided if drivers were assisted by an IVSS. In addition, this paper provides fundamental models and statistics on crash severity in support of this novel methodology, which address rear-end, lane change, and single vehicle roadway departure (SVRD) crash types. Simple mathematical models are described based on fundamental physics, which estimate Delta v (Δv) in each of the three crash types. The parameter Δv, a preferred measure of the severity of a collision, denotes the change in speed that a vehicle undergoes as a consequence of crashing. Crash statistics are utilized to correlate the values of Δv to the injuries suffered in each of the three crash types. The *Crashworthiness Data System* (CDS) crash database provides such statistics [2]. The CDS contains detailed data on a representative, random sample of thousands of minor, serious, and fatal crashes. There are 24 field research teams that study about 5,000 crashes a year involving passenger cars, light trucks, vans, and utility vehicles. These researchers perform a site investigation on the crash scene to provide a detailed description of the collision, and follow up by interviewing the crash victims and review medical records to determine the nature and severity of their injuries.

First, this paper describes the fundamental equations of a novel methodology that estimates the impact of IVSSs on the number of collisions and the number and severity of crash-related injuries. Next, mathematical models and relevant data are provided in support of estimating the injury reduction factor. This is followed by mathematical models that derive Δv from vehicle speeds at impact in rear-end, lane change, and SVRD collisions. After, the distribution of vehicle types and weights in the United States (US) vehicle fleet is supplied. The 1994 and 1995 CDS crash statistics are then presented to show the relationship between injury severity and Δv for a range of speed bins in each of the three collision types. Finally, this paper concludes with a summary.

SAFETY BENEFITS ESTIMATION METHODOLOGY

Intelligent vehicle safety systems (IVSSs) have the potential to reduce the number of motor vehicle collision and to alleviate the number and severity of crash-related injuries. The number of collisions that might be avoided with the assistance of an IVSS is estimated as follows:

$$B = N_{wo} \times SE \tag{1}$$

N_{wo}: Number of relevant collisions *without* the assistance of an IVSS.

SE: Total IVSS effectiveness in mitigating relevant collisions.

Relevant collisions refer to vehicular crashes that the IVSS is designed to address. Simple queries of crash databases, such as the *General Estimates System* (GES) database, yield values for N_{wo}. The parameter SE is defined by the following equation:

$$SE = \sum_{i=1}^{n} P_{wo}(S_i) \times E(S_i) \tag{2}$$

n: Total number of distinct safety-critical driving conflicts leading to N_{wo}.

S_i: A distinct safety-critical driving conflict.

$P_{wo}(S_i)$: Ratio of relevant collisions preceded by S_i over all relevant collisions *without* the assistance of an IVSS.

$E(S_i)$: IVSS effectiveness in mitigating relevant collisions preceded by S_i.

A set of five pre-crash variables in the GES and CDS crash databases provides a basis for the identification of distinct safety-critical driving conflicts (pre-crash scenarios) leading to different crash types. This set of variables outlines the sequence of events leading from a normal driving scenario into a collision. The following equation determines the value of $P_{wo}(S_i)$:

$$P_{wo}(S_i) = \frac{N_{wo}(S_i)}{N_{wo}} \tag{3}$$

$N_{wo}(S_i)$: Number of relevant collisions preceded by S_i *without* the assistance of an IVSS.

Values for $N_{wo}(S_i)$ can be obtained from the GES [3]. The parameter $E(S_i)$ is estimated by:

$$E(S_i) = 1 - \frac{p_w(S_i) \times \overline{M}_w(S_i)}{P_{wo}(S_i) \times \overline{M}_{wo}(S_i)} \tag{4}$$

$p_w(S_i)$: Probability of a collision when encountering a safety-critical driving conflict S_i *with* the assistance of an IVSS.

$\overline{M}_w(S_i)$: Number of encounters with a safety-critical driving conflict S_i per vehicle-miles traveled (VMT) *with* the assistance of an IVSS.

$p_{wo}(S_i)$: Probability of a collision when encountering a safety-critical driving conflict S_i *without* the assistance of an IVSS.

$\overline{M}_{wo}(S_i)$: Number of encounters with a safety-critical driving conflict S_i per VMT *without* the assistance of an IVSS.

Instrumented vehicles in field operational tests collect data that can be used to estimate the values of $\overline{M}_w(S_i)$ and $\overline{M}_{wo}(S_i)$. An estimate of $p_{wo}(S_i)$ is computed as follows:

$$p_{wo}(S_i) = \frac{\overline{N}_{wo}(S_i)}{\overline{M}_{wo}(S_i)} \tag{5}$$

$\overline{N}_{wo}(S_i)$ denotes the number of collisions preceded by a safety-critical driving conflict S_i per VMT *without* the assistance of an IVSS. This parameter can be estimated using driving exposure statistics and crash data from national crash databases. Consequently, the estimation of $E(S_i)$ solely relies on estimating $p_w(S_i)$. Potential estimation methods include Monte Carlo computer simulation [4] [5], principles of reliability theory [4], application of "extreme value" theory [6], or other suitable techniques.

The number and severity of crash-related injuries that might be alleviated with the assistance of an IVSS is measured by an injury reduction factor, R, defined as follows:

$$R = 1 - \frac{H_w}{H_{wo}} \tag{6}$$

H_w: Total harm caused by crash injuries *with* the assistance of an IVSS.

H_{wo}: Total harm caused by crash injuries *without* the assistance of an IVSS.

Generally, the "harm" is defined as the sum of injuries of crash victims, with each injury weighted in proportion to the economic cost of the outcome of such injury whether fatal or not. The total harm, H, can be formulated as:

$$H = N \times \overline{H} \tag{7}$$

N: Number of collisions.

$\overline{H}$: Average harm per collision.

Using above equations, the injury reduction factor, R, is expressed as follows:

$$R = 1 - \frac{\sum_{i=1}^{n} P_{wo}(S_i) \times \{1 - E(S_i)\} \times \sum_{k=1}^{\ell} P_w(S_i, \Delta v_k) \times \overline{H}(\Delta v_k)}{\sum_{i=1}^{n} P_{wo}(S_i) \times \sum_{k=1}^{\ell} P_{wo}(S_i, \Delta v_k) \times \overline{H}(\Delta v_k)} \tag{8}$$

ℓ: Range of Δv values.

Δv_k: Crash severity measure, denoting the change in speed bin k that a vehicle undergoes as a consequence of crashing.

$\overline{H}(\Delta v_k)$: Average harm per collision resulting in Δv_k for a specific manner of collision (e.g., rear-end impact).

$$P_w(S_i, \Delta v_k) = \frac{N_w(S_i, \Delta v_k)}{N_w(S_i)} \qquad (9)$$

$N_w(S_i, \Delta v_k)$: Number of relevant collisions that were preceded by a safety-critical driving conflict S_i and resulted in Δv_k *with* the assistance of an IVSS.

$N_w(S_i)$: Number of relevant collisions that were preceded by a safety-critical driving conflict S_i *with* the assistance of an IVSS.

$$P_{wo}(S_i, \Delta v_k) = \frac{N_{wo}(S_i, \Delta v_k)}{N_{wo}(S_i)} \qquad (10)$$

$N_{wo}(S_i, \Delta v_k)$: Number of relevant collisions that were preceded by a safety-critical driving conflict S_i and resulted in Δv_k *without* the assistance of an IVSS.

$N_{wo}(S_i)$: Number of relevant collisions that were preceded by a safety-critical driving conflict S_i *without* the assistance of an IVSS.

Equation (8) assumes that the vehicle crashworthiness (e.g., crash protection offered by vehicles), distribution of vehicle weights, and vehicle occupancy remain the same *with* and *without* the assistance of an IVSS. Therefore, the reduction of injury severity would occur due to lower closing speeds at impact (smaller Δv) if drivers were assisted by an IVSS. The values of $\overline{H}(\Delta v_k)$ can be derived from available crash databases. The ratios $P_w(S_i, \Delta v_k)$ and $P_{wo}(S_i, \Delta v_k)$ are possible to obtain when exercising Monte Carlo simulation to estimate the probabilities of a crash in safety-critical driving conflict S_i *with* and *without* the assistance of an IVSS. These simulations yield a number of collisions along with vehicle speeds at impact. These speeds can then be converted to values of Δv using simple models. An analysis of crash databases such as the CDS produces values of $P_{wo}(S_i, \Delta v_k)$. Such estimates can then be utilized as desired values to calibrate the models and techniques that estimate both $P_{wo}(S_i, \Delta v_k)$ and $P_w(S_i, \Delta v_k)$. Next, mathematical models and relevant data are presented to enable the estimation of the parameter R as expressed in Equation (8).

Δv ESTIMATION

Simple mathematical models based on Newtonian mechanics can be utilized to estimate Δv in rear-end, lane change, and single vehicle roadway departure (SVRD) collisions. Such models represent inelastic collisions among vehicles; i.e., the vehicles remain in contact as one combined object after the impact resulting in equal post-crash speeds.

REAR-END COLLISION – In a rear-end collision, the post-crash speed is expressed mathematically as [7][8]:

$$vf = \frac{(m_F \times v_{cF} + m_L \times v_{cL})}{(m_F + m_L)} \qquad (11)$$

m_F: Mass of following vehicle.

m_L: Mass of lead vehicle.

v_{cF}: Velocity of following vehicle at time of impact.

v_{cL}: Velocity of lead vehicle at time of impact.

vf: Final velocity of the coupled vehicles after impact.

The changes in speed as a result of the impact for the following vehicle, Δv_F, and for the lead vehicle, Δv_L, are then computed by:

$$\Delta v_F = v_{cF} - vf \qquad \Delta v_L = v_{cL} - vf \qquad (12)$$

LANE CHANGE COLLISION – Figure 1 illustrates one vehicle (Number 1) steering toward another vehicle (Number 2) that is traveling straight in an adjacent lane, prior to a lane change collision.

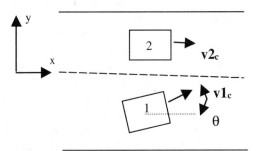

Figure 1.   Lane Change Crash Model Schematic

The longitudinal components of Δv for vehicle 1, $\Delta v1_x$, and vehicle 2, $\Delta v2_x$, are computed by:

$$\Delta v1_x = v1_c \times \cos(\theta) - vf_x \qquad \Delta v2_x = v2_c - vf_x \qquad (13)$$

$v1_c$: Velocity of vehicle 1 at time of impact.

$v2_c$: Velocity of vehicle 2 at time of impact.

θ: Angle of vehicle 1 impacting vehicle 2 measured clockwise relative to road direction (x-axis).

vf_x: Longitudinal component of final velocity of coupled vehicles after impact:

$$vf_x = \frac{(m_1 \times v1_c \times \cos(\theta) + m_2 \times v2_c)}{(m_1 + m_2)} \qquad (14)$$

m_1: Mass of vehicle 1.

m_2: Mass of vehicle 2.

Similarly, the lateral components $\Delta v1_y$ and $\Delta v2_y$ are calculated using the following equations:

$$\Delta v1_y = v1_c \times \sin(\theta) - vf_y \qquad \Delta v2_y = -vf_y \qquad (15)$$

vf_y: Lateral component of final velocity of coupled vehicles after impact:

$$vf_y = \frac{m_1 \times vl_c \times \sin(\theta)}{(m_1 + m_2)} \qquad (16)$$

In a lane change collision, total Δv for vehicle 1, $\Delta v1$, and total Δv for vehicle 2, $\Delta v2$, are then calculated as follows:

$$\Delta v1 = \sqrt{\Delta v1_x^2 + \Delta v1_y^2} \quad \Delta v2 = \sqrt{\Delta v2_x^2 + \Delta v2_y^2} \qquad (17)$$

SINGLE VEHICLE ROADWAY DEPARTURE (SVRD) COLLISION – The total Δv in SVRD collision type may be expressed as:

$$\Delta v = \alpha \times v_c \qquad (18)$$

α: Impact severity attenuation factor.

v_c: Speed of vehicle at time of impact.

This model assumes that the parameter v_c is equal to the travel speed of the vehicle when it leaves the roadway. Moreover, the final speed of the vehicle is assumed to be zero after the impact. The parameter α is a positive random number (≤ 1) that takes into account the effects of objects collided with in SVRD collisions. The distribution of α depends on the category of objects collided with. Table 1 presents the distribution of objects collided with in SVRD collisions based on 1994 and 1995 CDS crash statistics. Fixed objects include fences, walls, ground, curbs, fire hydrants, and other. Breakaway objects encompass trees, bushes, embankments, and posts designed to break away upon impact. Non-breakaway objects consist of poles or posts that are not designed to break away upon impact. Fixed traffic barriers comprise concrete barriers, impact attenuators, and guardrails. Non-fixed objects include passenger cars, trucks, pedestrians, animals, trailers, and other. Detailed analyses of SVRD collision cases are needed to statistically describe the distributions of α for the different object categories.

Table 1.   Distribution of "Objects Collided With" in SVRD Crashes

Object	%*
Fixed	30.1
Breakaway	24.3
Non-breakaway	22.6
Fixed traffic barrier	19.4
Non-fixed	3.6
Total	100.0

*: Based on 1994 & 1995 CDS

VEHICLE TYPE AND WEIGHT DISTRIBUTION

Distributions of vehicle types and weights on US roadways are needed to compute vf as indicated in Equations (11), (14), and (16). Figure 2 shows the distribution of vehicle types based on 1996 US registration statistics that counted approximately 203 million registered vehicles [9]. The "truck" type encompasses light trucks ($\leq$ 4,500 Kg), medium trucks, and heavy trucks (> 15,000 Kg). Light trucks accounted for about 93.3 percent of all trucks in 1996. Figure 3 illustrates the distribution of light vehicles, autos plus light trucks, which comprised about 95.3 percent of all registered vehicles in 1996. Table 2 indicates that the average weight of automobiles varied between 1,139 Kg for subcompact autos and 1,657 Kg for large autos over 1992-1997 period. The distribution of the truck population by gross vehicle weight, excluding light trucks, is shown in Figure 4 based on 1992 statistics.

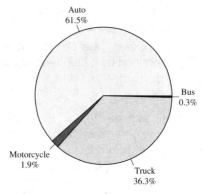

Figure 2.   Distribution of Vehicle Types in the US Based on 1996 Statistics

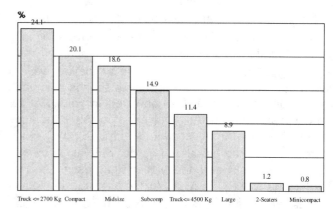

Figure 3.   Distribution of Light Vehicles Based on 1996 Statistics

Table 2. Weight of Automobiles by Size

Auto Type	Weight (Kg)*
Minicompact	1,227
Subcompact	1,139
Compact	1,211
Midsize	1,442
Large	1,657
2 Seaters	1,328

*: Average weight of new autos from 1992 through 1997.

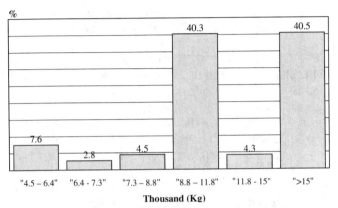

Figure 4. Distribution of Medium/Heavy Trucks by Weight Based on 1992 Statistics

AVERAGE HARM PER COLLISION

The average harm per collision resulting in Δv_k, $\overline{H}(\Delta v_k)$ in Equation (8), is obtained for rear-end, lane change, and single vehicle roadway departure (SVRD) collisions in 10 Km/h speed bins as follows:

$$\overline{H}(\Delta v_k) = \overline{H}_o(\Delta v_k) \times \{\text{Average number of occupants per vehicle}\} \times \{\text{Average number of vehicles involved in a specific collision}\} \quad (19)$$

$\overline{H}_o(\Delta v_k)$: Average harm per occupant as a result of Δv_k in a specific manner of collision, estimated by:

$$\overline{H}_o(\Delta v_k) = \frac{1}{N_o(\Delta v_k)} \times \sum_{i=0}^{6} w(i) \times I(i) \quad (20)$$

$N_0(\Delta v_k)$: Number of occupants of a specific crash type resulting in Δv_k.

w(i): Weighting coefficient corresponding to maximum injury severity i based on the Maximum Abbreviated Injury Scale (MAIS).

I(i): Total number of persons involved with a maximum injury severity i based on MAIS resulting from Δv_k collisions.

Table 3 defines the injury scale of MAIS and presents the "fatal equivalent" values that correspond to each MAIS severity level based on its crash economic cost [10] [11]. MAIS values are converted to "fatal equivalent" values so that injuries of different severity can be measured on a single ratio scale [12]. Table 4 presents statistics on the average number of occupants per vehicle and the average number of vehicles involved in three collision types for both light vehicles and trucks.

Table 3. MAIS levels and their "Fatal Equivalent" values

I	MAIS	w*
0	Uninjured	0.0014
1	Minor	0.0087
2	Moderate	0.0417
3	Serious	0.1250
4	Severe	0.2765
5	Critical	0.8483
6	Fatal	1

*: Based on 1994 economic cost [10]

The 1994 and 1995 CDS crash databases were queried to estimate the values of $\overline{H}_o(\Delta v_k)$ for rear-end, lane change, and SVRD collisions. A total of 1,235 crash cases or about 28 percent of all examined cases contained valid Δv information. Each crash case was weighted using its *Ratio Inflation Factor* in the CDS so as to approximate the national profile. Moreover, injuries are scored in the CDS database using the police-reported injury score KABCO which is converted to the MAIS scale by a translator derived from the 1982-1986 National Accident Sampling System (NASS) data.

Table 4. Average Values of Vehicles Involved and Vehicle Occupancy in Collisions^

Crash Type	Vehicle Type	# Vehicles per Crash	# Occupants per Vehicle
Rear-End	Light Vehicles	2.2	1.3
	Trucks	2.2	1.2
Lane Change	Light Vehicles	2.0	1.5
	Trucks	2.0	1.3
SVRD	Light Vehicles	1	1.4
	Trucks	1	1.1

^: Based on annual statistics from 1989-93 GES Crashes [11].

AVERAGE HARM PER OCCUPANT IN REAR-END COLLISIONS – A total of 455 rear-end crash cases or about 38 percent of all rear-end crash cases in the 1994 and 1995 CDS databases had valid total Δv information for both the lead and following vehicles. These cases were examined to estimate $\overline{H}_o(\Delta v_k)$ based on the highest total Δv experienced in each collision. Figure 5 illustrates the relationship between Δv and average harm per occupant in rear-end collisions, expressed in terms of fatal equivalent values. The average harm per occupant should increase with increasing values of Δv. However, there appears to be an anomaly in the distribution due to the low number of cases available in some Δv bins, especially in the "15 – 25" and "> 65" Km/h Δv bins. It is recommended that a fatal equivalent value of 0.028 be assigned to $\overline{H}_o(\Delta v_k)$ at Δv above 55 Km/h (using the weighted average of fatal equivalent values at the last two Δv speed bins). Moreover, a value of 0.006 is suggested for the "15 – 25" Km/h Δv bin based on the increasing rate of $\overline{H}_o(\Delta v_k)$ between "25 –35" and "45 – 55" Km/h Δv bins.

AVERAGE HARM PER OCCUPANT IN LANE CHANGE COLLISIONS – The 1994 and 1995 CDS crash databases contained 67 cases or about 22 percent of all lane change crash cases with valid total Δv information for both vehicles involved in the collision. The values of $\overline{H}_o(\Delta v_k)$ are estimated using the highest total Δv experienced by either vehicle in each collision. Figure 6 illustrates the relationship between Δv and average harm per occupant in lane change collisions as estimated from the CDS database and displays a modified distribution to adjust the observed anomaly. As seen in Figure 6, an anomaly exists in the distribution of observed injury severity data due mainly to the very low number of lane change crash cases available in the two-year period under consideration.

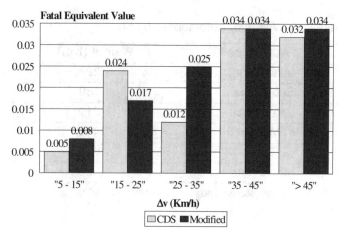

Figure 6. Distribution of Average Harm Per Occupant in Lane Change Collisions by Highest Δv

AVERAGE HARM PER OCCUPANT IN SVRD COLLISIONS – A total of 713 cases or about 24 percent of all SVRD crash cases with valid total Δv information were extracted from the 1994 and 1995 CDS databases. Figure 7 illustrates the distribution of the average harm per occupant in SVRD collisions by 10 Km/h Δv bins. The average harm per occupant is expected to increase with Δv as evident in Figure 7. This harm increased dramatically between "45 – 55" and "55 – 65" Km/h Δv bins, and then was reduced slightly at Δv over 65 Km/h. This may be attributed to the small number of valid cases available for the last two speed bins. It is recommended that a fatal equivalent value of 0.376 be assigned to $\overline{H}_o(\Delta v_k)$ at Δv above 55 Km/h (using the weighted average of $\overline{H}_o(\Delta v_k)$ at the last two Δv bins).

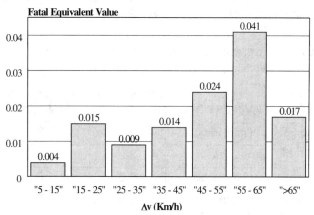

Figure 5. Distribution of Average Harm Per Occupant in Rear-End Collisions by Highest Δv

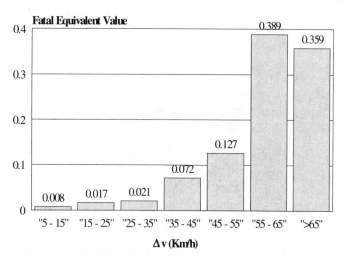

Figure 7. Distribution of Average Harm Per Occupant in SVRD Collisions by Δv

SUMMARY AND CONCLUDING REMARKS

This paper delineated a novel methodology and fundamental supporting data to assess the impact of intelligent vehicle safety systems on the number of collisions and the number and severity of crash-related injuries using computer simulations. Simple mathematical models with representative parameters were described which derive the parameter Δv from speeds at impact and weights of vehicles involved in rear-end, lane change, and SVRD collisions. The SVRD collision model requires data on the impact severity attenuation factor, α, which are not readily available. Data are needed to better describe α for different object categories. The 1994 and 1995 crash databases were queried to identify the relative frequency of striking five object categories in SVRD collisions. In support of Δv estimation, statistics were presented on vehicle types and weights based on 1996 vehicle registration in the U.S.

Statistical data on the number of vehicles involved per crash and the number of occupants per vehicle were provided in order to estimate the average harm per collision. The 1994 and 1995 CDS crash databases were examined to obtain relationships between Δv and average harm per occupant in each of the three collision types. Unfortunately, a small number of crash cases in the CDS database contained valid information on Δv. Consequently, the results obtained had anomalies in the distribution of the average harm per occupant with respect to Δv. To remedy this problem, at least 5 years of CDS data might be needed to generate the appropriate distributions.

ACKNOWLEDGMENT

Mr. John Smith of the Volpe Center's Accident Prevention Division is recognized for his contribution to this research work by providing data from the 1994 and 1995 CDS crash databases.

REFERENCES

1. Najm, W.G. and daSilva, M.P., "Benefits Estimation Methodology for Intelligent Vehicle Safety Systems Based on Encounters with Safety-Critical Driving Conflicts". Submitted for Publication at Intelligent Transportation Society of America's Tenth Annual Meeting & Exposition, Boston, MA, May 2000.

2. National Center for Statistics and Analysis, "National Automotive Sampling System, 1995 Crashworthiness Data System, Data Collection, Coding and Editing Manual". United States Department of Transportation, National Highway Traffic Safety Administration, Washington, D.C., 20590.

3. Najm, W.G., Wiacek, C.J., and Burgett, A.L., "Identification of Precrash Scenarios for Estimating the Safety Benefits of Rear-End Collision Avoidance Systems". Fifth World Congress on Intelligent Transport Systems, Seoul, Korea, October 1998.

4. NHTSA Benefits Working Group, "Preliminary Assessment of Crash Avoidance Systems Benefits". Version II, United States Department of Transportation, National Highway Traffic Safety Administration, December 1996.

5. Najm, W.G. and Burgett, A.L., "Benefits Estimation for Selected Collision Avoidance Systems". Fourth World Congress on Intelligent Transport Systems, Berlin, Germany, October 1997.

6. Campbell, K.L., Joksch, H.C., and Green, P.E., "A Bridging Analysis for Estimating the Benefits of Active Safety Technologies". NHTSA Contract No. DTNH22-93-D-07000, UMTRI-96-18, April 1996.

7. Evans, L., "Driver Injury and Fatality Risk in Two-Car Crashes Versus Mass Ratio Inferred Using Newtonian Mechanics". GMR-7928, April 1993.

8. Khadilkar, A.V., Redmond, D., and Ausherman, V.K., "Collision Avoidance System Cost-Benefit Analysis". Volume I, United States Department of Transportation, National Highway Traffic Safety Administration, DOT HS 806 242, September 1981.

9. Davis, S.C., "Transportation Energy Data Book: Edition 18". ORNL-6941, U.S. Department of Energy, DE-AC05-96OR22464, September1998.

10. Blincoe, L.J., "The Economic Cost of Motor Vehicle Crashes, 1994". United States Department of Transportation, National Highway Safety Administration, DOT HS 808 425, July 1996.

11. Wang, J.-S., Knipling, R.R., and Blincoe, L.J., "Motor Vehicle Crash Involvements: A Multi-Dimentional Problem Size Assessment". ITS America Sixth Annual Meeting, Houston, TX, April 1996.

12. Knipling, R.R., Mironer, R., Hendricks, D.L., Tijerina, L., Everson, J., Allen, J.C., and Wilson, C., "Assessment of IVHS Countermeasures for Collision Avoidance: Rear-End Crashes". Appendix B, United States Department of Transportation, National Highway Traffic Safety Administration, DOT HS 807 995, May 1993.

2001-01-0462

Forward Collision Warning: Preliminary Requirements for Crash Alert Timing

David J. LeBlanc
University of Michigan Transportation Research Institute

Raymond J. Kiefer and Richard K. Deering
General Motors Corporation

Michael A. Shulman, Melvin D. Palmer
Ford Motor Company

Jeremy Salinger
Veridian/ERIM International

ABSTRACT

Forward collision warning (FCW) systems are intended to provide drivers with crash alerts to help them avoid or mitigate rear-end crashes. To facilitate successful deployment of FCW systems, the Ford-GM Crash Avoidance Metrics Partnership (CAMP) developed preliminary minimum functional requirements for FCW systems implemented on light vehicles (passenger cars, light trucks, and vans). This paper summarizes one aspect of the CAMP results: minimum requirements and recommendations for when to present rear-end crash alerts to the driver. These requirements are valid over a set of kinematic conditions that are described, and assume successful tracking and identification of a legitimate crash threat. The results are based on extensive closed-course human factors testing that studied drivers' last-second braking preferences and capabilities. The paper reviews the human factors testing, modeling of results, and the computation of FCW crash alert timing requirements and recommendations.

INTRODUCTION

Over 1.8 million rear-end crashes occur annually in the United States with approximately 2,000 associated fatalities and over 800,000 injuries. [1][2]. Forward Collision Warning (FCW) systems are intended to provide alerts to assist drivers in avoiding or mitigating rear-end crashes. In 1996, the Crash Avoidance Metrics Partnership (CAMP) – a partnership between Ford Motor Company and General Motors Corporation – began a project to define and develop key pre-competitive enabling elements of FCW systems. This project was conducted under a cooperative agreement with the U.S. National Highway Transportation Safety Administration (NHTSA). The motive was to enhance consistent countermeasure system implementation across manufacturers, which will improve customer understanding and acceptance and help to accelerate the implementation of FCW systems.

This paper addresses one aspect of the requirements proposed by CAMP: guidelines for when FCW alerts should be presented to a driver. These alert timing requirements are applicable to a set of objective test procedures developed in the project, and to circumstances with similar initial values of inter-vehicle range, vehicle speeds, and vehicle accelerations. This paper addresses alert timing only; the prediction of vehicle paths and identification of objects that may pose threats are not discussed here.

OEMs and suppliers alike are active in FCW research and development. Other available research relevant to this paper's subject includes the extensive CAMP project final report [3], an earlier project by Frontier Engineering [4], as well as FCW draft standards under consideration by ISO and SAE committees.

The paper begins by presenting the crash problem that the FCW system requirements address. Next, the two phases of human factors testing are reviewed – "baseline" testing to investigate lay drivers' last-second braking decisions for use in determining when to issue crash alerts, and "interface" testing, to develop alert modality requirements and to validate the results of the baseline testing. Requirements for when to issue FCW system crash alerts are presented, and a separate section provides an algorithm for computing the required and recommended crash alert timing for a wide set of approach conditions. See [3] and [5] for other results from this project.

THE FCW CRASH PROBLEM

The specific crash problem which an FCW system should address is described in terms of the prioritized list of six rear-end crash scenarios in Table 1. These scenarios were selected from previous analysis work [6], which combined crash outcome statistics (1991 General Estimates System, 1990 Michigan and 1991 North Carolina police reports) with causal factors [7]. These scenarios satisfy three conditions: the scenarios are observable by an autonomous, line-of-sight FCW system; a warning may have helped a driver brake to avoid or mitigate the impending crash; and the crash types are high-frequency, high-severity events. In this analysis, severity comprehends both the direct costs of crashes and the functional years lost due to death or incapacitating injury. The six scenarios selected represent 19.5% of all crashes and account for 16.2% of the direct costs and 9.2% of functional years lost from motor vehicle crashes in the U.S. annually.

Table 1. Prioritized list of relevant rear-end crash scenarios (percentages refer to annual U.S. motor vehicle crashes)

Scenario	Frequency (%)	Functional years lost (%)	Direct Cost (%)
Inattentive driver	12.0	4.9	10.2
Distracted driver	2.0	1.7	1.9
Poor Visibility	2.0	1.6	1.7
Aggressive driver	1.5	0.5	1.1
Tailgate	1.0	0.3	0.8
Cut-in	1.0	0.2	0.5
Totals	**19.5**	**9.2**	**16.2**

The most common conditions associated with rear-end crashes are straight roads during the daytime under clear weather conditions. Driver inattention is the major causal factor in these rear-end crash scenarios, as shown in Table 1. Driver distraction is also significant . Distraction and inattention are used here in the sense of [7], where a driver is *distracted* by "some event, activity, object or person" and a driver is *inattentive* when choosing to "direct his attention elsewhere for some non-compelling reason," such as "unnecessary wandering of the mind, or a state of being engrossed in matters not of immediate importance to the driving task."

It is possible that FCW systems may provide some benefit in other crash scenarios. While pedestrian and animal crashes may also be mitigated by FCW systems in some instances, these are typically very different scenarios from rear-end crashes and are not considered in the performance requirements set developed.

HUMAN FACTORS TESTING

Requirements for crash alert timing and crash alert modality (auditory, visual and/or haptic alerts) were developed by conducting a series of closed-course human factors studies using a "surrogate target" methodology developed in this program (see [5] for a summary of this, which is provided in detail in [3]). The "surrogate target" consists of a molded composite mock-up of the rear half of a passenger car mounted on an impact-absorbing trailer that is towed via a collapsible beam, as shown in Figure 1.

The surrogate target provides a realistic crash threat to drivers, yet absorbs impacts with a 10 mile per hour velocity differential without permanent damage [3]. This "surrogate target methodology" allows experimenters to safely place naive drivers in realistic rear-end crash scenarios on a proving ground and observe their behavior.

BASELINE TESTING – In the first phase of human factors testing, drivers were asked to perform last-second braking maneuvers while approaching a slowing or stopped vehicle (surrogate target) without FCW alerts. This was called "baseline" testing. Drivers were asked to wait to brake until the last possible moment in order to avoid colliding with the surrogate target. In performing these judgments, subjects were instructed to use either "normal", "comfortable hard," or "hard" braking pressure.

These last-second braking judgments were made on approaches to both a stationary and a moving surrogate target. Approaches to a stationary surrogate target were performed at 30, 45, and 60 mph. Approaches to a moving surrogate target involved initially following the target at steady state, at a fixed speed (30, 45, and 60 mph). The surrogate lead vehicle then began to decelerate at a computer-controlled constant braking level. For each of the three initial speeds, three lead vehicle decelerations were used: -0.15, -0.28, and –0.39 g. When the surrogate target was moving, drivers were instructed to follow at their normal headways.

The different braking instructions enabled the identification and modeling of drivers' perceptions of "aggressive normal braking" and "hard braking". Thirty-six younger (20-30 year old), 36 middle-aged (40-51 year old) and 36 older (60-71 year old) drivers were tested, with an equal number of males and females in each age group. Over 3,800 last-second braking trials were performed; the test location shown in Figure 1 is a straight, one-mile, two-lane roadway in the GM Milford Proving Ground.

Figure 1. Experimental vehicles with surrogate lead vehicle

BASELINE STUDY RESULTS – The data obtained in the baseline study was analyzed statistically to build a model that could be used by an FCW system to predict the onset of drivers' last-second braking. Several dependent measures were explored to find the most predictive model. Results suggest that drivers' "last-second" braking decisions are modeled better using deceleration-based models, rather than time-based ones suggested in previous research (e.g., headway or time to collision) [8].

Within deceleration-based models, two metrics appeared relevant. The "required deceleration" measure is defined as the constant deceleration level required (at the moment of the driver's chosen onset of deceleration) to avoid the crash, given the lead and following vehicles' speeds at braking onset and an assumption that the lead vehicle continues to slow at the prevailing deceleration rate until stopped. Hence, the required deceleration measure is a hypothetical rather than observed driver braking behavior. The "actual deceleration" measure is defined as the constant deceleration level required (at the moment of the driver's chosen onset of deceleration) to yield the observed braking distance.

Figure 2 shows the average required deceleration over all trials, arranged in the graph according to the different kinematic conditions and the different braking instructions. These deceleration measures vary with initial driver speed and lead vehicle deceleration. Drivers braked harder at higher speeds and as the lead vehicle braked harder. This finding also sharply contrasts with some previous FCW system crash alert timing approaches, in which driver deceleration is assumed to be a single fixed value across kinematic conditions [4] [9]. The deceleration-based measures were relatively uninfluenced by driver age or gender.

Statistical analysis of the baseline study data was used to create a model of last-second braking decisions. The objective was to determine whether a single crash alert timing could provide alerts that occur *after* the vast majority of drivers' last-second "normal" braking onsets occurred, and yet also occur soon enough to allow almost all drivers to respond and successfully avoid an impact. This objective attempts to define FCW system timing that provides reduction in rear-end crash harm,

while acknowledging that the FCW system's potential for harm reduction can be realized only if drivers are not subjected to excessive occurrences of unwanted, "too-early" alerts.

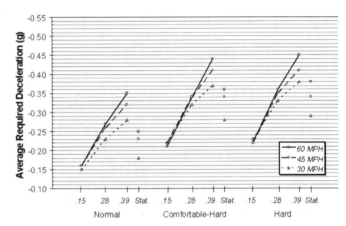

Figure 2. Average required deceleration at braking onset by following vehicle (as function of braking instruction, POV (lead vehicle) braking, and initial vehicle speed; stationary trials are labeled "Stat.")

The 50th percentile required deceleration measure obtained under "hard" braking instructions appeared promising as an estimate of how hard drivers prefer to brake in response to an alert. Figure 3 shows percentile values of the deceleration measures for data collected from the 108 subjects for trials in which the drivers are initially following the surrogate lead vehicle at 45 mph, and the lead vehicle brakes at a constant –0.28 g. The leftmost curve of percentile values is the required deceleration parameter for the last-second "normal" braking instruction. This curve may represent drivers' preferred braking onset behavior for normal last-second braking. Any alert for this kinematic condition requiring driver deceleration magnitudes less than that of this curve's rightmost endpoint (approx. –0.34 g) might be perceived as "too early" by the vast majority of drivers.

The middle curve in Figure 3 is the required deceleration parameter calculated for the "hard" braking instruction. This data may represent the preferred braking onset behavior for drivers executing a last-second hard braking maneuver. For this kinematic condition, an alert that assumes driver decelerations equal to a given point along this curve could be perceived as an acceptable avoidance braking maneuver for those drivers to the right of the point on the curve. Conversely, for drivers to the left of that point, the alert may result in braking that is uncomfortably hard. The rightmost curve in the figure is the actual deceleration parameter for the "hard" braking instruction. This curve models the level of (constant) deceleration which drivers actually used to avoid the crash. As the deceleration level required to avoid the crash increases, this distribution shows the percentage of drivers remaining (to the right) who demonstrated that they were able to brake at this level or harder. Braking at a lower level (to the left in the figure) in an actual collision situation would still realize benefit from reduced impact speed.

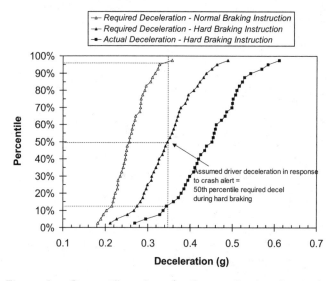

Figure 3. Percentile values for the required and actual deceleration measures, for trials with 45 mph initial speed and –0.28g lead vehicle braking

Thus by accommodating driver preferences for hard braking it appears possible to minimize "too early" alerts for a high percentage of drivers while still allowing sufficient distance for most drivers to avoid the crash by hard braking. Because these three distributions exhibit a similar relationship across all 12 distinct kinematic conditions studied, the 50th percentile "required deceleration" parameter for "hard" braking was modeled across all test conditions and used for crash alert timing purposes in the interface studies. The model for the 50th percentile required deceleration is termed the CAMP "required deceleration parameter (RDP)", and consists of four terms:

RDP: Acceleration (g's) = -0.165 + 0.685 x (lead vehicle acceleration, in g's) + 0.080 (if lead vehicle is moving) – 0.00877 x (closing speed, in m/sec).

Acceleration is considered to be negative for a vehicle that is slowing. The model reflects the data shown in the previous figure: drivers' deceleration magnitudes increased as lead vehicle deceleration magnitude increased; drivers were more aggressive with stopped lead vehicles (for a given closing speed); and drivers braked so that their required deceleration was greater as the closing speed increased.

This model is significantly different from previously developed alert criteria that are based on time-based measures (e.g., headway-time or time-to-collision [8]) or assume a single fixed deceleration level across all kinematic conditions [4][9]. The difference may be attributed in part to the surrogate target methodology, which is believed to present a more realistic crash threat than previously available.

INTERFACE STUDIES – The second set of experiments are collectively referred to as the "interface" studies. These were conducted to (a) define how to present FCW alerts to drivers, and (b) evaluate and validate the crash alert timing strategy developed from the baseline study. Item (b) is the focus of the remainder of this paper; item (a) (alert modality) is summarized in [5] and addressed in detail in [3].

For the interface studies, younger, middle-aged and older drivers (108 total) were asked to brake in response to FCW crash alerts while approaching the surrogate target under the same speed and lead vehicle deceleration conditions examined in the baseline study. Both alerted and unexpected (or surprise) braking events were executed, each with naive drivers and drivers experienced with the alerts. In two of the three studies, drivers were unaware the vehicle was equipped with an FCW system crash alert prior to the surprise braking event. Several strategies were employed to create an inattentive/distracted driver during this surprise event, including engaging the driver in natural conversation, asking the driver to respond to some background-type questions, and asking the driver to search for a (non-existent) indicator light on the conventional instrument panel.

These experiments used a one-stage approach to FCW crash alerts, as opposed to a multi-stage alert, in which different stages of an alert would occur as the situation becomes more threatening. Reasons for using a one-stage approach in these experiments include the discovery of a driver-acceptable single-point crash alert timing approach (represented by the RDP model); compatibility with adaptive cruise control (ACC) alerts under consideration; simplicity of customer education (mental model) and production implementation;

minimizing alerts that are too early for some drivers; and avoiding the rapid sequencing of alert stages that would occur in some conflict scenarios.

The alert timing for the interface studies was built upon the RDP model described earlier. A last-second approach was used: a crash alert is given when the assumed driver response to the alert, along with the assumed future motion of the lead vehicle would lead to the bumpers just touching (i.e., the range and closing speeds go to zero simultaneously). The assumed driver response consists of a reaction time, followed by a step change in the acceleration of the FCW-equipped vehicle to the RDP model-predicted value described above. The lead vehicle acceleration is assumed to be constant throughout the event, unless it comes to a stop, in which case it remains at rest.

The driver response model uses two key parameters – driver brake reaction time and driver deceleration response. The driver deceleration response parameter is computed from the 50th percentile required deceleration model developed from the "hard braking" instruction from the baseline study, presented earlier. Two reaction times were used in the interface studies. Some trials were "alerted" trials, in which the driver was told to wait for the FCW alert. The reaction time used was 0.52 s, based on pilot studies which were later validated with outside test participants. (Additional delay times are added for brake actuation and alert presentation delays). Other trials were "surprise" trials, in which drivers were intentionally distracted by the experimenter while the lead vehicle (surrogate target) began braking without brake lights at – 0.37 g from a 30 mph, steady-state following situation. Surprise moving trials used a 1.5 sec driver brake reaction times (Ref. [10] states that for "reasonably" straightforward situations, 85-95% of drivers will respond with a perception-reaction time of 1.5 sec or less after the first appearance of the object/condition of concern). Driver reaction times measured during this study were used later to recommend appropriate values for FCW systems.

Because a major objective of the interface studies was to investigate alert modality strategies, the testing included six different combinations of alert presentation to the driver. A visual component selected from several candidates using an ANSI testing procedure was included in all combinations (icon shown in Figure 4); speech and non-speech tones were included in five combinations; and a brake pulse was employed in two combinations. The visual icon was presented on either a high head-down display or a heads-up display (HUD). See [3] for modality descriptions and observed differences in reaction time and driver subjective responses.

The alert timing approach used was subjectively rated by the drivers (on average) as "just right" timing under a wide range of driver speed and lead vehicle deceleration conditions, under both expected and surprise braking event conditions. Figure 5 shows drivers' subjective ratings for expected braking event conditions for two of the interface studies. The crash alert timing approach employed also allowed drivers to respond to the crash alert in a manner which allowed them to avoid impacts with the surrogate lead vehicle during surprise braking conditions. Across all surprise braking event trials (one per subject), 104 of 108 drivers avoided impacting the surrogate target without assistance from the passenger-side experimenter.

WARNING

Figure 4.   Visual icon used in interface studies

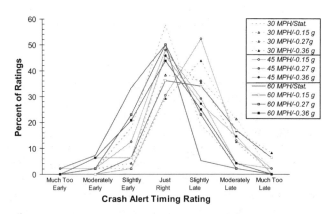

Figure 5.   Crash alert timing ratings during alerted trials

RECOMMENDED AND REQUIRED CRASH ALERT TIMING

Based on the human factors testing described, a number of preliminary requirements for an FCW system driver interface were developed. Detailed alert modality requirements are given in [3] and [5]. Some of the key alert timing requirements are now summarized. A single-stage crash alert is recommended, but earlier stages are acceptable if they do not degrade effectiveness of the latest, most imminent stage whose timing is addressed here. A driver-adjustable alert timing is allowed, but the most imminent (latest) alert stage must meet the requirements described here.

A crash alert timing approach was recommended, and requirements were also given to determine whether the most imminent crash alert stage is "too early" or "too late," as a function of range, vehicle speeds, and vehicle accelerations at the alert onset. These recommendations and requirements are limited to a domain of validity that is determined by the kinematic conditions of the human factors testing, and by the manner in which the timing is expressed. That is, the models developed from the baseline study use data from twelve specific kinematic scenarios. The model's ability to predict driver preferences and performance may not generalize to situations much different than those represented in the testing.

The three alert timing results (recommended, too-early, and too-late timings) are presented by first listing the two driver behavior parameters that define the timings, then providing an algorithm that uses these parameters to compute the alert onsets in terms of a range between the vehicles. The parameter sets are shown in Table 2.

Table 2. Driver response parameters that define recommended and limiting crash alert timings

Parameter	"Too early"	Recommended	"Too late"
Assumed reaction time	1.52 sec	1.18 sec	1.18 sec
Assumed driver acceleration	use RDP	use RDP	use ADP

The two reaction time values in the table – 1.18 and 1.52 sec – are the 85th and 95th percentile (i.e., slower) reaction times observed in those surprise interface tests that used the distraction technique that led to the highest upper-percentile values. In these tests, 24 drivers were asked to find a non-existent telltale indicator on the instrument panel.) These subjects were unaware that the vehicle was equipped with an FCW system. These higher percentile reaction times were close to those found for the two other interface studies conducted here. These values correspond well to the percentiles recommended by Olson (1996) for "reasonably" straightforward situations.

For assumed driver deceleration responses in Table 2, "RDP" refers to the required deceleration parameter described earlier. The "ADP" denotes the "actual deceleration parameter," which is a regression fit over host vehicle speed to the set of each driver's average actual braking performance during trials in which the lead vehicle braked at –0.39 g (the test condition with the highest level of lead vehicle braking). The ADP, which is used as an upper limit to drivers' deceleration performance, is shown below:

ADP: Acceleration, g's = -0.260 - 0.00727 x (FCW vehicle speed, m/sec)

COMPUTING ALERT TIMING REQUIREMENTS

Given the parameters of Table 2 above, an algorithm is now presented to compute the recommended and the required alert onset range for a given kinematic situation.

To test compliance of a particular FCW system with these requirements, the range at which the FCW system's alert begins is observed. The kinematic conditions at that moment are then used to compute the recommended and required values using the algorithm below.

For purposes of this discussion, the FCW-equipped vehicle will be called the "subject vehicle" (SV), and the preceding vehicle that represents a potential collision threat will be called the "principal other vehicle" (POV). Let V_{SV} and V_{POV} denote the initial speeds of the SV and the POV, respectively, as shown in Figure 6. Let aSV and aPOV denote the decelerations of the SV and the POV, respectively, at the moment of interest. (Acceleration is negative for braking). Let DT denote the total delay time between the crash alert onset and when the driver is expected to decelerate the vehicle in response to the crash alert. The total delay time includes both the driver's reaction time and the nominal brake system lag. The driver's expected acceleration response to the alert is denoted a_{SVR}.

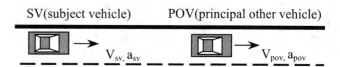

Figure 6. Initial situation of vehicle pair

DOMAIN OF VALIDITY – The required and recommended CAMP timing values are valid over a restricted domain of initial conditions which are dictated by two limiting factors. The first and most important is the kinematic conditions addressed by the human factors testing. In conditions much different than those represented in these tests, the models (RDP and ADP equations) may not be good approximations for driver braking decisions. The second factor is the algorithm presented below. For approach situations similar to those in the human factors tests or the objective test procedures in [3], the algorithm does provide an exact computation of the range needed to bring range and closing speed to zero simultaneously, using the last-second crash alert timing approach and the driver response models described earlier. For some less common kinematic conditions, however, the algorithm does not provide an exact solution for the last-second approach (e.g., computing alert onset range for a FCW-equipped vehicle accelerating hard toward a lead vehicle). An algorithm that provides an exact computation over all possible kinematic situations includes much more involved logic, and would be needlessly complex for computing these alert timing requirements.

Listed below are conditions that define a domain of validity sufficient to guarantee exact computations of the last-second approach. Enforcing these conditions avoids the extensive logic needed to handle all possible kinematic situations:

- SV speed is expected to be greater than the POV speed at the end of the total delay time.
- SV speed is not expected to go to zero during the delay.
- SV speed is initially at least 16 kph.
- SV acceleration at crash alert onset has an absolute magnitude that is no greater than 0.1g.
- POV speed is non-negative.
- If the POV is initially moving, it will not come to rest during the delay.
- The POV is either decelerating or not accelerating at more than 0.08g.

If any of these conditions do not hold, the equations that follow may not be applicable for computing the exact requirements for alert timing.

COMPUTING REQUIRED AND RECOMMENDED ALERT TIMING – To compute crash alert timing requirements or recommendations, five steps are used. These compute the minimum range needed between vehicles to avoid a collision, assuming (1) the lead vehicle acceleration remains constant (unless the vehicle comes to a stop, in which case it is assumed to remain stopped), and (2) the following vehicle acceleration remains constant for a delay time period, then instantaneously changes to an assumed value. *This algorithm is not intended to represent a general-purpose crash alert timing algorithm, but is rather a computation of alert timing requirements or recommendations for a limited set of kinematic conditions.*

Five steps for computing timing requirements:

1. Consider whether the algorithm is applicable, given the domain of validity constraints discussed earlier. If applicable, continue.

2. Project vehicle speeds from current values to the end of the total delay time. The total delay time includes the sum of the assumed driver reaction time (from Table 2), plus a typical delay time between a rapid brake pedal application and deceleration of the vehicle (e.g., 0.20 sec). The predicted speeds at the time of SV acceleration onset are:

$$V_{SVP} = V_{SV} + a_{SV}*DT$$

$$V_{POVP} = V_{POV} + a_{POV}*DT$$

3. Evaluate the driver acceleration response, a_{SVR}, that is assumed to occur after the delay time. Table 2 shows that the recommended crash alert timing, as well as the "too early" timing, uses the RDP equation. The "too-late" timing uses the ADP equation. If the RDP equation applies, evaluate a_{SVR} as shown in Figure 7. If the ADP equation applies, evaluate a_{SVR} as shown in Figure 8. Notice that computations of both use speeds that are predicted for the end of the delay time, i.e., the time of driver deceleration onset.

$$
\begin{aligned}
&\text{If } V_{POVP} > 0 \text{ and } a_{POV} < 0, \\
&\quad a_{SVR}(\text{g's}) = -0.165g + (0.685g/g)*a_{POV} + 0.080g \\
&\qquad\qquad + (-0.00877g/m/s)*(V_{SVP} - V_{POVP}) \\
&\text{Else if } V_{POVP} > 0 \text{ and } a_{POV} >= 0, \\
&\quad a_{SVR}(\text{g's}) = -0.165g + 0.080g \\
&\qquad\qquad + (-0.00877g/m/s)*(V_{SVP} - V_{POVP}) \\
&\text{Else} \\
&\quad a_{SVR}(\text{g's}) = -0.165g \\
&\qquad\qquad + (-0.00877g/m/s)*(V_{SVP} - V_{POVP})
\end{aligned}
$$

Figure 7. Computing the assumed driver acceleration response using the RDP

$$a_{SVR}(\text{g's}) = -0.260g - (0.00727g/m/s)V_{SVP}$$

Figure 8. Computing the assumed driver acceleration response using the ADP

4. Compute the minimum range at which an alert would be needed so that the model of driver response would just bring the closing speed to zero as the range went to zero. (Derivations of the following equations are not presented. The equations follow from a straightforward application of kinematics using the simple models presented, and assuming the conditions above apply.)

The alert range, R, is the sum of two terms: the amount that the range will decrease during the total delay time ("delay time range," or DTR), plus the minimum range (at SV deceleration onset) needed to avoid a collision ("braking onset range," or BOR):

R = DTR + BOR

DTR is the difference in distances covered during the delay time, DT:

$$DTR = (V_{SV} - V_{POV})*DT + \tfrac{1}{2}(a_{SV} - a_{POV})*(DT)^2$$

BOR is computed using one of two possible expressions. These correspond to whether the POV is expected to be moving or stopped when the "contact" occurs (contact is the moment at which the models predict the range rate and range both go to zero). The following conditional determines which of these two cases is expected:

If $V_{POV}/a_{POV} < DT + V_{SVP}/a_{SVR}$,

Then: Contact expected when POV is stopped,

Else: Contact expected while POV is still moving.

A more robust conditional statement that gracefully handles special cases (e.g., POV stationary or at constant-speed), is derived from the above:

If $a_{POV}*V_{SV} <= a_{SVR}*V_{POV} - a_{POV}*DT*(a_{SV} - a_{SVR})$,

Then Contact expected when POV is stopped,

Else: Contact expected while POV is still moving.

Contact with a stopped POV: This includes cases in which the POV is initially stopped as well as cases in which the POV decelerates to a stop during the delay time or during the braking maneuver of the SV. For these cases, BOR is the difference between the expected stopping distances of the POV and SV. Expected stopping distances are computed using conditions occurring at the end of the total delay time:

If $a_{POV} = 0$ (i.e., stationary POV),

Then $BOR = (V_{SVP})^2/(-2*a_{SVR})$

Else,

$BOR = (V_{SVP})^2/(-2*a_{SVR}) - (V_{POVP})^2/(-2*a_{POV})$

Contact while POV is still moving: The case in which contact is expected when the POV is moving includes cases in which the POV is not decelerating, and in fact is accelerating within the conditions assumed earlier. It also includes cases in which the POV is decelerating, but conditions are such that contact is still expected before the SV deceleration can occur quickly enough to avoid an impact. One common situation leading to this case is when the SV is tailgating at higher speeds and the POV begins braking at significant levels. If contact is expected when the POV is moving the braking onset range is:

$BOR = (V_{SVP} - V_{POVP})2/(-2*(a_{SVR} - a_{POV}))$.

Regardless of which braking onset range equation is used, the alert onset range R can now be computed as the sum of DTR and BOR.

5. Apply other applicable requirements that may affect the required range at alert onset (see [3], Chapter 4, Section 4.7). For example, if the steps above yield an alert onset range greater than the upper limit on required warning ranges (100 meters), then the range computed is held to this value. The reader is advised to apply the requirements summarized in this paper with a familiarity of the context of FCW system operation, such as described in [3].

CONCLUSION

This paper presents preliminary CAMP minimum requirements and recommendations for when FCW crash alerts should be presented to drivers. These are valid over a set of approach situations that are similar to the testing situations described herein. These results are just one aspect of FCW performance addressed in a recent CAMP project intended to accelerate successful deployment of FCW systems. Included here is an algorithm for computing the required and recommended inter-vehicle range at which a crash alert should begin, as a function of vehicle speeds and accelerations. These computations are based on statistical modeling of the results of two phases of human factors tests that are reviewed. This modeling predicts a driver's deceleration response to a crash alert, as a function of vehicle speeds

and accelerations. Neither time-based measures (e.g., headway, time to collision) nor models assuming a single value for driver deceleration over all approach conditions were as predictive of the test observations as the deceleration-based model developed here. Other test results provide a basis for selecting a driver reaction time under surprise or unexpected braking conditions.

There is no claim that these requirements can be met with currently available technology. It is also possible that countermeasure systems which do not meet all of the proposed requirements may still provide drivers with some level of crash avoidance / mitigation benefit.

These results are subject to a number of limitations. First is the domain of initial conditions evaluated in the human factors testing was finite. Second, all human factors testing was conducted during clear weather daylight conditions on a straight, dry, level road. Third, instrumentation-quality data was used in testing: "instantaneous" knowledge of lead vehicle behavior (including deceleration) was obtained from on-board instrumentation via vehicle-to-vehicle communications. Fourth, the crash scenario evaluated was an in-lane approach to a stopped vehicle or a lead vehicle exhibiting constant deceleration levels. Fifth, drivers were instructed to use only braking as an avoidance maneuver, and not steering.

While the scenarios evaluated represent the majority of rear-end crashes, further testing is necessary to establish driver acceptance of the proposed alert timing and interface modality requirements under different operating conditions using autonomous sensor data. Other conditions that should be considered are nighttime, bad weather, and non-constant lead vehicle deceleration profiles. Finally, extensive field operational testing is necessary, at a minimum, to better understand what levels of nuisance alerts ("too-early" alerts) are acceptable to drivers.

ACKNOWLEDGMENTS

The authors wish to acknowledge key contributions that made these results possible. Marie Cassar (now at GM) supervised test execution, Carol Flannagan conducted statistical modeling, and GM Milford Proving Ground staff participated in all vehicle testing. The project was conducted under a cooperative agreement with the U.S. NHTSA (Jack Ference, COTR), with helpful feedback from several groups within US DOT.

REFERENCES

1. Knipling, R., Wang, J.-S., and Yin, H.-M. (1993). *Rear-end crashes: problem size assessment and statistical description.* Report No. DOT-HS-807-994. Washington, D.C.: U.S. Department of Transportation, National Highway Traffic Safety Administration.

2. Najm, W., Wiacek, C., and Smith, J. (1998). *Rear-end crash and precrash scenario analysis.* Presentation at NHTSA Symposium on Rear-end Collision Avoidance, Laurel MD, October 1998.

3. Kiefer, R., LeBlanc, D., Palmer, M., Salinger, J., Deering, R., and Shulman, M. (1999). *Development and validation of functional definitions and evaluation procedures for collision warning/avoidance systems.* DOT HS 808 964. Office of Crash Avoidance Research, National Highway Transportation Safety Administration. Washington, DC.

4. Wilson, T.B., Butler, W., McGehee, D.V. and Dingus, T.A. (1997). *Forward-looking collision warning system performance guidelines.* SAE paper #970456. Society of Automotive Engineers. Warrendale, PA.

5. Kiefer, R., (2000). *Developing a forward collision warning system timing and interface approach by placing drivers in realistic rear-end crash situations.* Proc. 2000 Human Factors & Ergonomics Society Conference, San Diego, CA, 2000.

6. General Motors (1996). *44 crashes, v.3.0.* Warren, MI: North American Operations, Crash Avoidance Department.

7. Treat, Tumbas, McDonald, Shinar, Hume, Mayer, Stansifer and Castellan (May 30, 1979). *Tri-level study of the causes of traffic accidents: Executive summary.* Report No. DOT-HS-805-099. Washington, D.C.: U.S. Department of Transportation, National Highway Traffic Safety Administration.

8. Horst, A. R. A. van der (1990). *A time-based analysis of road user behavior in normal and critical encounters.* (Doctoral dissertation, Delft University of Technology, Delft, The Netherlands). (Available from the TNO Human Factors Research Institute, PO Box 23, 3769 Soesterberg, The Netherlands.)

9. Burgett, A., Carter, A., Miller, R., Najm, W., and Smith, D. (1998). *A collision warning algorithm for rear-end collisions.* 16th International Technical Conference on Enhanced Safety of Vehicles Abstracts, 98-S2-P-3 1, Washington, D.C., May 1998.

10. Olson, P.L. (1996). *Forensic aspects of Driver Perception and Response Time.* Lawyers & Judges Publishing Company, Inc.: Tuscon, AZ.

2001-01-0461

Simulating Rear-End Collision Warnings Using Field Operational Test Data

Richard A. Glassco and Daniel S. Cohen
Mitretek Systems

ABSTRACT

This paper discusses the definition and performance of the warning algorithm in a potential Rear-End Collision Avoidance System (RECAS). Given vehicle range, velocity, and leader deceleration data, the warning algorithm computes whether imminent braking is required to prevent a collision with a vehicle directly ahead, and determines whether a warning should be given to alert the driver. The result of the study is an estimate of the number of warnings per hour under normal driving conditions for various types of drivers.

INTRODUCTION

The study of the warning algorithm in a RECAS used data collected in a Field Operational Test (FOT) conducted for the National Highway Traffic Safety Administration (NHTSA) by the University of Michigan Transportation Research Institute (UMTRI)(References 1 and 2). The data collected by UMTRI contains information for over 114,000 miles (186,300 km) of naturalistic driving by 108 drivers, spanning more than 3,000 hours of driving. The FOT was designed to study Adaptive Cruise Control (ACC), so the database is called the ACC database in this paper.

The researchers developed computer procedures for reading the ACC database, checking the values for consistency, and passing them to the warning algorithm. The warning algorithm uses four input variables: the follower speed, the range between the two vehicles, the closing rate between the two vehicles, and the leading vehicle deceleration rate. The study included an exploration of what assumptions the algorithm should make and which "lockout" conditions would prevent a warning from being issued.

The project successfully integrated a RECAS warning algorithm into a large collection of naturalistic driving data. It gives an answer to the question "How often may a rear-end collision warning be issued during normal driving?" by calculating the number of times per hour the RECAS warning alarm would have sounded if it had been installed during the UMTRI FOT.

The full version of the project report documenting the RECAS algorithm, the data processing procedures, and the results is listed as Reference 3.

THE ACC DATA BASE

UMTRI conducted the FOT in 1996 and 1997 to study the potential performance of an ACC system. Ten Chrysler Concordes were equipped with a prototype ACC system. These cars were driven by 108 volunteer drivers, evenly distributed by age and gender. Most driving occurred in southeastern Michigan, but some drivers took longer interstate trips. In all, there were 11,092 trips recorded, covering 114,084 miles (183,600 km) and lasting 3,049 hours. The data collected constitutes an immensely valuable record of "naturalistic" driving under normal conditions by normal drivers (1).

The prototype ACC system consisted of:

- A forward-looking optical range and range-rate sensing device in the car front grille
- A Global Positioning System (GPS) for tracking vehicle location
- A computer system for recording these measurements and other driving/location information about the instrumented car.
- A computer algorithm for providing ACC by managing the throttle. The FOT did not include any automated control of the brake.
- A forward-pointing video camera for recording visual images.

Numerical data items were collected or computed every 0.1 seconds, stored in the onboard computer, and later stored in a table of a Microsoft Access® database. The video camera was operational all the time, collecting images every 0.1 seconds. However, most of the video images were discarded, since the volume of video data would have been too overwhelming to store and analyze. Video images were retained when the ACC system determined there was either a "near encounter," (an approach to another vehicle closer than a defined distance) or a "brake intervention," (application of the brakes at or exceeding a defined deceleration level while traveling faster than a specified speed).

Half of the 108 drivers were male and half were female. One-third of the drivers were between the ages of 20 and 30, one-third between the ages of 40 and 50, and one-third between the ages of 60 and 70. Within each age band the male/female ratio was 50/50.

As part of its analysis, UMTRI categorized the 108 drivers into five classes, representing different driving styles. UMTRI characterized each driver on two scales. The first scale was whether the driver tended to drive faster or slower than the driver in front. The second scale was whether the driver tended to drive closer or farther away than the average car-following distance from the vehicle in front. A driver's relative position on these scales determined his or her driver type:

- Drivers on the faster/closer end of the scales were labeled "Hunter/Tailgaters."
- Drivers on the faster/farther end of the scales were labeled "Planners" (it requires planning to drive faster than surrounding traffic yet maintain long following distances).
- Drivers on the slower/farther end of the scales were labeled "Ultra-Conservative."
- Drivers falling into more than one of these categories were called "Extremists." Since their behavior is mixed, little can be said about them as a group.
- Drivers falling into none of these categories (i.e. average speed and following distance) were called "Flow conformists."

The numeric data for each driver is in the format of a separate Microsoft Access database. Each trip taken is a table in the database. Trips may range from a few seconds to a few hours in duration. Each table contains a value every tenth of a second for each of 36 variables. Some of these were taken directly from the sensor systems such as range, follower velocity, brake on or off, ACC on or off. Some were derived from those primary values, such as follower or leader deceleration and rate of change of curvature, and some values were added by later analysis.

The ACC FOT was not designed to collect data all the time. Data recording was turned off when the leader speed was less than thirty percent of the follower's speed. In addition, the speed of the lead vehicle was dropped when the speed of the following vehicle was less than 25 miles per hour. These data dropouts were not considered to be a limitation because the FOT was designed as a study of cruise control at moderate to high speeds. However, the missing data made it impossible to estimate the frequency of collision warnings at low leader or follower speeds. Data procedures oriented toward the RECAS analysis are described in a later section.

DEFINITION OF THE RECAS WARNING ALGORITHM

This section describes the collision warning algorithm used in this study. It is derived from an equation presented by Burgett et al. (Reference 4) in 1998, which made the following assumptions:

- If the leader is decelerating, the leader will continue to decelerate at the same constant rate to a stop. Otherwise the leader will maintain constant velocity.

- The following vehicle maintains the same constant speed until the driver responds to the warning by applying the brake.
- The time required for an inattentive driver to respond to a warning by applying the brake pedal is 1.5 seconds. This time was suggested by a study performed using the Iowa Driving Simulator.
- The alerted driver will brake at a constant deceleration rate of 0.75 g (about 24.1 feet per second per second).
- The minimum distance between the two vehicles during or after braking will be two meters.

Given these assumptions, the warning algorithm computes a "time to warning", starting from the current instant. The warning algorithm uses four input variables:

- Follower speed
- Range between the two vehicles
- The closing rate between the two vehicles (also called range-rate or Rdot)
- Leading vehicle deceleration rate

In Phase I of the study (Reference 5), the original equation in reference 4 was evaluated only at the instant the leader began to decelerate, with the assumption that lead and following vehicles were traveling at the same speed. In Phase II of this study with continuous vehicle data available, the equation was evaluated every 0.1 second whenever either the lead vehicle was decelerating or the following vehicle was moving faster than the lead vehicle. Reference 3 provides a full statement and derivation of the warning equation.

The time to warning is the number of seconds to pass before a warning should be given, assuming a 1.5 second reaction time and 0.75g deceleration. Given these assumptions, the following vehicle would come safely to a stop two meters behind the lead vehicle.

If the warning algorithm returns a time-to-warning that is greater than zero, no warning is necessary at the current moment. Either hard braking is not necessary or the following vehicle may wait more than 1.5 seconds to begin braking. In this case the analysis program passes on to read the next configuration values 0.1 seconds later.

If the warning algorithm returns a time-to-warning less than or equal to zero, then a warning would have been issued if it had been installed in the ACC FOT, provided that a lockout would not have prohibited the warning, as described in the next few paragraphs.

Although the configuration of the two vehicles may be such that a warning would be issued, there may be reasons why a warning should not be issued (i.e., locked out). In general, a warning should be locked out if there is reason to suspect (a) the situation is momentary or a result of erroneous data, or (b) the driver is alert and is capable of responding appropriately without a warning.

A lockout condition must be able to be determined in real time. A perfect lockout system would be able to tell when the following driver is in command of the situation. Given the data collected by the ACC FOT, only a few indicators of driver responsiveness are possible.

Over a series of iterations, six conditions emerged as lockouts to evaluate in this study. Even if a warning is indicated by the warning equation, no warning will be issued if any of the following conditions occur:

- The follower is traveling slower than 10 miles per hour (16 km per hour)
- The warning situation has existed less than 0.2 seconds
- The follower's brakes are already on
- The follower's brakes have been on at any time within the last two seconds
- The follower has just completed a lane change maneuver to end behind a new leader
- The warning is within five seconds of a previous warning

Evaluating the warning algorithm every 0.1 seconds means that the algorithm can respond in real time to fluctuations in vehicle dynamics. If the lead vehicle slows down briefly, but speeds up again before a critical situation is reached, the RECAS system will not issue a warning. The system can also respond to changes in the level of leader deceleration. For example, a warning might not be necessary if the leader begins braking with a deceleration of 0.1 g, but may become necessary if the leader's braking level increases to 0.2 g.

Although the algorithm can follow changes to vehicle dynamics in real time, it cannot predict the future. For example, the algorithm assumes that if the lead vehicle is decelerating, it will maintain the same deceleration until it stops. Experience shows that frequently the lead vehicle will subsequently speed up, or it may increase its deceleration. Rarely is constant deceleration found outside of controlled experiments. However, the algorithm has no way to predict what will happen, so it assumes constant deceleration.

PROCEDURES FOR ANALYSIS OF ACC DATA

The researchers developed a Visual Basic® computer program to read, filter, and analyze the numeric ACC database files. This section outlines the major steps in the process of extracting data values from the ACC database and cleaning up the data for further processing.

DATA DROPOUTS AND SPIKES. As is common for data obtained from live sensors, data spikes and data dropouts occur frequently in the ACC FOT database. A data dropout is an instance when one of the measured values suddenly drops to zero. A data spike is any case where a data value assumes a value significantly different from the preceding and following values for a brief interval.

Data dropouts and spikes occur most frequently in the range data, but there are also occasional data dropouts in vehicle velocity and deceleration data. Since the study derived range-rate, lead vehicle speed, and lead vehicle deceleration from the range and following vehicle speed data, the researchers attempted to detect and correct dropouts and spikes only in the range and following vehicle velocity data. The use of a secondary procedure for smoothing data, the Butterworth filter, is described in a later subsection.

CORRECTING RANGE DATA. The range from the following vehicle to the leading vehicle should be a continuous variable, except when the follower changes lane or the lane leader ahead leaves the lane or a new lane leader cuts in ahead.

The analysis program tries to distinguish the times when there is a legitimate sudden change in the range from the times when a sudden range change is a sensor error that should be corrected. If there is a sudden jump in range, the algorithm looks ahead to see if there is a range value within the next second that is consistent with the previous good range and range-rate values. If such a value is found, the program interpolates across the gap of bad or missing data. If interpolation is not possible, the program assumes a new leader has been found or there is no longer a leader, and starts from the new range.

Another problem encountered frequently in the range data happened immediately after a new leader cut in front of the instrumented vehicle. As the new leader angled across the field of view of the range sensor, the first reflection came from the front bumper, so the reported range was actually the distance to the front of the new leader. As the leader straightened out in the lane in front of the follower, the sensor swept back along the side of the vehicle to finish in the normal configuration measuring the range to the back of the leading vehicle. The resulting rapid decrease in reported range makes it appear to the analysis program that the lead car is decelerating rapidly – a situation very likely to trigger an inappropriate warning. The researchers' approach to solving this problem involved "back-extrapolation." If the range to the leader settled out to a fairly constant value 0.5 seconds after an abrupt change in the range data (suggesting a cut-in), then the program extrapolated backwards in time from the fairly constant range data to the instant after the appearance of the new target. The reported range values for the first 0.5 seconds were discarded.

Additional evidence of erroneous range values was the "backscatter" flag in the database. The ACC data collection system set the value of a backscatter variable greater than 50 whenever it suspected that the range reported by the sensor was the result of reflection from anything other than a vehicle in the lane ahead. Whenever the analysis program read a backscatter flag value greater than 50 in the database, it ignored all data for the next six seconds. The most frequent cause of high backscatter was snow, rain, or fog particles in the air.

RANGE RATE AND LEADER VELOCITY. The researchers bypassed problems with lead vehicle velocity and range-rate by calculating them directly from the range and following vehicle velocity data provided in the ACC database.

The range-rate at each point was determined by computing the slope of the best linear fit to the current range plus and minus one-half second. Since data is reported every

0.1 seconds, the algorithm used up to 11 points for the calculation. However, if there was a sudden change in range because of target loss or acquisition of a new target, the non-matching range values were not used in the calculation.

Because future range values as well as previous range values were used at each point, the calculation of range-rate is "in synch" with the range. For example, the range-rate becomes negative the instant that range begins to decrease. This method is not suitable for real-time response, however. It can only be used for after-the-fact data analysis.

Leader velocity was computed by adding the range-rate to the follower velocity.

LEADER AND FOLLOWER ACCELERATION. The researchers computed the leader deceleration as the slope of the best linear fit to the current leader velocity plus and minus one-half second. Since data is reported every 0.1 seconds, the algorithm used up to 11 points for the calculation. However, if there was a sudden change in leader velocity because of target loss or acquisition of a new target, the non-matching range values were not used in the calculation. Again, the "center-based" calculation for deriving the change in velocity could be performed for after-the-fact analyses, but is not possible for real time warning system operation.

The follower acceleration/ deceleration were computed by the same method described above. Since follower velocity is always continuous, the determination of follower deceleration is not subject to sudden changes.

USE OF A BUTTERWORTH FILTER. Before data collected by sensors is used in further calculations, data-smoothing techniques are often employed to reduce random noise. The derivation of range rate and acceleration as slopes of best linear fits as described in previous sections are one way of reducing noise. Another common method is the use of digital filters.

The Butterworth filter is a commonly-used digital filtering technique for reducing the amount of noise arising from high-frequency fluctuation, while allowing low-frequency fluctuation corresponding to actual trends in the data to pass through.

It is not appropriate to employ a smoothing technique across a data discontinuity caused by target loss or new target acquisition. For the same reason, a smoothing filter will not correct errors from data spikes or dropouts; in fact, they would make these errors harder to detect and correct. Corrections for spikes and dropouts must be made first, and the smoothing filter applied to segments between discontinuities.

The Butterworth filter was used in a post-processor to smooth data for four-second segments surrounding potential warnings for selected drivers.

APPLYING THE RECAS ALGORITHM The RECAS algorithm has been described in the third section of this paper. The algorithm was applied to compute a time to warning every 0.1 second whenever there was a leader and the follower's speed was greater than ten feet per second.

If the warning algorithm returns a time-to-warning less than or equal to zero, then a warning would have been given,

if the RECAS system had been installed in the FOT vehicle and if the warning had not been precluded by a lockout criterion. When a potential warning was generated, the program wrote a line to a summary file containing the configuration at the time of warning and information that allowed the lockouts to be determined and applied.

The program also wrote out vehicle information to a detail file for further study. For each potential warning, the program wrote out vehicle configuration data every 0.1 seconds for two seconds before and two seconds after the time the warning would have occurred.

Figure 1 is a graph of a sample set of values recorded before and after a potential warning

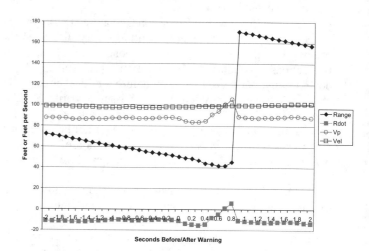

Figure 1. Sample Range, Rdot, and Velocities Surrounding a Warning

For this graph the time of the potential warning is set at zero. The range is decreasing steadily (i.e. the range rate is negative) because the follower speed (Vel) is greater than the leader speed (Vp). The warning would have been issued at time 0. However, at time 0.4 the leader's speed increases, and at time 0.9 the range jumps significantly, indicating that either the leader or follower changed lanes, and the follower has a new leader.

POST PROCESSING. After all the potential warnings for a driver were computed, the program closed the database file and returned to the summary and detail files written for each potential warning. It then determined which of these potential warnings should lead to a warning after smoothing with the Butterworth filter and discarding warnings that would have been precluded by a lockout.

After all the lockouts were applied, the remaining warnings would have been given if a RECAS system had been installed in the ACC FOT. The total number of warnings each driver would have received divided by the total number of hours that driver was on the road results in the average number of warnings per hour for each driver.

VIDEO ANALYSIS. The video records of braking episodes were immensely valuable for understanding various configurations and behaviors that were not clear from the numeric data. In many cases involving lane changes, transient targets, or crossing traffic, the numeric data could not tell the whole story. In cases where video episodes were available, the sequence of events became clear. Dozens of individual events were viewed in the process of fine-tuning the data cleanup algorithms. However, most potential warnings did not have a matching video episode, and there was no automated analysis of video files.

RESULTS

This section presents the results of the analysis program run for 105 drivers in the ACC FOT database. Data for the remaining 3 drivers was unreadable.

AVERAGE WARNINGS PER HOUR. Figure 2 presents a graph showing the average number of warnings per hour for each driver, sorted in ascending order. Table 1 shows selected percentiles and the average. Only three drivers would have average warning rates significantly greater than one per hour. Five drivers would have had no warnings at all during the FOT.

| | Percentile | | | | | |
	10	25	50	75	90	Average
All drivers	0.03	0.08	0.20	0.42	0.72	0.31
Gender						
Female	0.00	0.05	0.14	0.32	0.50	0.22
Male	0.06	0.11	0.27	0.49	0.74	0.40
Age						
20-30	0.07	0.14	0.29	0.50	0.77	0.38
40-50	0.04	0.07	0.22	0.37	0.61	0.36
60-70	0.00	0.05	0.12	0.27	0.50	0.19
Type						
Extremist	0.04	0.06	0.12	0.29	0.48	0.21
Flow conformist	0.07	0.16	0.27	0.36	0.50	0.28
Hunter-tailgater	0.04	0.28	0.50	0.77	0.92	0.61
Planner	0.00	0.06	0.11	0.16	0.49	0.23
Ultra-conservative	0.00	0.05	0.09	0.16	0.37	0.12

Table 1. Percentiles and Averages for Selected Groupings of Drivers

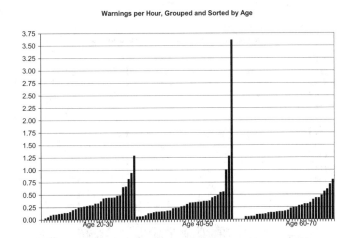

Figure 3. Warning Rate by Age

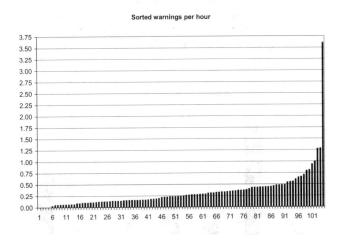

Figure 2. Number of Warnings Per Hour

Figure 3 presents the sorted number of warnings per hour separated by driver age category. Although a wide range of warning rates can be found in each group, the differences among the warning rates are evident in the graph and are borne out by the averages and percentiles presented in Table 1. The youngest drivers had the highest average warning rate and the highest percentile values at each point. The oldest drivers had the lowest average and percentile warning rate.

Figure 4 presents the results by driver gender. Male drivers had a much higher average warning rate than female drivers in the FOT, and the percentiles were higher at each point, as shown in Table 1. There were the same number of drivers in each age group and gender.

Figure 5 presents the same information separated by driver style as defined by UMTRI and described in the second section. This figure shows that "Hunter/tailgater" drivers had the highest warning rate, followed by "Flow Conformists". The "Ultra-Conservative" drivers had the smallest warning rate, followed by the "Planners." The "Extremists," actually a mixture of other driving styles, are in the middle.

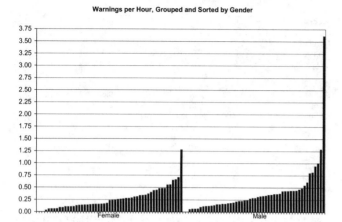

Warnings per Hour, Grouped and Sorted by Gender

Figure 4. Warning Rate by Driver Gender

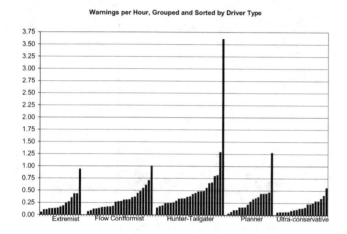

Warnings per Hour, Grouped and Sorted by Driver Type

Figure 5. Warning Rate by Driving Style

Table 1 and Figures 3 through 5 agree with the popular perceptions of groups that tend to drive more aggressively, and thus are likely to receive more warnings. Whether the warning rates are high enough to constitute an annoyance to drivers or whether the drivers would modify their driving style in reaction to the warnings are issues that could be explored in a future FOT.

Lockouts. This section discusses the effects of the lockouts employed to reduce the number of inappropriate warnings. Figure 6 illustrates the effect on warnings per hour of each lockout, applied individually and in combination. Each lockout condition is applied independent of the other five. The seventh bar shows the number of warnings remaining if all six lockouts were applied in combination. The final bar shows the number of these warnings that remained after smoothing the data two seconds before and after the potential warning with a Butterworth filter. This number is 0.31 warnings per hour, as reported in Table 1.

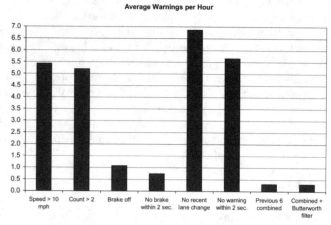

Average Warnings per Hour

Figure 6. Average Warnings per Hour Remaining After Lockouts

It is clear that locking out a potential warning if the follower's brakes are on or have been on within the last two seconds has the greatest effect in reducing the number of warnings. The second bar shows that approximately 40% of the warning situations last less than 0.2 seconds; they are triggered either by a transitory situation or bad sensor data.

While examining the effect of the lockout titled "Speed > 10 mph," the researchers derived a histogram of the follower's speed at the time of the potential warning. Figure 7 shows the proportion of potential warnings falling in each speed category after lockouts.

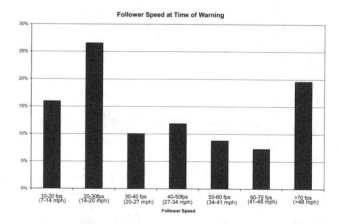

Follower Speed at Time of Warning

Figure 7. Follower Speed at Time of Warning

Another potential lockout situation exists if an ACC system with braking capability is engaged at the time of a potential warning. If the RECAS determines that the ACC is engaged and can achieve the required deceleration without driver intervention, then no warning is necessary. For each potential warning in the FOT, the researchers determined:

- Whether the ACC system was actively engaged
- Whether a deceleration of 0.3 g beginning immediately would be sufficient to stop the vehicle behind the leader

If both conditions are true, then a warning might not be necessary. Table 2 presents the number of potential warnings falling into this category for each set of lockouts. The result is that a very small percentage of the potential warnings would not have been required. Most of the time that a warning might have been required, the ACC system was not active. The result suggests that the overlap between ACC systems and RECAS is small.

Lockout	Potential Warnings	ACC on & sufficient Number	Percent
No lockout	18,983	115	0.6%
Speed > 10 mph	15,567	108	0.7%
Count > 2	10,674	70	0.7%
Brake off	2,886	49	1.7%
No brake within 2 sec.	2,194	52	2.4%
No recent lane change	19,004	108	0.6%
No warning within 2 sec.	15,446	102	0.7%
All combined	798	21	2.6%

Table 2. Number and Percent of Time ACC Could Have Provided Sufficient Braking

WHAT HAPPENED AFTER WARNINGS. In all the FOT, there were no collisions. In each case where the RECAS indicated that hard braking was necessary to avoid a collision, a collision was avoided. It should be noted that if the following driver began braking sooner than 1.5 seconds after the warning algorithm would have issued a warning (which was often the case), more gentle braking than 0.75 g was required to avoid a collision. In the majority of cases, however, there was no braking at all; instead the situation changed so that braking was not necessary.

Four types of events happened following the time a RECAS warning would have been issued:

- The following driver applied the brakes within two seconds
- Either the leader or the follower changed lanes, so that the follower was no longer behind the leader
- The leading vehicle speeded up so that it was no longer a potential problem
- The lead vehicle decreased its deceleration rate so that its deceleration was no longer high enough to constitute a threat

More than one of these events may have happened at a time. In fact, the third set is a subset of the fourth set. Figure 8 shows the proportions of warnings that were followed by each type of resolution.

Event type one (follower braking) is the resolution anticipated by the warning algorithm. However, at least for the type of driving situations captured in the FOT data, this was the least likely resolution for a warning situation.

A lane change resolved a warning situation more frequently than follower braking. Unfortunately the available data frequently could not reveal whether the following car or the leading car changed lane. The question can be resolved by inspection of the corresponding video if a video episode was recorded at the time of the warning, but (a) videos were not recorded for the majority of potential warning times, and (b) manual inspection of the videos that do match is very time-consuming. A lane change by the follower can be anticipated by the driver, but not by the warning algorithm, unless the driver uses the turn signal and the warning system is able to detect the turn signal. A lane change by the leading car cannot be anticipated by warning algorithm and may not be anticipated by the following driver.

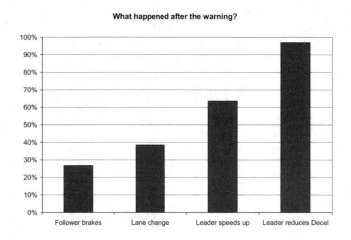

Figure 8. Warning Episode Resolution

The most frequent event to resolve a warning episode is the leading vehicle speeding up, or at least decreasing the rate of deceleration. This event cannot be anticipated by the warning algorithm and may or may not be anticipated by the following driver. The driver has access to more visual information than the collision warning system.

In each of the cases where one of the vehicles changes lane or the lead vehicle does not continue decelerating at the same speed, the following driver may consider that a warning would be unnecessary or even a nuisance. However, the warning algorithm has no way of predicting that one of these events will take place, and thus must act on a conservative basis.

OTHER ANALYSES. Mitretek conducted two other analyses that are not described in this paper. They are:

1. A study of the sensitivity of the results to the assumptions of 1.5 seconds reaction time and 0.75 g deceleration. The results were re-calculated with a reaction time of 1.0 seconds and again with a follower deceleration of 0.4 g.
2. An analysis of the road type (freeway, arterial, ramp, etc.) on which warnings would have occurred.

Results from these investigations are presented in Reference 3.

CONCLUSIONS

This section presents conclusions about the value and the application of an automatic collision warning system that may be drawn from the study.

THE APPLICABILITY OF FOT DATA TO RECAS ANALYSIS. The project has shown the usefulness of naturalistic driving data collection and analysis in assessing the impact of rear-end collision avoidance warning systems. Prior to this project, the formulation of the collision warning algorithm was a hypothetical exercise – there was no way to estimate how frequently a condition would arise in normal driving that would trigger such a warning. A field operational test directly employing RECAS will be the best source of data. In preparation for such a test, this project has successfully integrated the warning algorithm into a large collection of naturalistic driving data and has answered the question "How often would a collision warning have been issued if it had been installed in the FOT?"

Several FOTs have been conducted to collect naturalistic driving data for various purposes. Additional insights may be achieved by applying the techniques of this study to these FOT databases.

Naturalistic driving field operational tests and analyses similar to the ACC FOT should continue to be pursued to help in evaluating rear-end collision avoidance and other collision avoidance systems. Naturalistic driving studies need to be carefully designed to obtain the appropriate data to assess a wide variety of collision avoidance concepts. Analysis of this type of data can be very resource-intensive, and the study design must consider the need for each data element and how it will be used in the analysis. Studies should include a wide range of driving conditions (roadway type, speed, urban/rural, light conditions, surface conditions, etc.) and drivers.

Video coverage during critical events was very helpful to the success of this analysis, and it is recommended that video documentation of critical events be included in all FOTs.

AUTOMATED DATA COLLECTION AND PROCESSING. The body of the ACC data is a significant achievement with significant time spent by the FOT designers to collect clean and consistent data. As discussed in the body of this report, UMTRI and then Mitretek have developed procedures to address issues of consistency and data irregularity – problems inherent in any large repository of data collected by real-time sensors. These procedures should be considered as lessons learned by others designing data collection and analysis plans for field operational tests.

It should be stressed that the analyses performed as part of this work were based on post-processing of the field data. Obviously, the actual success of the on-board RECAS warning algorithm is dependent on the ability of the sensing system to provide accurate and consistent real time data output.

A collision warning algorithm that runs in real time must be able to "keep up." It must be able to perform its calculations at least as quickly as new data is presented to it.

THE RECAS ALGORITHM AND LOCKOUTS. The collision warning equations are based on the physics of vehicle motion and deceleration. There is little room for changes in these equations to tune the performance of the system, other than the assumed deceleration rate and the reaction time. However, if the RECAS system issued a warning every time the algorithm indicated it, there would be an unacceptably high rate of warnings per hour. The lockouts identified for use with the RECAS algorithm significantly reduce the number of unnecessary warnings. For the ACC data, the chosen lockouts reduced the average number of warning per hour from 7.1 warnings per hour to 0.31 warnings per hour. The lockouts should be considered an integral part of the RECAS package along with the warning algorithm. The lockouts need to be defined so that they can be evaluated in real time along with the warning algorithm.

Findings suggested by these analyses are consistent with popular perceptions about which groups tend to drive more aggressively, and thus are likely to receive more warnings. This correlation between driving style and the results of the study tends to confirm that the RECAS warning algorithm and filters are behaving as they should.

ACKNOWLEDGMENTS

The authors acknowledge August Burgett, David Smith, and other NHTSA staff for writing the original warning algorithm and providing ideas and encouragement throughout the project. The authors thank NHTSA contractor Robert Miller for much work developing and extending the warning algorithms.

REFERENCES

1. P. Fancher, R. Ervin, J Sayer, M. Hagan, S. Bogard, Z. Bareket, M. Mefford, J. Haugen, "Intelligent Cruise Control Field Operational Test", University of Michigan Transportation Institute (UMTRI), May 1998 (UMTRI-98-17).

2. "Evaluation of the Intelligent Cruise Control System, Volume I", Volpe National Transportation Center, U.S. Department of Transportation, November 1998. (DOT-VNTSC-NHTSA-98-3)

3. R. Glassco and D. Cohen, "Analysis of a Rear-End Collision Avoidance System (RECAS) Warning Algorithm Using Field Operational Test Data," Mitretek Systems, Washington, DC, December 2000.

4. A.L. Burgett, A. Carter, R.J. Miller, W.G.Najm, and D.L. Smith, "A Collision Warning Algorithm for Rear-End Collisions," 16th International Technical Conference on Enhanced Safety of Vehicles (ESV), Windsor, Canada, June 1-4, 1998, Paper Number 98-S2-P-31.

5. R. Glassco, A. Burgett and A. Chande, "Evaluation of a Rear-End Collision Avoidance Warning System using Data Collected in an Intelligent Cruise Control field Operational Test," Proceedings of 6th Annual ITS World Congress, Toronto Canada, November 1999

2000-01-0346

An Integrated Approach to Automotive Safety Systems

Stephen N. Rohr, Richard C. Lind, Robert J. Myers,
William A. Bauson, Walter K. Kosiak and Huan Yen
Delphi Automotive Systems

ABSTRACT

The industry strategy for automotive safety systems has been evolving over the last 20 years. Initially, individual passive devices and features such as seatbelts, airbags, knee bolsters, crush zones, etc. were developed for saving lives and minimizing injuries when an accident occurs. Later, preventive measures such as improving visibility, headlights, windshield wipers, tire traction, etc. were deployed to reduce the probability of getting into an accident. Now we are at the stage of actively avoiding accidents as well as providing maximum protection to the vehicle occupants and even pedestrians. Systems that are on the threshold of being deployed or under intense development include collision detection / warning / intervention systems, lane departure warning, drowsy driver detection, and advanced safety interiors.

In this paper, we will discuss the concept of *the safety state diagram*, a unified view of the automotive safety system, and the technologies that are required to implement this vision. Advanced ideas such as pre-crash sensing, anticipatory crash sensing, X-by-wire systems, advanced safety interiors, integrated vehicle electrical/electronics systems, data networks, and mobile multimedia (telematics) will also be addressed.

INTRODUCTION

The expanded use of electronics, microcontrollers, sensors, actuators, high-speed data busses, X-by-wire technologies, etc. in the automotive industry will have a major impact on the architecture of future safety systems. Many traditional safety technologies are beginning to merge. Looking at the vehicle as a personal safety system, one can describe five vehicle driving scenarios: normal driving state, warning state, crash avoidable state, crash unavoidable state, and post event state. The first three states focus on accident avoidance while the last three states focus on damage mitigation (with an overlap of the third state). Using the state diagram, it is apparent that automotive safety concerns should be addressed with an integrated system approach.

THE INTEGRATED SAFETY SYSTEM STATE DIAGRAM

As depicted in Figure 1, an integrated automotive safety system may be thought of as a series of interdependent safety states.

Figure 1. Integrated Safety System State Diagram

THE AVOIDANCE ZONE – The three states that comprise the avoidance zone include: normal driving state, warning state, and collision avoidable state. It is important to note that all the safety actions occurring in these states reduce the probability of a collision. This is also known in the industry as *active safety*.

Normal Driving State – Under the normal driving state, a driver enjoys many of the comfort and convenience features afforded by modern automotive electronics. For instance, a millimeter wave or laser-based adaptive cruise control (ACC) system maintains either a constant vehicle cruising speed or a constant headway between vehicles. In addition to the conventional AM/FM radio broadcasts, the on-board telematics system provides cell phone and wireless data capability, as well as GPS, map, and navigational aids. If desired, real-time traffic information can also be accessed through either the Internet or a preferred Call Center.

These features not only provide the needed information in a timely manner, but also offer added safety and security protection for vehicle occupants. For example, the navigation system with its turn-by-turn instructions lets the driver concentrate on the task of negotiating the traffic because there is no need to take his eyes off the road to look for his destination. Additionally, the telematics system allows the vehicle occupants to stay in touch with the dealer or repair shop should anything go wrong with the car.

The next generation of ACC systems will be able to handle low-speed stop-and-go traffic situations as well. This is especially significant since a large percentage of accidents take place under these driving conditions. Some vehicles are already equipped with the night vision system that provides enhanced vision at nighttime for the driver. The system is implemented in such a way that the infrared (IR) image of the scene is projected via the head up display (HUD) unit and superimposed on the driver's natural field of view. With this implementation, the driver enjoys enhanced vision without having to take his eyes off the road.

Some driver monitoring systems have made their way into commercial fleet vehicles because of the desire to ensure public safety and to minimize costly accidents. Most likely, these systems will soon appear in passenger cars as well. Using a combination of biological sensors, eye tracking devices, and vehicle steering information, it is possible to infer the degree of driver alertness. Appropriate countermeasures then can be employed to stimulate the driver or warn the driver that it is time to pull off the road and rest.

A roadway condition sensor that can reliably indicate to the driver whether the roadway ahead is wet, dry, icy, rough, etc. is a valuable safety tool especially during the wintertime. With this sensor information, the vehicle could issue an advisory to the driver to adjust his speed accordingly. Ultimately, the vehicle may automatically adjust its parameters to help the driver maintain control of the vehicle. In short, the vehicle development allows drivers and passengers under the normal driving state to be well protected by a large array of safety, comfort and convenience features. The interior of the vehicle will not only provide the maximum level of protection possible but will adapt according to each occupant's preference. The telematics functions will be working in concert with the safety features of the vehicle through a well designed Human Machine Interface (HMI).

Warning State – Sensing systems are the key in the warning state. The Integrated Safety System (ISS) must maintain full awareness of the driving situation in order to detect potential crash situations. Sensing needs range from external object detection (other vehicles, trees, signs, etc.) to internal vehicle states (tire pressure, vehicle stability, etc.). Sensor and data-fusion algorithms will combine information from various sensors to form a model of the current situation. Ultimately, the ISS needs to be able to sense objects around the entire (360 degrees) vehicle.

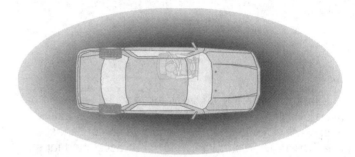

Figure 2.   ISS 360 degree sensing

Once the situation is understood by ISS, the appropriate warning can be delivered to the driver. At this point, the driver must take action to avoid the collision.

Practical vehicle applications involved in the warning state include:

- Low tire pressure warning
- Impending rollover warning
- Lane / Roadway departure warning
- Parking assistance/warning
- Back-up assistance/warning
- Blind-spot warning
- Rear-end collision warning
- Lane change warning

One particularly difficult problem arises when sensing objects at longer ranges: the issue of path prediction. If the ISS is unable to accurately project the host vehicle's path, it will be impossible to determine which objects in the field of view represent a threat to the host vehicle. If, in the scenario shown below, the ISS vehicle is unaware that the road is curving, then there are several objects that would appear to be directly in the path of the host vehicle. The ISS would incorrectly warn the driver of all the supposed threats. For the ISS to be effective, it is important to provide warning or control signals at the appropriate time and to minimize false alarms and nuisance alerts.

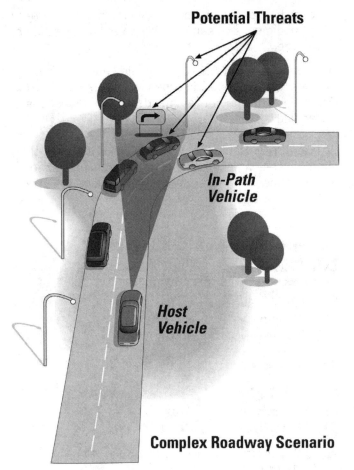

Potential Threats

In-Path Vehicle

Host Vehicle

Complex Roadway Scenario

Figure 3.   Complex Roadway Scenario

Assuming the driver follows the road, the only object actually in the path of the host vehicle is the vehicle around the curve in the same lane. Depending on the vehicle spacing and relative speeds, this vehicle may or may not represent a threat. Correct detection of road geometry would enable the ISS to correctly ignore all objects except the vehicle in the same lane as the host vehicle.

The solution to the path estimation problem is highly dependent on roadway geometry, inter-vehicle kinematics, driver reaction times, and braking behaviors. Yaw-rate sensors can be used to determine roadway curvature, but only when the host vehicle is already in the curve. For scenarios such as curve entry or curve exit, other sensors such as solid-state cameras will be needed. GPS, digital road maps, and roadside transponders can also be used to increase the accuracy and robustness of the ISS path estimation algorithms.

Several other issues surface in the warning state:

- Threat Assessment: The ISS must determine whether a particular situation poses a threat and merits warning the driver. Driver preferences must be taken into account as well. One driver may not want to be warned until the threat is severe while another driver may want to be notified at the slightest hint of trouble.

- False alarms and nuisance alerts: It is important to eliminate or minimize false alarms to win the driver's confidence in the system.
- Human Factors: In a critical situation, the ISS must notify the driver in a way that is quickly recognized and that encourages the driver to take the proper action. It is vital that warnings not distract or confuse the driver during an impending collision.

Collision Avoidable State – The collision avoidable state is our last opportunity to avoid an accident and return to the normal driving state. Everything that is done in this state is with the intention to completely avoid the accident. Reaction time as well as vehicle stability and control are extremely important in this state. The ability to stop or steer clear of harm is the most important. Examples of features included in this state are:

- Automatic stopping
- Automatic lane change
- Lane keeping
- Chassis and suspension control
- Vehicle rollover prevention
- X-by-wire (steer, brake, throttle)

Figure 4.   Future "by wire" vehicle control

The implementation of the collision avoidable state is primarily dependent upon two things: a suite of sensors and sensor fusion algorithms that provide information about the state of the vehicle and its surroundings as defined in the warning state; and a suite of X-by-wire products (steering, braking, throttle, and suspension) that de-couple the actuation from the mechanical input provided by the driver as shown in the illustration above. This de-coupling is a key enabler as it allows the vehicle to be commanded to perform various maneuvers without direct driver input.

Consider the current anti-lock brake systems that are prevalent on passenger vehicles and light trucks today. In these systems, the braking function is augmented by a computer controlled brake release and then re-applied to mitigate the effect of wheel slippage. It is important to note that the initiation of the brake function today still requires the driver's input.

Collision avoidance features will evolve into three modes:

- Driver initiated
- Vehicle initiated
- A blend of both

Consider again a simple braking maneuver in a vehicle equipped with the necessary sensors and a brake-by-wire system. Under any driving conditions, the vehicle will know its speed, closing speed of approaching objects, road surface conditions, driver intended path, and vehicle attitude (pitch and yaw). Once the driver requests braking, the system will provide the appropriate level of braking effort to effect a normal stop.

What will happen if the ISS vehicle has detected a slowed or stopped object in its path and the driver has ignored all of the warnings or is unable to command the brake function by stepping on the brake pedal? Because of the brake-by-wire product the vehicle will have the ability to initiate braking without input from the driver.

The above examples can all be accomplished with only a brake-by-wire system in the ISS vehicle. The options to enhance vehicle performance and stability are greatly increased when by-wire steering, throttle, and suspension are added. If the ISS vehicle detects that it is closing upon an object too quickly, then a simple vehicle commanded throttle reduction might be a viable response.

Or, perhaps the driver has initiated a very quick turn that will take the vehicle into an unstable condition. This could be from lack of driving experience, or over reaction to a driving situation. With an effective ISS system, the steering angle may be reduced, throttle reduced, independent differential braking applied and the suspension stiffened all simultaneously without input from the driver to keep the vehicle stable.

THE MITIGATION ZONE – The three states that comprise the mitigation zone include: collision avoidable state, collision unavoidable state, and the post event state. All safety actions that occur in these states focus on reducing the effects of a collision.

Collision Avoidable State – Note that this state appears in both the avoidance and mitigation zones. It's obvious that the best way to protect an occupant, pedestrian, or property is to avoid an accident. If the accident cannot be avoided, then the goal is to reduce its effects. Many of the systems used in collision avoidance come into play for damage mitigation. With automatic braking, for example, we can slow down the vehicle as much as possible to minimize injury and damage.

Using the same sensors and fusion algorithms described in the warning state, the ISS vehicle could prevent the driver from directing the vehicle to cross the roadway centerline and into the path of an oncoming vehicle. Or, it could inhibit the driver from a collision with a utility pole or bridge abutment that can possibly be at the right side of the vehicle.

In essence, when all the vehicle control authority of braking, steering, throttle, and suspension has been used, the ISS vehicle will attempt commands for the "softest possible landing."

Collision Unavoidable State – Since this is the point of no return, everything should be done immediately before as well as immediately after the crash to reduce the effects of the accident.

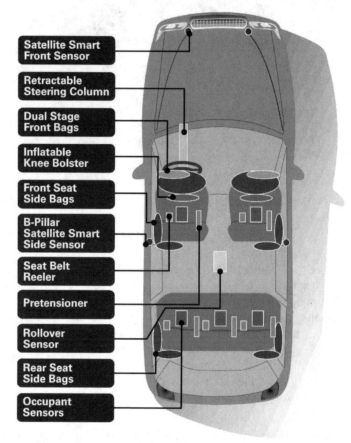

Figure 5.   Near-term, high-content occupant restraint system.

The collision unavoidable state has traditionally encompassed the realm of occupant protection. These features include everything from crashworthy vehicle structures and interior padding to seatbelts and airbags. Interest in "smart" or advanced airbag systems has intensified recently in order to provide improved occupant protection under a variety of real-world accidents, as well as to minimize the potential adverse effects caused by airbag deployments. "Smart" restraint systems are intended to be more adaptable to various real-world factors such as crash type, crash severity, seat belt usage, and occupant type and position.

As occupant protection countermeasures increase in number and sophistication, electronic sensing requirements continue to grow. The finer the sensing resolution of both vehicle dynamics and occupant kinematics, the "smarter" the complete occupant protection system becomes.

Although the above technologies provide significant benefits in the area of occupant protection, the advent of collision avoidance technologies has now made it possible to incorporate new functionality into the vehicle.

Anticipatory or pre-crash sensing is a key enabler not only to post-impact countermeasures (such as variable stage airbags and seatbelt tensioners), but also for resettable, pre-impact countermeasures (such as adaptable interior and exterior structures and pedestrian protection countermeasures).

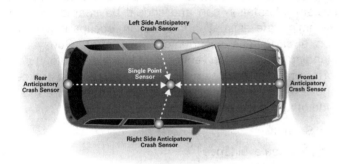

Figure 6.   Joint discrimination using anticipatory crash sensors

Post Event State – After an accident has occurred, the ISS vehicle will automatically assess the severity of the event based on a number of sensory indications:

- Did the airbags deploy?
- Did the vehicle roll over?
- What is the rest position of the vehicle?
- How many occupants are there?
- What are the vital signs of the occupants?
- Is there a fuel leak?
- Is there a fire?

In the case of a severe accident (i.e., airbags were deployed), the vehicle's telematics system will automatically dial 911 (or equivalent) and summon help. If the occupants are still able to communicate with the dispatcher, the extent of injuries can be obtained and passed on to the paramedics. If the occupants suffered severe injuries and could not communicate with the dispatcher, then on-board biosensors can be used to assess the situation. If the vehicle is equipped with a video camera in the passenger compartment, video images can be used to aid in situation assessment.

In addition, the vehicle will have enough intelligence to detect and extinguish fires, release seatbelts, and unlock car doors (allowing easy egress as well as greater ability to be reached by rescuers).

To prevent fires caused by a ruptured fuel line, the fuel pump can be shut off automatically, the engine can be turned off remotely, and unnecessary electrical power can be disconnected. If the event takes place at night, the vehicle can also provide illumination and road hazard flashing warning lights.

In the post event state, there are many possibilities for enhancing the survivability of the victims. Many of these features are enabled by the telematics system. In order for these features to be available, the telematics system must survive the accident. Therefore, there is an implied level of robustness that must be built into the system.

A CASE STUDY – COLLISION AVOIDANCE AND OCCUPANT PROTECTION

An interesting phenomenon occurred as we looked at the future of pieces to the integrated safety systems puzzle. No matter how we approached the problem, it was evident that feature/function sets, as well as technology building blocks, tended to merge as time passed. In the following example, we'll discuss how collision avoidance and occupant protection tend to *blend together*.

FOREWARN® COLLISION AVOIDANCE – Collision avoidance systems depend on short- and long-range sensors to characterize the location and motion of objects around the vehicle. A typical system determines object attributes with a suite of sensors, develops a model of the scene around the vehicle, and issues a set of vehicle control commands depending on the desired system function.

In its simplest form, the function would be to issue a warning to the driver so that the driver can take the appropriate avoidance action. In a full collision avoidance system, the desired function involves automatic lateral or longitudinal control of the vehicle.

To develop this capability, the industry has been following a path similar to the one shown below. Several enabling technologies are needed to achieve the vision of a true collision avoidance system: long- and short-range object detection sensors, x-by-wire systems, appropriate human-machine interfaces, etc.

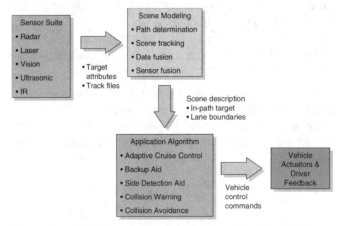

Figure 7.   Collision Avoidance System Mechanization

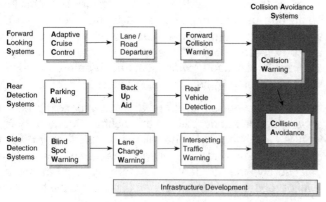

Figure 8.   Collision Avoidance Systems Development

SHORT-RANGE PROXIMITY SENSING NEEDS – Collision avoidance systems are being developed to do as the name implies – avoid accidents. As we've shown, short-range proximity sensing is one of the key enablers. We know, however, that there is a long list of needs for short-range sensors as well. Some of these potential applications include:

- Power doors and liftgates
- Express close window and sunroof systems
- Occupant position sensing
- Security system applications
- Pre-crash sensing systems
- Active pedestrian protection systems
- Parking / Back-up aid
- Close cut-in detection
- Anti-trap trunks

The main point to understand is that features that traditionally have been looked at separately by product-focused teams basically need the same fundamental technical solutions.

ANTICIPATORY (PRE-CRASH) SENSING – By integrating long-range cruise control and collision warning sensors with short-range sensors needed for urban automatic cruise control (stop-and-go driving), pre-crash sensing systems will have the ability to perform object detection and tracking up until the actual time of impact. This ability now provides the capability to calculate angle-of-impact and region-of-impact, both critical parameters to understand crash type and to deploy the appropriate occupant protection countermeasures.

The logical next step is to integrate the vision systems utilized for lane tracking/lane departure into the system to provide object classification. This feature will provide the remaining information needed to truly predict crash severity.

CONCLUSION

No matter how you approach it, the future of automotive safety systems is certainly an integrated, vehicle systems-level approach.

Safety technology roadmaps are beginning to look alike. Collision avoidance sensors and occupant recognition sensors employ basically the same technologies. The same can be said about vehicle dynamic control sensors and vehicle crash sensors, as well as distributed safety architectures and distributed mobile multimedia architectures.

Safety features and functions are blending. The best way to protect an occupant is to avoid the accident. Subsystem information can and should be shared (vehicle dynamic state estimation information, occupant information, airbag status, scene information, etc.). Subsystem blending can enhance vehicle (and integrated safety) performance (e.g. a collision threat can mute a radio and cell phone, etc.). Proximity sensors can be used for a multitude of applications including:

- Security
- Collision avoidance
- Occupant position and recognition
- Pre-crash sensing
- Pedestrian protection systems

As a result, a systems approach to integrated safety is driving our future developments.

ACKNOWLEDGMENTS

The authors would like to acknowledge the Delphi Automotive Systems Advanced Vehicle Systems, Collision Avoidance, AED, Chassis, and Restraints Electronics teams for their assistance and hard work in making the contents of this paper a reality.

REFERENCES

1. W.G. Najm, "Comparison of Alternative Crash Avoidance Sensor Technologies," SPIE Vol. 2344 Intelligent Vehicle Highway Systems, 1994.

2. W.K. Kosiak and G. Nilson, "Benefits and Issues of Anticipatory Crash Sensing for Enhancing Occupant Protection," 3rd International Symposium of Sophisticated Car Occupant Safety Systems, 1996.

3. W.K. Kosiak and S.N. Rohr, "Future Trends in Restraint Systems Electronics," SAE Brazil, 1999.

4. Y. Shimizu, T. Kawai, J. Yuzuriha, "Improvement in driver-vehicle system performance by varying steering gain with vehicle speed and steering angle: VGS (Variable Gear-ratio Steering system)," SAE 1999-01-0395, 1999.

5. H. Kuzuya, H. Nakashima, K. Satoh, "Development of Robust Motor Servo Control for Rear Steering Actuator Based on Two-Degree-of-Freedom Control System," SAE 1999-01-0402, 1999.

6. S.M. El-Demerdash, A.M. Selim, "Vehicle Body Attitude Control Using an Electronically Controlled Active Suspension," SAE 1999-01-0724, 1999.

7. G.R. Widmann, "Delco Electronics' Activities in Collision Avoidance – The Path to Deployment of a 'Smart Vehicle'," SAE Toptech, 1997.

2000-01-1301

Development of an Intersection Collision Warning System Using DGPS

Zhengrong Yang, Takashi Kobayashi and Tsuyoshi Katayama
Japan Automobile Research Institute

ABSTRACT

In this paper, an intersection collision warning system using DGPS (Differential Global Positioning System) will be proposed. The system is developed to prevent collisions of vehicles crossing at intersections, especially at well visibility intersections without traffic lights. Two GPS receivers are installed on two vehicles on the move towards the same intersection from different directions. The position and velocity information of the vehicles are measured by on-board GPS receivers, and then transmitted from one vehicle to another by inter-vehicle communication (IVC). Therefore, the relative position and direction of each vehicle and collision judgment coefficient (CJC) which is defined by using the relative position of vehicles are calculated. After taking the crossing position of directions and the variation of CJC into account, the position of the intersection and the possibility of collision can be predicted in advance. Warning will be given to drivers with a prearranged timing.

INTRODUCTION

A possible explanation of causative factors in intersection accidents was proposed by N.Uchida, et al. [1]. It is gen-erally believed that a vehicle which may collide remains unchanged in the visual field, as shown in Figure 1. If the shape of the triangle defined by two vehicles and an intersection does not change with time, it becomes diffi-cult for two drivers to find each other unless they come close to the intersection; a different triangle configuration shows that it is easy for one driver to see the other well before reaching the intersection, as shown in Figure 2. These two examples have shown that the human visual search system has an inherent weakness which is a risk factor for certain types of accident. A support system for overcoming this inherent weakness will be needed for traffic safety in the near future.

In order to eliminate or mitigate traffic accidents at inter-sections, many detection systems have been developed. In this paper, a collision warning system using DGPS for preventing collision accidents at intersections is described. The system can be used in many traffic situa-tions, especially at intersections with good visibility with-out traffic lights and under the condition of bad weather. As a support system for drivers, it would be effective. The collision avoidance algorithm will be described below.

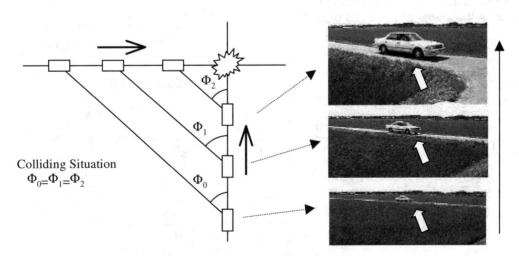

Figure 1. Relative Vehicle Position and View of Crossing Vehicle

Colliding Situation
$\Phi_0 = \Phi_1 = \Phi_2$

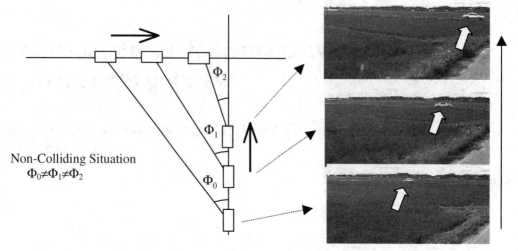

Non-Colliding Situation
$\Phi_0 \neq \Phi_1 \neq \Phi_2$

Figure 2. Relative Vehicle Position and View of Crossing Vehicle

SYSTEM ELEMENTS AND EXPERIMENT

The system sketch map is shown in Figure 3, which consists of a base station and two vehicles with GPS receivers. The base station (see Figure 4) sends necessary information to automobiles installed with modems (transmitter and receiver, see Figure 5) for inter-vehicle communication (IVC), so that vehicle position can accurately be measured.

The base station and the automobiles receive the signals from GPS satellites. A signal to improve the coordinate accuracy of the vehicles A and B is calculated at the base station, and sent to vehicles A and B at the same time. All this information (namely, three dimension coordinate, or latitude/longitude, velocity and time, etc.) for one vehicle will then be transmitted to another vehicle by IVC to provide vehicle position in real time. The positions of both vehicles will be stored in the computer, and the relative geometrical position and vehicle directions will be calculated during movement. Therefore, based on the velocities and directions of both vehicles, If the directions of the vehicles cross, the intersection position and the possibility of vehicle collision at the intersection can be predicted in advance.

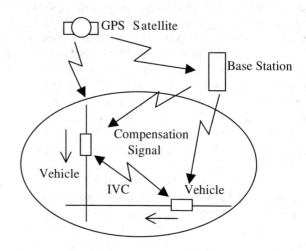

Figure 3. Basic Elements of the System

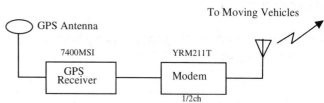

Figure 4. Function of the Base Station

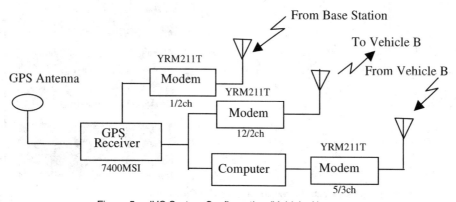

Figure 5. IVC System Configuration (Vehicle A)

The experiment was performed on a village road. Two vehicles (A and B) with the system shown in Figure 5 were used. The velocities of the vehicles remained constant at 10km/h-40km/h. The starting point of each vehicle was about 100 or 200m away from the intersection. And, the sampling frequencies of the GPS receivers at the base station and vehicles were set to 1Hz.

DETECTION METHOD

As shown in Figure 1, it is clear that we will have a relationship of $\Phi 0=\Phi 1=\Phi 2$ if the shape of the triangle formed by the vehicles and intersection doesn't change with time. In other words, if both vehicles run at constant speed and will collide at the intersection, the triangle shape remains the same. And this can be defined by the following coefficient,

$$S =(Ya-Yb)/(Xa-Xb) \qquad (1)$$

where Xa, Ya are the coordinate values of vehicle A, Xb, Yb are the coordinate values of vehicle B. In the case of Figure 1, S will be constant.

For this reason, whether vehicles will collide or not at an intersection can be judged by the variation of this coefficient. We thus call the coefficient S the Collision Judgment Coefficient (CJC).

As an example, the relative position between vehicles is shown in Figure 6. We assume that vehicle A is located at Xa,Ya (Xa=80m,Ya=0m) at a certain time and has velocity Va, and vehicle B at Xb, Yb (Xb=0m,Yb=160m) respectively. Xa and Yb are also the distances of vehicles A and B to the intersection respectively. And assume also that the velocity of vehicle B is Vb=2Va. When these values are substituted in Equation 1, S (CJC)= -2. If vehicles A and B run towards the same intersection at the constant velocity shown above, the line that connects a and b will move in parallel towards the intersection. CJC will be a constant value (absolute value) as shown in Figure 7. In this case, it would be certain that the vehicles will collide at the intersection. However, under the same velocity condition, let us assume that vehicle A is 90m or 70m away from the intersection to vehicle B at a certain time (see Figure 6). In this case, this means that vehicle A will pass the intersection earlier or later than vehicle B respectively. Thus, collision would not occur. The lines ac and ad are not in a parallel movement. The CJC variance (absolute value) in this case is shown in Figure 7. It is shown that, when vehicles run at constant velocity and collide at intersection, CJC is a constant value. But when vehicles move at uncertain or changing velocities, CJC will fluctuate.

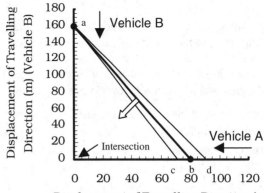

Figure 6. Relative Position between Vehicles at Intersection

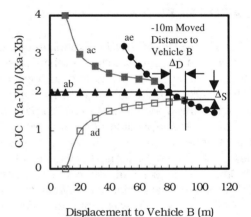

Figure 7. Variation of CJC for Relative Position of Vehicles

Under the same velocity condition shown above, if the starting position of vehicle A is shifted in the longitudinal direction at every five meters in respect to vehicle B position (Xb=0m,Yb=160m), a line ae as shown in Figure 7, the CJC will vary with the relative positions between the vehicles. ΔD is a length range that shows vehicle A is displaced 10m with respect to vehicle B, namely a length of about twice of a passenger car. ΔS is the variance of CJC for the length range ΔD. As shown in this figure, when one vehicle has about 10m displacement from the initial position, CJC will fluctuate about 10%. In this time, the fluctuation of CJC is evaluated by comparing it with the mean value of the past n points (here, n=3). If the CJC value fluctuates more than 10% of the preceding mean value of CJC, the automobiles (passenger cars) travelling towards the same intersection do not collide at the intersection. However, if the fluctuation of CJC is less than 10% of the preceding mean value, the possibility that the vehicles collide at the intersection may exist. Hence, in order to use CJC for judging collision probability, we suggest a 10% fluctuation rate of CJC with respect to the mean value of the past n points for passenger cars.

WARNING TIMING

Warning timing is strongly related to driver perception responding time, vehicle velocity and traffic situation. It is also an important subject on human interface design and must be examined in detail. However, under the present circumstances, we make the following assumptions:

1. Driver's forward view prediction time (a kind of margin time)[2]: 2 (sec)
2. Warning perception time ranged from about 1.0 to 1.2 second [3]: 2 (sec)
3. Braking response time ranged from about 0.4 to 0.8-second [4]: 1.0 (sec)
4. Communication system delay: 0.8 (sec)

Summing up, we suppose a warning time of 6 seconds before reaching an intersection.

SIMULATION RESULTS

INTERSECTION PREDICTION – Experiment and computer simulations for the system algorithm were carried out. The simulation result of the intersection prediction based on the experimental data is shown in Figure 8. The travelling directions were calculated by the past m (m=3) positions and present position of each vehicle. The following parameter values were used: A and B vehicle velocities were 10km/h, their distances to the intersection were 100m. It is shown that, when the vehicles travel towards the same intersection for 4 seconds, the intersection that both vehicles will cross at is found to be about 90m ahead and is well predicted in advance by the direction of each vehicle.

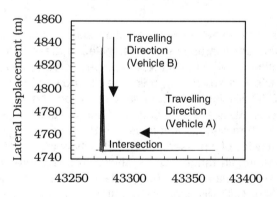

Figure 8. Intersection Prediction using Travelling Direction

COLLIDING SITUATION – The simulation results using the experimental data are shown in Figure 9. Figure 9(a) is a situation where vehicles A and B had a velocity of 20km/h, both vehicles were 100m from the intersection. Figure 9(b) shows a case of different velocities. The distances to the intersection were 200m (vehicle A) and

100m (vehicle B), and the velocities of vehicles A and B were 40km/h and 20km/h, respectively.

The judgment signal was defined as the one that presents the possibility of an intersection in front of a vehicle. When the value of the judgment signal is zero, it means that there is not an intersection; if it takes on a certain value, there exists a possibility of an intersection ahead in the travelling direction. As shown in these figures, when the vehicles travelled towards the same intersection for about 2 - 3 seconds, the judgment signal changed from zero to a certain level, and an intersection about 100m-200m ahead was predicted in advance by the direction of each vehicle. In the proposed system algorithm, as the judgment signal becomes a certain value, collision judgment calculation will be performed. The calculation results of CJC are plotted in Figure 9(a) and (b). Because the fluctuation of the coefficient is less than 10% in the neighborhood of the intersection, a warning is issued to the drivers when the vehicles approach the intersection and are only 6 seconds away from the intersection. The warning will continue until the vehicles pass the intersection.

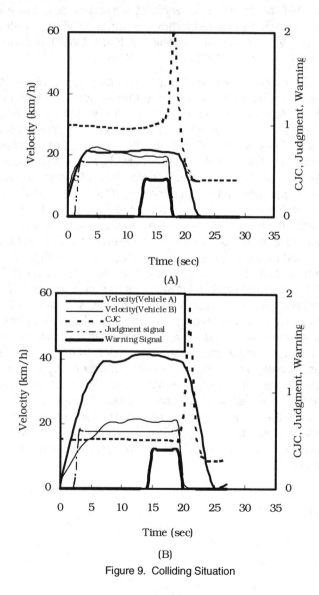

(A)

(B)

Figure 9. Colliding Situation

NON-COLLIDING SITUATION – Figure 10 shows a non-colliding situation, where the distance of both vehicle A and B from the intersection is 100m. The velocities are 15km/h (vehicle A) and 10km/h (vehicle B), respectively. On the basis of the velocity and distance information, the vehicles would not collide at the intersection, thus the warning should not be issued to the drivers. Nevertheless, the intersection that is approached is correctly predicted in advance.

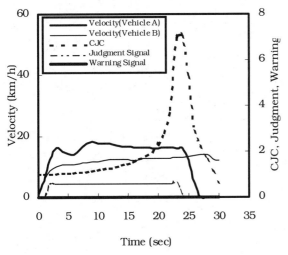

Figure 10. Non-colliding Situation

CONCLUSION

An intersection collision warning system using DGPS was proposed. The effectiveness of the system was confirmed by experiment and computer simulation. Using the relative position between automobiles and vehicle direction, the presence of an intersection, and the possibility of inter-vehicle collision can be predicted in advance. Warning will be issued to drivers at a timing that has been set up when the mean value of the collision judgment coefficient (CJC) fluctuates less than 10% in comparison with the preceding mean value of CJC for passenger cars.

Although the experiment and simulation were carried out on general passenger cars, the system may be adapted to all kinds of automobiles, if the CJC coefficient corresponding to the length of vehicle is appropriately modified.

In the study, the judgment of collision was performed solely based on the CJC value. The future direction of study will be focused on more effective collision judgment algorithm by integrating the predicted time to reach the intersection and human interface.

REFERENCES

1. N.Uchida, K.Fujita and T.Katayama. Detection of Vehicle Crossing Path at Intersection. Journal of Japan Automobile Research Institute, Vol.19, No.2, February 1997.
2. I.Kageyama and Y.Nozaki. Lateral Control for Autonomous Vehicle with Risk Level. The 4th Transportation and Logistics Conference, Japan. No.2204. December 5, 1995.
3. T.Fujioka, K.Muramatsu and M.Aso. ITS Study about Driver's Behavior Using Driving Simulator -Driver's Behavior in Emergency Situation-. JSAE Review, No.72. September 1995.
4. P.L.Olson. Driver Perception Response Time. SAE Paper, 890731(1989).

CONTACT

Zhengrong YANG
E-mail: zyang@jari.or.jp
http://www.jari.or.jp

1999-01-1238

An Experimental Investigation of a CW/CA System for Automobiles

Sung Ha Kim, Pil Soo Moon, Woon Sung Jang, Kun Sang Kim and Seong-chul Lee
KIA Motors Corp.
Seoul, KOREA

Minsu Woo and Kyongsu Yi
School of Mechanical Engineering, Hanyang University
Seoul, KOREA

ABSTRACT

CW/CA(Collision Warning /Collision Avoidance) Systems have been an active research and development area as interests and demands for the advanced vehicle increase. A CW/CA 'Hardware-in-the-Loop Simulation (HiLS)' system has been designed and used to test a CW/CA algorithm, radar sensors, and warning displays under realistic operating conditions in the laboratory. A CW/CA algorithm has two parts. One is a distance decision algorithm that determines the critical warning and braking distance and the other is a brake control algorithm for collision avoidance. The CW/CA HiLS system consists of a controller in which a DSP chip is installed, a preceding vehicle simulator, a radar sensor and a warning display. The controller calculates velocities of the preceding and following vehicles, relative distance and relative velocity of the vehicles using vehicle simulation models. The relative distance and velocity are applied to the vehicle simulator that is controlled by a DC motor. The relative distance of the vehicle simulator is measured by the radar sensor and is fed back into the controller. Finally, the controller gives warnings and applies the brakes according to the CW/CA algorithm.

INTRODUCTION

Most Accidents are caused by driver's unconsciousness of a dangerous situation. So, it would be greatly helpful for drivers if they perceive the danger ahead of the accident in any situation. The CW/CA system is the advanced safety device that gives an advance notice to the driver and makes him meet that situation. In even more dangerous situation it applies the brakes to avoid the accident. To develop the system like this, many kinds of simulations are necessary for a driver's safety and saving expenses and a period of time. In this point of view, the CW/CA HiLS system is very useful.

The CW/CA HiLS system is implemented with a controller, a vehicle simulator, a radar sensor, a warning display, and so on. A 24GHz radar sensor only gives information of the range to an object. Kalman filter is used for the range rate. A vehicle simulator is controlled by DC motor and simulates the motion of a preceding vehicle. Two 6-DOF models for a preceding and a following vehicle will be operated in the controller to generate the reference signal of the simulator. The controller has high speed of process. It determines the level of the danger by a distance decision algorithm and applies the brakes by a brake control law. Simultaneously, it controls a simulator and a warning display [7].

The warning index that indicates the level of the danger is the function of the warning distance, the braking distance and the relative distance. The warning and the braking distance are mainly defined with the relative distance and velocity. The brake controller will operate while the warning index is less than zero. It allows the brakes to be applied to slow down the vehicle in constant deceleration.

A CW/CA HARDWARE-IN-THE-LOOP SIMULATION SYSTEM

The CW/CA HiLS system is shown in fig. 1. It consists of a simulator that simulates the motion of a preceding vehicle, a radar sensor, a controller, a warning display and a DC motor. The control program is written in C code and downloaded to a controller through the ether net. The controller communicates with a personal computer while running and calculates velocities of the preceding and following vehicles, relative velocity and relative distance of the vehicles using vehicle simulation models. It gives the signal to the motor driver in accordance with relative velocity and relative distance. And a DC motor driver operates the vehicle simulator.

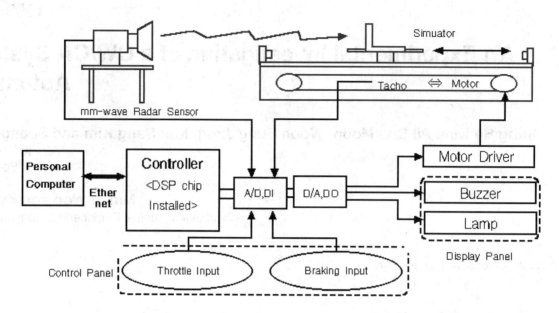

Figure 1.The schematic diagram of the CW/CA HiLS system

The relative distance of the vehicle simulator is measured by a radar sensor and is fed back to the controller. The controller estimates the relative velocity from the relative distance that is measured by the radar sensor at every sampling time and judges the level of danger by calculating the warning index. For the last step, the warning display device displays how much dangerous the present situation is.

KALMAN FILTER DESIGN FOR RELATIVE DISTANCE AND VELOCITY

A radar sensor gives only relative distance information of the vehicle simulator. Relative velocity as well as relative distance is very important information in a CW/CA algorithm. So, it has to be calculated in the controller. There is several ways to get the velocity from the distance. Among them, Kalman filter is applied to ensure noise-rejection and to get the accurate velocity signal as follows.

The basic relations of the distance, the velocity and the acceleration are:

$$v = \frac{dx}{dt}, \quad a = \frac{dv}{dt} \tag{1}$$

Let $x_1 = x$, $x_2 = v$, the state equation is:

$$\begin{aligned}\dot{x}_1 &= x_2 \\ \dot{x}_2 &= a\end{aligned} \quad \dot{x} = \begin{pmatrix} 0 & 1 \\ 0 & 0 \end{pmatrix} x + \begin{pmatrix} 0 \\ 1 \end{pmatrix} a \tag{2}$$

A distance signal comes out of the radar sensor and is measured:

$$\tag{3}$$

where n is random noise.

Figure 2. Comparison of Kalman filter and the numerical method

Assuming a desirable value for noise variance, the estimate of velocity could be gained from the following equation [5].

$$\tag{4}$$

where   mean estimates of

Fig. 2 is one of the experimental results of estimating range rate signal from a radar sensor. Notice that Kalman filter works better than the numerical method in estimation of a range rate.

VEHICLE MODEL

The objective of simulating two vehicle models in the controller is to generate more realistic signal for a reference of the simulator. The 6-DOF vehicle model includes dynamics of engine, torque converter, shaft torque, front wheel, rear wheel and vehicle body [1]. And gear ratio and tire slip to get longitudinal force with are also considered. Results of fig. 3 are acquired by experiments and computer simulation. They validate the use of vehicle models.

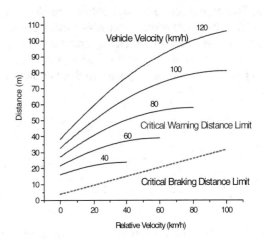

(a)   critical distances

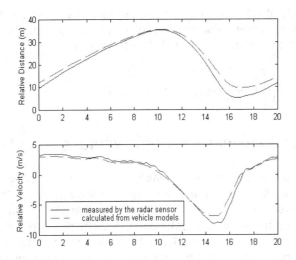

Figure 3.   Validation of vehicle models

CW/CA ALGORITHM

DISTANCE DECISION ALGORITHM – A CW/CA algorithm consists of a distance decision algorithm and a brake control algorithm for collision avoidance. The first algorithm determines the level of warning and on/off of braking according as the relative distance and velocity vary. All warning and braking decisions are based on the following non-dimensional warning index [1]:

$$w = \frac{d - d_{br}}{d_w - d_{br}} \qquad (5)$$

$$d_w = \frac{1}{2}\left(\frac{v^2}{a} - \frac{(v - v_{rel})^2}{a} \right) + v \cdot (\tau_{sys} + \tau_{hum}) + d_0 \qquad (6)$$

$$d_{br} = v_{rel} \cdot (\tau_{sys} + \tau_{hum}) + 0.5 \cdot a \cdot (\tau_{sys} + \tau_{hum})^2 \qquad (7)$$

where d is actual vehicle spacing, d_{br} is the critical braking distance and d_w is the critical warning distance. Also, v is the vehicle velocity, v_{rel} is the relative velocity between vehicles, a is the maximum deceleration of both vehicles, t_{sys} is the system delay, t_{hum} is the driver delay and d_0 is a headway offset.

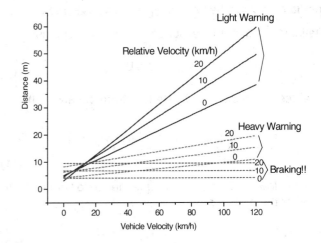

(b)   warning points

Figure 4.   The characteristics of a distance decision algorithm for warning and braking

If w is greater than 1, it means that the vehicle is driving safely and no warning occurs. If w lies between 1 and a (where a is distinction between light warning and heavy warning. It is restricted within 0 to 1 and closer to 0 than 1 in most cases.), then warning display device warns the driver lightly. In that case only visual warning occurs. If w is between a and 0, d is very close to d_{br}. Strong warnings that include audio and visual warnings are given to the driver.

Finally, if w is less than 0, then d become less than d_{br}. So, the system applies the brakes to avoid prevent the collision. This algorithm could be modified by scaling all distances based on driving styles and environmental conditions such as tire-road friction [2], the weather, day or night, and so on. Two figures of fig. 4 show the characteristics of a distance decision algorithm. From them, it is easily noticed where the controller will give warnings to a driver and apply the brakes.

BRAKE CONTROL ALGORITHM – As mentioned before, The object of the brake control is to slow down the vehicle in constant deceleration when the warning index is less than 0. A two-surface sliding controller will be used to control the vehicle with desired deceleration. First, the vehicle dynamics can be reduced to the simplified model as follows.

$$\dot{v} = \frac{T_e - R_g R_d (F_a r + T_{rr} + T_b)}{\beta} \tag{8}$$

$$\beta = \frac{I_e + I_g + (R_g R_d)(I_w + mr^2)}{r} \tag{9}$$

where T_e is the engine torque, T_b is the brake torque, R_g is the gear ratio, R_d is the final speed reduction ratio, F_c is the aerodynamic drag force, r is the wheel radius, T_{rr} is the rolling resistance moment, I_e is the inertia of the engine, I_g is the inertia of the gear, I_w is the inertia of driven wheels and m is the vehicle mass.

Defining the upper level surface for acceleration as:

$$S_a = a_{des} - a = 0 \tag{10}$$

The desired torque can be derived from equation (8):

$$T_{bdes} = \frac{1}{R_g R_d}(T_e - \beta a_{des} + R_g R_d r F_a) \tag{11}$$

Next, following the desired torque, define lower level surface for the brakes as:

$$S_b = T_b - T_{bdes} \tag{12}$$

To force the surface to zero, we set:

$$\dot{S}_b = -\lambda_b S_b \tag{13}$$

So, in the development of this brake torque controller, T_b and $_{bdes}$ are to be differentiated. Notice that the derivative of T_{bdes} results in many terms. To avoid this explosion of terms, the following first-order filter is used to get z_b:

$$\tau_{bz}\dot{z}_b + z_b = T_{bdes} \tag{14}$$

Moreover, from the brake dynamics, we can get [3,4]:

$$\tau_b \dot{T}_b + T_b = kP_{mc} - 150 \tag{15}$$

Finally, the pressure of master cylinder to guarantee the desired deceleration of the vehicle is derived from above equations, (13), (14) and (15) as follows:

$$P_{mc} = \frac{1}{k}\left[T_b + \frac{\tau_b}{\tau_{bz}}(T_{bdes} - z_b) - \lambda_b(T_b - z_b)\tau_b + 150 \right] \tag{16}$$

COLLISION WARNING TEST VEHICLE

A test vehicle has been established as shown in fig. 5 and fig. 6 to verify the CW system on the real road. The radar sensor was installed at the front side of the engine compartment. Spatial restriction moves its position a little bit left from the center. The CW controller embedded in the trunk space hooked up the speed signal from the instrument panel as well as the relative distance signal from the radar sensor. It estimates the relative velocity applying the Kalman filter and gives warnings depending on the situation.

A warning device was designed as in fig. 7. The warning indicator indicates the degree of danger. The number of lamps is lit up as the situation goes more dangerous. The buzzer and the warning light operate periodically. The period has three steps.

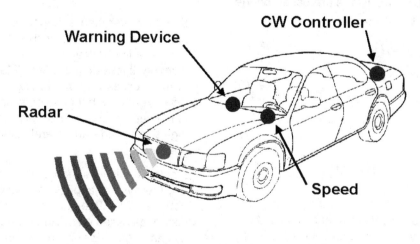

Figure 5. Installation of Collision Warning system

Figure 6. A front view of the CW test vehicle

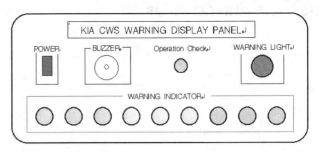

Figure 7. The warning display panel

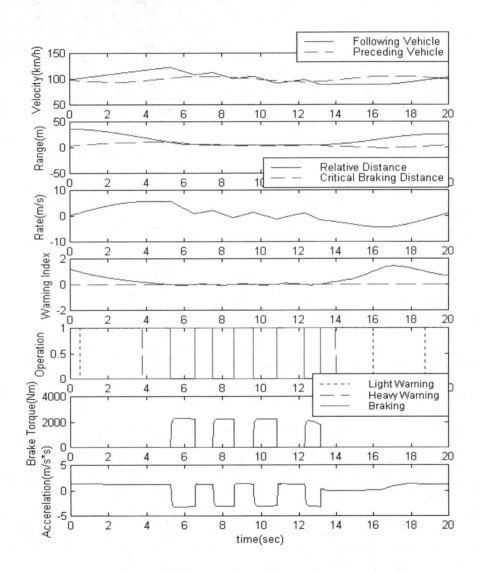

Figure 8. Experimental Results of the CW/CA HiLS

RESULTS AND CONCLUSION

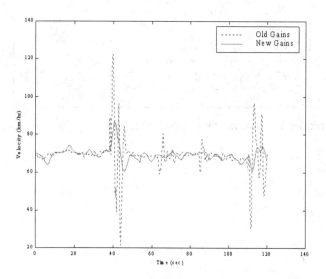

Figure 9. Tuning of Kalman filter gains in real road test

The CW/CA algorithm was simulated on the CW/CA HiLS system that consists of a radar sensor, controller, a preceding vehicle simulator, warning displays, and so on.

Two vehicles were realized in the controller and their relative behavior was applied to the simulator. Undergoing various realistic driving conditions, the controller made decisions of warning and braking based on the signal from a radar sensor. When the warning index became less than zero, the controller generated brakes torque to avoid the accident. By means of these procedure the performances of CW/CA algorithm were verified.

Fig. 8 shows experimental results of CW/CA algorithm using HiLS in laboratory. When the following vehicle goes over the preceding vehicle's speed, the range and the warning index drop. As the range drops below the critical warning distance, warnings are timely given to a driver. In spite of warnings, if a driver would not do anything and the range drops below the critical braking distance, the brakes are automatically applied. In this experiment the desired brakes torque was set to -3 (m/s^2). Notice that the acceleration of the following vehicle (7th figure from the top) is almost -3 (m/s^2) while braking. This causes the velocity of the following vehicle to reduce to that of preceding one and the driver to be safe. After that, if the warning index increases over the critical warning distance all the warnings and the braking will disappear.

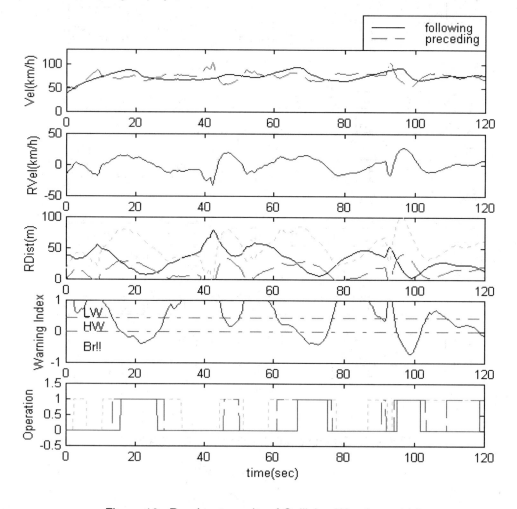

Figure 10. Road test results of Collision Warning vehicle

Fig. 9 shows the result of estimating the preceding vehicle velocity using the Kalman filter. In experiment, the preceding velocity was supposed to run at 70 (km/h). The result looks fairly good. Old gains acquired in HiLS were not directly applicable to the case in real road test. The reason is considered as the change of the radar sensor frequency. The gains were tuned several times and adopted properly as shown. In that figure the radar sensor switching of preceding vehicles caused two spiky parts. For the radar sensor was installed left a bit, it was occasionally sensitive to vehicles on the left lane.

One of the results of the road test is shown in fig. 10. As the velocities of the following and the preceding varies, the relative distance goes up and down. Accordingly the warning index is calculated and corresponding warnings occur timely. A distinct value of the warning index between light and heavy warning was set several times to find the optimum. The Collision Warning system operated successfully.

From the road test of CW system two things were noticed. The first thing is that the light warning might not be necessary for most healthy drivers. One type of warning is enough to refresh a driver's attention before braking. Heavy warning will cover up the region of light warning if the warning criterion of warning index value changes appropriately. Next, gains of the critical warning and braking distance should be scheduled depending on the typical driving style. A driver could choose one of them.

In this paper, as described above, the job of the brake controller is to reduce the vehicle velocity in constant deceleration to prevent the collision. However, due to abrupt generation of the brake torque the driver may feel discomfort. Unfortunately, the road test of a CW system has not been done because the automatic braking system is not installed completely. Changing the desired acceleration according to the vehicle condition could be a way to improve driver comfort. Various way of braking that makes the driver feel comfort and prevent an accident simultaneously shall be studied additionally.

CONTACT

KIA Motors Corp.,
Seoul, 153-030, KOREA
Tel: +82-2-890-2624, Fax:+82-2-890-2666
e-mail : kmktc1@kia.co.kr
Hanyang University,
Seoul, 133-791, KOREA
Tel: +82-2-290-0455 Fax: +82-2-295-9970
e-mail : kyongsu@email.hanyang.ac.kr

REFERENCES

1. J. Karl Hedrick, "Driving Systems for Optimal Vehicle to Vehicle Distance Control," Technical Report, University of California at Berkeley, Berkeley, CA, 1997

2. Kyongsu Yi, "Estimation of Tire-Road Friction for Collision Warning Algorithm Adaptation," Technical Report, School of Mechanical Engineering, Hanyang University, 1997

3. J. K. Hedrick, J. C. Gerdes, D. B. Maciuca, D. Swaroop, "Brake System Modeling, Control and Integrated Brake/ Throttle Switching," California PATH Research Report, University of California at Berkeley, Berkeley, CA, 1997.

4. J.C. Gerdes, "Decoupled Design of Robust Controllers for Nonlinear System: As motivated by and Applied to Coordinated Throttle and Brake Control for Automated Highways," PH.D. Thesis, University of California at Berkeley, Berkeley, CA, 1996.

5. Frank L. Lewis, Optimal Estimation with an Introduction to Stochastic Control Theory, A Wiley-Interscience Publication, JohnWiley & Sons, 1986

6. TMS320C3x/4x Optimizing C Compiler, Texas Instruments, 1997.

7. S.K. Kim, C.K. Song, K.S. Kim, S.C. Lee, M.S. Woo, K.S. Lee,"An Experimental Investigation of a Collision Warning System for Automobiles Using Hardware in-the-Loop Simulations," KSAE Conference, vol. II, pp. 820 – 825, Dong-A University, Pusan, Korea, May 1998.

Driver/Vehicle Characteristics in Rear-End Precrash Scenarios Based on the General Estimates System (GES)

Christopher J. Wiacek and Wassim G. Najm
Volpe National Transportation Systems Center

ABSTRACT

Dynamically-distinct precrash scenarios in rear-end collisions were identified in a recent study conducted by the Volpe National Transportation Systems Center, of the United States Department of Transportation, Research and Special Programs Administration, in conjunction with the National Highway Traffic Safety Administration (NHTSA) using NHTSA's General Estimates System (GES) crash database from 1992 through 1996. Precrash scenarios represent vehicle dynamics immediately prior to a collision. This paper provides a statistical description of the five most frequently-occurring rear-end precrash scenarios in terms of vehicle and driver characteristics, using the 1996 GES database. The statistics presented in this paper encompass driver characteristics of the following vehicle including avoidance maneuver attempted before impact, crash contributing factors, driver age, and gender; vehicle body types involved in these rear-end precrash scenarios; and initial travel speeds of the following vehicle under various posted speed limits. The results of this study will be useful in estimating the safety benefits of advanced-technology rear-end collision avoidance systems in terms of both crash number reduction and severity mitigation.

INTRODUCTION

The rear-end crash type has the highest frequency of occurrence among all vehicular crashes, which accounts for approximately one quarter of all police-reported crashes [1]. Vehicle-based rear-end collision warning systems and adaptive cruise control systems have the potential to alleviate the rear-end crash problem. The potential safety benefits of such systems can be best assessed by estimating the effectiveness of these systems in different rear-end precrash scenarios, taking into consideration crash statistics associated with each of these scenarios. In a recent study, dynamically-distinct rear-end precrash scenarios have been identified using the National Highway Traffic Safety Administration's (NHTSA) General Estimates System (GES) crash database [2]. This database provides the largest nationally-

representative crash sample available, which includes about forty-eight thousand police-reported crashes annually. Moreover, the GES database enables the identification of precrash scenarios based on vehicle dynamics and critical events that occur immediately prior to a collision in conjunction with roadway geometry [3].

This paper provides a statistical description of major rear-end precrash scenarios based on approximately 12,000 rear-end crash cases in the 1996 GES database. Specifically, this paper presents statistics on drivers in the following vehicle and on body types of vehicles involved in these rear-end precrash scenarios. The *Vehicle/Driver File* as well as the *Person File* of the GES database were used to identify the characteristics of both the following vehicle driver and vehicles involved. Driver characteristics include driver response to critical events, crash contributing factors, driver age and gender. Vehicle characteristics encompass body types of both striking and struck vehicles and following vehicle travel speed prior to the driver's realization of an impending danger. It should be noted that crash statistics on maximum injury severity, roadway surface condition, posted speed limit, and light and atmospheric conditions can be found in Reference [2] for the five most frequently-occurring rear-end precrash scenarios. Finally, the results of this study will be useful in estimating the safety benefits of advanced-technology rear-end collision avoidance systems in terms of both crash number reduction and severity mitigation.

PRECRASH SCENARIO IDENTIFICATION

Rear-end precrash scenarios were identified by examining combinations of two precrash variables in conjunction with the *Accident Type* and *Vehicle Role* variables, all in the GES *Vehicle/Driver File*, along with the *Roadway Alignment* and *Traffic Control Device* variables in the GES *Accident File*. The first precrash variable, *Movement Prior to Critical Event*, describes a vehicle's activity prior to the driver's realization of an impending critical event or danger. This variable discerns vehicle maneuvers, such as passing or turning, and dynamic states such as stopped or decelerating. Our analysis of rear-end crashes listed the lead vehicle as "decelerating" to a stop shortly before impact if its dynamic state was coded as *Stopped In Traffic Lane* on a straight road either due to a traffic control device or in order to make a turn. This assumption was based on a previous study conducted under an NHTSA-sponsored project to develop performance guidelines for advanced-technology rear-end crash countermeasure systems [4]. It was assumed that a forward-looking sensor on the following vehicle would have the lead vehicle within plain view while decelerating to a stop on a straight road. If the same conditions occurred on a curved road, the lead vehicle was listed as "stopped" shortly before impact, as initially coded in the GES, based on the assumption that a forward-looking sensor would not have the lead vehicle in view until it came to a complete stop. The second precrash variable, *Critical Event*, identifies the critical event which made the crash imminent (i.e., something occurred which made the collision possible). This variable does not refer to culpability. The *Accident Type* variable contains a category on same trafficway/same direction, rear-end crashes which is further defined by various crash configurations. The *Vehicle Role* variable indicates vehicle role in single or multi-vehicle crashes, such as striking or struck. The *Roadway Alignment* variable points to whether the road is straight or curved. Finally, the *Traffic Control Device* variable indicates the type and whether or not a traffic control device was present at the crash site.

A set of about twenty distinct rear-end precrash scenarios was identified by examining relevant GES cases that were coded in the database from 1992 through 1996. This set accounted for an average of 98 percent of all rear-end crashes over this five-year period. Table 1 defines and ranks the top five rear-end precrash scenarios in a descending order based on the weighted relative frequency of occurrence. These five precrash scenarios comprise an average of about 89 percent of all rear-end crashes.

DRIVER CHARACTERISTICS

The characteristics of drivers involved in rear-end collisions are obtained from the *Vehicle/Driver* and *Person Files* in the GES database. Such characteristics include avoidance maneuver attempted in response to the imminent crash, driver factors contributing to the crash, driver age, and gender. The statistics on major driver characteristics of the following vehicle are presented in this paper for the top five rear-end precrash scenarios using the 1996 GES database. Next, driver characteristics are statistically described in terms of avoidance maneuver attempted, contributing factors, driver age and gender, respectively.

Table 1. Definition and Relative Frequency of Top Five Rear-End Precrash Scenarios (Based on 1992- 1996 GES) [2]

No.	Scenario Definition	Relative* Frequency, %
1	Both following and lead vehicles are traveling at constant speed on a *straight road* and lead vehicle then decelerates.	37.0
2	Following vehicle is traveling at constant speed on a *straight road* and encounters a lead vehicle stopped in traffic lane ahead.	30.2
3	Following vehicle is traveling at constant speed on a *straight road* and encounters a lead vehicle traveling at a constant, lower speed ahead.	14.1
4	Both following and lead vehicles are decelerating on a *straight road* and lead vehicle then decelerates at a higher rate.	4.5
5	Following vehicle is traveling at constant speed on a *curved road* and encounters a lead vehicle stopped in traffic lane ahead.	3.0
	Sum	88.8

*: Relative frequency represents the average value from 1992 through 1996.

AVOIDANCE MANEUVER ATTEMPTED – The statistics on the avoidance maneuver attempted by the driver in the following vehicle are obtained from the *Corrective Action Attempted* variable in the *Vehicle/ Driver File* of the 1996 GES database. This variable describes the actions taken by the driver in response to the impending danger. Table 2 presents the statistical distribution of failed driver attempts to avoid a rear-end crash in the top five precrash scenarios. As seen in Table 2, there was no action attempted by the following driver in over 78 percent of the cases in rear-end precrash scenarios 1, 2, 3, and 5. This percentage is lower in precrash scenario 4 at about 69 percent. Driver inattention could explain the lack of corrective action taken by the driver in the following vehicle, which was estimated to cause about 76 percent of all rear-end collisions [1]. In case of corrective action taken, braking was the most likely response to the rear-end crash threat among the top five scenarios. Such information might be useful in the development of driver decision models in computer simulation of rear-end precrash scenarios.

Table 2. Percent Distribution of Attempted Avoidance Maneuvers (Based on 1996 GES)

Action Attempted	No. 1	No. 2	No. 3	No. 4	No. 5
No Action	81.4	78.4	83.8	68.6	86.2
Braked	12.2	15.5	8.1	25.7	11.1
Steered	1.1	2.2	1.7	1.4	0.7
Braked & Steered	0.5	1.0	0.4	1.4	0.2
Accelerated	0.1	0.0	0.0	0.0	0.0
Other/No Details	0.3	0.1	0.4	0.1	0.0
Unknown	4.5	2.8	5.6	2.8	1.8
Sum	100.1*	100.0	100.0	100.0	100.0

*: Rounding Error

DRIVER CONTRIBUTING FACTORS – The GES database contains a number of variables referring to circumstances, conditions, and events that may have contributed to the crash. From the GES *Vehicle/Driver File*, this study examined the variables *Driver Distracted By*, *Driver's Vision Obscured By*, *Driver Drinking in Vehicle*, and *Violations Charged*. The variable *Driver Distracted By* attempts to capture distractions which may have influenced driver performance while the variable *Driver's Vision Obscured By* identifies visual circumstances which may have contributed to the cause of the crash. In addition, the variable *Person's Physical Impairment* from the GES *Person File* was also examined in this study.

Table 3 presents the statistics on contributing factors of the driver in the following vehicle for each of the top five rear-end precrash scenarios. Among the five scenarios, the driver was distracted more in precrash scenarios 2 and 3 at about 7 percent and 8 percent of their cases, respectively. Internal distraction was mostly cited in distraction cases. Audible or haptic warnings issued by a rear-end collision avoidance system could be effective in cases where a driver was distracted by alerting the driver to the impending danger. Table 3 also shows that driver's vision was obscured in about 2 percent of the cases in scenarios 2, 3, 4, and 5. Reflected glare and bright sunlight were the dominant factors in rear-end precrash scenarios 2, 3, and 5; while rain, snow, or fog was mostly cited in vision obstruction cases in scenario 4. The forward-looking sensor(s) of a rear-end collision avoidance system would have to detect the lead vehicle and warn the driver when appropriate in cases where the driver's visibility is compromised. Alcohol use by the driver of the following vehicle was reported in about 5 percent and 6 percent of the cases in scenarios 1 and 3, respectively. This statistic is approximately 3 percent in the remaining three scenarios. It should be noted that alcohol use by the driver is generally much higher in fatal crashes.

According to Table 3, the driver of the following vehicle was charged with a violation in as high as 51 percent of the cases. Violations charged to the driver included alcohol or drugs, speeding, alcohol or drugs and speeding, reckless driving, failure to yield right-of-way, hit and run, revoked license, running a traffic signal or stop sign, and other violations/no details. Unfortunately, the "other violation/no details" codes were commonly cited under the variable *Violations Charged* in the GES. Of the known violations, speeding was the dominant charge. Finally, the driver was physically impaired in about 2 percent of the cases in scenario 3, the highest among the five rear-end precrash scenarios. Drowsy, ill, or blackout was mostly indicated in these cases. Generally, rear-end collision warning systems would not be effective in cases involving drivers drinking/drunk in the vehicle or physically-impaired.

Table 3. Percent Distribution of Contributing Factors (Based on 1996 GES)

Driver Factor	No. 1	No. 2	No. 3	No. 4	No. 5
Distracted	4.6	7.1	7.8	1.7	4.2
Vision Obstructed	0.7	1.6	1.5	1.5	2.1
Drinking Alcohol	4.6	2.6	5.7	2.6	2.8
Violation Charged	48.5	51.0	48.3	26.9	36.1
Physically Impaired	0.8	0.7	1.7	0.2	0.9

Note: Driver factor statistics are not mutually exclusive and numbers in columns do not add up to 100.

DRIVER AGE AND GENDER – The *Age* and *Sex* variables in the GES *Person File* were used to categorize both age and gender of following vehicle drivers involved in the top five rear-end precrash scenarios. Tables 4 and 5 present, respectively, the percent distributions of driver age and gender in each of the five scenarios. Drivers between the age of 16 and 24 years are over-represented in all five scenarios at approximately 30 percent of rear-end collision cases, given that this age category constitutes about 21 percent of all licensed drivers [5]. Conversely, drivers over 64 years of age are under-represented at about 6 percent of rear-end collision cases, given that this age category accounts for about 13 percent of all licensed drivers.

Table 4. Percent Distribution of Following Driver Age (Based on 1996 GES)

Driver Age	No. 1	No. 2	No. 3	No. 4	No. 5
16 – 24 years	30.9	35.0	30.5	31.6	29.9
25 – 64 years	63.9	58.4	64.3	61.7	65.1
65 + years	5.2	6.7	5.2	6.7	5.0
Sum	100.0	100.1*	100.0	100.0	100.0

*: Rounding Error

Table 5 shows that male drivers are slightly over-represented in rear-end collisions at about 60 percent, considering that licensed male drivers constitute about 53 percent of the driving population [5].

Table 5. Percent Distribution of Following Driver Gender (Based on 1996 GES)

Driver Gender	No. 1	No. 2	No. 3	No. 4	No. 5
Male	62.5	61.2	65.4	55.9	59.9
Female	37.5	38.8	34.6	44.1	40.1
Sum	100.0	100.0	100.0	100.0	100.0

VEHICLE CHARACTERISTICS

The statistical characteristics of vehicles involved in rear-end collisions, in terms of body type and travel speed, can be obtained from the GES *Vehicle/Driver File*. Specifically, this paper presents statistics on body type of both striking and struck vehicles and on travel speed of the following vehicle for each of the top five rear-end precrash scenarios.

VEHICLE BODY TYPE – The variable *Body Type* indicates the type of the vehicle involved in the crash, including automobiles, utility vehicles, van-based light trucks, light conventional trucks, buses, medium or heavy vehicles, motored cycles, and 'other' vehicles such as snowmobiles, farm equipment, street sweepers, etc. Table 6 shows the percent distribution of vehicle types involved in rear-end collisions for both striking and struck vehicles. Moreover, the statistics are presented for each of the top five rear-end precrash scenarios, defined in Table 1, under four vehicle categories. The "light vehicle" category consists of automobiles, light trucks, utility vehicles, and vans. The "truck" category encompasses both medium and heavy trucks. The "other" category includes motored cycles and 'other' vehicles.

Table 6.   Percent Distribution of Vehicle Types Involved (Based on 1996 GES)

| Striking | Scenario | Struck | | | |
		Light Vehicle	Truck	Bus	Other
Light Vehicle	1	95.2	1.2	.7	.2
	2	95.8	1.4	.9	.1
	3	91.0	3.4	.6	.7
	4	85.2	.5	0.0	.1
	5	96.2	1.5	0.0	0.0
Truck	1	2.2	.2	.2	0.0
	2	1.5	.1	0.0	0.0
	3	2.9	.2	0.0	0.0
	4	12.1	.8	.4	0.0
	5	1.6	.1	0.0	0.0
Bus	1	.1	0.0	0.0	0.0
	2	0.0	0.0	0.0	0.0
	3	.4	0.0	0.0	0.0
	4	.4	0.0	0.0	0.0
	5	.5	0.0	0.0	0.0
Other	1	.2	0.0	0.0	0.0
	2	.4	0.0	0.0	0.0
	3	.7	0.0	0.0	0.1
	4	.4	0.0	0.0	0.0
	5	0.0	0.0	0.0	0.0

As seen in Table 6, a light vehicle struck another light vehicle in over 90 percent of rear-end collisions classified under precrash scenarios 1, 2, 3, and 5. This statistic is lower for precrash scenario 4 at about 85 percent. It should be noted that light vehicles accounted for about 95.4 percent of the 1995 vehicle fleet in the United States; trucks, buses, and "other" vehicles comprised respectively 2.4, 0.3, and 1.9 percent of the 1995 vehicle fleet [6]. In rear-end precrash scenario 4 (Where both following and lead vehicles were initially decelerating before the critical event), a truck struck a light vehicle in 12.1 percent of the cases, compared to only 3 percent or below in the other scenarios. This statistic is exceptional given that a light vehicle struck a truck in only 0.5 percent of the cases in rear-end precrash scenario 4. Among the five rear-end precrash scenarios, the truck is most likely to hit a lead vehicle in rear-end precrash scenario 4 (13.3 percent of scenario cases).

FOLLOWING VEHICLE TRAVEL SPEED – Statistics on travel speed of the following vehicle were obtained from the *Travel Speed* variable, which indicates the actual travel speed of the vehicle in miles per hour (MPH) prior to the driver's realization of an impending danger. This paper presents statistical distributions of following vehicle travel speed with respect to various posted speed limits (*Speed Limit* variable in the GES *Accident File*). This was an initial attempt to quantify the relationship between actual vehicle speed and certain posted speed limit. Figure 1 illustrates the distribution of actual speed data with respect to 35, 45, and 55 MPH posted speed limits. The data presented in Figure 1 were an aggregate of the 'known' data on vehicle speeds for the top five rear-end precrash scenarios. It should be noted that about 70 percent of the actual speed data in the GES database were coded as unknown. The average vehicle speed was about 23, 27, and 39 MPH respectively in the 35, 45, and 55 posted speed limit zones. These results show that a significant majority of rear-end collisions occur below the posted speed limit. This finding might be attributed to congested traffic conditions that restrict drivers from maintaining the posted speed.

CONCLUSION

This paper presented a statistical description of driver and vehicle characteristics for the five most frequently-occurring precrash scenarios in rear-end collisions, based on the 1996 GES database. Driver characteristics were described for the driver of the following vehicle in terms of failed avoidance maneuver attempts, factors contributing to the crash, age, and gender. The data showed that the driver did not attempt any avoidance maneuver in over 78 percent of the cases. Braking was the most likely avoidance action taken when the driver reacted to the impending threat. Moreover, the driver was charged with a violation in as high as 51 percent of the cases. In addition, drivers between the ages of 16 and 24 were over-represented in rear-end collisions at about 30 percent of all cases. Vehicle characteristics were obtained for body type of all vehicles involved and actual travel speed of the following vehicle. A light vehicle struck another light vehicle in over 90 percent of rear-end collisions. Also, trucks were most likely to hit another vehicle in rear-end precrash scenario where both vehicles were initially decelerating. Finally, the available known data showed that the following vehicle was traveling at an average speed below the posted speed limit before colliding with a lead vehicle.

DISCLAIMER

The conclusions and opinions expressed in this paper are those of the authors and do not represent the position of the U.S. Department of Transportation, with respect to the matters discussed.

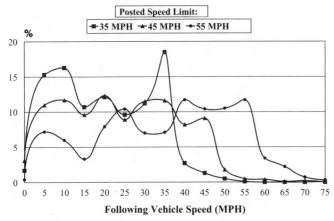

Figure 1. Distribution of Following Vehicle Speed vs. Posted Speed Limit (Based on 1996 GES)

REFERENCES

1. Najm, W., Mironer, M., Koziol, J.,Wang, J., & Knipling, R., Synthesis Report: Examination of Target Vehicular Crashes and Potential ITS Countermeasures, DOT HS 808 263, June 1995.

2. Najm, W., Wiacek, C., Burgett, A., Identification of Precrash Scenarios for Estimating the Safety Benefits of Rear-End Collision Avoidance Systems, 5th ITS World Congress, Seoul, Korea October 1998.

3. U.S. Department of Transportation, National Highway Traffic Safety Administration, National Accident Sampling General Estimates System (GES) - Users's Manual - 1995 File, National Accident Sampling Systems General Estimates System Technical Note, DOT HS 807 796.

4. Wilson, T., IVHS Countermeasures for Rear-End Collisions, Task 1, Volume IV: 1992 NASS CDS Case Analysis, Interim Report, DOT HS 808 564, February 1994.

5. Teets, M.K., Editor, Highway Statistics Summary to 1995, United States Department of Transportation, Federal Highway Administration, FHWA-PL-97-009, July 1997.

6. Davis, S.C., Transportation Energy Data Book: Edition 17, Oak Ridge National Laboratory, ORNL-6919, Oak Ridge, TN, August 1997.

COLLISON AVOIDANCE SYSTEMS

Performance of a Rear-End Crash Avoidance System in a Field Operational Test

Wassim G. Najm and Jonathan Koopmann
Volpe National Transportation Systems Center

ABSTRACT

This paper characterizes the capability of a rear-end crash avoidance system based on data collected from a field operational test. The system performs forward crash warning and adaptive cruise control functions. The test consists of 66 subjects who drove 10 equipped vehicles on public roads over 157,000 km. System characterization addresses the ability of the forward-looking sensor suite to maintain in-path target tracking and discern between in-path and out-of-path targets; the efficacy of the alert logic in warning the driver to driving conflicts that may lead to rear-end crashes; and the visibility, audibility, and readability of information displayed by the driver-vehicle interface.

INTRODUCTION

This paper presents the results of an analysis to characterize the performance and capability of an automotive rear-end crash avoidance built for light vehicles (e.g., passenger cars, vans, minivans, sport utility vehicles, and light trucks). This rear-end crash avoidance system is known as the Automotive Collision Avoidance System (ACAS), which consists of both forward crash warning (FCW) and adaptive cruise control (ACC) functions [1]. The FCW detects, assesses, and alerts the driver of a potential hazard in the forward region of the host vehicle. The ACC utilizes automatic brake and throttle to maintain speed and longitudinal headway control. According to the 2002 General Estimates System crash database, light vehicles were involved in approximately 1.8 million police-reported rear-end crashes in the United States or about 29% of all light vehicle crashes. These rear-end crashes resulted in about 850,000 injured persons.

The ACAS underwent a field operational test (FOT) that was conducted with ten equipped vehicles from March 2003 to November 2004 [2]. Sixty-six subjects were selected from the state of Michigan as FOT participants. They were split equally by gender and three age groups: younger (20 – 30 years old), middle age (40 – 50 years old), and older (60 – 70 years old). Each subject drove

the ACAS-equipped vehicle as his or her own personal car for a test period of four weeks, unsupervised and unrestricted. The first week was dedicated to collecting baseline driving data, i.e., *without* the assistance of the ACAS. During this week, FOT subjects drove with manual control and also had the option of using conventional cruise control. During the remaining three weeks, driving was performed *with* the assistance of the ACAS. In that period, subjects drove the FOT vehicles with either manual control or manual control augmented with the FCW function, and they also had the option of engaging ACC. Two hours of training was provided for FOT participants prior to starting the FOT.

An independent evaluation was conducted to address three goals in support of the decision process to deploy rear-end crash avoidance systems: characterize system performance and capability, achieve a detailed understanding of safety benefits, and determine driver acceptance of the system [3]. The FOT generated objective data gathered by on-board data acquisition systems and subjective data obtained from test subject interviews, surveys, and focus group sessions. Numerical and video data were collected from 157,000 km distance traveled. A system characterization test was independently conducted to supplement the FOT data and acquire additional data on the performance of system sensors and automatic controls from controlled, predetermined on-road routes. This test was executed on a wide variety of roadway configurations and environmental conditions for a total distance of 392 km.

This paper reports on the results of the system capability goal of the independent evaluation that examines the operational performance of ACAS in the driving environment. It addresses the main components of the system: sensor suite, alert logic, and driver-vehicle interface (DVI). The characterization of the forward-looking sensor suite examines how well the system rejects out-of-path targets, and detects and tracks closest in-path targets. This analysis was based in part on observations from 8-second FOT episodes of video and numerical data, which were triggered by crash imminent alerts during the ACAS disabled and ACAS enabled test periods. It should be noted that the system was operating in the background during the ACAS disabled test period, where FOT subjects did not see or hear the alerts. System characterization test data were used to determine the rejection ratio of out-of-path targets and the rates of missed, lost, or intermittent detection of in-path targets. In addition, FOT surveys provided a subjective evaluation of the missed and false target rates by the forward-looking sensor suite. The ability of the alert logic component of the system to issue a correct signal (efficacy) was examined using data from FOT episodes triggered by crash imminent alerts and FOT surveys. The ACAS issues a "true" signal (warning) when the host vehicle is on a rear-end crash course with an in-path obstacle (i.e., situations requiring a signal). On the other hand, a "false" signal is issued in situations not requiring a signal such as out-of-path targets or the host vehicle not on a collision path with a lead vehicle in its lane. The degree of nuisance generated by ACAS alerts was qualitatively measured using FOT surveys. Drivers would most likely perceive out-of-path target alerts as nuisance. Moreover, "true positive" signals issued by the ACAS might also be considered as "nuisance" if drivers subjectively judged them as too early or not necessary. The ability of the DVI to properly convey system information to the driver was qualitatively evaluated using FOT surveys. In particular, the opinions of FOT subjects are reported on how well they were able to see the HUD while driving, read the displayed information, and hear auditory alerts.

Next, this paper describes FCW and ACC functions of the rear-end crash avoidance system. This is followed by a general characterization of crash imminent alerts received during the FOT. After that, performance results are discussed separately for the sensor suite, alert logic, and DVI. This paper concludes with recommendations on system improvement areas.

SYSTEM DESCRIPTION

The rear-end crash avoidance system consists of both FCW and ACC functions. A suite of sensors supports these two functions and comprises a combination of vehicle original equipment manufacturer sensors with forward-looking radar, forward-looking camera, differential global positioning system with map matching, and a yaw-rate sensor. The radar measures range, range-rate, and azimuth angle to a maximum of 15 targets from 1 to 150 m with a sampling frequency of 10 Hz. The detection range of the radar is limited on curves with a radius of curvature below 500 m. Moreover, the maximum horizontal field of view of the radar is 15° with an azimuth resolution angle of 2°; the vertical beam width is 4.1°. The forward-looking camera determines the road geometry ahead of the host vehicle from 15 to 75 m, vehicle heading angle, and vehicle lateral position within the lane of travel. A color head-up display (HUD) presents visual information to the driver by projecting an image on the windshield, which subtends a visual angle of 1.5° vertical and 3.0° horizontal.

FORWARD CRASH WARNING

The FCW function provides drivers with cautionary and crash imminent alerts that assist them in avoiding or reducing the severity of rear-end crashes between the front of their vehicle and the rear of a lead vehicle moving or stationary. Cautionary alerts are presented visually to the driver in a graded scale by vehicle icons on the HUD. Crash imminent alerts are conveyed visually on the HUD and audibly by a vehicle speaker. FCW is enabled when the vehicle ignition is turned on, and cannot be disabled by the driver. This function does not activate until the speed of the host vehicle exceeds 40 km/h and will remain active until the vehicle slows to below 32 km/h. The warning function is disabled under certain conditions specified by system designers such as dirty radar, sharp curve, or heavy precipitation. The range of the warning function is set to a maximum of 100 m. The driver can adjust the sensitivity of the visual cautionary alerts with a sensitivity adjustment control (six settings) but the driver cannot change the timing of the auditory crash imminent alert. The factors that determine when to issue a crash imminent alert include, but not limited to, range and range rate between the host and lead vehicles, host vehicle speed, lead vehicle acceleration, and brake pedal press. The HUD provides a graded visual display that reflects the degree of the closing gap between the host vehicle and the lead vehicle based on the FCW sensitivity setting. The most sensitive setting of FCW produces the most cautionary alerts because FCW responds to the host vehicle closing in on obstacles ahead at farther distances with lower range rates.

ADAPTIVE CRUISE CONTROL

The ACC function maintains both a selected cruise speed (speed control mode) when there is no lead vehicle limiting its forward motion, and a selected headway (headway control mode) with a lead vehicle that is traveling slower than the selected cruise speed. The headway adjustment control consists of six discrete steps that vary from a minimum of one to a maximum of two seconds. This same control also sets the desired cautionary alert timing of the FCW function. The ACC is engaged by the driver and becomes active when the speed of the host vehicle exceeds 40 km/h. At first ACC engagement, the initial headway setting is set to the maximum value. In headway control mode, the ACC can slow the host vehicle by throttle application or brake to pace a lead vehicle moving slower than the set speed. Once vehicle speed falls below 32 km/h, the driver is alerted to take manual control of the vehicle. The ACC does not respond to stopped vehicles ahead – unless the stopped vehicle was initially being tracked as a moving vehicle. The maximum automatic braking capability of the ACC is limited to 0.3g (2.9 m/s^2). The brake lights of the host vehicle turn on when vehicle brakes are automatically applied. The ACC goes into a standby mode when the brakes are manually applied. The ACC does not automatically accelerate the host vehicle until the driver manually accelerates above 40 km/h and then initiates the resume function or the set speed function. The ACC function has a crash imminent warning capability if the ACC braking authority is inadequate to prevent a rear-end crash. When ACC is engaged, the driver does not receive visual cautionary alerts.

CHARACTERISTICS OF CRASH IMMINENT ALERTS

Sixty-six subjects received a total of 980 crash imminent alerts during the field operational test, or about 0.62 crash imminent alert per 100 km traveled. The breakdown of these alerts by the various driving conditions is as follows:

- Road type: 80% or 1.08 alerts per 100 km traveled on non-freeways, and 20% or 0.23 alert per 100 km traveled on freeways.
- Weather: 92% or 0.62 alert per 100 km traveled in clear conditions, and 8% or 0.62 alert per 100 km traveled in adverse weather.
- Ambient light: 82% or 0.69 alert per 100 km traveled in lighted conditions, and 18% or 0.42 alert per 100 km traveled in dark conditions.
- Traffic: 39% or 0.35 alert per 100 km traveled in low traffic, 53% or 1.19 alerts per 100 km traveled in moderate traffic, and 8% or 1.58 alerts per 100 km traveled in heavy traffic.
- Road junction: 55% of all alerts were received at non-junctions, 35% at intersections, 7% at driveways, and 3% on ramps.
- Host vehicle speed: 67% of all alerts were triggered at host vehicle speed over 56 km/h. Moreover, a relatively high rate of 2.18 alerts per 100 km traveled occurred at travel speeds between 40 and 56 km/h. This speed bin had about 28% of all alerts.

Out-of-path target alerts accounted for 44% of all alerts, which warned of objects that are not in the path of the vehicle and thus posing no safety risk. On the other hand, alerts triggered by in-path targets amounted to 56% of all alerts that may or may not warn of an impending crash. Moving, in-path or out-of-path, targets caused 62% of all alerts. About 92% of these alerts fall under two categories: moving in-path targets and stationary out-of-path targets. Some alerts due to moving in-path targets could be a source of nuisance to drivers who judge that these situations do not pose any immediate rear-end crash threat.

SENSOR SUITE

The capability of the forward-looking sensor suite is examined in terms of its ability to discriminate between in-path and out-of-path targets, and to detect and track closest in-path targets.

IN-PATH TARGET DETECTION AND TRACKING

In-path targets triggered 0.35 crash imminent alert per 100 km traveled. The majority or 97% of these alerts was attributed to moving targets. About 87% of moving in-path target alerts were triggered when both the host and lead vehicles were traveling on a straight road. Only 7% of these alerts were issued when both vehicles were on a curve. Changing lanes, turning, or passing by the host vehicle behind an in-path moving vehicle triggered 12% of all moving in-path target alerts. On the other hand, about 47% of moving in-path target alerts were caused by the lead vehicle changing lanes, turning, or making a left turn across the path of the host vehicle. In these cases, the lead vehicle posed no danger to the host vehicle. About 62% of the alerts due to moving in-path targets occurred in the vicinity of intersections or driveways. Moreover, 60% of the moving in-path target alerts were triggered at host vehicle speed greater than or equal to 56 km/h.

A system characterization test, independent of the FOT, was conducted to examine the detection and tracking of in-path vehicles by the forward-looking sensor suite under different roadway curvatures and environmental conditions. This analysis focused on late, intermittent, and lost detections of targets that were entirely within the same lane as the host vehicle, and were moving within 100 m or stopped within 70 m from the host vehicle. Detection here refers to lead vehicles being declared as closest in-path targets by the system, and not crash imminent alerts. Late detection was marked if the sensor suite did not detect the lead vehicle under these conditions. Intermittent detection was noted if the sensor suite first detected the target under these conditions and then lost it for up to 3 seconds before target reacquisition. Any detection lost for 3 seconds or more was recorded as lost detection. The results show that the forward-looking sensor suite was late in detecting about 17% of the vehicles encountered on curves below 500 m radius, as opposed to 14% on curves with higher radius. Moreover, late detections were observed in 18% and 15% of the targets respectively in rain and clear weather. Intermittent detection was noted in 28% of the targets on curves below 500 m radius, as opposed to 24% on curves with higher radius. The results of late and intermittent detections do not show a significant impact by sharp curves (radius < 500 m). In contrast, sharp curves accounted for lost detection in 22% of the targets as opposed to only 8% on curves with higher radius. A higher rate of lost detections was observed in rain than in clear weather during the day, and much higher rate at night than the day in clear weather.

OUT-OF-PATH TARGET DETECTION AND REJECTION

Out-of-path targets caused a rate of 0.27 crash imminent alert per 100 km traveled. About 83% of all out-of-path target alerts were due to stationary objects. A total of 42%, 10%, and 8% of these objects were located respectively on curve, curve entry, and curve exit. On the other hand, the host vehicle received 36%, 16%, and 4% of these alerts while located respectively on curve entry, curve, and curve exit. Thus, the majority of stationary out-of-path target alerts was associated with curved roadways. Objects on straight roads triggered 32% of stationary out-of-path target alerts when the host vehicle was simply traveling straight. This could be caused by radar misalignment in case of roadside objects, or a weakness in the bridge rejection algorithm in case of a bridge or overhead sign.

The breakdown of crash imminent alerts due to stationary out-of-path targets by the various driving conditions is as follows:

- Road type: 77% or 0.38 alert per 100 km traveled on non-freeways, and 23% or 0.10 alert per 100 km traveled on freeways.
- Weather: 96% or 0.24 alert per 100 km traveled in clear conditions, and 4% or 0.10 alert per 100 km traveled in adverse weather.
- Ambient light: 82% or 0.25 alert per 100 km traveled in lighted conditions, and 18% or 0.16 alert per 100 km traveled in dark conditions.
- Road junction: only 9% of alerts occurred in the vicinity of intersections and driveways as opposed to 62% of moving in-path target alerts. About 87% of stationary out-of-path target alerts happened at non-junctions.
- Host vehicle speed: only 25% of all stationary out-of-path target alerts were triggered at host vehicle speed below 56 km/h.

The rate of stationary out-of-path target alerts on non-freeways is higher than on freeways due to sharper curves and more abundant roadside furniture on non-freeways. It is important to note that the adverse weather does not appear to affect the radar sensor in creating false targets as apparent in the lower alert rate in adverse weather. The higher alert percentage at non-junctions can be explained by the presence of more curves than at road junctions. Finally, 75% of all stationary out-of-path target alerts were triggered at host vehicle speeds over 56 km/h. Thus, suppressing this type of alert at lower speeds would not impact the percentage of these false alerts.

The capability of the forward-looking sensor suite to reject out-of-path targets was assessed using data from the system characterization test. The results show that the rejection ratio of overhead objects (e.g., bridges, overhead signs and walkways) was 100% for crash imminent alerts, given that 308 such objects were encountered during the system characterization test. Moreover, the threat assessment algorithm rejected 97.9% of out-of-path stationary objects, and 99.6% of lead vehicles in the adjacent lane when the host vehicle was negotiating a curve. Out-of-path stationary targets comprised mailboxes, signs, guardrails, light poles, and other stationary roadside objects, excluding overhead objects. The lead vehicle turning caused 5 crash

imminent alerts; in 3 of these cases, the lead vehicle was more than half way out of the host vehicle lane. Overall, the threat assessment algorithm suppressed 97.7% of crash imminent alerts that might have been triggered by either host or lead vehicle maneuver to change lanes. Environmental conditions did not appear to have any impact on crash imminent alert rejection due to out-of-path targets.

Suppression of visual alerts due to out-of-path targets by roadway curvature and environmental conditions was also examined using system characterization test data. Visual alerts only displayed warning icons to the driver on the HUD. The icons range from small green vehicle indicating target detection, to large yellow vehicle indicating a crash imminent alert would be issued if no action were taken. All icons were accounted for regardless of their level or length of time displayed. The system characterization test was conducted with FCW at an intermediate sensitivity setting. The system suppressed the display of visual warning icons in 96.4% of overhead objects encountered during the system characterization test. Moreover, the system did not issue visual alerts to 82.8% of out-of-path stationary objects, and 73.6% of lead vehicles in the adjacent lane when the host vehicle was negotiating a curve. Visual alerts were also suppressed during 92.9% of host or lead vehicle lane changing maneuvers. In contrast to crash imminent alerts, environmental conditions appear to have some impact on the rejection ratios of visual alerts due to out-of-path targets. The lowest rejection ratios were observed during nighttime driving, perhaps due to limitations of the vision sub-system that depends on outside lighting or the lack of traffic ahead that is used in part to predict the forward path of the host vehicle. Moreover, rejection ratios were lower in rain than in clear weather during daytime driving due to lower visibility and heavy precipitation.

The following question in the post-drive survey provided a subjective assessment of the forward-looking sensor suite to deal with out-of-path targets:

- How often, if ever, did FCW give you a warning that was false (*false target*)?

The question was scaled from 1 (very frequently) to 7 (very infrequently) with 0 for never. Only 3% of the subjects replied that FCW never had a warning when there were no other vehicles to warn about. About 56% of the subjects indicated infrequent (scales 5-7) false warnings as opposed to 21% who reported frequent (scales 1-3) false warnings.

ALERT LOGIC

This part of the analysis assesses the efficacy of crash imminent alerts based on FOT triggered episodes and evaluates their nuisance on drivers using FOT survey data.

The efficacy of the warning logic is judged by driver maneuver response to crash imminent alerts and driver state (distraction and eyes-off-road) during alert-triggered video episodes. Figure 1 compares driver maneuver response before and after in-path target alerts during the ACAS enabled test period in which FOT subjects heard the auditory alerts. The percentage of off-throttle response was significantly reduced, from 45% before the in-path alert to 9% after the alert. On the other hand, the brake response jumped from 8% before the alert to 56% after the alert. The percentage of no response declined from 43% before the alert to 29% after the alert during the ACAS enabled test period. Figure 2 compares driver response after the in-path target alert between the ACAS disabled and ACAS enabled test periods. FOT subjects had higher response rates and braked more when they heard the alerts during the ACAS enabled test period. This would suggest that heard in-path target alerts elicited drivers to respond, but this is greatly dependent on the driving situations.

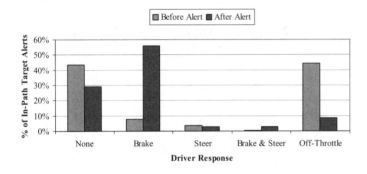

Figure 1. Driver Action Before and After In-Path Target Alerts during ACAS Enabled Test Period

Distraction was observed in about 39% of the in-path target alerts. Driver distraction includes dialing phone, talking/ listening to the phone, singing/whistling, grooming, adjusting controls, scratching face, yawning, drinking/eating/smoking, talking to passenger, reading, searching interior, scanning back adjacent lanes, scanning rear-view mirror, looking to the side/outside car, reaching for items, and other distractions. FOT subjects had their eyes off the road in 3% of the in-path target alerts. An FOT subject was noted to have his/her eyes off the road if the driver glanced away from the road ahead for a time period greater than or equal to 1.5 seconds before the alert.

Subjects were asked to rate the timing of the FCW auditory alert with the following statement:

- Overall, evaluate the timing of the auditory alert when FCW was responding to a vehicle ahead.

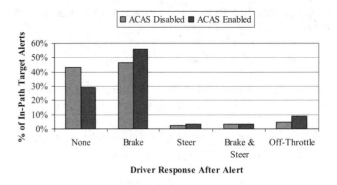

Figure 2. Comparison of Driver Response After In-Path Target Alerts between ACAS Disabled and Enabled Test Periods

Based on a scale from 1 (too early) to 7 (too late) with 4 being just right, FOT subjects responded with a mean score of 4.71. About 36% of the subjects thought that FCW alert timing was just right. About 52% of the subjects judged FCW timing as late (scales 5-7) as opposed to only 12% who judged it as early (scales 1-3). An inquiry was made into the subjects' opinion about changing the design of FCW alert timing by two survey items translated below into one statement:

- If I were designing an FCW system, I would add an alert timing setting that allowed me to receive alerts *sooner/later* than the *most/least* sensitive alert timing setting that I experienced with this FCW system.

About 67% of the subjects disagree (scales 1-3) to having *later* alert setting as apposed to 45% who disagree to *sooner* setting. On the other hand, 21% of the subjects agree (scales 5-7) to *later* setting as opposed to 47% for *sooner* setting. The usefulness of FCW alerts was investigated by asking the subjects to "rate the extent to which FCW alerts were useful in providing a warning about a driving situation that might result in a collision", at a scale between 1 (not at all useful) and 7 (very useful). About 59% of the subjects rated it to be useful (scales 5-7) as opposed to only 27% who judged it not to be useful. Two more questions were asked of subjects, which address the efficacy of FCW alerts:

- How often, if ever, did FCW not give you an alert when you felt that one was necessary (*missed alert*)?
- How often, if ever, did FCW give you an alert in a situation that you felt was appropriate (*appropriate alert*)?

Subjects were asked to rate their response to these two questions using the following scale: 1 (never), 2 (once or twice), 3 (once or twice per week), 4 (several times per week), 5 (once per day), and 6 (several times per day). About 61% of the subjects felt that they never had a missed alert, and 17% experienced 1 or 2 missed alerts in total. On the other hand, only 12% indicated that they never had an appropriate alert. About 64% of the subjects felt that they experienced 1 to 6 appropriate alerts in total.

NUISANCE ALERTS

Few questions were asked of subjects, which indirectly refer to alerts that might be the source of nuisance:

1. How often, if ever, did FCW give you an alert where you could not identify the source of the alert (*unidentified source*)?
2. How often, if ever, did FCW give you an alert in a situation that you felt was not appropriate (*inappropriate alert*)?
3. How frequently did "*driving situations listed below*" result in FCW alerts that you felt were not necessary?
   a. When a vehicle ahead of me turned
   b. When I passed a moving vehicle
   c. When a vehicle ahead changed lanes
   d. When my vehicle changed lanes
   e. When a vehicle cut in front of me
   f. When I cut in behind another vehicle
   g. When I passed a sign, light post, or guard rail
   h. When I passed a parked vehicle

Table 1 lists the mean values for the questions above. One third of the subjects indicated that they never got an alert that they could not identify its source. About 62% acknowledged that they received between 1 and 6 alerts in total with an unknown source. About 56% indicated that they received 1 to 6 inappropriate alerts in total, with 15% indicating never to receive an inappropriate alert.

Table 1. Subjective Evaluation of Inappropriate FCW Alerts

Q	Topic	Mean
1	Unidentified source	2.0
2	Inappropriate alert	2.9
3a	Lead vehicle turning	2.8
3b	Host vehicle passing	2.0
3c	Lead vehicle changing lane	2.2
3d	Host vehicle changing lane	1.9
3e	Lead vehicle cut in	2.3
3f	Host vehicle cut in	1.7
3g	Host vehicle passing sign, etc	2.9
3h	Host vehicle passing parked car	1.8

1 never, 2 once or twice, 3 once or twice/week, 4 several times/week, 5 once/day, 6 several times/day

DRIVER-VEHICLE INTERFACE

HUD readability and sound audibility of the driver-vehicle interface are evaluated based on subjective judgment of FOT subjects in response to survey items. The DVI capability to effectively convey information to the driver in terms of HUD readability and alert sound audibility was subjectively captured by the following six questions:

1. How easy was it to drive and see the HUD at the same time?
2. How easy was it to see the HUD?
3. Overall, how easy was it to see all the information shown on the HUD?
4. How easily were you able to detect the visual crash alerts?
5. How easily were you able to recognize alerts from FCW?
6. How easily were you able to detect the audio crash alert?

Table 2 displays the mean values for driver evaluation of the DVI. An overwhelming majority of subjects were able to see and hear text and audio messages transmitted by the DVI. All subjects (100%) easily (scales 5-7) drove and saw the HUD at the same time, 98.5% easily recognized alerts from FCW, 97% easily saw the HUD and detected the visual and audio crash alerts, and 96.5% saw all the information shown on the HUD overall.

Table 2. Subjective Evaluation of DVI Information Display Capability

Q	Topic	Mean
1	Drive & see HUD	6.73
2	See HUD	6.67
3	See HUD information	6.66
4	Detect visual alerts	6.65
5	Recognize alerts	6.73
6	Detect audio alerts	6.80

1 very difficult ... 7 very easily

CONCLUSION

The FCW function of the rear-end crash avoidance system incorporates state-of-the-art sensor technologies for short-term deployment plans. However, improved signal processing and threat assessment algorithms would enhance FCW alert efficacy by recognizing slower lead vehicles transitioning from the path of the host vehicle to out of its path. This event generated numerous unnecessary crash imminent alerts during the FOT, and even forced the ACC to automatically brake in response to lead vehicles exiting the freeway. Stationary out-of-path targets were mostly the source of false crash imminent alerts. Remedies to deal with this particular problem have been suggested, including the disregard of the closest in-path stationary target. The suggestion is for the threat assessment algorithm to rely completely on the closest in-path moving target that accounts only for moving vehicles and for stopped vehicles tracked by the radar to be moving prior to stopping. The examination of video episodes revealed few cases where stationary-tagged vehicles triggered the crash imminent alerts, mainly at intersections. Thus, a concern is raised regarding the elimination of stationary-tagged targets from the threat assessment algorithm. Luckily, subjects were attentive when these alerts were issued in the FOT. Alerts triggered by stationary-tagged targets would be very helpful in preventing a rear-end crash if drivers were inattentive.

The analysis of crash imminent alerts also showed that increasing the threshold operating speed of FCW over 40 km/h would not make any significant impact on false and nuisance alerts. To boost driver acceptance of FCW at the expense of some limited safety benefits, it is recognized that a tradeoff must be made between alert rates and the operating envelope and sensitivity of FCW. The ACAS incorporated many subsystems to track targets at long ranges in the path of the host vehicle. One of these subsystems is the global positioning system/geographic information system mapping capability to figure out the road geometry ahead of the host vehicle. It appears that this feature had little impact on target tracking and crash imminent alerts as evident from the system characterization test that was conducted in an area where map information was not available and the alert rate did not seem to differ from the rates observed by FOT subjects with available map data. Given the cost of such a feature, the ACAS could perform without it unless, of course, this feature is also a part of a navigation device or a curve speed warning system. Moreover, it is recommended that human factors tests be conducted to obtain user feedback on the usability of some of the HUD icons presented to FOT subjects by the ACAS. Only the cautionary and crash imminent alert icons of FCW were tested prior to building the pilot vehicle for the FOT. The results of the independent evaluation suggest marginal acceptance of FCW as well as some positive safety indicators that warrant deployment at least at low-level market penetration.

Additional research may be necessary to reduce the rates of false and nuisance alerts of FCW and to enhance the timing of crash imminent alerts for mid-term deployment plans. Proceeding with further FCW enhancement activities may depend on successful results (driver satisfaction, units sold, and positive safety impact) from short-term deployment and good market penetration levels. The recognition of the driver state would improve FCW alert timing, ranging from low complexity to identify the location of driver face (facing forward or sideways), medium complexity to track the eyes of the driver, to high complexity to measure the cognitive load of the driver. This research could build on current efforts undertaken in the SAVE-IT program [4]. Another FCW improvement might be achieved with the use of digital image processing of the forward scene to discern the objects that the radar is tracking. This would greatly reduce the rates of crash imminent alerts due to stationary out-of-path targets.

Vehicle to vehicle communications are suggested to improve the forward-looking sensing capability of FCW for long-term deployment plans. This research would build upon prior work in vehicle safety communications [5]. This enhancement would call upon lead vehicles to transmit information about their dynamic state to following vehicles, given wider deployment of FCW in the vehicle fleet. This would improve the timing of crash imminent alerts by reducing the rates of "too late" alerts (increasing crash prevention potential) as well as "too early" alerts (decreasing nuisance alert rate). Proceeding with such system improvement activity might depend on significant market penetration rates of FCW in the vehicle fleet during the next five to ten years.

REFERENCES

1. General Motors Corporation, "*Automotive Collision Avoidance System Field Operational Test (ACAS FOT) – Final Program Report*". DOT HS 809 886, National Highway Traffic Safety Administration, Washington, DC, 2005.
2. University of Michigan Transportation Research Institute and General Motors, "*Automotive Collision Avoidance System Field Operational Test – Methodology and Results*". DOT HS 809 900, National Highway Traffic Safety Administration, Washington, DC, 2005.
3. Najm, W.G., Stearns, M.D., Howarth, H., Koopmann, J., and Hitz, J., "*Evaluation of an Automotive Rear-End Collision Avoidance System*". Draft Final Report, Volpe National Transportation Systems Center, Cambridge, MA, 2005.
4. Witt, G.J., Zhang, H., and Smith, M., "*Phase 1 Research Summary and Phase 2a Planning Document. SAfety VEhicle(s) Using Adaptive Interface Technology (SAVE-IT)*". Volpe National Transportation Systems Center, Cambridge, MA, 2004.
5. CAMP Vehicle Safety Communications Consortium Consisting of BMW, DaimlerChrysler, Ford, GM, Nissan, Toyota, and VW, "*Vehicle Safety Communications Project, Task 3 Final Report – Identify Intelligent Vehicle Safety Applications Enabled by DSRC*". DOT HS 809 859, National Highway Traffic Safety Administration, Washington, D.C., March 2005.

CONTACT

Wassim G. Najm, General Engineer, Advanced Safety Technology division, Tel (617) 494-2408, Fax (617) 494-2995, e-mail: Najm@volpe.dot.gov.

DEFINITIONS, ACRONYMS, ABBREVIATIONS

ACAS: Automotive Collision Avoidance System
ACC: Adaptive Cruise Control
DVI: Driver-Vehicle Interface
FCW: Forward Crash warning
FOT: Field Operational Test
HUD: Head-Up Display

A Geodesics-Based Model for Obstacle Avoidance

Jason C. Olmstead Muhs and Jingzhou Yang
The University of Iowa

ABSTRACT

This paper presents a path prediction model for obstacle avoidance. A geodesics model is used to obtain the desired path in Cartesian space. The distance between the start target point (and end target point) and the surface of an obstacle is minimized to determine the boundary points of a geodesic across the surface of the obstacle. The model then numerically solves for a geodesic curve between the two boundary points of the geodesic on the surface of the obstacle. The model offsets the resulting discrete points on the geodesic in the positive normal direction (outside of the obstacle) to form a path of motion around the obstacle.

INTRODUCTION

Trajectory planning of human upper body movement is one of the most challenging problems in digital human simulation. Many tasks require the arm to move from its initial position to a specified target position without any constraints, or via a point for a curved path in case of obstacle avoidance. Obstacle avoidance is an important part of digital human simulation because in order for digital humans to simulate real-life behavior, they must be able to maneuver in a manner consistent with real human motion. Digital humans must be able to avoid obstacles in their path in order to be useful in modeling real-life tasks. This is the case for Santos™, the digital human that is the focus of the Virtual Soldier Research (VSR) program. The purpose of Santos™ is to be able to perform prototype testing for external organizations and provide feedback on his physical status; such as muscle stress, muscle fatigue, joint torques, heart rate, etc. In order for Santos to be able to perform prototype testing, Santos™ must be able to reach around obstacles that inhibit his ability to directly reach for a target and must be able to do so in a manner that a real person would.

The case of obstacle avoidance is very important for trajectory planning in virtual environments. A lot of research has gone into obstacle avoidance, especially in robotics. However, applying obstacle avoidance to real-time virtual simulations is an area that has not been looked into as much. Flash and Hogan (1985) presented a mathematical model for obstacle avoidance that used a via-point to control the path of an end-effector around an obstacle between an initial point, *A*, and a destination point, *B*, as seen in Figure 1.

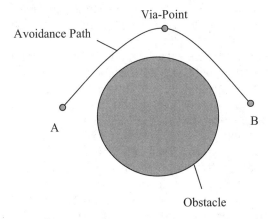

Figure 1: The via point is located above the obstacle and the path of the end-effecter moves from point A to point B through the via-point

The path of motion was optimized using a minimum-jerk model, which minimized the third derivative of position in all three directions of Cartesian space.

$$\xi = \frac{1}{2} \int_0^{t_f} \left(\left(\frac{d^3 x}{dt^3} \right)^2 + \left(\frac{d^3 y}{dt^3} \right)^2 + \left(\frac{d^3 z}{dt^3} \right)^2 \right) dt \qquad (1)$$

This model was applied to motion prediction in a virtual environment by Mi (2003). The downside to using this model in a virtual environment is that the user has to input the via-point for the end-effector (finger tip) to move through. This necessitates some simple trial and error and is very tedious. The need for user input also makes it impossible for any simulations to be done in real-time. Instead, a method should be created to determine the location of the via-point automatically in order to facilitate simulation in real-time.

Collision avoidance has been implemented into Jack™ by The University of Pennsylvania. A discrete 3-D space is used to compute optimal, shortest distance paths that avoid collisions. An iterative process is used to determine postures at each consecutive node in the 3-D workspace determined by the algorithm as it searches for a shortest distance path. While this method may be fast, it can still be inefficient to make an iteration and then make multiple checks for the next node in the process. It is more efficient if a path can be created without having to make checks at every discrete point in the path.

This paper presents a methodology for predicting and simulating the path around an obstacle generated by humans in a natural motion of the torso and upper extremity.

PATH IN CARTESIAN SPACE

Designing an algorithm to automatically determine a via-point is a minimum requirement for real-time obstacle avoidance. It is not necessarily sufficient, however. Moving around objects with complicated surfaces, such as those with long extremities, with one via-point does not necessarily produce a realistic motion. The via-point has to be placed in just the right position to ensure the hand and arm get around the obstacle without colliding with it. This is very difficult to do because of the complexity of obstacle surfaces. This problem can be simplified, however, by the addition of extra via-points, or control points, as they will now be called. The problem of getting around the obstacle is simplified because the use of multiple control points can ensure that the path does not collide with the obstacle. The points must be correctly spaced, though. If the control points were bunched together in a small region of the path, it would be similar to having just the one via-point. The question, now, becomes how to determine the control points. One possible solution to these questions is to offset the surface of the obstacle outward in the normal direction and find key points on the offset to use as control points. The offset would ensure there is ample space between the path and the obstacle. This works, but there are more efficient ways of accomplishing this. One such method uses geodesics and is a small part of a larger motion prediction algorithm, shown in Figure 2. Figure 2 shows that if an obstacle is not present, collision avoidance is not necessary. If an obstacle is present, a geodesics-based model, Figure 3, is used to move around the obstacle. The path determined is sent to the motion prediction algorithm and the path of motion is determined based on optimal joint movement.

BOUNDING SURFACES

Obstacles with complicated surfaces will often arise in situations where collision avoidance is required. Any

method developed to avoid collisions must be able to successfully avoid such complicated surfaces. One such way to accomplish this is to simplify the obstacles with a simple surface. The surface used to simplify obstacles must have certain characteristics, though. The surface must be complete so geodesics may be computed, it must be convex so motion paths are not following the contour of a surface inside a concavity, it must not have edges so an offset of the geodesic points is always possible, and a consistent definition must exist because designing a program to compute derivatives analytically for a surface is very difficult and time consuming. Using surfaces with consistent definitions avoids this problem. Surfaces such as the torus meet most of the conditions but they don't reasonably fit complicated surfaces since a torus is shaped like a ring. NURBS can represent the most complicated surfaces but do not have a consistent definition and are far from being simple. Ellipsoids are the only simple surfaces that meet all of the listed conditions above. The only variances in the definition of any ellipsoid are the radii, which determine the size of the ellipsoid. Ellipsoids are defined as follows:

$$\mathbf{x}_{\text{Ellipsoid}} = \{R_x \sin u \cos v, R_y \sin u \sin v, R_z \cos u\} \quad (2)$$

R_x is the radius in the x-direction, R_y is the radius in the y-direction, and R_z is the radius in the z-direction. The definition of the ellipsoid is a function of the surface parameters u and v. The method of predicting a collision-free path around obstacles is based on the computation of geodesics.

GEODESICS 3D MODEL

Geodesics is a topic discussed in differential geometry. "A geodesic on a surface μ is the shortest curve among all piecewise-differentiable curves on μ connecting two points." (Gray, 1999) What this means is that a geodesic represents the shortest distance between two points on a complete three-dimensional surface. The reason this is useful is that a function describing the curve between the initial and final points can be determined numerically. This is accomplished by forming two second-order differential equations based on the parameters u and v of the surface function $\mathbf{x}(u,v)$. The formulation of these equations, however, starts with the coefficients of the first fundamental form of the surface $\mathbf{x}(u,v)$.

The first fundamental form is used to determine arc lengths on surfaces. The coefficients are:

$$E = \mathbf{x}_u \cdot \mathbf{x}_u \quad (3)$$
$$F = \mathbf{x}_u \cdot \mathbf{x}_v \quad (4)$$
$$G = \mathbf{x}_v \cdot \mathbf{x}_v \quad (5)$$

where

$$\mathbf{x}_u = \frac{\partial \mathbf{x}}{\partial u} \qquad (6)$$

$$\mathbf{x}_v = \frac{\partial \mathbf{x}}{\partial v} \qquad (7)$$

These coefficients can than be used to determine the Christoffel symbols, which are functions of the coefficients of the first fundamental form. The Christoffel symbols are then used to formulate the differential equations to be solved for geodesics. The values of the Christoffel symbols are:

$$\Gamma^1_{11} = \frac{GE_u - 2FF_u + FE_v}{2(EG - F^2)} \qquad (8)$$

$$\Gamma^2_{11} = \frac{EF_u - 2EE_v + FE_u}{2(EG - F^2)} \qquad (9)$$

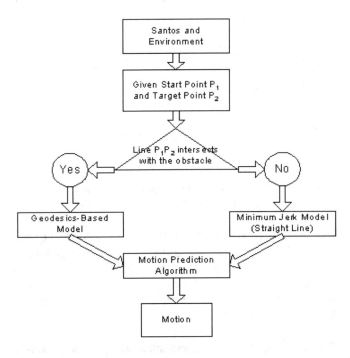

Figure 2: A simple motion algorithm of the upper extremities of a virtual human

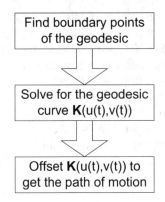

Figure 3: A simple diagram highlighting the main steps of the geodesics-based obstacle avoidance model

$$\Gamma^1_{12} = \frac{GE_v - FG_u}{2(EG - F^2)} \qquad (10)$$

$$\Gamma^2_{12} = \frac{EG_u - FE_v}{2(EG - F^2)} \qquad (11)$$

$$\Gamma^1_{22} = \frac{2GF_v - 2GG_u + FG_v}{2(EG - F^2)} \qquad (12)$$

$$\Gamma^2_{22} = \frac{EG_v - 2FF_v + FG_u}{2(EG - F^2)} \qquad (13)$$

where

$$E_u = \frac{\partial E}{\partial u} \qquad (14)$$

$$E_v = \frac{\partial E}{\partial v} \qquad (15)$$

$$F_u = \frac{\partial F}{\partial u} \qquad (16)$$

$$F_v = \frac{\partial F}{\partial v} \qquad (17)$$

$$G_u = \frac{\partial G}{\partial u} \qquad (18)$$

$$G_v = \frac{\partial G}{\partial v} \qquad (19)$$

The two second-order differential equations for the geodesics can now be written as:

$$u'' = -\Gamma^1_{11} u'^2 - 2\Gamma^1_{12} u'v' - \Gamma^1_{22} v'^2 \qquad (20)$$

$$v'' = -\Gamma^2_{11} u'^2 - 2\Gamma^2_{12} u'v' - \Gamma^2_{22} v'^2 \qquad (21)$$

Solving for the function of the curve numerically, using equations (20) and (21), allows the curve to be discretized so multiple points exist on the curve. These points can then be used as control points to control the path around the obstacle after the curve has been offset from the surface. The start and end points have to be determined before the function of the geodesic can be determined. Figure 4 shows what a geodesic on a sphere would look like, as an example. Notice that the curve does not follow the latitude in this case as some may guess with respect to the two endpoints shown. A geodesic on a sphere is always a segment of a great circle (a circle with the same diameter as the sphere).

The initial and final points of the geodesic are determined by the initial and final locations of the end-effector. NPOPT, a Fortran-based optimization software, was used to accomplish this task with constraints used from an algorithm designed by Zhihua Zou and Jing Xiao at The University of North Carolina at Charlotte (Zou and Xiao, 2003). The program requires an initial guess at the solution and minimizes the squared distance $F_i(u,v)$, represented by (22), between the obstacle surface and the initial/final location of the end-effector.

$$F_i(u,v) = \|w_i\|^2 = \min_{(u,v)}\|\mathbf{x}(u,v) - \mathbf{P}_i\|^2 \qquad (22)$$

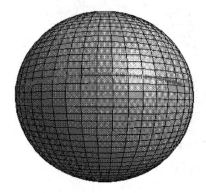

Figure 4: Geodesic on a sphere

The initial and final locations of the end-effector are, respectively, represented by i=1,2.

The closest point on the surface is defined by $F_i(u^*,v^*)$ and for this to be a local minimum, the necessary condition is that

$$\nabla F_i(u^*,v^*) = 0 \qquad (23)$$

According to Zou and Xiao, the sufficient condition for $\mathbf{x}(u^*,v^*)$ to be the closest point on $\mathbf{x}(u,v)$ to $\mathbf{P}_i$ is:

$$A = \mathbf{x}_{u^*}\bullet\mathbf{x}_{u^*} + \mathbf{x}_{u^*u^*}\bullet(\mathbf{x}(u^*,v^*) - \mathbf{P}_i)$$
$$B = \mathbf{x}_{u^*}\bullet\mathbf{x}_{v^*} + \mathbf{x}_{u^*v^*}\bullet(\mathbf{x}(u^*,v^*) - \mathbf{P}_i)$$
$$C = \mathbf{x}_{v^*}\bullet\mathbf{x}_{v^*} + \mathbf{x}_{v^*v^*}\bullet(\mathbf{x}(u^*,v^*) - \mathbf{P}_i) \qquad (24)$$
$$A > 0$$
$$B^2 - A\cdot C < 0$$

If the necessary and sufficient conditions are satisfied for a particular closest point, then the point is acceptable. If the point does not satisfy the conditions, a new guess must be made for the surface parameters u and v (Zou and Xiao, 2003). A diagram of what this procedure

entails is shown in Figure 5. Arc $\overline{AB}$ is the geodesic and arc $\overline{CD}$ is the offset of the geodesic and the path of motion for the end-effector. Points C and D are the boundary points of the geodesic and points C and D are the offsets of A and B. Segment $\overline{CH}$ shows that if there is no obstacle, the shortest distance between the initial point and the desired point is a straight line and obstacle avoidance is not necessary.

Once the boundary points of the geodesic are determined, the geodesic path across the surface can be determined. This was accomplished by using EleGeodesic software created by Anderson (1996) at Elements Research. The software uses a relaxation method to determine a numerical solution for the geodesic curve across the obstacle surface. It requires

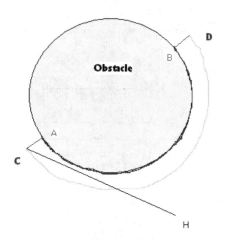

Figure 5: A geometric representation of how an offset geodesic would look on a sphere

the two boundary points of the curve determined from the NPOPT software and a guess at the initial direction of the curve (the gradient of the obstacle surface at the initial point of the geodesic). It then uses an iteration method to minimize the distance and the curve relaxes to the shortest distance curve.

The offset direction is accomplished by normalizing the cross product of $\mathbf{x}_u$ and $\mathbf{x}_v$ at all the discrete points on the curve using:

$$U = \frac{\mathbf{x}_u \times \mathbf{x}_v}{\|\mathbf{x}_u \times \mathbf{x}_v\|} \qquad (25)$$

The offset distances of each point depend on the offset distances of the boundary points and the number of discrete points on the geodesic. If the offset distance of the initial point of the geodesic is d_o and the offset distance of the final point of the geodesic is d_n, then the offset distance of the ith discrete point, p_i, will be determined by:

$$d_i = d_o + (d_n - d_o)\left(\frac{i}{n+1}\right), 0 \le i \le n \qquad (26)$$

where d_i is the offset distance of the ith point on the geodesic and n is the number of points on the geodesic exclusive of the initial point. This transition between the offset distances of the boundary points of the geodesic will provide a smooth path around the obstacle.

RESULTS

A simple example demonstrating this method begins with an end-effector located at the point {1.1,0,0}. An obstacle, represented by a sphere of radius one, is located at the origin. The final desired point of the end-effector is {0,0,-2.4}. This results in the following initial and final boundary points, respectively, of the geodesic on the sphere.

$$\begin{aligned}\{u_0,v_0\} &= \{0,0\}\\ \{u_f,v_f\} &= \{3.14159,0\}\end{aligned} \qquad (27)$$

These points are used to determine the surface coordinates of the discrete points of the geodesic on the sphere. The surface coordinates are converted back into Cartesian coordinates so the cross product of the surface gradient components can be determined in order to offset the points. Figure 6 shows what the geodesic looks like on the sphere. The path is offset by moving each discrete point off the obstacle in the normal direction with the boundary points of the offset path defined as the initial and final points of the end-effector. Figure 7 shows the obstacle with the offset geodesic defining the path of avoidance. The offset distance of the initial point of the path is 1.1 units since the

distance of the initial point of the end-effector from the obstacle is 1.1 units. Likewise, the offset distance of the final point of the path is 2.4 units since the distance of the final point of the end-effector from the obstacle is 2.4 units. Each subsequent point after the initial point is offset an equal amount more than the previous point so that a smooth path is obtained. Figure 7 shows what the offset geodesic looks like. The path determined from this algorithm will be sent to the motion prediction algorithm, also developed in the VSR program, which will ensure that the joint motion of the virtual human is optimized. The motion prediction will enable reaching motions to be performed based on human factors measures that provide a basis for enabling Santos™ to make realistic motions. Examples of human factors measures are minimizing discomfort or joint displacement. This combined algorithm will provide efficient obstacle avoidance within virtual environments as has been described.

CONCLUSION

The proposed method for predicting hand movement around obstacles is very efficient and will consequently fit very well with the virtual human model because of the real-time capability of the model. It was shown that determining the boundary points of a curve on a surface could be found simply by determining the closest points on the surface to the initial and final positions of the end-effector. It is important to note that the sphere in the example previously presented is only an example of a surface surrounding an obstacle. Ellipsoids are the surfaces that will be used to surround any complicated obstacle. Spheres are merely special forms of ellipsoids. The geodesics algorithm can then be used to determine the minimum distance between the two boundary points. Once the curve is determined as a function of the obstacle surface parameters and is discretized through the numerical solution, it can be offset in the normal direction to provide a path around the obstacle. The path can then be used to optimize joint motion during obstacle avoidance.

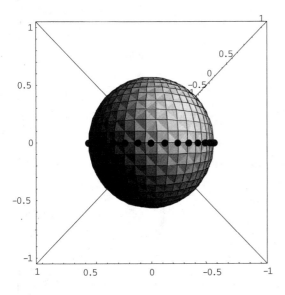

Figure 6: Spherical obstacle viewed from the right side with the geodesic represented by the black dots

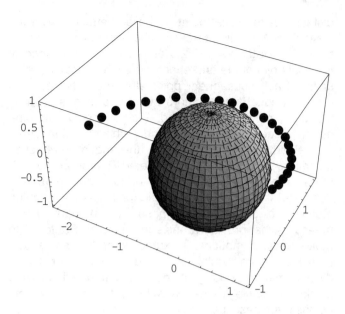

Figure 7: Obstacle with path of avoidance

ACKNOWLEDGMENTS

This research was funded by the US Army TACOM project: Digital Humans and Virtual Reality for Future Combat Systems (FCS).

REFERENCES

1. T. Flash and N. Hogan (1985), "The coordination of arm movements: an experimentally confirmed mathematical model", The Journal of Neuroscience, 5(7): 1688-1703.
2. Z. Mi (2004), *Task-Based Prediction of Upper Body Motion*, Ph.D. Dissertation, The University of Iowa, Iowa City, IA.
3. Y. Liu and N. I. Badler, "Real-Time Reach Planning for Animated Characters Using Hardware Acceleration", Computer Animation and Social Agents, IEEE Computer Society, New Brunswick, NJ, May 2003, pp. 86-93.
4. A. Gray (1999), *Modern Differential Geometry of Curves and Surfaces with Mathematica*, 2nd ed., CRC Press LLC, Boca Raton, FL.
5. Z. Zou and J. Xiao (2003) "Tracking Surfaces in Real Time," *Proceedings of the 2003 IEEE/RSJ International Conference on Intelligent Robots and Systems*, October, Las Vegas, NV, 2692-2698.
6. W.L. Anderson, Elements Research, http://www.elesoft.com/geodes.html, Software Copyright 1996, Software Updated April 2005.

CONTACT

Jason C. Olmstead Muhs, Graduate Research Assistant, Virtual Soldier Research program, Center for Computer Aided Design, The University of Iowa, Email: jolmstea@engineering.uiowa.edu

Advanced Protection for Vulnerable Road Users

Eddie Moxey
University of Surrey

Neil Johnson, Michael G. McCarthy and William M. McLundie
InfraRed Integrated Systems (IRISYS) Ltd., Transport Research Laboratory (TRL) Ltd. and Jaguar Cars Ltd.

ABSTRACT

There is increasing international interest in collision avoidance and more recently in safety systems for 'Vulnerable Road Users' (VRUs). The aim of the 'Advanced Protection for Vulnerable Road Users' (APVRU) project was to develop a sensor system capable of detecting VRUs, distinguishing them from the road environment, tracking their position and predicting potential impacts with the vehicle. Our fusion of sensor data from multiple, short range, high accuracy, pulse radar units and a low-resolution passive infrared sensor array serves to eliminate the majority of clutter by negating false triangulation combinations, whilst ensuring that only thermally distinct moving targets are considered. The intention is that a derivative of the sensor system *may* eventually provide the technological link between VRU detection and a driver warning, collision avoidance and/or the activation of a safety system on the vehicle (external airbags). The performance of the APVRU system was demonstrated during real-world accident scenarios, impacting with an anthropometric dummy, at speeds up to 25kph.

INTRODUCTION

Pedestrians and cyclists, referred to collectively as 'Vulnerable Road Users' (VRUs), account for approximately 1 in 4 of all road deaths in the European Union [1]. According to accident statistics over 6,000 pedestrians are killed annually on EU roads [1]. In the United Kingdom, 71% of pedestrian fatalities occurred at impact speeds under 40 mile/h and 85% of fatalities were injured by the front of the vehicle[1] [2, 3]. These statistics indicate the scale of the VRU problem and the proportion of casualties to which an effective warning and/or active safety system may offer benefit.

Previous research by the Transport Research Laboratory (TRL)[4], suggested that the design of vehicle fronts could afford adequate protection to pedestrians in low speed impacts by using soft bumpers and pop-up bonnets. However, further to a series of adult and child pedestrian impact tests, this report concluded that by using external airbags on the bumper and bonnet of a vehicle, there was potential for a significant reduction in head injury (which accounts for around 60% of all fatalities). While it was suggested that an active pedestrian safety system could also be used to activate the vehicle's brakes prior to impact, the activation of an external airbag system should be explored, and therefore provided an initial basis for the APVRU project. This paper details the design of the *proof of concept* sensor system developed under the APVRU project.

PEDESTRIAN AIRBAG INVESTIGATION

Although some patents have been filed [5, 6], there is little work concerning the experimental investigation of external airbags for pedestrian protection. The previously mentioned work on external airbags [4] found significant reductions of biomechanical injury values for specific impacts. However, since no external pedestrian airbag or pedestrian impact dummies existed, the testing used Occupant Protection Assessment Test (OPAT), Anthropometric Test Dummies (ATD's) and off-the-shelf passenger airbag systems mounted externally on the bonnet of a vehicle. Thus, this purely experimental approach does not allow these results to be applied to the wide range of impact conditions that would occur in real-world VRU accidents. As a result of this, a different approach was taken within the APVRU project.

An explicit finite element model was constructed using LS-DYNA[TM]. This model incorporated a specifically-created pedestrian model (as described by Howard, *et al* [7]), a correlated C-class (small family hatchback) vehicle and a staged airbag system, consisting of a bumper airbag and a scuttle airbag, based on best estimates of how a possible external safety system may operate. A possible design for this, *proof-of-concept*, pedestrian airbag system is depicted in Figure 1.

---

[1] UK Fatals database, populated by TRL on behalf of the UK Department for Transport

A review of recent accident statistics and an investigation into both driver and pedestrian behavior, during simulated emergency situations, was conducted for this project and subsequently published [3].

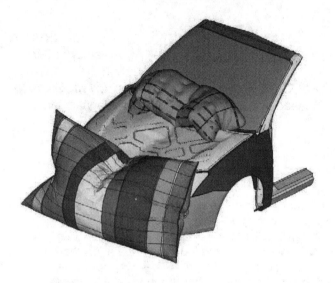

Figure 1: Proof of concept pedestrian airbag system.

A set of experiments was constructed allowing several fundamental variables to be examined:

- Timing implications for the required response of a pedestrian sensor system.
- Examination of the biomechanical effects on the pedestrian if the airbag was to fire correctly and too early or too late.
- Examination of the effects on pedestrian location and height (i.e. 90th percentile compared with a 6-year old child).

The concept behind the staged bumper and scuttle airbag system is that, following a trigger signal from the pre-impact sensing system predicting an unavoidable impact, the bumper airbag will deploy first, reducing initial impact, bending and shear on the ankle and knee. The scuttle airbag will then fire in sequence allowing reduction in head acceleration, and protecting the head and upper body from large decelerations. Full details of the FE analysis is given in the main project report [8].

A number of scenarios were simulated with both a validated 50 percentile adult male FE model and a 6 year old child FE model [7]. It was found that in perfect circumstances the system may reduce adult lower leg and head injuries. However, it was also found that potential late activation of the system adds energy into the event and could lead to an increase in the severity of injury to the pedestrian. In addition, the scuttle airbag had no beneficial effect for the head impact for the child humanoid.

The FE model was able to prove that an elementary solution of fixing a simple occupant-based system on the exterior of the vehicle is not effective. It may be possible

to engineer a much more complex system that includes adaptive venting. However, the problems with airbag geometry, durability and reliability in operation over several years, may prove to be a very difficult issue. A more effective solution may be to combine an active and passive system, taking the best and most applicable aspects from both.

ANTHROPOMETRIC DUMMY

In order to demonstrate the APVRU tracking system under real-world conditions, an anthropometric dummy with infrared and radar profiles representative of a human adult was developed. Apart from being representative of a "hot-bodied" human [8], the dummy had to be lightweight and yet able to withstand numerous low-speed impacts without damaging the vehicle or itself. The dummy was constructed from a wire-frame skeleton covered with stiff foam. This was then wrapped with heating wire elements and tin foil before being clothed and mounted onto a small wheeled trolley. Both the thermal and radar signatures of the dummy can be modified to suit. Comparative measurements of the dummy against several human targets showed no significant difference in the infrared profile and an acceptable comparison with the radar profile of human subjects.

SENSORS AND OVERVIEW OF THE APVRU SYSTEM

Most of the previous attempts to solve VRU tracking have relied on radar and/or video-based, CCTV, systems [9, 10, 11, 12]. Visual image processing to identify VRU's by either their shape or motion is both computationally expensive and unreliable in complex urban environments. Fusion with the radar data improves the positional accuracy for tracking, however complex methods of data association tend to limit the reaction times of such systems.

Our review of current sensor technologies [13] found that while a number of alternatives are possible, the technically superior option, given the requirement for real-time operation in a heavily cluttered environment, was the combination of short range radar and passive infrared sensors. The use of a low resolution infrared detection array provides the ability to distinguish VRU's from the environment and track them across the field of view. In addition, the use of high resolution, short range, radar provides the ability to accurately track the position of targets over the ground plane. While the use of infra-red cameras has been proposed [14], the fusion of radar and infrared sensor data within a target tracking system has demonstrated, for the first time, the ability to track "hot-bodies" in front of a vehicle in real, non-simulated, road environments using a relatively low-cost system.

The radar unit selected was the 'High Resolution Radar' (HRR) developed by M/A-COM, a subsidiary of Tyco Electronics Inc. This unit, designed originally for proximity sensing for the American automotive market, is a short range (0.2 - 20m), radar operating in the 24GHz Industrial, Scientific and Medical (ISM – radio spectrum) Band. The unit consists of a microwave front-end, with an integrated processor and a 'Controller Area Network' (CAN) protocol interface. Although capable of continuous wave operation, the unit is normally operated in pulse mode and reports, via CAN, the range and amplitude of the signal reflected back from each of a maximum of ten targets every 20 ms.

The infrared pedestrian detection and tracking head is an experimental detector platform based around a low element count infrared detector array. This novel, low-cost, thermal sensing technology, developed by IRISYS, has already been applied to a diverse range of applications. Infrared radiation is focused onto a 16x16 pyroelectric detector array using a germanium lens giving a 60° field of view, while the array is scanned at just over 30 frames per second. For this application, a long wave pass infrared filter (approximately 6.5 to 15 μm) was employed so that the device was optimized for the detection of humans. The detector platform includes a DSP which enables all low-level signal processing and target tracking to be handled locally. The tracking system is based around an elliptical contour tracker capable of concurrently tracking multiple thermally distinct moving targets with sub-pixel accuracy. This provides estimates of the position, shape/size, and velocity of multiple, uniquely identified, targets. Since this sensor is only sensitive to changes in incident radiation and the tracker will only consider smoothly changing elliptical responses, the system is effective at minimizing clutter and noise.

One of the principle factors influencing radar range accuracy is the amplitude of the reflected signal that varies with the radar cross section of the particular target. The use of more than two radar units allows a ground plane position measurement to be derived for each combination of two or more radars, resulting in multiple measurements that can be combined to give a more accurate range and position estimate. It was found that three units, configured as described below, gave sufficiently accurate results to achieve a *proof of concept* system. A practical system is likely to require more radar units, as suggested in Klotz and Rohling's [15] work.

In a production system, the most convenient location for mounting the radar units would probably be behind the front bumper with the IR Head/s at the upper edge of the windscreen or incorporated into the headlamps. Adding the complication of such packaging issues was considered an unnecessary distraction from the development of the core technology. The most straightforward approach was therefore adopted with the IR mounted in the centre and the three HRR units mounted equidistant across the width of the vehicle on a modified roof bar, as shown in Figure 2.

Figure 3 shows the block diagram of the hardware elements of the final APVRU system. Three HRR units, the synchronization box (Synch) and the IR Head are connected to a Central Processing Unit (CPU), via a dedicated CAN bus (CAN1). The synchronization box provides "soft synchronization" by transmitting a time offset that is used to align the data transmission of the HRR units. The second, vehicle bus (CAN 2) is used to receive the vehicle's ground speed and transmit a confidence measure update and deployment indication to a display and interface unit called a MicroGen[2]. This is a programmable, multi-port, prototyping interface unit that provided both interrogation and isolation from the vehicle CAN bus.

Although this was done mainly for safety reasons, it is likely that a production system would need a dedicated CAN bus for the sensors, employed for this and similar collision avoidance applications, in order to accommodate the required bandwidth.

Figure 2: IR Head and 3 HRR sensors mounted on a modified roof bar.

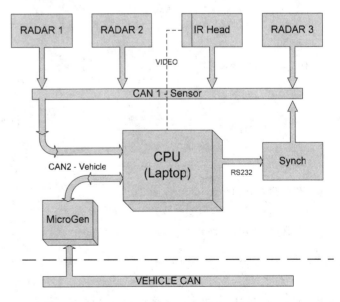

Figure 3: System-level block diagram with sensors, CAN busses and communication.

---

[2] Product details may be found at: www.add2.co.uk

Full details of all the equipment used in this project can be found in the project's final report [8].

SOFTWARE AND HOT-BODY TRACKING

The APVRU system is required to track "targets of interest," VRU's, in real-time and make predictions regarding the likelihood of potential impacts. The system must be able to distinguish VRUs from other objects which reflect the transmitted radar pulses or emit infrared radiation similar to that of humans. Only when tracks from both these signals have a common trajectory can the target be treated as a possible VRU. Essentially the system must perform as a "hot-body" tracker.

SOFTWARE STRUCTURE AND DATA PROCESSING

As suggested above, this hot-body tracking system must perform two separate tracking tasks. The first filtering stage stabilizes range measurements from the individual radar units by combining multiple responses and minimizing the effect of missing data. The second stage uses the multiple ground plane measurements from the triangulation and data fusion stages to update target position and impact prediction estimates. Both levels of tracking in this system are based on Kalman filtering techniques which are described below. Figure 4 shows a block diagram representation of the two levels and the data processing structure. Since the IR Head has built-in tracking, a local trajectory store (a list structure containing all data associated with each active target) is simply updated directly. A similar store is used within the range filtering blocks and the combined tracking block.

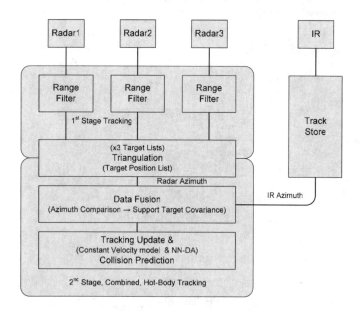

Figure 4: Conceptual block diagram of radar range filtering and hot-body tracking.

The lower half of Figure 4 contains three processing blocks; radar triangulation, data fusion, and tracking

update & collision prediction. These are drawn within a single block labeled "2nd Stage Combined, Hot-body, Tracking" which adjoins the upper background block surrounding the three radar filtering blocks. This representation illustrates the two levels of tracking, their close cooperation and the implementation of the data fusion and impact prediction in the second level of the Kalman filtering. The radar triangulation sub-routine, initiated from within the main tracking filter, uses the data from the three separate range stabilization filters to create a single list of target position measurements that forms the basis of the algorithms developed to track targets across the ground plane and make predictions about any potential impact.

RADAR DATA FILTERING AND TRIANGULATION

The Kalman filter is a popular mechanism for system state estimation and control. The technique involves the iterative update of the system state estimate and associated error covariance. The normal operation involves two phases. The first 'time update' or 'prediction' phase updates these quantities to reflect the effect of an elapsed period of time with no new information. The second 'measurement update' or 'correction' phase updates the quantities to reflect the effect of a new measurement.

For each radar unit, a range tracking module uses the noisy list of radar responses to update a small number of consistently tracked target range estimates using a 'Nearest-Neighbour' data association (NN-DA) mechanism.

Since target range is likely to be approximately locally linear with respect to the radar sampling rates, a constant velocity model was used to describe target range dynamics. Given that range is scalar valued, this model results in a two-dimensional state vector (range and rate of change of range). This yields a simplification of the Kalman filtering equations, avoiding the computationally expensive matrix manipulation encountered when the filter is applied to problems involving higher dimensional state and measurement vectors. Since the radar units return no information relating to the "fitness" of a range measurement, a constant measurement noise variance is used in the correction phase.

Since the HRR units only report the range and amplitude of the reflected signals, the actual position on the ground plane must be computed by triangulation from radar pairs. Although the worst case complexity of the triangulation procedure is $O(n^2)$ operations (when both radars give rise to n measurements), many of the possible combinations of range estimates may be discarded as they are impossible given the particular radar baseline.

Despite this constraint and the reduction in the number of range estimates achieved by range filtering, false ground plane measurements will still be generated. The vast majority of these responses will either be short lived, appear to move in an erratic manner, or will have no support from similarly moving thermal targets. The effect of these "ghost" responses are minimized by the subsequent stage of filtering.

SENSOR FUSION AND HOT-BODY TRACKING

A primary consideration when implementing any form of data, or sensor, fusion is the problem of registration. The data from each sensor must have a common frame of reference. Unfortunately, it can be difficult to accurately determine a target's ground plane position from its projection onto the view plane. An alternative approach was therefore adopted.

Since we are seeking a method to track the ground plane motion of hot bodies, the aim of the data fusion stage is to ensure that only radar targets corresponding to thermally distinct objects are considered. The covariance associated with ground plane position measurements from the triangulation phase is essentially a measure of confidence in the estimate. Thus we can modify the measurement covariance such that it also reflects the confidence in the object being a hot body. This is achieved by adjusting the covariance using a pseudo probability which reflects the degree of support for the target given by the infrared system. This method allows infrared support to be effectively integrated over time by the Kalman filtering used in the ground plane tracking; hence continuous infrared support is required for the ground plane track to be maintained. In addition to providing an effective solution to achieving hot body tracking, this approach helps to ensure that "ghost" tracks and noisy measurements from the triangulation phase are, in general, ignored.

The infrared support probability is derived by considering the similarity in target azimuth (relative to the direction the vehicle is facing), since this can be easily and accurately obtained in both ground plane and image plane coordinate systems. For each triangulated ground plane measurement, the thermal target with most similar azimuth is selected and the azimuth similarity used to calculate the degree of infrared support. Although this approach fails to discriminate between targets with the same azimuth, this is only significant if the targets' azimuths remain identical over an extended period of time.

Using the above fusion mechanism to produce sets of modified target position measurements ground plane target tracking is achieved using a similar process to that used in filtering range estimates.

IMPACT PREDICTION

The final stage of the APVRU system is perhaps the most important and is likely, in the long term, to prove the most difficult to achieve with the levels of performance required for a commercial system. In general, the future trajectory of the target relative to the vehicle may be highly non-linear. While the principal contribution is from the motion of the vehicle itself (which can be affected by many factors), the actions of the VRU, such as any attempt to avoid the vehicle, cannot be ignored. The robust prediction of future relative motion would therefore need to differentiate both vehicle and target motion, classify VRU type and be able to generate the required stochastic predictions based on detailed models of vehicle and VRU dynamics.

Since the generation of such detailed non-linear predictions was beyond the scope of the APVRU project, it was decided to focus on constrained scenarios in which the relative trajectory of the target is linear. Given this simplifying assumption, the Kalman filter employed within the ground plane tracker provides a powerful mechanism for predicting the future target position and, crucially, the probability distribution of the position. This is achieved by using the prediction phase of the Kalman filter update equations in isolation. In this mode of operation, the filter uses its model of the process to estimate the most likely system state at a given point in the future, together with the variance associated with that state estimate at that point in time. When this approach is applied to the filter used within the ground plane tracking mechanism, a sequence of predictions shows the target position moving linearly from its initial point with a constant velocity, whilst its elliptical confidence region (as derived from the estimate covariance) remains centered on the predicted position but expands over time. Using this evolving distribution, the point when the highest probability of an impact with the vehicle will occur and the value of that probability can be found.

Figure 5 is a screen shot of the hot-body tracking system in operation. The left-hand half is the ground plane with the vehicle's bonnet shown as the dark grey 2m square block in the bottom centre. The right-hand half shows a 16 x 16 pseudo-image from the IR Head with a video image superimposed in the centre. The latter was only crudely aligned for test purposes and none of the image data is used by the tracking system. The individual radar reflections are shown along the 3 horizontal lines as the bottom of both grids. The pedestrian's thermal track is shown on the right-hand grid and the combined hot-body tracking along the projection of the azimuth from the IR track is shown on the left-hand image. The system is indicating a 77% probability of impact in 0.52 seconds and has triggered a deployment signal for the "imaginary" external airbags.

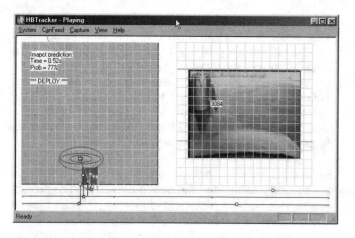

Figure 5: Screen shot image of the hot-body tracking system in operation.

TESTING AND PERFORMANCE

The APVRU system was tested with a number of simulated, real-world, accident scenarios, each more challenging than the last. In order to convey the results of these test scenarios herein the following 14 groups, covering 43 individual scenarios was derived. Groups 1 to 10 were conducted with human volunteers walking or running at various trajectories toward the vehicle, groups 11 to 13 were conducted with a moving vehicle and drive-by scenarios and the final group with our anthropometric dummy.

1. A pedestrian crossing the vehicle left to right or right to left at varying distances.
2. A pedestrian walking along various parallel trajectories toward the vehicle.
3. A pedestrian running along various parallel trajectories.
4. A pedestrian walking along various diagonal trajectories.
5. Two pedestrians walking along various parallel trajectories
6. Two pedestrians walking along various intersecting trajectories toward the vehicle.
7. A pedestrian running past a second pedestrian walking toward the vehicle on parallel trajectories.
8. A pedestrian walking/running at varying speeds across the front of the vehicle and then changing direction toward the vehicle (simulating a pedestrian running into the road).
9. As (8) with a second pedestrian walking parallel toward the vehicle along the pavement.
10. As (8) from behind a parked car.
11. The vehicle driving past a pedestrian standing, at varying distances, at the side of the road.
12. The vehicle driving past two pedestrians walking, at varying distances, at the side of the road.
13. The vehicle driving past pedestrians and vehicles parked at the side of the road.
14. Live impact tests with the dummy at varying vehicle speeds up to 25kph (15.5mph).

The limit in the last was initially due to safety concerns, since both the system operator and driver had to be inside the vehicle during the tests. Further comment on this limit will be given later.

Based on a 5-point scale (1 being "poor" and 5 being "good") the performance of both the system's ability to acquire a valid track and to maintain a tracking lock for each of the test scenarios was graded by subjective analysis. In each case the inspector also graded the appropriate deployment indication for the system with a "Yes/No" tick-box. The mean and standard deviation of each group was calculated together with the percentage of appropriate deployment and is presented in Figure 6. Plus and minus one standard deviation is illustrated. The numbers in brackets on the central graph indicate the total number of individual tests in each grouping. This number is double the number on separate test scenarios, since each was conducted twice.

Apart from only a 50% appropriate deployment in test group 13 (from only 4 tests and probably due to lower level of prediction accuracy from tracking multiple pedestrians plus parked and oncoming vehicles), these results suggest a good to excellent level of performance. However, it should be noted that these are simplified test scenarios in largely controlled environment and should only be regarded as evidence of *proof of concept* for the sensor system.

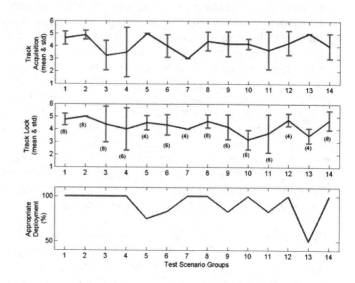

Figure 6: Mean and standard deviation graphs for the system's track acquisition and tracking lock plus the percentage of appropriate deployments.

DISCUSSION

There are a number of points which should be considered before any conclusions can be made about this project and/or any suggestions for future work. There are three areas for discussion; development and production costs (against consumer expenditure), technical and regulatory limitations and reliability and possibility of litigation. All of these are interrelated and will be addressed collectively in the following paragraphs.

First, consider similar target tracking problems in cluttered environments and the level of system development and research expenditure. Comparing the sensor technology used in military and/or civil aviation

232

against that of automotive applications we can note that the degree of difficulty of the problem is relatively low, due to the low number of targets in a largely uncluttered environment (there are relatively few aircraft in a largely empty sky), and yet the level of expenditure and development has been considerable. Conversely, the degree of difficulty in tracking targets on the ground is very high, due to the considerable amount of clutter from buildings, road furniture and other road users, yet the level of development and expenditure is still very low.

Due to the considerable cost of development, system integration and packaging, if pedestrian tracking and/or active protection systems are to be implemented on production vehicles, they are likely to only be incorporated as part of a wider suite of collision avoidance applications. Sensor systems, more specifically radar systems, are generally tuned for their particular application. Developing a sensor suite capable of detecting and tracking targets across a number of applications invariably requires design compromise and the development of a system with high level logic, data sharing capability, and fast, real-time, processing. As with all engineering problems within a commercial marketplace, there is a conflict of interest between the level of performance and the overall cost. However, it is particularly apparent in this case.

The radar unit selected for this project was, at the time, considered to be the best available choice. However, there are a number of problems with the use of this particular unit for this application. Firstly, although the design of the HRR units yields a good compromise between modern technology and low-cost for "parking aid" and/or "stop & go" applications, their performance falls short of that required for APVRU. The range is far too short and the update rate is too slow for tracking pedestrians from a vehicle traveling at standard urban speeds. At 30mph a vehicle covers 13.41m/s, or 1m every 74.57ms. From this speed and given the estimate of inflation for an external airbag, the maximum time window for the tracking algorithm is less than 2s, with a minimum update rate, based on the results from our system, of at least 100Hz. This is assuming the VRU is already stationary in the middle of the road. Unfortunately the HRR unit's update rate is 50Hz and the maximum usable range is much less than the specified maximum of 20m. Furthermore, if the output of the system were to actively apply the brakes in order to slow or stop the vehicle prior to impact with a VRU then the system would need to be able to detect the VRU at a much earlier point in time than that demonstrated by this system.

Excluding increased initial cost, the concept of a combined system of multiple sensors for a suite of applications presents a number of advantages. Firstly, the use of longer range, automatic cruise control and collision warning sensors could be used to initialize pedestrian tracking at a much earlier point in time and offer the real possibility of being able to actively apply the vehicle braking before impact. An increased number of sensors, across and/or around the vehicle, would offer an improved field of view and greater positional accuracy in the triangulation algorithms.

Further to the results from the FE analysis, if the airbag were to fire and stabilize to early, then any potential benefit would be reduced However, more importantly, if the system were to fire late and the point of impact be prior to airbag stabilization, then the airbag could add energy into the event, subsequently causing greater injury. Further study is required to quantify the real operational risks and benefits of such a system. Ultimately, correct operation cannot be guaranteed in 100% of cases due to the non-deterministic nature of the problem. This leads us to the question of quantifying system reliability.

These comments should not be taken to imply that the tracking system or that the activation of an active, pyrotechnic, protection system has no merit. An alternative protection system could consist of a passive system in the vehicle bumper with external airbags up the bonnet and/or around the scuttle and windscreen pillars. Such a system would have no active system that would extend beyond the area that the vehicle would normally occupy and as such could not add energy into an impact that was not already present.

POSSIBLE IMPROVEMENTS

While the *proof of concept*, "hot-body," tracking system developed for this project was successfully demonstrated at speeds up to 25kph, there are a number of possible areas for future research and development.

Although there are now new suppliers of automotive radar equipment, more work is required to improve both the performance of individual units and the combination of long-, medium- and/or short-range radar systems.

The use of two IR Heads would be an advantage, both to positional accuracy, by using the range information from stereo processing algorithms, and the identification and removal of false, non-pedestrian, targets. Additional improvements to the IR tracking system could include improving the sensitivity of the device, particularly to looming targets; minimizing incorrect target initializations, and providing target classification capabilities.

While the APVRU system has shown reasonable performance in a number of constrained test scenarios, the system is still relatively simple and there are a number of significantly more advanced (and more computationally expensive) tracking filters (such as more advanced methods of data association and range gating and filters involving split/merge reasoning) that could be investigated in order to deal with increased levels of clutter in the tracker's field of view. Furthermore, considerable work is required to provide a more realistic and robust impact prediction capability.

CONCLUSION

The over-riding conclusion from the results of our FEA of a possible design for an external, pedestrian, airbag is that substantially more work is required before such a system could be recommended for incorporation into production vehicles. Further investigation in the use a combination of active and passive safety systems are recommended and likely to yield a more practical solution.

Based on the update rate of the radar and considering the subjective analysis of the success of the triggering event at 25 km/h, tests at higher speeds were considered likely to be beyond the capabilities of the current system. Further hardware improvements, specifically the update rate of the HRR units, would be required to achieve effective performance at higher impact speeds.

ACKNOWLEDGMENTS

This project was sponsored by the UK's Engineering and Physical Sciences Research Council (GR/M87054/01), the Department for Transport and the Department for Trade and Industry under the Foresight Vehicle Programme.

REFERENCES

1. **Organisation for Economic Co-operation and Development (OECD) / International Road Traffic and Accident Database (IRTAD),** Summary Report 2001, Federal Highway Research Institute (BASt), Bergisch, Germany (http://www.oecd.org/ and http://irtad.bast.de).
2. **Department for Transport** (DfT), UK. Road Accidents Great Britain: The Casualty Report 2000. The Stationary Office (TSO), London 2001.
3. **McCarthy, M.G.** Foresight Vehicle: Using accident data to develop Advanced Protection for Vulnerable Road Users. SAE World Congress and Exhibition, Detroit, MI, USA, March 2002. Book No: SP-1695, Doc. No: 2002-01-1122. SAE Tech.Pub.
4. **Holding, P.N., Chinn, B.P. and Happian-Smith J.** Pedestrian protection – An evaluation of an airbag system through modelling and testing. 17th Enhanced Safety of Vehicles (ESV) Conference, Amsterdam, Holland, June 2001. PaperNo.330.
5. **UK Patent application: GB2397559,** Airbag Arrangement for Windscreen & A-Pillar (H LAND, YNGVE (SE); FREDRIKSSON, RIKARD) AutoLiv
6. **US Patent application: 20040232663,** Externally developed Airbag Device (Takimoto, Takayuki) TAKATA
7. **Howard, M.S., Thomas, A.V., Koch, W., Hardy, R. and Watson, J.** Validation and applications of a finite element pedestrian humanoid model for use in pedestrian accident simulations. Int. Conference on the Biomechanics of Impact (IRCOBI). September 2000, Montpellier, France.
8. **McCarthy, M.G., Moxey, E., Johnson, N. and McLundie, W.M.** The Advanced Protection of Vulnerable Road Users (APVRU) Project: Final Report. Rpt.No. PR/SE/970/04, prepared for the LINK Foresight Vehicle Programme by the Transport Research Laboratory (TRL) Ltd. (available online at: http://www.apvru.com/)
9. **Broggi A., Bertozzi, M., Fascioli, A. and Sechi M.** Shape- based pedestrian detection. IEEE Intelligent Vehicles Symposium. Pp:215-220 2000.
10. **Yamada, Y., Okuda, K. and Shiraishi, N.** Pedestrian protection and rear-end collision injury reduction technologies on ASV-2. 7th Intelligent Transport Systems World Congress, Turin, Italy, Nov 2000.
11. **Gavrila, D.M., Kunet, M. and Lages U.** A multi-sensor approach for the protection of vulnerable traffic participants – the PROTECTOR project. IEEE Instrumentation and Measurement Conference, Budapest, Hungary, May 2001.
12. **Tons, M., Doerfler, R., Meinecke, M. and Obojski M.** Radar sensors and sensor platform used for pedestrian protection in the EC-funded project SAVE-U. IEEE Intelligent Vehicles Symposium, pp:813-818, 2004.
13. **Kerr, D.S., Parker, G.A. and McLundie, W.M.** Conflict simulation and sensor evaluation for the Advanced Protection of Vulnerable Road Users – A Foresight Vehicle project. In SAE World Congress and Exhibition, Detroit, MI, USA, March 2002. Book No: SP-1695, Doc. No: 2002-01-0826. SAE Tech.Pub.
14. **Bertozzi, M., Broggi, A., Carletti, M., Graf, T., Grisleri, P. and Meinecke M.** IR Pedestrian Detection for Advanced Driver Assistance Systems. 25th Pattern Recognition DAGM Symposium, Magdenberg, Germany, 2003
15. **Klotz, M. and Rohling, H.** A 24GHz short range radar network for automotive applications. Proc. CIE International Conference on Radar, Beijing, China, 15-18 October 2001. pp.115-119.

CONTACT

The APVRU consortium consisted of the Transport Research Laboratory Ltd., InfraRed Intetgrated Systems (IRISYS) Ltd., Jaguar Cars Ltd., and the University of Surrey. Although the main author resides at the latter the project management and point of contact should be via Mr Michael McCarthy at TRL Ltd. It is intended that video files of the test data being processed by the APVRU tracking system will be made publicly available – contact TRL for more information at: www.trl.co.uk

2005-01-1479

On Automatic Collision Avoidance Systems

Thorsten Brandt, Thomas Sattel and Jörg Wallaschek
L-LAB: Public Private Partnership of University of Paderborn and Hella KGaA Hueck & Co.

ABSTRACT

In future automotive collision avoidance systems (CAS) collision-free path planning will be required. For this reason, two possible path planning techniques are compared. The first method is based on the mathematical theory of differential games, the second method focuses on so-called elastic bands. However, the method of elastic bands is essentially modified to provide solutions in complex driving situations. Furthermore, it is shown how path planning, path following control and driving condition estimation can be embedded in a possible overall CAS-setup. Therein, driving conditions are estimated based on the characteristic velocity. Finally, simulation results for an emergency maneuver are presented.

INTRODUCTION

The future development of today's driver assistance systems strongly depends on the efficiency of automotive actuators and environmental sensors such as radar, lidar, laser scanners, video systems and appropriate image processing software. However, sensor fusion of these systems promises a prolonged innovation thrust [1]. Thus, in the most far-reaching scenarios future vehicles will become completely autonomous. These intelligent vehicles require collision avoidance systems (CAS) to handle unexpected emergency situations.

Within this framework it seems reasonable to consider technological extensions of already existing driver assistance systems, for example lane departure warning or emergency braking, in order to provide evasion functionalities. For the time being, these CAS can take over the driving task in emergency situations that a human driver cannot handle.

If the estimated collision risk exceeds a prescribed threshold value the CAS initiates an emergency maneuver. Depending on the complexity of the situation, different actions can be taken: braking, evading or evading in combination with braking or accelerating. Today's emergency systems are only capable of sheer braking. This work mainly focuses on evading maneuvers at constant velocity to provide the foundation for systems that combine evasion maneuvers with velocity changes.

Besides further developments in sensor and actuator technologies as well as in the identification of emergency situations, the following fundamental problems have to be solved and are addressed in this paper:

1. <u>Path planning</u>: generation of possible collision-free evasion paths,

2. <u>Path following</u>: determination of control signals such as the steering angle to guide the vehicle on the planned evasion path,

3. <u>Driving condition estimation</u>: model based prediction whether the planned path is drivable or not.

First, a possible overall CAS-setup is suggested. Next, two methods for collision-free path planning are presented and compared. The first method is based on the mathematical theory of differential games. Path planning in situations with *one* oncoming vehicle can be accomplished by numerically solving the "*Modified Game of Two Cars*" as shown by Lachner et al. [2,3] in the *1990*s. The second approach employs so-called elastic bands which were introduced for robotic path planning by Quinlan and Khatib [4] in the *1990*s and recently adapted to automotive applications by Gehring and Stein [5] and Hilgert et al. [6]. Therein, the main focus lay on vehicle following and lane change maneuvers. However, for collision avoidance further modifications need to be implemented. It is shown how the borders of the road and moving obstacles can be incorporated in the framework by use of repulsive potential fields.

Subsequently, path following control and model based driving condition estimation are discussed. An adequate vehicle model and driving condition estimation based on the *Characteristic Velocity Stability Indicator* are mentioned. Finally, simulation results for an emergency maneuver incorporating a static obstacle, an oncoming vehicle as well as the borders of the road are presented.

CAS: AN OVERALL SYSTEM SETUP

An overall setup of CAS combining, among others, the detection of emergency situations, path planning, path following and driving condition estimation, respectively, is illustrated in Fig. 1. The connecting lines between the blocks symbolize transferred sensor data as well as logical values. Therein, logically true values are signified by bold face tags.

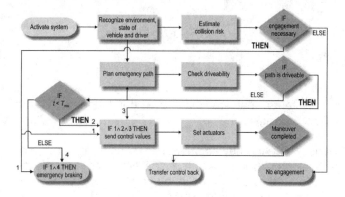

Fig. 1: Possible overall CAS-setup: logical framework of an initiated emergency maneuver

Once the CAS is activated it permanently evaluates environmental sensor data, the driving conditions of the CAS-equipped vehicle and eventually the state of the driver to identify emergency situations. Thereby, emergency situations are characterized by a high *collision risk* with other vehicles, pedestrians or any kind of obstacle, respectively. If the estimated collision risk exceeds a prescribed threshold value, the CAS initiates an emergency maneuver as shown in Fig. 1.

In doing so, possible emergency paths are planned based on the available sensor data. Then, the planned emergency paths are checked for drivability, for example by the *Characteristic Velocity Stability Indicator*. This iterative procedure searches for a drivable emergency path until a valid solution is found or the time limit T_{max} is exceeded. In case that no valid solution can be found within the given time limit emergency braking without steering should be initiated, as depicted in Fig. 1.

However, if in a detected emergency situation a drivable emergency trajectory is generated fast enough, the necessary control signals to guide the vehicle on that trajectory are sent to the actuators of the vehicle. In this, *x-by-wire* technology seems suitable. When the emergency maneuver is finished the control is transferred back to the driving instance. This can be a human driver or a system that autonomously controls the vehicle.

PATH PLANNING

Since path planning is an essential feature of future CAS two promising automotive path planning techniques are presented. The first is based on differential games. Second, a method using elastic bands is outlined and a comparison of both methods is given.

DIFFERENTIAL GAMES

A possible method to generate emergency trajectories that avoid collisions with *one* oncoming vehicle is to formulate the collision avoidance problem as differential game. This section gives a path planning approach incorporating the *Modified Game of Two Cars* based on the work of Lachner et al. [2,3].

The basic idea of the *Modified Game of Two Cars* is to consider the CAS-equipped vehicle E (Evader) and an oncoming vehicle P (Pursuer) as parts of one dynamical system, bounded by the borders of the road.

The configuration of the system is depicted in Fig. 2. The states of this system are the lateral distances of the centers of mass to the middle of the road x_E and x_P, the distance between the two vehicles in road direction y, the orientations of the velocity vectors ϕ_E and ϕ_P and the velocity v_E of the CAS-equipped vehicle E. For clarity, the states can be combined in a state vector

$$\mathbf{z} := \left(x_P, x_E, y, v_E, \phi_P, \phi_E\right)^T.$$

The velocity of the oncoming car, v_P, is assumed to be constant.

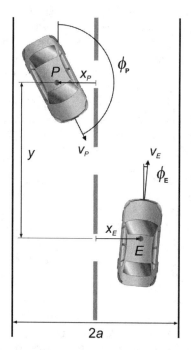

Fig.: 2: Modified Game of Two Cars

Lachner et al. [2,3] modeled both vehicles as rigid bodies with normalized mass in the plane of the road, neglecting the rotational moment of inertia. Therein, it is notable that the rotational motion is kinematical constrained

$$\dot{x}_P = v_P \sin\phi_P \ , \qquad (1)$$

$$\dot{x}_E = v_E \sin\phi_E \ , \qquad (2)$$

$$\dot{y} = v_P \cos\phi_P - v_E \cos\phi_E \ , \qquad (3)$$

$$\dot{v}_E = b_E \, \eta_E \ , \qquad (4)$$

$$\dot{\phi}_P = w_P \, u_P \ , \qquad (5)$$

$$\dot{\phi}_E = w_E(v_E) u_E \ . \qquad (6)$$

The maximum deceleration is marked by the constant value b_E. After normalization, the control variables consist of the nondimensional steering-wheel deflections u_P and u_E of both vehicles as well as of the nondimensional drive or brake force η_E of E.

The scaling factors are the velocity depending maximum angular velocities

$$w_P = \frac{\mu_0 g}{v_P} \ , \qquad (7)$$

$$w_E = \frac{\mu_0 g}{v_E} \ . \qquad (8)$$

Therein, g denotes the gravitational constant and μ_0 the friction coefficient between the tires and the surface of the road.

Inequality constraints result from the requirement not to leave the road of width $2a$,

$$-a \le x_P \le a \ , \qquad (9)$$

$$-a \le x_E \le a \ , \qquad (10)$$

as well as from the bounds of the nondimensional control variables

$$u_{P,min} \le u_P \le u_{P,max} \ , \qquad (11)$$

$$u_{E,min} \le u_E \le u_{E,max} \ , \qquad (12)$$

$$\eta_{E,min} \le \eta_E \le \eta_{E,max} \ . \qquad (13)$$

The characteristic property of this coupled nonlinear state space model is that not all control variables are accessible. E only has control over the variables u_E and η_E, while P can only access u_P. Combining the control

variables of E in a control vector $\mathbf{u_E}$, the state space model (1)-(6) can be written as

$$\dot{\mathbf{z}} = \mathbf{f}(\mathbf{z}, u_P, \mathbf{u_E}) \ . \qquad (14)$$

The objective of E is to avoid a collision with P. Therefore, a minimum distance between E and P has to be maintained over the time period of the evasion maneuver

$$d_{min} := \min_{0 \le t \le \infty} \sqrt{(x_P(t) - x_E(t))^2 + y(t)^2} \ . \qquad (15)$$

In order to design a control strategy for collision avoidance for E, maximization of d_{min} is desired. It is assumed that E and P cannot communicate. Thus, E has no knowledge about possible driving strategies of P and vice versa. For this reason no common strategy can be planned.

However, a frequently used approach to solve differential games is to assume a worst-case scenario: for the oncoming vehicle P an optimal strategy for hypothesized collision intension is presumed. In doing so, the task is to identify the corresponding strategy of the collision avoiding vehicle E, that is the normalized steering angle, $u_E(t)$, and the nondimensional drive or brake force $\eta_E(t)$.

In order to solve this maximization problem of the minimal distance between both vehicles, Lachner et al. [2,3] simplify the problem by considering the squared lateral distance between both vehicles at the instant of mutual passing ($t=t_f$)

$$d_{final} := (x_P(t_f) - x_E(t_f))^2 \ . \qquad (16)$$

The applied numerical technique is very time consuming and needs a sufficiently accurate estimate of the start solution because convergence cannot be guaranteed for all start solutions. Furthermore, the numerical procedure is very complex and the time consumed for the solution varies for different initial conditions. For this reason, the method of differential games can only be used offline for path planning. For real time synthesis, e.g. neural networks can be trained with solutions, which are generated offline for different initial conditions as illustrated in Lachner et. al [2,3].

ELASTIC BANDS

Another possible approach to generate emergency trajectories for vehicle collision-avoidance maneuvers is given by the concept of elastic bands. The basic idea how this concept can be employed for vehicle path planning is illustrated in Fig. 3. An elastic band consists of nodes connected by springs. One of the terminating nodes is fixed to the host vehicle, the other can arbitrarily be placed on the road along the planned path. The elastic band moves virtually with the host vehicle and follows the planned path. For autonomous path planning, information about the environment solely rely on sensor data of the host vehicle. Hence, the configuration of the nodes of the elastic band is described with respect to a reference frame (x_V, y_V), fixed to the host vehicle.

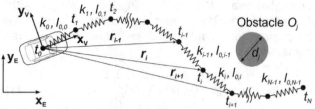

Fig. 3: Elastic band described in a vehicle fixed reference frame (x_V, y_V), earth fixed reference frame (x_E, y_E) and obstacle O_j

Without environmental disturbances the elastic band is in its initial equilibrium configuration. Environmental changes cause external forces acting on the nodes. These external forces shall repel the elastic band from environmental borders or obstacles. This enforces a new equilibrium configuration of the nodes of the elastic band, resulting in a deformed shape, as shown for example in Fig. 3. When the path planning procedure is finished, the elastic band is detached from the CAS-equipped vehicle at time t_0. Then, the elastic band virtually lying on the road is driven through by the CAS-equipped vehicle. Noteworthy, that a copy of the vehicle fixed reference frame (x_V, y_V) remains on the surface of the road at time t_0 and is denoted (x_P, y_P).

Compared to the original concept, applied by Gehring and Stein and Hilgert et al. [5,6], two modifications are introduced. First, the influence of the borders of the road on the equilibrium configuration of the elastic band is incorporated. Second, the influence of moving obstacles is taken into account. Both effects are described by suitable virtual potential fields.

In the following, the internal and external forces acting on each node are described based on internal and external potentials, respectively. Subsequently, force equilibrium is formulated and some comments on the computation of the equilibrium solutions are given. Finally, an interpolation method for smoothing the transition between the nodes of the elastic band is mentioned.

Internal Potentials of an Elastic Band

Adjacent nodes of the elastic band are coupled by linear springs. The nodes are numbered in temporally ascending order with $i=0,...,N-1$: $t_{i+1} > t_i$, as they are successively driven through by the CAS-equipped vehicle. Using the position vectors $\mathbf{r}_i$ to each node, as illustrated in Fig. 3, the linear coupling between neighbouring nodes can be formulated by internal potentials

$$V_i^{int} := \frac{1}{2} k_i \left(\left\| \mathbf{r}_{i+1} - \mathbf{r}_i \right\| - l_{0,i} \right)^2 \quad , \qquad (17)$$

where $l_{0,i}$, k_i denote the length of the undeformed spring i and the corresponding spring stiffness, respectively. The distance between adjacent nodes is given by the Euclidean norm

$$\left\| \mathbf{r}_{i+1} - \mathbf{r}_i \right\| = \sqrt{\left(r_{i+1,x} - r_{i,x} \right)^2 + \left(r_{i+1,y} - r_{i,y} \right)^2} \quad . \quad (18)$$

The total internal potential, governing the behavior of the elastic band, is

$$V^{int} := \sum_{i=0}^{N-1} V_i^{int} = \frac{1}{2} \sum_{i=0}^{N-1} k_i \left(\left\| \mathbf{r}_{i+1} - \mathbf{r}_i \right\| - l_{0,i} \right)^2 \quad . \quad (19)$$

The directional derivative of the total internal potential with respect to the position vector of the corresponding node yields the internal forces

$$\mathbf{F}_i^{int} = -\frac{\partial V^{int}}{\partial \mathbf{r}_i} = -k_{i-1} \left(\left\| \mathbf{r}_i - \mathbf{r}_{i-1} \right\| - l_{0,i-1} \right) \frac{\mathbf{r}_i - \mathbf{r}_{i-1}}{\left\| \mathbf{r}_i - \mathbf{r}_{i-1} \right\|}$$
$$+ k_i \left(\left\| \mathbf{r}_{i+1} - \mathbf{r}_i \right\| - l_{0,i} \right) \frac{\mathbf{r}_{i+1} - \mathbf{r}_i}{\left\| \mathbf{r}_{i+1} - \mathbf{r}_i \right\|}$$

$$(20)$$

acting on each node.

External Potentials of the Borders of the Road

Likewise, the influence of the borders of the road on the nodes of the elastic band is described by repulsive external potentials

$$V^{B_q}(\mathbf{r}_i) := -k^{B_q} \ln \left\| \mathbf{r}_i - \mathbf{r}_i^{B_q} \right\| \quad \text{with} \quad q \in \{l, r\}, \quad (21)$$

where the potential of the left and right border are defined separately. The effect of the borders of the road on the elastic band can be illustrated by springs, having spring stiffness k^{B_q}, acting between the borders and the nodes of the elastic band, as shown in Fig. 4.

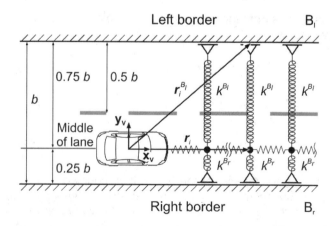

Fig. 4: Neutral position of an elastic band

The position vector $\mathbf{r}_i^{B_q}$ marks the point of the border, having the shortest distance to the corresponding node i. According to (21), the external potentials decay logarithmically towards the center of the road. Analogously to the procedure carried out to derive the internal forces, the external forces, acting on the nodes of the elastic band become

$$\mathbf{F}_i^{B,q} = -\frac{\partial V_i^{B_q}}{\partial \mathbf{r}_i} := \frac{k^{B_q}}{\left\|\mathbf{r}_i - \mathbf{r}_i^{B_q}\right\|} \cdot \frac{\mathbf{r}_i - \mathbf{r}_i^{B_q}}{\left\|\mathbf{r}_i - \mathbf{r}_i^{B_q}\right\|} \quad . \quad (22)$$

To adjust the neutral position of the elastic band, the spring stiffnesses k^{B_l} and k^{B_r} in Fig. 4 must be chosen appropriately. This is accomplished by assuming force equilibrium at each node in the neutral position of the elastic band. As a consequence a relation between the spring stiffnesses and geometric parameters of the road, see Fig. 4, follows

$$\mathbf{F}_i^{B_l} = \mathbf{F}_i^{B_r} \Rightarrow \frac{k^{B_l}}{k^{B_r}} = \frac{0.75b}{0.25b} = 3 \quad . \quad (23)$$

For finding equilibrium solutions of the elastic band at small distances to the borders of the road limited external forces

$$\mathbf{F}_i^{B_q} = k^{B_q} e^{-\left\|\mathbf{r}_i - \mathbf{r}_i^{B_q}\right\|^2} \cdot \frac{\mathbf{r}_i - \mathbf{r}_i^{B_q}}{\left\|\mathbf{r}_i - \mathbf{r}_i^{B_q}\right\|} \quad (24)$$

turned out to be useful.

External Potentials of Moving Obstacle

To meet the requirements of a collision-avoidance approach, it is important to incorporate the influence of moving obstacles on the deformation of the elastic band. As sketched in Fig. 3, obstacles are modeled by safety circles having diameters d_j being larger than the sum of the maximal extension of both the obstacle and the CAS-equipped vehicle. It is presupposed that information on position and velocity of obstacles are provided by the sensor system of the CAS-equipped vehicle at the beginning of the path planning, t_0. In a first approach, the velocity of obstacles during the evasion maneuver is assumed to be constant. Thus, the path of the obstacle during the evasion maneuver is linearly extrapolated as illustrated in Fig. 5.

Assuming that the CAS-equipped vehicle drives through the emergency path, given by the elastic band, at a planned longitudinal velocity $v_x(t)$, the time t_i when the vehicle passes through the corresponding node i can be determined.

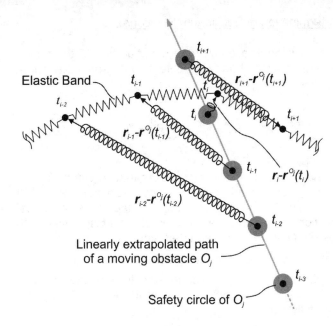

Fig. 5: Incorporation of a moving obstacle O_j

The key idea behind the construction of potentials for the moving obstacle is as follows. To each node i at time t_i along the elastic band corresponds a position of the obstacle along its own path. The correlating nodes of the elastic band and the positions of the obstacle O_j at times t_i, $i = 1, ..., N$ are coupled by a repelling potential field

$$V^{O_j}(\mathbf{r}_i) := -k^{O_j} \ln\left(\left\|\mathbf{r}_i - \mathbf{r}^{O_j}(t_i)\right\| - \frac{d_j}{2}\right) \quad , \quad (25)$$

where k^{O_j} denotes the stiffness parameter and $\mathbf{r}^{O_j}(t_i)$ is the position vector to the obstacle at time t_i. The coupling of the elastic band with the obstacle is illustrated in Fig. 5 by coiled springs. The external forces repelling the elastic band from the obstacle are given by

$$\mathbf{F}_i^{O_j} = -\frac{\partial V^{O_j}(\mathbf{r}_i)}{\partial \mathbf{r}_i} := \frac{k^{O_j}}{\left\|\mathbf{r}_i - \mathbf{r}^{O_j}(t_i)\right\| - \frac{d_j}{2}} \cdot \frac{\mathbf{r}_i - \mathbf{r}^{O_j}(t_i)}{\left\|\mathbf{r}_i - \mathbf{r}^{O_j}(t_i)\right\|} \quad . \quad (26)$$

Also for the moving obstacles bounded external forces

$$\mathbf{F}_i^{O_j} = k^{O_j} e^{-\left(\left\|\mathbf{r}_i - \mathbf{r}^{O_j}(t_i)\right\| - \frac{d_j}{2}\right)^2} \cdot \frac{\mathbf{r}_i - \mathbf{r}^{O_j}(t_i)}{\left\|\mathbf{r}_i - \mathbf{r}^{O_j}(t_i)\right\|} \quad (27)$$

turned out to be useful.

Determination of Equilibrium Solutions

The collision-free equilibrium configuration of the nodes of the elastic band is determined from the force equilibrium

$$\mathbf{F}_i^{sum} = \mathbf{F}_i^{int} + \mathbf{F}_i^{B_l} + \mathbf{F}_i^{B_r} + \sum_j \mathbf{F}_i^{O_j} = 0 \quad , \quad (28)$$

resulting in a nonlinear algebraic equation which can for example be solved using a Newton-Raphson method. It is noteworthy that at each iteration step of the numerical solution procedure, the coupling of the elastic band with the obstacle must be updated since the position vector $\mathbf{r}^{O_j}(t_i)$ changes with the deformation of the elastic band. In doing so, it is suggested to vary the stiffness parameters k_i in order to prevent the elastic band from start twisting.

Smooth Transition of the Equilibrium Solution

After determining the equilibrium configuration of the elastic band, the transition between adjacent nodes is smoothed by a time-parameterized cubic-spline interpolation. The time-parameterized emergency path expressed in the road fixed reference frame is then given by $(x_P(t), y_P(t))$.

COMPARISON OF DIFFERENTIAL GAMES AND ELASTIC BANDS

A comparison of the two given path planning methods is summarized in Tab. 1. It can be followed that elastic bands are more suitable for automotive collision avoidance. The major drawbacks of the method based on differential games are the restricting assumptions about the vehicle model and the fact that only one obstacle can be incorporated in a 2-player-zero-sum game. Furthermore, collision-free maneuvers can only be guaranteed if the oncoming vehicle acts exactly as assumed in the differential game.

Table 1: Comparison of differential games and elastic bands for path planning

	Modified Game of Two Cars [2,3]	Elastic bands
Vehicle model	The vehicles are modeled as particles with kinematic constraints.	The method does not require a particular vehicle model.
Velocity of CAS-equipped vehicle	Changes in the velocity of the CAS-equipped vehicle are part of the solution.	The velocity of the CAS-equipped vehicle is not part of the method yet.
Return to the initially planned path	Path planning terminates at the instant of mutual passing.	The last node of the elastic band can arbitrarily be placed on the road.
Number of obstacles	The method is limited to one oncoming vehicle.	The number of obstacles is variable.
Assumptions on the obstacles	Worst-case behaviour for the oncoming vehicle is assumed. Collision avoidance cannot be guaranteed in case of different behavior!	The path of a moving obstacle is linearly extrapolated yet.
Real time capability	This method is not real time capable. Real time synthesis can for example be performed by training of neural networks with solutions that are generated offline.	The method of elastic bands is real time capable in case of limited velocities. At higher velocities, real time capability can be gained by limiting the number of iterations or by training of neural networks as well.

PATH FOLLOWING

Having planned the final emergency path, a path following procedure is required that brings out the steering angle of the CAS-equipped vehicle along the emergency path for the planned longitudinal velocity $v_x(t)$. In Hilgert et al. [7] a least-square procedure is used to compute the steering angle along planned paths for lane change maneuvers. It turned out that this method is computationally cost intensive. The path following procedure outlined here, is based on a closed-loop control system. The components of the control system consisting of a vehicle model, combined with a linear lateral guidance model and a PID-controller are described in the following.

The used reference frames are depicted in Fig. 6. The equations of motion of the CAS-equipped vehicle are formulated in the vehicle fixed reference frame (x_V, y_V). The emergency trajectory is described in the reference frame (x_P, y_P) which was detached from the vehicle at time t_0. Furthermore, a reference frame $(\tilde{x}_V, \tilde{y}_V)$ moving on the emergency trajectory is introduced for the lateral guidance procedure. As a global reference the earth fixed reference frame (x_E, y_E) is used.

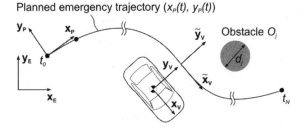

Fig. 6: Reference frames with unit vectors

VEHICLE MODEL

For performance reasons, the complexity of the vehicle model should be as low as possible. However, an appropriate level of accuracy must be retained. In collision avoidance maneuvers the lateral vehicle dynamics are of most interest. A widely used model in lateral vehicle dynamics is the bicycle model, see Fig. 7.

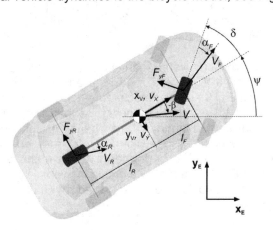

Fig. 7: Bicycle model

This planar model combines the left and right track into a single one. Often the equation of motion are linearized assuming constant velocity V of the center of gravity c.g.. Up to lateral accelerations of approximately 4 m/s^2, the linear model provides reliable results. The linear equations of motion can be written as [8]:

$$\begin{bmatrix} \dot{\beta} \\ \dot{\psi} \end{bmatrix} = \mathbf{A} \begin{bmatrix} \beta \\ \psi \end{bmatrix} + \mathbf{B} \delta \qquad , \qquad (29)$$

with

$$\mathbf{A} = \begin{bmatrix} -\left(\dfrac{c_{\alpha F} + c_{\alpha R}}{Vm} \right) & \left(\dfrac{-c_{\alpha F} l_F + c_{\alpha R} l_R}{V^2 m} - 1 \right) \\ \left(\dfrac{-c_{\alpha F} l_F + c_{\alpha R} l_R}{J_z} \right) & -\left(\dfrac{c_{\alpha F} l_F^2 + c_{\alpha R} l_R^2}{V J_z} \right) \end{bmatrix} \quad (30)$$

and

$$\mathbf{B} = \begin{bmatrix} \dfrac{c_{\alpha F}}{Vm} \\ \dfrac{c_{\alpha F} l_F}{J_z} \end{bmatrix} \qquad , \qquad (31)$$

where β and $\dot{\psi}$ denote the state variables *side slip angle* and *yaw rate*, respectively. The control variable δ indicates the steering angle. The parameters of the model are described and quantified in Tab. 2.

However, it should be noted that for higher lateral accelerations than 4 m/s^2 or changing velocity the nonlinear bicycle model should be employed. Equations of motion including a simple power train model can for example be found in [9].

Table 2: Vehicle parameters [9]

Symbol	Description	Value	Unit
m	Vehicle mass	1280	[kg]
J_z	Moment of inertia about z_V-axis	2500	[kgm^2]
l_F	Distance of front axle to c.g.	1.203	[m]
l_R	Distance of rear axle to c.g.	1.217	[m]
c_{aF}	Front tire cornering stiffness	100000	[N/rad]
c_{aR}	Rear tire cornering stiffness	100000	[N/rad]

241

STEERING ANGLE DETERMINATION

For the determination of the steering angle $\delta(t)$ along the planned time-parameterized emergency path $(x_P(t), y_P(t))$ a lateral guidance model is used, as depicted in Fig. 8. Its state variables are the lateral deviation y and the angular deviation $\Delta\psi$. The lateral deviation y describes the shortest distance between the center of gravity of the vehicle and the path.

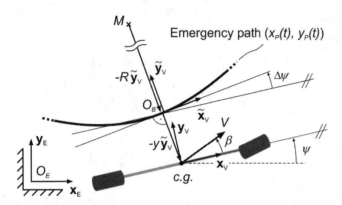

Fig. 8: Guided bicycle model with emergency trajectory

The projection of the velocity V of the vehicle onto the unit vector $\tilde{\mathbf{y}}_V$, pointing towards the instantaneous center of curvature M, describes the lateral deviation rate $\dot{y} = V\sin(\beta + \Delta\psi)$. For small angles the following relation holds for the lateral deviation rate

$$\dot{y} = V(\beta + \Delta\psi) \quad . \tag{32}$$

For small angles, the angular deviation rate is given by

$$\Delta\dot{\psi} = V\kappa - \dot{\psi} \quad , \tag{33}$$

where κ denotes the curvature, which is evaluated by

$$\kappa(t) = \frac{\dot{x}_P\ddot{y}_P - \dot{y}_P\ddot{x}_P}{\left(\sqrt{\dot{x}_P^2 + \dot{y}_P^2}\right)^3} \quad , \tag{34}$$

for the planned time-parameterized emergency path $(x_P(t), y_P(t))$. The coupled equations of the laterally guided bicycle model are

$$\begin{bmatrix} \dot{\beta} \\ \ddot{\psi} \\ \dot{y} \\ \Delta\dot{\psi} \end{bmatrix} = \begin{bmatrix} A_{11} & A_{12} & 0 & 0 \\ A_{21} & A_{22} & 0 & 0 \\ V & 0 & 0 & V \\ 0 & -1 & 0 & 0 \end{bmatrix} \begin{bmatrix} \beta \\ \dot{\psi} \\ y \\ \Delta\psi \end{bmatrix} + \begin{bmatrix} B_1 & 0 \\ B_2 & 0 \\ 0 & 0 \\ 0 & V \end{bmatrix} \begin{bmatrix} \delta \\ \kappa \end{bmatrix} \quad , \tag{35}$$

with A_{ij} and B_i according to (30) and (31), respectively.

This model is embedded into a simple closed-loop structure with a PID-controller, where the lateral deviation $y(t)$ is regulated at zero and the curvature $\kappa(t)$ acts as disturbance function, which is prescribed by the planned path. The steering angle $\delta(t)$ is the control variable. The controller parameter are adjusted to avoid oscillatory behaviour of the closed-loop system.

DRIVING CONDITION ESTIMATION

The driving condition estimation can for example be performed based on the characteristic velocity

$$v_{ch}^2(t) := \frac{v_x^2(t)}{1 - \dfrac{\delta(t)\,v_x(t)}{\dot{\psi}(t)\,(l_F + l_R)}} \quad . \tag{36}$$

A suitable method is the so-called *Characteristic Velocity Stability Indicator* CVSI [10]. In this method the longitudinal and the characteristic velocity of the CAS-equipped vehicle are compared. The advantage is that only few data are needed, which are uncomplicated to compute or measure. The corresponding driving conditions to the CVSI values are given in Tab. 3.

Table 3: Interpretation of CVSI values

CVSI	Driving condition
1	Under-steering, stable
2	Neutral-steering, stable
3	Over-steering, stable
4	High-over-steering, indifferent
5	Braking away, unstable

SIMULATION RESULTS

In the following simulations, the path planning is performed in presence of two obstacles. Therefore, the method of differential games cannot be applied and the method of elastic bands is employed. The initial setting is as follows:

The CAS-equipped vehicle drives in the middle of the right lane on a 7 m wide road at a speed of 15 m/s. Then, at time t_0 the CAS detects an emergency situation and initiates an evasion maneuver. At that time a non-moving obstacle appears 67 m ahead. An oncoming vehicle travels at a speed of 25 m/s from an initial distance of 65 m. The performed evasion maneuver is depicted in Fig. 9. Therein, the oncoming vehicle and the non-moving obstacle are represented by their safety circles.

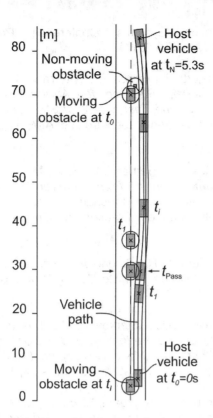

Fig. 9: Simulation result of the evasion maneuver

The parameters that are used in the path planning are summarized in Tab. 4. The forces of the borders of the road and the obstacles were modeled according to (24) and (27), respectively. The computation of the steering angle for the path following was performed by use of a PID-controller. The steering angle, the lateral acceleration as well as the yaw rate, computed based on the linear bicycle model are given in Fig. 10. It can be seen that the lateral acceleration for this maneuver stays below 4 m/s^2. Therefore, the use of the linear bicycle model is sustainable. Furthermore, the CVSI is displayed and indicates that the vehicle moves stably on the planned evasion path.

CONCLUSIONS

An overall setup for CAS was suggested. Especially, it was shown how path planning, path following and driving condition estimation can be embedded in this framework. Subsequently, a method for emergency path planning based on differential games was compared to a method based on elastic bands. Therein, the method of elastic bands was modified for automotive collision avoidance problems. The comparison showed that the latter method seems to be more suitable for automotive path planning. Furthermore, an adequate vehicle model was extended by a guidance model. The CVSI was used for model based driving condition estimation. Finally, sample simulation results for an emergency maneuver employing the described setup were given.

Table 4: Path planning parameters

Symbol	Description	Value	Unit
d_1	Safety diameter of oncoming vehicle	4	[m]
d_2	Safety diameter of non-moving obstacle	3.6	[m]
k^{O_1}	Force scaling factor of oncoming vehicle	8	[-]
k^{O_2}	Force scaling factor of non-moving obstacle	6	[-]
k^{B_l}	Force scaling factor of left border	10	[-]
k^{B_r}	Force scaling factor of right border	$10 \cdot e^{-24.5}$	[-]
k_i	Internal spring stiffness of interval i (Initial value equal for all intervals)	53	[N/m]
$l_{0,i}$	Initial length of internal spring of interval i (equal for all intervals)	0.77	[m]
n	Number of nodes	53	[-]

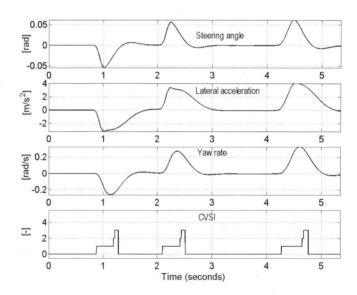

Fig. 10: Steering angle, yaw rate, lateral acceleration and CVSI of the CAS-equipped vehicle following the emergency trajectory based on a linear bicycle model

ACKNOWLEDGEMENTS

Many thanks to the *L-LAB* and the *International Graduate School of Dynamic Intelligent Systems* in Paderborn for supporting this work with facilities and a research scholarship.

REFERENCES

[1] Vukotich, A., Kirchner, A.: *Sensor fusion for driver-assistance-systems*. Elektronik im Kraftfahrzeug, Baden-Baden, Germany, 2001 (in German)

[2] Lachner, R.: *Echtzeitsynthese optimaler Strategien für Differentialspiele mit schneller Dynamik mit Anwendungen bei der Kollisionsvermeidung*. PhD Thesis, University of Clausthal-Zellerfeld, Germany, Papierflieger, 1997

[3] Lachner, R., Breitner, M.H., Pesch, H.J.: *Real Time Computation of Differential Games with Applications to Collision Avoidance*. International Series of Numerical Mathematics, Vol. 124, Birkhäuser, Basel, 1998

[4] Quinlan, S., Khatib, O.: *Elastic Bands: Connecting path planning and control*. Proceedings of the IEEE Conference in Robotics and Automation 93, 1993

[5] Gehring, S.K., Stein, F.J.: *Elastic Bands to Enhance Vehicle Following*. IEEE Intelligent Transportation Systems Conference Proceedings, Oakland (CA), USA, 2003

[6] Hilgert, J., Hirsch, K., Bertram, T., Hiller, M.: *Emergency Path Planning for Autonomous Vehicles Using Elastic Band Theory*. IEEE/ASME Conference on Advanced Intelligent Mechatronics AIM, Kobe, Japan, 2003

[7] Hilgert, J., Bertram, T., Hiller, M.: *Path Planning for Autonomous Driving from a Vehicle Dynamics Point of View*. 4. VDI Mechatroniktagung Innovative Produktentwicklungen, Frankenthal, Germany, 2001 (in German)

[8] Isermann, R.: *Diagnosis Methods for Electronic Controlled Vehicles*. Vehicle System Dynamics 36, No. 2-3, pp. 77-117, 2001

[9] Smith D.E., Starkey, J.M.: *Effects of Model Complexity on the Performance of Automated Vehicle Steering Controllers: Model Development, Validation and Comparison*. Vehicle System Dynamics 24, 1995

[10] Börner, M. Andreani, L., Albertos, P., Isermann, R.: *Detection of Lateral Vehicle Driving Conditions Based on the Characteristic Velocity*. IFAC World Congress 2002, Barcelona, Spain, 2002

Driver's Degree of Dependence on Collision Avoidance Systems

Akira Kurosaki, Wai Cheong Choy,
Hiroaki Kosaka, and Hirokazu Nishitani
Graduate School of Information Science, Nara Institute of Science and Technology

ABSTRACT

The authors performed an experiment on a collision avoidance system (CAS) by using a driving simulator. A driver is exposed to a variety of events designed with the support system. In each event, based on whether the brakes were operated, we proposed two indices, namely the alarm ignorance ratio and the fail-safe ratio, to indicate the degree of the driver's dependence on the support system. In addition, from the simplified NASA-TLX and the post-experiment interview, the driver's mental workload was measured. Based on the degree of dependence, the authors plan to design a support system that collaborates with the driver.

1 INTRODUCTION

In recent years, due to the development of information-communication technology, much progress has been made in the Intelligent Transport Systems (ITS) used in cars. Currently, much research is focused on technologies for the advanced safety vehicle (ASV), such as the lane deviation-prevention system and the vehicular-gap control system, as well as the advanced cruise-assist highway system (AHS) [1,2] which includes a transportation infrastructure, in-vehicle sensors and other systems. Recent years have also seen an increased number of cars on the market featuring electronic toll collection (ETC) and adaptive cruise control (ACC). Hence, there is a need to improve the safety and reliability of these support systems and to reduce the driver's workload while driving.

However, if the driver is too dependent on a collision avoidance system, there is the possibility that the driver could become unable to avoid an accident without a support system, even under normal situations. In addition, if the timing for the information to be provided is inappropriate, the driver might find the system unnecessary or even useless. In this research, we perform an experiment on a collision avoidance system by using a driving simulator. Based on an analysis of the driver's behavior and conduct, we evaluate the driver's degree of dependence on the support system. We also investigate the correlation between the driver's degree of dependence and the driver's conduct and behavior.

2 DRIVING SIMULATOR

It is nearly impossible to perform an experiment with a real car in order to analyze a driver's condition during the encounter of danger, since it could obviously lead to serious accidents. Therefore, a driving simulator must be used to analyze a driver's behavior and conduct. In this research, we used the HONDA Driving Simulator (DS) shown in Figure 1 to perform the experiments [3].

Figure 1 : Honda Driving Simulator

Using the scenario editor function of this DS, we can design obstacles and create events according to our objectives. To evaluate the driver's degree of dependence on the support system, we collected data on the driver's behavior and conduct and then performed a subjective evaluation of the driver.

3 EXPERIMENT OF COLLISION AVOIDANCE SYSTEM

3.1 EXPERIMENTAL OVERVIEW

The experiment is based on a collision avoidance support system that provides information when there is a vehicle crossing an intersection. The subjects are required to travel along the straight advancing course as shown in Figure 2. As for the test course, there is an intersection every 40 meters along the straight road in the urban area. Because the crossing vehicles are hidden behind buildings, the subjects won't be able to notice them until they get close to the intersections. As for the support system used in this research, when there is a crossing vehicle, an alarm sounds at about 27.5 meters from the intersection. When the support system is in operation, the alarm sounds without fail whenever there is a crossing vehicle to inform the subject of the danger. The subjects were required to perform on this course for 8 rounds a day over 2 days (i.e. 16 rounds).

In this experiment, ten graduate students were used as subjects (S1 - S10). Before the experiment began, the collision avoidance support system was explained to the subjects and they were informed that they would also answer a subjective evaluation questionnaire after the course. They were required to maintain an approximate speed of 30 km/h. The experiment was performed using the automatic transmission.

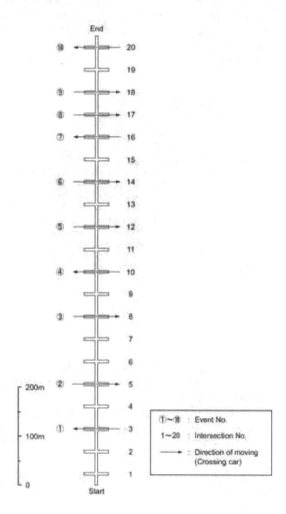

Figure 2 : Test course

3.2 EXPERIMENTAL PROCEDURE

The experimental conditions are listed in Table 1, where the columns represent the intersection number at which a crossing vehicle exists and the rows represent each round of the test course as shown in Figure 2. The symbols used in the table represent the events occurring on the course during the round. For example, R4, indicates that it is the fourth round the subject is performing the course and there are events on the 8th and 20th intersections. ● indicates that the support system is not in operation, which means there is no alarm regardless of whether there is a crossing vehicle (without system). O indicates that the support system operates normally, which means an alarm sounds when there is a crossing vehicle (system normal). X indicates that there is a defect in the support system, which means there is no alarm even when there is a crossing vehicle (system defect). ▲ indicates that the support system operates erroneously, which means an alarm is sounded even though there is no crossing vehicle (system error).

Table 1 : Experimental Conditions

Day	phase		3	5	8	10	12	14	16	17	18	20
1	I	R1			●				●			
		R2						●				
		R3							●			
		R4			●							●
	II	R5		O								O
		R6			O				O			
		R7					O					
		R8		O					O			
		R9		O					O			
		R10			O					O		
		R11				O					O	
2	III	R12	O				X				O	
		R13			O						O	
		R14						▲				
		R15			▲			▲				▲
		R16					X				O	

● : without system (with crossing vehicle, no warning)

O : system normal (with crossing vehicle, with warning)

X : system defect (with crossing vehicle, no warning)

▲ : system error (no crossing vehicle, with warning)

The experiment was performed over a period of 2 days. First, the subjects were asked to perform the test course without the support system (phase I). After that, they were asked to perform the test course while using the support system (phase II). Once they become used to the support system, they were required to perform the test course with the support system sometimes operating abnormally or erroneously (phase III). If an accident occurs during the experiment, the experiment is disrupted.

3.3 MEASUREMENT INDICES

The measurement indices used for the experiment are shown below:

- Vehicle's position coordinates(X,Y), Vehicle's speed [km/h]
- Brakes opening, acceleration opening [%]
- Subjective evaluation of driver (NASA-TLX questionnaire, interview)
- DS's replay data and video images
- Video images of the driver's movements

From the data on acceleration and brake opening, we can obtain the time of changing pedals and the starting time of braking. The interval between the sampling time for these 2 openings, the vehicle's position coordinates, and the vehicle's speed is 10 [ms].

4. EXPERIMENTAL RESULTS AND DISCUSSION

4.1 BRAKE OPERATION

It is the authors' preference to discuss about the braking operation and the subjective evaluation for each and every subject in every round. However, due to the limited space and the difficulty in doing so, we decided to categorize the results of the calculation into 3 groups. A representative for each group is chosen and a discussion is done based on these representatives. The 3 groups are categorized based on the braking ratio which is the percentage of the brake pedal being stepped (number of times the brake pedal is stepped/number of intersections). Group A has a percentage of 90% or more, Group B has a percentage of 50% or more and less than 90%, and Group C has a percentage of less than 50%. S3 is chosen to represent Group A, S8 to represent Group B, and S2 to represent Group C. The discussion will be based mainly on the results of these 3 subjects from here onwards. The number of accidents and the braking ratio in each phase, and the braking ratio for R1 – R16 are shown in Table 2.

It is anticipated that with or without a crossing vehicle, the operation of the brakes while heading toward an intersection changes. From the brakes opening, we investigated whether the brake pedal was stepped just before going through an intersection. Some of the results are shown in Table 3 – Table 5. The column numbers in the table corresponds to the intersection number in Figure 2. O represents the operation of the brakes while heading toward the intersection, thus making the vehicle slow down or come to a complete halt. X represents the occurrence of an accident. From Table 3 - Table 5, we conclude the following:

Table 2 : Number of accidents and the braking ratio in each phase, and the braking ratio for R1 – R16

	Group A				Group B		Group C			
	S3	S9	S7	S5	S8	S10	S1	S6	S2	S4
Overall braking ratio (%)	100.0	100.0	99.7	93.4	81.9	51.3	49.3	46.5	41.3	22.3
Number of accidents in Phase I	0	0	1	0	0	0	0	0	0	1
Number of accidents in Phase II	0	0	0	0	0	0	0	0	0	0
Number of accidents in Phase III	0	0	0	0	0	2	1	1	1	2
Braking Ratio (%) in R1	100.0	100.0	100.0	100.0	100.0	100.0	100.0	100.0	100.0	25.0
Braking Ratio (%) in R2	100.0	100.0	92.9	100.0	100.0	100.0	100.0	100.0	100.0	80.0
Braking Ratio (%) in R3	100.0	100.0	100.0	100.0	100.0	100.0	100.0	100.0	100.0	55.0
Braking Ratio (%) in R4	100.0	100.0	100.0	100.0	100.0	100.0	100.0	100.0	100.0	50.0
Braking Ratio (%) in R5	100.0	100.0	100.0	25.0	100.0	100.0	25.0	10.0	15.0	10.0
Braking Ratio (%) in R6	100.0	100.0	100.0	70.0	100.0	20.0	10.5	10.0	30.0	10.0
Braking Ratio (%) in R7	100.0	100.0	100.0	100.0	100.0	15.0	5.3	5.0	5.0	10.0
Braking Ratio (%) in R8	100.0	100.0	100.0	100.0	100.0	10.0	10.5	10.0	30.0	15.0
Braking Ratio (%) in R9	100.0	100.0	100.0	100.0	75.0	100.0	25.0	10.0	10.0	15.0
Braking Ratio (%) in R10	100.0	100.0	100.0	100.0	35.0	10.0	15.0	10.0	15.0	15.0
Braking Ratio (%) in R11	100.0	100.0	100.0	100.0	20.0	10.0	15.0	10.0	10.0	10.0
Braking Ratio (%) in R12	100.0	100.0	100.0	100.0	50.0	8.3	8.3	8.3	16.7	8.3
Braking Ratio (%) in R13	100.0	100.0	100.0	100.0	65.0	30.0	30.0	15.0	65.0	10.0
Braking Ratio (%) in R14	100.0	100.0	100.0	100.0	80.0	30.0	55.0	40.0	5.0	15.0
Braking Ratio (%) in R15	100.0	100.0	100.0	100.0	85.0	30.0	80.0	100.0	20.0	15.0
Braking Ratio (%) in R16	100.0	100.0	100.0	100.0	100.0	33.3	95.0	100.0	30.0	0.0
Braking Ratio (%) in Phase I	100.0	100.0	98.6	100.0	100.0	100.0	100.0	100.0	100.0	57.4
Braking Ratio (%) in Phase II	100.0	100.0	100.0	85.0	75.7	37.9	15.3	9.3	16.4	12.1
Braking Ratio (%) in Phase III	100.0	100.0	100.0	100.0	76.0	27.4	57.6	56.5	28.3	10.7

Phase I

All the subjects slowed down or came to a complete halt regardless of whether there was a crossing vehicle except for S4.

However, for S4, there were some intersections where he did not step on the brakes. During the interview, S4 replied that he acknowledged the crossing vehicle by its sound and therefore passed through certain intersections without stepping on the brakes. In addition, S4 said that an accident occurred due to the difference in the braking systems between the DS and a real car.

However, for S7, an accident occurred in R2 despite having slowed down and performing safety confirmation at all the intersections. Based on the interview, the subject replied that the difference in the braking sensation between the DS and a real car making it difficult to brake on time and that caused the accident.

Phase II

In the cases of subjects of Group B and C, the brakes were operated at all intersections whenever a warning was provided. On the other hand, for the intersections without a warning, the rate at which they all passed through the intersection without braking was extremely high. This is due to the fact that the subjects depended on the support system, trusting that when there was a warning alarm, there was no crossing vehicle.

In contrast, subjects of Group A stepped on the brakes and performed safety confirmation for all intersections, regardless of whether there was a crossing vehicle. In the interview, three subjects of Group A mentioned that although the support system can be trusted, they chose not to depend on it. In addition, S8, who initially did not trust the system, started to trust the system gradually as the experiment progresses resulting in the decrease of the brake operation when entering an intersection.

Phase III

In the cases of S1, S2, S4, S6 and S10, an accident occurred at the 12th intersection in R12 due to the defect of the system. The fact that these subjects depended entirely on the support system is assumed to be the cause of the accidents. Moreover, for subject of Group A, who did not really trust the system, there was no accident in R12.

An interview with S8 revealed that the workload while driving decreased with the help of the support system in Phase II. S8 added that after the first system defect in Phase III, he started to distrust and stopped depending on the support system.

In the case of S2, he passed through many intersections after R12 without slowing down or bringing the car to a complete halt. This showed that S2 continued to depend on the support system even after the accident occurred.

As for S3, from the very beginning he neither trusted the support system nor depended on it, so he stepped on the brakes and performed safety confirmation at all of the intersections. In the interview, he replied that up until the system became defective, he started to depend a little on the support system. However, he stopped trusting the support system after the system defect.

The system defect occurred a second time in R16, also at the 12th intersection. For all the subjects in Group A, there was no accident because they all slowed down for each intersection. As for S2, although he trusted the support system throughout the test course, there was no accident because he drove with extreme caution and managed to apply on the emergency brake to avoid an accident. As for S1 and S6, they did not depend on the support system and managed to avoid an accident. However, S4 and S10 caused an accident for the second time. In the interview, S10, who is among the subjects who caused an accident, replied that he did not depend on the support system but somehow an accident happened.

Table 3 : Operation of brakes at intersection （Phase I）

		1	2	3	4	5	6	7	8	9	10	11	12	13	14	15	16	17	18	19	20
Sx *	R1	O	O	O	O	O	O	O	O	O	O	O	O	O	O	O	O	O	O	O	O
	R2	O	O	O	O	O	O	O	O	O	O	O	O	O	O	O	O	O	O	O	O
	R3	O	O	O	O	O	O	O	O	O	O	O	O	O	O	O	O	O	O	O	O
	R4	O	O	O	O	O	O	O	O	O	O	O	O	O	O	O	O	O	O	O	O
S7	R1	O	O	O	O	O	O	O	O	O	O	O	O	O	O	O	O	O	O	O	O
	R2	O	O	O	O	O	O	O	O	O	O	O	O	O	X	(disrupted by accident)					
	R3	O	O	O	O	O	O	O	O	O	O	O	O	O	O	O	O	O	O	O	O
	R4	O	O	O	O	O	O	O	O	O	O	O	O	O	O	O	O	O	O	O	O
S4	R1						O		X	(disrupted by accident)											
	R2				O	O		O	O	O	O	O	O	O	O	O	O	O	O	O	O
	R3			O		O				O	O		O	O	O		O		O		O
	R4			O		O			O		O				O		O	O	O	O	O

*Sx : All subjects except S4 and S7

Table 4 : Operation of brakes at intersection （Phase II）

		1	2	3	4	5	6	7	8	9	10	11	12	13	14	15	16	17	18	19	20
S2	R9					O											O				
	R10								O									O			O
	R11										O								O		
S3	R9	O	O	O	O	O	O	O	O	O	O	O	O	O	O	O	O	O	O	O	O
	R10	O	O	O	O	O	O	O	O	O	O	O	O	O	O	O	O	O	O	O	O
	R11	O	O	O	O	O	O	O	O	O	O	O	O	O	O	O	O	O	O	O	O
S8	R9	O	O	O	O	O	O	O	O	O	O	O			O		O	O			
	R10	O	O	O	O				O									O	O		
	R11										O	O							O	O	

Table 5 : Operation of brakes at intersection （Phase III）

		1	2	3	4	5	6	7	8	9	10	11	12	13	14	15	16	17	18	19	20
S2	R12			O	O								X	(disrupted by accident)							
	R13		O		O	O			O	O	O	O	O	O				O	O	O	O
	R14												O								
	R15								O						O		O				O
	R16			O									O		O				O	O	O
S3	R12	O	O	O	O	O	O	O	O	O	O	O	O	O	O	O	O	O	O	O	O
	R13	O	O	O	O	O	O	O	O	O	O	O	O	O	O	O	O	O	O	O	O
	R14	O	O	O	O	O	O	O	O	O	O	O	O	O	O	O	O	O	O	O	O
	R15	O	O	O	O	O	O	O	O	O	O	O	O	O	O	O	O	O	O	O	O
	R16	O	O	O	O	O	O	O	O	O	O	O	O	O	O	O	O	O	O	O	O
S8	R12			O									O	O	O	O	O	O	O	O	O
	R13				O			O	O	O	O	O	O			O	O	O	O	O	O
	R14	O		O	O		O			O	O	O	O	O	O	O	O	O	O	O	O
	R15			O	O	O	O		O	O	O	O	O	O	O	O	O	O	O	O	O
	R16	O	O	O	O	O	O	O	O	O	O	O	O	O	O	O	O	O	O	O	O

O : Braking   X : Accident

without system (with crossing vehicle, no warning)

system normal (with crossing vehicle, with warning)

system error (no crossing vehicle, with warning)

system defect (with crossing vehicle, no warning)

4.2 Alarm ignorance ratio and the fail-safe ratio

Based on the correlations between the occurrence of events and the operation of brakes described in 4.1, the correlations between the driver's behavior and the support system can be found and the measurement indices can be deduced.

The driver's behavior and conduct while using the support system can be classified into the following four categories:

(1) Alarm sounded and the brake pedal was stepped
(2) Alarm sounded but the brake pedal was not stepped
(3) Alarm did not sound but the brake pedal was stepped
(4) Alarm did not sound and the brake pedal was not stepped

In the case of (2), the driver ignored the support system, and this is found to be very dangerous driving conduct requiring attention. As for (3), the driver does not depend totally on the support system, making judgments and creating a margin for error. We can obtain valuable information on the lack of trust in the support system and the degree of dependence on the support system from both cases. Therefore, an index conversion of frequency of these two cases was considered.

In (2), the warning alarm was ignored. With $(n_1 + n_2)$ being the number of intersections where an alarm was sounded in the first round and n_1 being the number of cases where (1) and (2) are mutually exclusive events, the alarm ignorance ratio a can be derived from the equation below:

$$a = \frac{n_2}{(n_1 + n_2)} \quad , \quad (1)$$

Where a represents the ratio of the driver ignoring the warning alarm. Whenever the alarm sounds and the brake pedal is not stepped, the value for a increases. In this experiment, for all subjects, regardless of whether there was a crossing vehicle, whenever the alarm sounded, all of the subjects either slowed down or brought the car to a complete halt, giving us $n_2 = 0$, and thus the alarm ignorance ratio was $a = 0$.

In (3), the driver carried out the fail-safe function. With $(n_3 + n_4)$ being the number of intersections where an alarm did not sound in the first round and n_4 being the number of cases where (3) and (4) are mutually exclusive events, the fail-safe ratio, fs, can be derived from the equation below:

$$fs = \frac{n_3}{(n_3 + n_4)} \quad . \quad (2)$$

The fs before and after the system defect are shown in Figure 3.

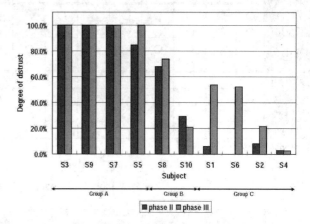

Figure 3 : fs for each subject

For all the subjects in Group A, the fs in Phase II and Phase III are close to 100% due to the fact that they slowed down and came to a complete halt at each and every intersection. As for the subjects in Group B, there were signs that the subjects didn't depend on the support system as they continued to drive with caution. This explains why the fs for the subjects in Group B is higher than Group C in Phase II. As for the subjects in Group C, due to the fact that they depended heavily on the support system, the fs in Phase II is low. In addition, we discovered that S1 and S6 returned to their normal driving behavior, which is not depending on the support system in Phase III.

The fs in Phase II (R5-R11) is shown in Figure 4.

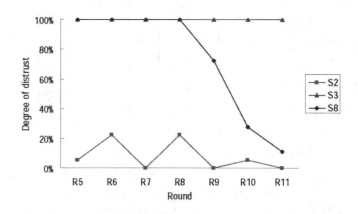

Figure 4 : fs for each subject in Phase II

250

From Figure 4, the *fs* value for S2 is low. As for S3, even in Phase II where the support system was used, regardless of whether there was an alarm, the brake pedal was stepped at all of the intersections, resulting in the *fs* for S3 to be 100% for all cases. In addition, S8, who initially did not trust the system and performed safety confirmation at every intersection, started to trust the system gradually as the experiment progresses resulting in the decrease of the brake operation when there is no alarm while entering an intersection.

The *fs* in Phase III (R12-R16) is shown in Figure 5.

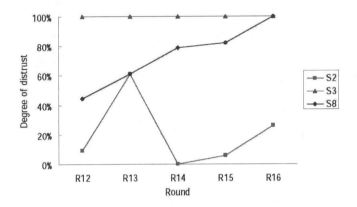

Figure 5 : *fs* for each subject in Phase III

The *fs* value represents the ratio when the brake pedal was not stepped in a case when the alarm did not sound. Since *fs* expresses the lack of trust in the support system, from Figure 5 we know that for S8, even when there was no crossing vehicle, the car slowed down or was brought to a complete halt, meaning that his distrust increased with occurrence of system defects and system errors. As for S2, after the accident at the 12th intersection in R12, there was an increase in his distrust once at R13, but since the system returned to normal after that, assuming that the system has recovered, *fs* decreased. In addition, for S3, who did not trust the system at all, the *fs* value was always 100%.

4.3 MENTAL WORKLOAD

There is the possibility that the occurrence or non-occurrence of an alarm or crossing vehicle can have an influence on the driver's mental workload. In order to measure the mental workload (MWL) of the driver every 2 rounds, the simplified NASA-TLX, which is the AWWL (adaptive weighted workload) [4], and the interview after the test course were used. The results for the AWWL analysis are shown in Figure 6.

Here, we focus on the mental workload (MWL) after the first system defect in R12. As for S8, his AWWL score increased after R12. This is due to an increase in frustration and effort as he was trying to return to his normal driving behavior, which is not depending on the support system. In the case of S2, due to frustration arising from the accident in R12, we saw an increase in the AWWL score as well as an increase in the distrustfulness and dissatisfaction in the support system. As for S3, the influence of frustration increases as the experiment goes on, and due to that the AWWL score increased.

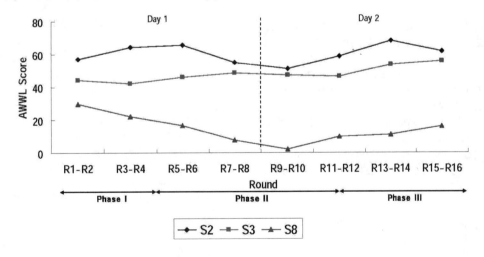

Figure 6 : AWWL score for each subject

5 CONCLUSION

In this research, a driver is exposed to a variety of events designed with the collision avoidance system of a DS. In each event, based on whether the brakes were operated, we proposed two indices to indicate the degree of the driver's dependence on the support system. In addition, from the AWWL score and the post-experiment interview, the driver's mental workload was measured.

From the AWWL results and whether the brakes were operated, we compared the ten subjects. Based on the operation of the brake pedal in Phase II, and whether an accident occurred during the system defect in Phase III, the authors found that there is a big difference among the subjects in their driving conduct and behavior after an accident. In addition, the proposed fail-safe ratio fs reflects on the ratio of defensive driving and the driver's distrust of the support system. The alarm ignorance ratio a and the fail-safe ratio fs give us information on the driver's degree of dependence on the support system.

The authors will continue to investigate the fundamental factors involved in the driver's degree of dependence using a, fs, and any other related parameter. In addition, based on the three performance elements of recognition, judgment, and operation, we plan to investigate these fundamental factors comprehensively on a support system in real-time. We intend to measure the physiological index data (line of sight, heart rate, etc.), analyze them, and then investigate their relations with the degree of dependence in real-time. Based on the degree of dependence, we plan to design a support system that collaborates with the driver.

REFERENCES

1. ITS HANDBOOK 2003-2004, The Road Bureau Ministry of Land, Infrastructure and Transport, http://www.its.go.jp/ITS/2003HBook/p44-49.pdf
2. Journal of Society of Automotive Engineers of Japan, Year Book, "Intelligent Transport Systems", Vol.58, No.8, pp166-172 (2004)
3. Honda Driving Safety Promotion Center, Manual of Honda Driving Simulator (2001)
4. Kenji Itoh, Sonoko Kuwano and Akinori Komatsubara, Handbook of Human Factors, pp138-151, Asakura Publishing, Tokyo (2002)

Driver Crash Avoidance Behavior: Analysis of Experimental Data Collected in NHTSA's Vehicle Antilock Brake System (ABS) Research Program

Graeme F. Fowler and Robert E. Larson
Exponent, Failure Analysis Associates

Laura A. Wojcik
Packer Engineering

ABSTRACT

As part of the National Highway Traffic Safety Administration's (NHTSA) Light Vehicle Antilock Brake System (ABS) Research Program a study was conducted to examine driver crash avoidance behavior and the effects of ABS on drivers' ability to avoid a collision in a crash-imminent situation. The test track study, described in detail in the SAE paper "Driver Crash Avoidance Behavior with ABS in an Intersection Incursion Scenario on Dry Versus Wet Pavement" [1], was designed to examine the effects of ABS versus conventional brakes, ABS brake pedal feedback level, and ABS instruction on driver behavior and crash avoidance performance.

Exponent has obtained the electronic data collected by NHTSA in the dry pavement study and analyzed the steering inputs to better understand how drivers respond to emergency avoidance situations. The results of this study can also be used to put the steering magnitudes, rates, and patterns used by test drivers performing idealized avoidance-type maneuvers on the skid-pad into the proper perspective by comparing them with what truly surprised drivers apply when confronted with an emergency situation requiring a steering response.

INTRODUCTION

As part of the National Highway Traffic Safety Administration's (NHTSA) Light Vehicle Antilock Brake System (ABS) Research Program a study was conducted into driver crash avoidance behavior in a simulated crash-imminent event. The protocol for the study, along with a summary of the results, is provided elsewhere [1 and 2]. According to NHTSA, the study was designed to examine the effects of ABS versus conventional brakes, ABS brake pedal feedback level, and ABS instruction on driver behavior and crash avoidance performance. The protocol for their investigation involved naïve subjects driving on a test track roadway under the guise that they were participating in a study of how average drivers steer and maintain speed in typical driving conditions. The driver traveled through an intersection three times where two vehicles were stationary at the stop lines of the crossing lane. Before the driver approached the intersection for a fourth time, the stationary vehicles were replaced with realistic looking stryofoam mockups. As the driver approached intersection, the styrofoam vehicle on the right intruded 6 feet into their lane of travel. The resulting driver steering and braking inputs as well as the vehicles' response were recorded with on-board instrumentation and video cameras. The protocol targeted an approach speed of 45 mph.

The experiment was designed to elicit emergency steering response under true surprise conditions. It was not intended to be representative of all crash avoidance-type scenarios or pre-crash on-road maneuvers. However, as pointed out by the NHTSA [1], the study is "useful in determining not only the extent to which drivers are able to maneuver the vehicle, but also drivers' physical capacity to supply control inputs to the vehicle. Insight into drivers' ability to maintain control during a panic maneuver and ability to avoid a collision can also be gained from this research." An understanding of typical drivers' response to an emergency avoidance is of interest to researchers in the areas of vehicular crash reconstruction as well as vehicle handling, controllability and rollover resistance. In fact, over the years numerous avoidance type maneuvers that purportedly model this situation have been proposed to investigate vehicles' limit handling and stability. The most notable being the International Standards Organization 3888 Part 2 Double Lane Change (ISO-DLC) and Consumer's Union Short Course (CU-SC) maneuvers. Both are discussed extensively in [3].

The NHTSA driver performance study represents the most extensive investigation into driver behavior in a crash-imminent situation using actual vehicles since the 1970's [4, 5, 6]. In the dry pavement experiment, 192 subjects were involved. Also of significance is the fact that the study utilized modern vehicles - a 1995 Chevrolet Lumina and a 1996 Ford Taurus - whose steering response is representative of passenger vehicles in the driving fleet today.

Exponent has obtained the electronic data collected by NHTSA in the dry pavement study for analysis of the drivers' behavior. In this paper we summarize the drivers' control responses obtained in the experimental study with a focus on their steering magnitudes and rates. Since data from this study are likely to be used by researchers as a reference for the capability of lay drivers to apply steering inputs in an emergency situation, the control inputs of those subjects who recorded the highest steering amplitudes are discussed in some detail. Next, the data from subjects who responded to the mock intersection collision by steering in both directions (reverse steered) were singled out for examination. Reverse steer inputs are intentionally generated in emergency lane change maneuvers using the ISO-DLC and CU-SC protocols with the intent of stressing vehicles' limit handling characteristics. It is of interest then, to compare the steering responses of typical drivers in this study to those of test drivers when performing the ISO-DLC and CU-SC maneuvers as a way of understanding the severity of these tests and their relevance to real-world crash situations. This comparison is provided in the final section.

ANALYSIS OF DRIVER BEHAVIOR

NHTSA's experimental design included multiple variables, such as different vehicles with unique ABS properties, response time allowed prior to the impending collision (time to intersection, TTI), conventional braking versus ABS, driver gender, and ABS driver instruction. With the exception of conventional versus ABS brake systems' impact on the steering and braking response, we have not addressed the impact of these additional factors. (These are discussed in detail in [1 and 2].)

An examination of the in-vehicle logs documenting the notes of the onboard observer indicated that 16 of the 192 subjects anticipated the vehicle crossing into the intersection. Of those 16 subjects, the in-vehicle logs also reported that only one (6%) failed to avoid impacting the incursion vehicle. In contrast, 41% of the subjects that did not anticipate the event impacted the styrofoam vehicle. Since the drivers who anticipated the incursion may have reacted differently to those who were truly surprised, they were not considered in our evaluation of driver behavior.

DEFINITION OF "EVENT"

A review of the electronic data files for the 176 subjects who did not anticipate the vehicle incursion led us to define an "event" window based on either time or vehicle speed during which the driver and vehicle data were to be analyzed. For the analysis presented here, data were analyzed for 5 seconds starting from the movement of the foam incursion vehicle or until the subject's vehicle speed dropped below 10 mph, if that occurred prior to the end of the 5 second period. It should be noted that movement of the incursion vehicle into the subject vehicle's lane began at either 2.5 or 3.0 seconds before the subject vehicle would have reached the intersection, based on the subject's actual approach speed. Five seconds was sufficient for the subject vehicle to either pass the incursion vehicle or, if heavy braking was applied, complete the primary avoidance maneuver. Steering inputs after the vehicle's speed had dropped below 10 mph were assumed to not be associated with the primary avoidance maneuver. Initial speeds ranged from 39 to 50 mph. It was noted that, within the defined event window, some subjects had begun to accelerate again and apply steering inputs which were probably associated with re-aligning the vehicle on the roadway after the avoidance maneuver. With respect to these variances in driver behavior, the extent of the event window was considered to be conservative.

SUMMARY OF DRIVER RESPONSES

Given the above definition of the "event," the data were processed to analyze the drivers' response. Initial review of the steering and braking data indicated that – even with this relatively simplistic crash-imminent scenario – a diverse range of outcomes was observed. This variability is clearly demonstrated in **Figure 1**, where the maximum absolute steer angle for each subject is plotted against the change in speed during the event. The change in speed was used as a surrogate for braking effectiveness since subjects implemented a wide range of brake forces and application times. Steering amplitudes ranged from 0 to 271 degrees and speed changes ranged from a small increase to a 37 mph reduction.

In order to broadly classify the subjects' braking and steering response, we assumed that a speed change of 6 mph or less corresponded to zero or minimal braking effort. (For these subjects, peak braking pedal forces were at or below 35 lbs). Peak steering magnitudes of 25 degrees or less were assumed to be indicative of minimal steering response. With these thresholds in mind, the majority of subjects (61%) steered and braked to avoid colliding with the obstacle (i.e., their speed change exceeded 6 mph and they steered in excess of 25 degrees). The next largest group – representing 28% of the subjects – principally braked in response to the emergency, without significant steering. Only 7% of the subjects applied steering only, which notably, is the control response modeled in both the ISO-DLC and CU-SC dynamic maneuvers. A number of subjects (3.4%) appeared to not respond in any meaningful way to the emergency situation. In the experiment, approximately 40% of the subjects "crashed" into the encroaching vehicle, which provides an indication of the severity of the emergency scenario simulated.

Figure 1: Maximum Absolute Steer Angle v. Change in Speed During Event, No Anticipation

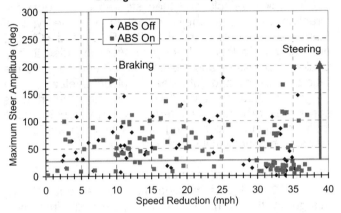

In **Figure 1**, vehicles with ABS active and ABS deactivated (simulating vehicles with conventional braking systems) are delineated. It can be observed from the Figure that the higher peak steering angles (for example, 150 deg and greater) were associated with heavier braking, regardless of the braking system. In their paper [1], Mazzae et al. reported the average avoidance steer magnitudes for subjects with ABS and those with conventional brake systems as 49 and 61 degrees, respectively. Although the difference in steer angle was not statistically significant, the authors did note that "drivers appeared to alter their steering behavior based on the degree to which they felt the steering inputs were affecting the motion of the vehicle in the desired direction. Subjects with ABS made smaller steering inputs and used lower steering input application rates than subjects with conventional brakes. The reason for this difference is believed to be that subjects made increasingly large steering with conventional brakes since, with locked wheels, their steering inputs were not effective in directing the vehicle's motion." This observation was confirmed when we looked specifically at the 60 subjects in the conventionally braked vehicles. Seven of these subjects applied the brakes with sufficient magnitude to lock the front wheels (based on the wheel speed data). The average peak steering amplitude for the 7 subjects was 122 degrees compared to 54 degrees with no lockup. Because of the potential confounding influence of ABS as apposed to conventional brakes, in the analysis that follows we will display the data with the different braking systems delineated.

Figure 2: Peak Left and Right Steer Angles

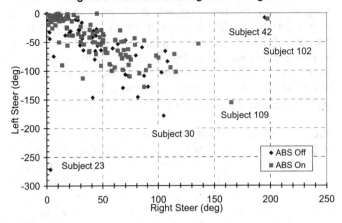

Figure 2 displays the peak left and peak right steer angles recorded for the 176 subjects during the event. The data reveal that, in the experimental crash-imminent scenario, 95% of the subjects input steering amplitudes of less than 130 degrees in an attempt to avoid the encroaching vehicle. Only 3% (5 subjects) exceeded 150 degrees of steering input. Three of these subjects were in conventionally braked vehicles (subjects 23, 42 and 30, representing 5% of that group) while two subjects (numbered 102 and 109) operated vehicles with the ABS activated (less than 2% of that sample).

TOP 5 STEERING EVENTS

The steering and braking responses of the five "large-steer" subjects, based upon the electronic data recorded during the event and information provided in the in-vehicle log, will be discussed below. Relevant steering, braking and vehicle response data for each subject are shown in **Appendix A, Figures A1 - A3 (conventional brakes)** and **A4 – A5 (ABS brakes).**

Conventional Brakes

Subject number 23 recorded the largest peak steer amplitude of 271 degrees. In response to the incursion vehicle, this subject first braked, which caused the front wheels to lock up. Approximately 0.75 seconds after braking – when the vehicle speed had decreased from approximately 43 mph to 27 mph – the subject steered to the left 271 degrees at a peak rate of 570 degrees/second.[1] At about the time of the maximum steer angle, the vehicle had almost come to a stop. During the steer, the subject continued to brake and, as a result of front wheel lockup, the vehicle did not respond to the steer input as evidenced by the negligible lateral acceleration generated (A_y). According to the in-vehicle log, the vehicle remained within the original lane of travel (right lane) in spite of the large left steer and did not collide with the obstacle.

---

[1] The steering rate was calculated by differentiating the data after implementing a 2 Hz low-pass filter.

To avoid a collision with the obstacle, subject 42 applied a right steer of 195 degrees at a peak rate of 356 degrees/second. Braking was initiated approximately at the same time as the steering input. Due to the braking, lateral accelerations of only 0.3 g's were developed during the "event." The vehicle migrated to the right to just straddle the road edge according to the in-vehicle log. This subject did impact the incursion vehicle.

Subject 30 braked hard and then steered to the left 178 degrees during his or her avoidance maneuver. Due to front wheel lock up at the beginning of the maneuver, the vehicle did not initially respond to the steering input. After the vehicle had decelerated to approximately 20 mph, the driver reduced braking and the vehicle responded with approximately 0.3 g lateral acceleration. The driver then steered to the right 105 degrees and began to accelerate. According to the in-vehicle logs, this subject did successfully avoid the obstacle by steering into the left lane.

ABS Brakes

Subject 102 braked heavily and steered right 197 degrees in response to the obstacle in his/her path. The peak steering rate was 320 degrees/sec and the vehicle left the roadway to the right. Although the speed approaching the intersection was 45 mph, the speed at the time of the peak steer input was 28 mph. This subject impacted the incursion vehicle. The peak lateral acceleration was 0.56 g's.

Subject 109 also initially steered right to avoid the incursion vehicle. The 165 degree right steer was followed by a 155 degree reverse steer, resulting in the largest peak-to-peak steering input of all the subjects (320 degrees). The peak steering rate was 870 degrees/sec. At the initiation of the event, the subject was traveling 44 mph. At the point of the maximum right steer the vehicle had slowed to 33 mph, at maximum left steer it was at 26 mph. As a result of the control inputs, the vehicle partially encroached the right shoulder and did impact the obstacle vehicle (according to the in-vehicle logs). The peak lateral acceleration generated in the maneuver was only 0.5 g's due to the heavy braking.

As can be established from **Figures A1 – A5** and the descriptions of the "large-steer" events above, all five of the subjects with steer amplitudes exceeding 150 degrees also braked heavily in response to the incursion vehicle. As a consequence of the braking, the maximum lateral accelerations developed during the avoidance maneuvers performed by the five "large steer" subjects was only 0.56 g's - substantially below the maximum lateral acceleration capability of the test vehicles. Three of the five subjects steered primarily in <u>one direction</u> during the course of the "event." In fact, the largest steer amplitudes - approximately 200 to 270 degrees - were from the subjects who steered in a single direction. The remaining 2 subjects steered in both directions to avoid the encroaching vehicle.

DRIVER AVOIDANCE INVOLVING REVERSE STEER INPUTS

We now restrict our attention to those subjects who steered in both directions during the "event" and whose peak steering amplitude in each direction is relatively comparable. It is assumed that included in this subset of drivers are those that made an attempt to steer around the incursion vehicle and then steered in the opposite direction to recover their original lane – the type of maneuver modeled in the ISO-DLC and CU-SC procedures. To this end, subjects were included if their peak-to-peak steer angle exceeded 50 degrees and the ratio of peak left and right steer amplitudes fell within the range of 0.33 to 3.0. This selection criterion is somewhat arbitrary, but removes from consideration those subjects who steered only a minimal amount or steered principally in just one direction.

Figure 3 documents the peak left and right steer angles for the 104 subjects meeting this requirement. Within this group, the average maximum steer angle input was 72 degrees, with more than 98% of the subjects utilizing peak steer amplitudes of less than 150 degrees.

Figure 3: Peak Left and Right Steer Angles - Reverse Steer Subjects

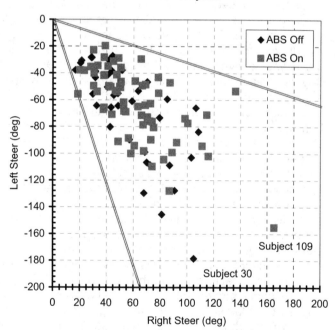

In **Figure 4** the distribution of the peak-to-peak steer angles for the subjects who input reverse steers is provided. We define the peak-to-peak steer angle as the sum of the peak left and right steer amplitudes recorded for each subject during the "event." The average peak-to-peak steer angle for these subjects is 125 degrees. Only two subjects (109 and 30) exceeded peak-to-peak angles of 230 degrees, using 320 and 283 degrees, respectively. Both of these subjects braked heavily during the avoidance maneuver, with one slowing to a stop in three seconds, and the other slowing to 11 mph in two seconds and then accelerating.

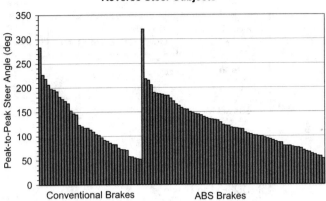

Figure 4: Peak-to-Peak Steer Angle
Reverse Steer Subjects

Figure 6: Maximum Steer Rate v Steer Amplitude
Reverse Steer Subjects

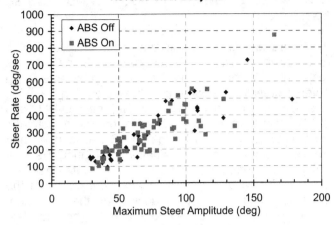

Figure 5 is a plot of the peak steer amplitude for each subject versus their speed approaching the intersection for the reverse steer subjects. For approach speeds above 45 mph, the maximum steer angles recorded did not exceed 120 degrees. The largest steer angle amplitude for these subjects was 179 degrees, associated with an initial speed of approximately 40 mph (subject 109).

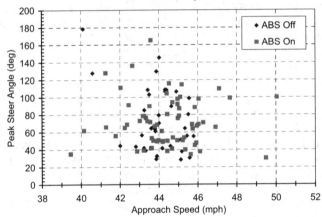

Figure 5: Peak Steer Angle v Approach Speed
Reverse Steer Subjects

Given the nature of the experiment, it is not surprising that an essentially linear relationship existed between the maximum steer amplitude and the steering rate for those subjects who input comparable left and right steering angles during their avoidance attempt. As shown in **Figure 6**, the average steering rate recorded for these subjects was approximately 290 deg/sec, with 98% of the subjects below 600 deg/sec. The maximum steer rate was 875 deg/sec.

The peak absolute value of lateral acceleration recorded during the "event" as a function of the maximum steer amplitude is plotted in **Figure 7**. Ninety four percent (94%) of the subjects generated lateral acceleration magnitudes of 0.6 g's or less. The highest lateral acceleration recorded for those subjects who input a reverse steer in the experiment was 0.73 g's[2]. Interestingly, the subjects experiencing the highest lateral accelerations did not apply the largest steering inputs due to the confounding influence of braking. Peak absolute values of lateral accelerations exceeding 0.6 g's were generated with steering amplitudes in the range of 90 to 110 degrees. Furthermore, the two subjects who developed the highest lateral accelerations (0.7 and 0.73 g's) both exited the roadway to the left (according to the in-vehicle logs). This loss of lane position suggests that lay drivers who steer aggressively in an avoidance maneuver and generate relatively high lateral accelerations are likely have difficulty keeping the vehicle within the driving lanes of a two-lane roadway.

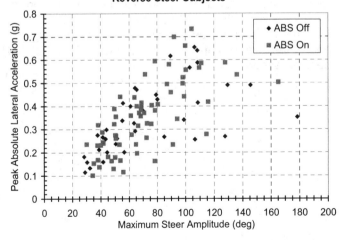

Figure 7: Peak Lateral Acceleration v Steer Amplitude
Reverse Steer Subjects

---

[2] The peak lateral acceleration of 0.73 g's was actually the highest recorded among all the subjects who did not anticipate the incursion vehicle.

COMPARISON WITH ISO-DLC AND CU-SC MANEUVERS

In **Figure 8** the peak left and right steering inputs recorded in the NHTSA driver study for those subjects responding to the simulated emergency by applying a reverse steer are compared to the steering amplitudes used by test drivers when performing the ISO-DLC and CU-SC maneuvers. The test driver data were obtained from testing performed by the NHTSA as part of their Phase IV experimental evaluation of maneuvers that may induce untripped rollovers [3]. The data points for the ISO-DLC and the CU-SC maneuvers represent the peak left and right steer angles at the maximum maneuver entrance speed for each of the four test vehicles in their nominal loading conditions[3].

The maximum peak-to-peak steering angles reported in the Phase IV tests were 454 degrees for the ISO-DLC and 713 degrees for the CU-SC. As discussed in the previous section, in the driver response study the maximum peak-to-peak steer angle recorded was 320 degrees, with 98% of the drivers applying steer reversals of 230 degrees or less. The steering magnitudes used by the test drivers to successfully negotiate the ISO-DLC and CU-SC courses cleanly at the maximum speeds well exceeded that which lay drivers were found to use in the simulated collision scenario.

We next looked at the steering inputs utilized by lay drivers who maneuvered the vehicle in a fashion that more closely followed the emergency lane change scenarios modeled by the ISO and CU maneuvers. In this case, we only considered subjects who avoided impacting the incursion vehicle by steering into the left lane according to the in-vehicle logs. As the ISO and CU lane change maneuvers do not involve braking, we further restricted the data set to include subjects whose speed change during the "event" was less than 15 mph - representing a relatively small amount of braking at most. Using these filters, the maximum steering amplitudes recorded by test subjects was 111 degrees and the maximum steering rate was 544 degrees/second.

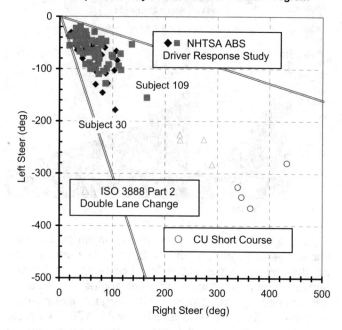

Figure 8: Comparison of Steering Inputs Recorded in NHTSA's Driver Response Study and Phase IV Research Program

In **Figure 9** the steering input, lateral acceleration, and vehicle speed for a Chevrolet Blazer driven by the three NHSTA's test drivers during their maximum speed clean CU-SC runs is compared to the two most severe steering reversals found in NHTSA's driver response study. As can be seen from the figure, the steering inputs used by the naïve test subjects (operating either a Chevrolet Lumina or Ford Taurus) were significantly smaller than those used by the test drivers in the CU-SC. The initial speeds in the driver response study were as high or higher than the CU-SC maximum speed clean runs, but in the driver response study the drivers who input the largest steering angles also utilized braking and their speed dropped off more rapidly. The lateral accelerations observed in the driver response study were also considerably lower than found from the CU-SC. This difference is due to a combination of factors, including the lower steering magnitudes, the reduced speed due to braking, and, in the case of Subject 30, a reduced lateral acceleration capability of the vehicle due to front wheel brake lock up.

CONCLUSIONS

This NHTSA study produced a diverse range of driver responses to a relatively simple accident avoidance situation. Most subjects steered and braked in response to the simulated emergency. The next largest group used their brakes as the principal method of avoidance. A relatively small number of subjects (7%) used only steering to avoid the incursion vehicle.

---

[3] 2001 Chevrolet Blazer, 2001 Toyota 4Runner, 1999 Mercedes ML 320, 2001 Ford Escape. Stability control enabled on vehicles with that that feature.

Figure 9: NHTSA ABS Driver Response study CU Short Course (Chevrolet Blazer) and Driver Response Study Test Data Comparison

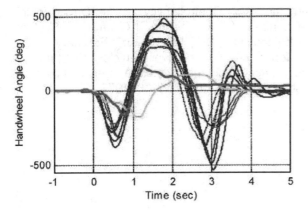

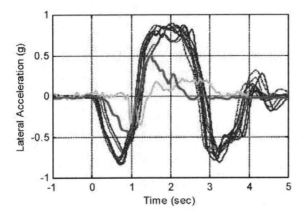

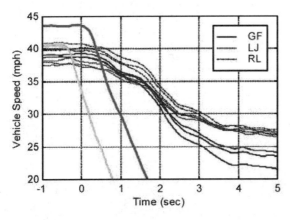

— CU Short Course Testing,
  NHTSA Figure 12.7 [3]
— Driver Response Subject 109,
  with ABS
— Driver Response Subject 30,
  with conventional brakes

Drivers of conventionally-braked vehicles that experienced front wheel lockup in the avoidance maneuver recorded, on average, higher maximum steering amplitudes (122 degrees) compared to those who did not lock up their front wheels (54 degrees). The average peak steering amplitude for the ABS vehicles was 49 degrees – almost identical to that of the subjects

in the conventionally-braked vehicles that did not lock up the front wheels in the avoidance maneuver. Consequently, drivers appeared to alter their steering input based on the degree to which they felt that their steering inputs were affecting the motion of the vehicle.

The largest steering amplitudes were recorded by subjects who steered primarily in one direction – often associated with front wheel lockup with the conventional brake system.

When we focused our attention on subjects who responded to the unexpected incursion with a reverse steer input, the average peak steer amplitude was 72 degrees, with the great majority of subjects (98%) inputting maximum steer amplitudes of less than 150 degrees. The average peak-to-peak steer angle was 125 degrees, with 98% of the subjects not exceeding 230 degrees. Only two subjects applied steering rates in excess of 600 degrees/second. The maximum steering rate measured in the experiment was 875 degrees/second.

For subjects who maneuvered the vehicle in a fashion that most closely represents that of the emergency lane change scenario modeled by the CU and ISO maneuvers, maximum steering amplitudes were less than 111 degrees.

Most subjects (94%) experienced lateral accelerations below 0.6 g's in response to their control inputs. The two subjects who recorded the highest lateral accelerations (0.7 and 0.73 g's) were not able to keep the vehicle within the driving lanes.

The ISO-DLC and CU-SC tests represent very severe vehicle handling maneuvers requiring steering angles and rates well in excess of what typical drivers implemented in the experimental collision avoidance scenario.

The peak steering amplitudes and rates applied by test drivers during skid-pad obstacle avoidance-type maneuvers should not be considered representative of what lay drivers might use when faced with a real-world collision scenario or when attempting to regain control on the highway.

ACKNOWLEDGMENTS

The research presented in this paper was sponsored by American Suzuki Motor Corporation. The authors gratefully acknowledge this support.

REFERENCES

1. "Driver Crash Avoidance Behavior with ABS in an Intersection Incursion Scenario on Dry Versus Wet Pavement," Mazzae, Barickman, Baldwin and Forkenbrock, SAE 1999-01-1288.

2. "NHTSA Light Vehicle AntiLock Brake System Research Program Task 5.2/5.3: Test Track Examination of Drivers' Collision Avoidance Behavior Using Conventional and Antilock Brakes," DOT HS 809 561, March 2003.

3. "A Comprehensive Experimental Evaluation of Test Maneuvers That May Induce On-Road, Untripped, Light Vehicle Rollover, Phase IV of NHTSA's Light Vehicle Rollover Research Program," Forkenbrock, Garrott, Heitz and O'Harra, DOT HS 809, October 2002.

4. "An Experimental Study of Automobile Driver Characteristics and Capabilities," Rice and Dell'Amico, Calspan Report No. ZS-5208-K-1, prepared for General Motors Corporation, March 1974.

5. "Automobile Controllability – Driver/Vehicle Response for Steering Control," Systems Technology Inc., DOT HS 801 407, February 1975.

6. "Performance of Driver-Vehicle System in Emergency Avoidance," Maeda, Irie, Hidaka, and Nishimura, SAE Paper 770130, 1977.

Appendix A: Response Data for the Five "Large-Steer" Subjects.

Figure A1: Subject 23 (Conventional Brakes)

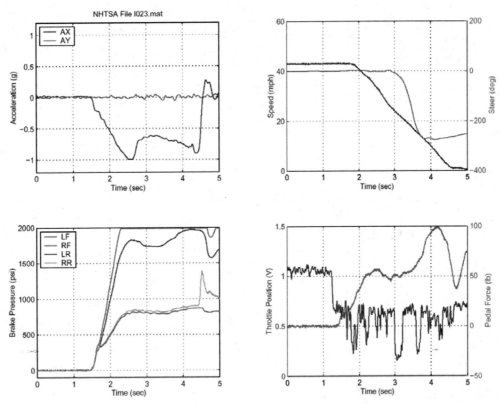

Figure A2: Subject 42 (Conventional Brakes)

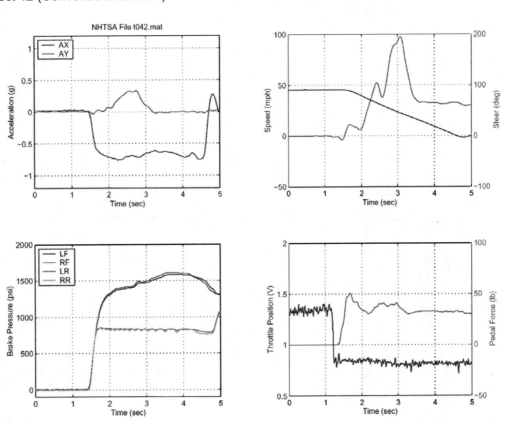

Figure A3: Subject 30 (Conventional Brakes)

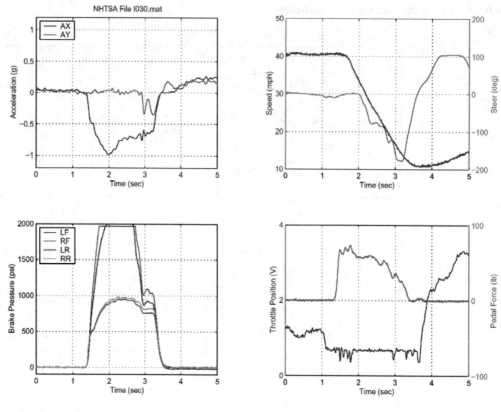

Figure A4: Subject 102 (ABS Brakes)

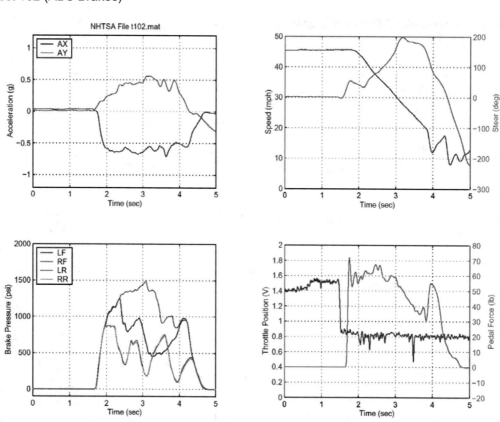

Figure A5: Subject 109 (ABS Brakes)

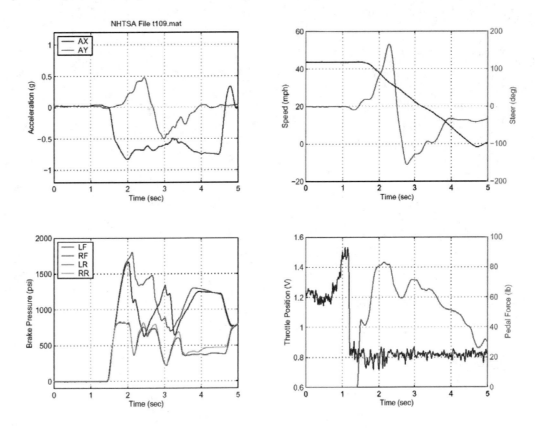

Predictive Safety Systems – Steps Towards Collision Mitigation

Peter M. Knoll and Bernd-Josef Schaefer
Robert Bosch GmbH, Driver Assistance Systems Business Unit

Hans Guettler and Michael Bunse
Robert Bosch GmbH, Restraint Systems Business Unit

Rainer Kallenbach
Robert Bosch GmbH, Automotive Electronics Division

ABSTRACT

Sensors to detect the vehicle environment are being used already today. Ultrasonic parking aids meanwhile have a high customer acceptance, and ACC (Adaptive Cruise Control) systems have been introduced in the market recently. New sensors are being developed at rapid pace. On their basis new functions are quickly implemented because of their importance for safety and convenience.

Upon availability of high dynamic CMOS imager chips Video cameras will be introduced in vehicles. A stereo capable Computer platform with picture processing capability will explore the high potential of functions. Finally, sensor data fusion will improve significantly the performance of the systems.

At the end of the 1980s this insight led to the vision of highly efficient street traffic, demonstrated in the "Prometheus" funded project. But at that time the electronic components necessary for these systems – highly sensitive sensors and extremely efficient micro-processors – were not yet ready for high-volume series production and automotive applications.

INTRODUCTION

Almost every minute, on average, a person dies in or caused by a crash. In 1998 more then 93,000 persons have been killed in the Triad (Europe, USA and Japan) in road traffic accidents leading to a socioeconomic damage of more then 600 bil. EUR.

The EU Commission has defined a demanding goal by cutting the number of killed persons to half until the year 2010. Bosch wants to contribute significantly to this goal by developing Driver Assistance Systems in close cooperation with the OEMs and thus, reduce the frequency and the severity of road accidents.

An important aspect of developing active and passive safety systems is the capability of the vehicle to perceive and interpret its environment, recognize dangerous situations and support the driver and his driving maneuvers in the best possible way.

TRAFFIC ACCIDENTS – CAUSES AND MEANS TO MITIGATE OR AVOID THEM

In critical driving situations only a fraction of a second may determine whether an accident occurs or not. Studies [1] indicate that about 60 percent of front-end crashes and almost one third of head-on collisions would not occur if the driver could react one half second earlier. Every second accident at intersections could be prevented by faster reactions.

DRIVER ASSISTANCE SYSTEMS FOR VEHICLE STABILIZATION

Only recently, statistic material has been published [2] showing that the accident probability for vehicles equipped with the ESP system (ESP=Electronic Stability Program) is significantly lower than for vehicles without ESP. Fig. 1 shows the high collision mitigation potential of this system.

Additional improvement is expected from systems like PRE-SAFE from DaimlerChrysler. PRE-SAFE combines active and passive safety by recognizing critical driving situations with increased accident possibility. It triggers preventive measures to prepare the occupants and the vehicle for possible crash by evaluating the sensors of the ESP and the Brake Assist.

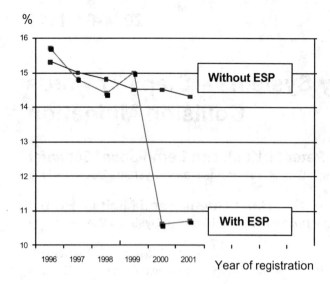

Figure 1: Influence of ESP on accident probability for vehicles without and with ESP [2]

To protect best passengers from a potential accident, reversible belt-pretensioners for occupant fixation, passenger seat positioning and sunroof closure are activated.

Like the vehicle interaction with the vehicles dynamic with ESP, the release of collision mitigation means can be activated only in the case when a vehicle parameter went out of control or when an accident happens. Today, airbags are activated in the moment when sensors detect the impact. Typical reaction times last 5 ms. In spite of the extremely short time available for the release of accident mitigation means, there is no doubt that airbags have contributed significantly to the mitigation of road accidents and, in particular, fatalities. But due to the extremely short time between the start of the event and the possible reaction of a system the potential of today's systems is limited.

Bosch as supplier of the ESP is proud of the achieved improvements but we want to achieve even more with a new generation of safety systems, being currently under development. These are the so called "predictive" driver assistance systems. They expand the detection range of the vehicle by the use of surround sensors. With the signals of these sensors objects and situations in the vicinity of the vehicle can be enclosed into the calculation of collision mitigating and collision avoiding means.

COMPONENTS OF PREDICTIVE DRIVER ASSIS-TANCE SYSTEMS

Today, the components for the realization of these systems – highly sensitive sensors and powerful microprocessors - are available or under development with a realistic time schedule, and the chance for the realization of the "sensitive" automobile is fast approaching. Soon sensors will scan the environment around the vehicle, derive warnings from the detected objects, and perform driving maneuvers all in a split second faster than the most skilled driver.

Electronic surround sensing is the basis for numerous driver assistance systems – systems that warn or actively intervene. Fig. 2 shows the detection areas of different sensor types.

Until now, due to the limited availability of sensors, only a few driver assistance systems could be established on the market up to now. One example is the Park Pilot from Bosch, which monitors objects at near range with the help of ultrasound technology. Sensors integrated in the bumper forward an acoustic or optical warning to the driver as soon as he approaches an obstacle. In the meantime, this system is widely used and has high acceptance with the customer. It is already in series-production in many vehicles [5].

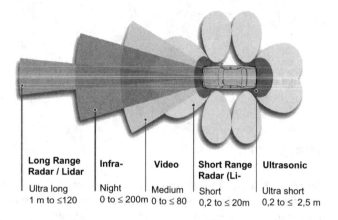

Fig. 2: Surround sensing: Detection fields of different sensors

Upon availability of appropriate sensors, new systems will be introduced in future vehicles. Their spectrum will range from warning systems to systems with vehicle interaction [4].

ULTRASONIC SENSORS

Reversing and Parking Aids today are using Ultra Short Range Sensors in ultrasonic technology. They have a detection range of approx. 1,5m. They have gained high acceptance with the customer and are found in many vehicles. The sensors are mounted in the bumper fascia. When approaching an obstacle the driver receives an acoustical and/or optical warning. The next generation of ultrasonic sensors will have a detection range of approx. 2.5m, and will thus, explore new applications like Parking Space Measurement and Semiautonomous Parking.

LONG RANGE RADAR 77 GHz

The 2nd generation Long Range Sensor with a range of approx. 120m is based on FMCW Radar technology. The narrow lobe with an opening angle of ± 8° detects obstacles in front of the own vehicle and measures the distance to vehicles in front. The CPU is integrated in the sensor housing. The sensor is multi target capable and can measure distance and relative speed simultaneously. The angular resolution is derived from the signals from 4 Radar lobes. Series introduction was made in 2001 with the first generation. Figure 3 shows the 2nd generation sensor. It will be introduced into the market in March, 2004. At that time this Sensor&Control Unit will be the smallest and lightest of its kind on the market.

The antenna window for the mm-waves is a lens of plastic material which can be heated to increase the availability during winter season. The unit is mounted in air cooling slots of the vehicle front end or behind plastic bumper material by means of a model specific bracket. Three screws enable the alignment in production and in service [4].

Fig 3: 77 GHz Radar sensor with integrated CPU for Adaptive Cruise Control

The information of this sensor is used to realize the ACC function (Adaptive Cruise Control). The system warns the driver from following too close or keeps automatically a safe distance to the vehicle ahead. The set cruise speed and the safety distance is controlled by activating brake or accelerator. At speeds below 30 km/h the systems switches off with an appropriate warning signal to the driver.

In future, additional sensors (Video, Short Range Sensors) will be introduced in vehicles. They allow a plurality of new functions.

SHORT RANGE SENSORS

Besides ultrasonic sensors, 24 GHz radar sensors (Short-Range-Radar (SRR)-Sensors) or Lidar sensors can be used in future to build a „virtual safety belt" around the car with a detection range between 2 and 20m, depending on the specific demand for the function performance. Objects are detected within this belt, their relative speeds to the own vehicle are calculated, and warnings to the driver or vehicle interactions can be derived.

Today, there is still a limitation for the introduction of the 24 GHz UWB (Ultra Wide Band) Radar imposed by the pending release of the frequency band for the mentioned applications. This release has been given in 2002 for the USA. In Europe this process is still going on and under intensive discussion, mainly opposed by the established services such as Earth Exploration Satellite Services, Radio Astronomy and Fixed Services. A worldwide harmonization is necessary.

VIDEO SENSOR

Figure 4 shows the current setup of the Robert Bosch camera module.

The camera is fixed on a small PC board with camera relevant electronics. On the rear side of the PC board the plug for the video cable is mounted. The whole unit is shifted into a windshield mounted adapter.

CMOS technology with non linear luminance conversion will cover a wide luminance dynamic range and will significantly outperform current CCD cameras. Since brightness of the scene cannot be controlled in automotive environment, the dynamic range of common CCD technology is insufficient and high dynamic range imagers are needed.

Fig. 4: Video camera module

SURROUND SENSING SYSTEMS AND DRIVER ASSISTANCE SYSTEMS

Based on the various sensor technologies, a plurality of application areas are possible for driver assistance systems.

LONG RANGE RADAR SYSTEM

Inattention is the cause of 68% of all rear end collisions! In 11% besides inattention following too closely is the cause, 9% of the rear end collisions are caused by following too closely alone. These statistics [6] show that 88% of rear end collisions can be influenced by longitudinal control systems. We assume a stepwise approach from convenience systems to safety systems where the first step has been made with the Adaptive Cruise Control.

- Step 1:
  Longitudinal Control (Adaptive Cruise Control)

  ACC and next ACC generations control the speed of a vehicle and control automatically the safety distance to a vehicle in front.

- Step 2:
  Predictive Safety Systems (PSS)

  Systems based on ACC interact with the vehicle in critical situations to avoid (in the best case) a potential accident or to mitigate the consequences of an unavoidable accident.

- Step 3:
  Systems for active collision avoidance

  Accidents are avoided by active interaction with the vehicle (longitudinal, lateral interaction and interaction with the engine management. This is still a vision.

THE FIRST STEP: ADAPTIVE CRUISE CONTROL (ACC)

Figure 5 shows the basic function of the ACC system. With no vehicle in front or vehicle in safe distance ahead, the own vehicle cruises at the speed which has been set by the driver (Fig. 5, up). If a vehicle is detected, ACC adapts automatically the speed in such a way that the safety distance is maintained (Fig. 5, middle) by interaction with brake and accelerator. In case of a rapid approaching speed to the vehicle in front, the system additionally warns the driver. If the car in front leaves the lane the own vehicle accelerates to the previously set speed (Fig. 5, below).

Fig. 5: Basis function of ACC

In order to avoid excessive curve speeds the signals of the ESP system are considered simultaneously. ACC will reduce automatically the speed. The driver can override the ACC system at any time by activating the accelerator or with a short activation of the brake.

The current system of the first generation is active at speeds beyond 30 km/h. To avoid too many false alarms, stationary objects are suppressed.

Bosch develops the system further: With the improved ACC this convenience function can be used also on smaller highways. The 2nd generation of the ACC will come on the market in early 2004. The next step in functionality will come with the Low Speed Following (LSF) function, with a fusion of the data of the long range radar with a short range sensor. This function will allow to brake the vehicle down to speed zero and to reaccelerate after a drivers confirmation. In a further step the fusion of ACC LSF with a Video camera will allow a complete longitudinal control at all vehicle speeds, and also in urban areas with a high complexity of road traffic scenery.

The today's ACC system is a convenience function supporting the driver to drive more relaxed. Starting from 2005 on, Bosch will extend the functionality of ACC to „Predictive Safety Systems", and enter, thus, into the field of safety systems.

THE SECOND STEP: PREDICTIVE SAFETY SYSTEMS

Predictive safety systems will pave the way to collision avoidance with full interference in the dynamics of the vehicle. They are partly based on signals derived from additional sensors, allowing to integrate the vehicle's surrounding. From the measurement of the relative speed between detected obstacles and the own vehicle, dangerous situations can be recognized in an early state. Warnings and stepwise vehicle interactions can be derived.

Figure 6 shows the shocking analysis of the braking behavior during collisions. In almost 50% of the collisions the drivers do not brake at all. An emergency braking happens only in 39% of all vehicle – vehicle accidents, and in 31% of the accidents with no influence of another vehicle, respectively.

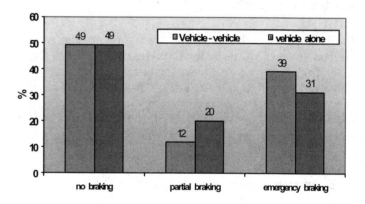

Fig. 6: Braking behavior during collision accidents [3]

This analysis confirms that inattention is the most frequent cause for collision type accidents and shows the high collision avoidance and collision mitigation potential of predictive driver assistance systems if the braking process of the driver can be anticipated or a vehicle interaction can be made by the vehicle's computer.

The introduction of predictive safety systems comes most likely with convenience systems where safety systems will use the same sensors. From 2005 on Bosch will extend ACC as the most important component of predictive safety systems to safety systems. If ACC recognizes a dangerous traffic situation the brake can be prefilled and the brake assist system can be prepared for a potential emergency braking. Future developments will incorporate functions to warn the driver very effectively and to perform automatic emergency braking.

In case of an emergency braking important fractions of a second can be used for a maximum reduction of kinetical energy.

VIDEO SYSTEM

The above mentioned Video technology will first be introduced for convenience functions that provide transparent behavior to and intervention by the driver. Fig. 7 shows the basic principle of operation for a video system.

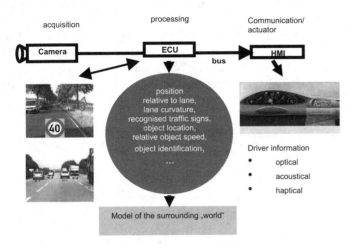

Fig. 7: Basic principle of a video sensor and functions being considered

The enormous potential of video sensing is intuitively obvious from the performance of human visual sensing. Although computerized vision has by far not achieved similar performance until today, a respectable plurality of information and related functions can readily be achieved by video sensing:

- lane recognition and lane departure warning, position of own car within the lane,
- traffic sign recognition (speed, no passing, ...) with an appropriate warning to the driver,
- obstacles in front of the car, collision warning
- vehicle inclination for headlight adjustments.

New methods of picture processing in conjunction with high dynamic imagers in future will further improve the performance of these systems [4]. Besides the measurement of the distance to the obstacle the camera can assist the ACC system by performing an object detection or object classification. Special emphasis is put on the night vision improvement function in the introduction phase of Video technology.

A high benefit can be achieved with a tail camera if objects approaching quickly from behind are detected, and the driver gets a warning signal when he intends to pass.

OUTLOOK

Figure 8 shows the enormous range of driver assistance systems on the way to the „Safety Vehicle". They can be sub-divided into two categories:

- Safety systems with the goal of collision mitigation and collision avoidance,

- Convenience systems with the goal of semiautonomous driving.

Driver support systems without active vehicle interaction can be viewed as a pre-stage to vehicle guidance. They only warn the driver or suggest a driving maneuver. One example is the parking assistant of Bosch. This system will give the driver steering recommendations when parking in order to park optimally in an automatically determined, prior measured parking space. Another example is the Night Vision Improvement system. As more then 40% of all fatalities occur at night this function has high potential for saving lives. Lane departure warning systems and systems detecting obstacles in the blind spot can also contribute significantly to the reduction of accidents as almost 40% of all accident are due to unintended lane departure.

ACC, which has already been introduced to the market, belongs to the group of active convenience systems and will further be developed to a functionality which allows driving in all speed ranges and in urban areas as well. If longitudinal guidance is augmented by lane-keeping assistance (also a video-based system for lateral guidance), and making use of complex sensor data fusion algorithms, automatic driving is possible in principle.

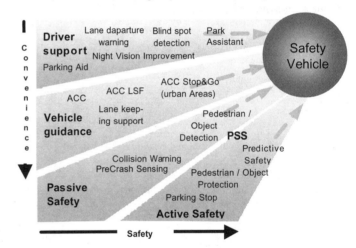

Fig. 8: Driver assistance systems on the way to the safety vehicle

Passive safety systems contain the predictive recognition of potential accidents and the functions of pedestrian protection.

The highest demand regarding performance and reliability is put on active safety systems. They range from a simple parking stop, which automatically brakes a vehicle before reaching an obstacle, to computer-supported control of complex driving maneuvers to avoid collisions. For example, the automatic emergency braking feature intervenes if a crash is unavoidable. In its highest levels of refinement, active systems intervene in steering, braking and engine management to avoid colliding with an obstacle. Here, the vision goes to the collision avoiding vehicle, making computer assisted driving maneuvers for crash avoidance.

CONCLUSION

The European Union has put the right emphasis on the e-Safety program with the vision to reduce fatalities to 50% until the year 2010. Car makers and suppliers have responded to the Commissions program and try to make their contributions to reach the goal [5].

The accident-free traffic, in our opinion, will remain a vision but we at Bosch see a plurality of means for a step wise introduction of convenience and safety systems for collision mitigation and future accident avoidance.

REFERENCES

[1] Enke, K.: „Possibilities for Improving Safety Within the Driver Vehicle Environment Loop, 7th Intl. Technical Conference on Experimental Safety Vehicle, Paris (1979)

[2] Anonymous statistics of accident data of the Statistisches Bundesamt (German Federal Statistics Institution), Wiesbaden, Germany (1998 – 2001)

[3] Statistics from the "Gesamtverband der Deutschen Versicherunswirtschaft e.V." (Association of the German Insurance Industry) (2001)

[4] Seger, U.; Knoll, P.M.; Stiller, C.: "Sensor Vision and Collision Warning Systems", Convergence, Detroit (2000)

[5] Knoll, P.M.: Fahrerassistenzsysteme – Realer Kundennutzen oder Ersatz für den Menschen? VDI, Deutscher Ingenieurtag, Münster, Germany (2003)

[6] NHTSA Report (2001)

CONTACT

peter.knoll@de.bosch.com

2003-01-0503

Rear-End Collision Velocity Reduction System

Kenji Kodaka, Makoto Otabe, Yoshihiro Urai and Hiroyuki Koike
Honda R&D Co., Ltd.

ABSTRACT

In Japan, rear-end collisions occur at higher frequency than many other kinds of traffic accident. The causes of rear-end collisions were therefore investigated. Accident statistics was used to conduct a statistical traffic accident analysis and a questionnaire survey was used to conduct a detailed traffic accident analysis. Simulation was then used to perform an accident analysis on the basis of those studies. The results suggested that many of these accidents were caused by momentary inattention during daily driving. Research was therefore carried out to determine what kind of collision avoidance assist system would be effective for use at such times. Tests were carried out to measure the obstacle avoidance characteristics of drivers using actual cars, and control timing parameters were established. In this process, the warning timing was set so that it would not lose its impact as a warning and also so that it would not interfere with the driver. The system was configured to make up for delays in recognition and judgment by means of brake control. The method of brake operation to be used was studied using both computation and testing with actual vehicle. The brake operation timing was set not to interfere with ordinary driver operations, and the deceleration was set to assist the driver's avoidance maneuvers. The result was creation of a system capable of contributing to the reduction of rear-end collisions.

1. INTRODUCTION

The Advanced Safety Vehicle (ASV) project promoted by the Ministry of Land, Infrastructure and Transport uses external sensing technology such as radar sensors to determine conditions in the vicinity in order to reduce traffic accidents. The corporations involved have proposed a variety of systems that function in this way to either avoid accidents before they occur or reduce accident damage, and there are expectations for accelerated movement toward practical application of these systems.

Systems designed to avoid collisions using brake control and external sensing technology such as radar sensors and cameras to predict collisions with the vehicle ahead had been proposed in the first phase of the ASV project.

They did not reach the point of practical application, due to technical and conceptual difficulties.

The present system concept was defined to be as accident avoidance support on the driver. The focus was also narrowed down to the rear-end collisions as the type of accident targeted. This report describes the creation of a system aimed to achieve early practical application.

2. SURVEY OF ACCIDENT SUMMARIES

Overall surveys[1][2] of traffic accidents involving injury or fatality in Japan in 1997 indicate that approximately 30% of all accidents were rear-end collisions, an extremely high percentage (Fig. 1).

A survey was next conducted of the distinctive features of rear-end collisions. Fig. 2 presents the frequency of rear-end collisions in relation to the speed at which the drivers recognize the danger. This indicates that most accidents occur when vehicles are cruising at low to medium speeds. Fig. 3 shows the human factors of the persons principally concerned who caused the accidents. It was determined that inattention in the forward direction and other such failures in recognition were a frequent factor. Fig. 4 shows the scenes of rear-end collisions classified by road geometry. It was determined that accidents on straight roads make up approximately 94% of the total, while extremely few accidents occurred on curved roads. This can be explained by the fact that inattention in the forward direction is a major factor in accidents, and since drivers are looking forward when they drive on curved roads, there is a lower incidence of inattention and therefore fewer accidents occur. Fig.5 shows the status of the other vehicle involved during a collision. It is apparent that rear-end collisions with stationary vehicles are in the great majority, accounting for approximately 87% of the total. The above accident surveys suggest that the scenes of rear-end collisions are likely to involve cruising relatively slowly on a straight road. Then, when the driver is slow to recognize the vehicle ahead due to inattention in the forward direction or other reasons (Fig.3), for example, condition mistaken, safety not-checked and miss-handling, the rear-end collision occurs.

271

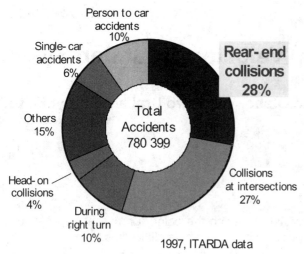

Fig.1 Composition of traffic accidents per year (1997)

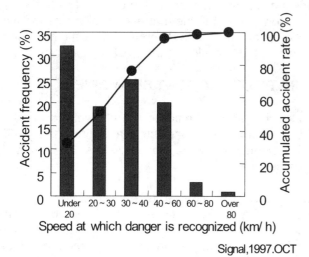

Fig.2 Frequency and accumulated rate of rear-end collisions in relation to the speed at which danger is recognized

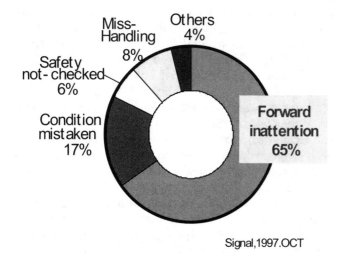

Fig.3 Factor in rear-end collisions

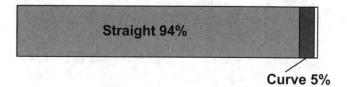

Signal, 1997.OCT

Fig.4 Road situation during rear-end collisions

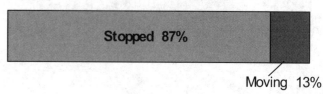

Signal,1997.OCT

Fig.5 State of vehicle ahead during rear-end collision (according to statistics)

3. ANALYSIS OF ACCIDENT CAUSES

3.1. QUESTIONNAIRE SURVEY

A questionnaire survey was conducted within Honda to examine the circumstances under which actual accidents occur.

The survey targeted people who had actually experienced rear-end collisions and those who had almost experienced such accidents. The status of the vehicle ahead was examined first, and this showed that the majority of vehicles subject to these accidents were moving vehicles that were being followed from behind (Fig. 6). This finding contradicts Fig. 5, discussed above. Further investigation of the main causes showed, as in Fig. 7, that delay in recognizing deceleration and unexpected deceleration by the other vehicle were factors in approximately half of the accidents.   In other words, the drivers had not been looking to the side for prolonged periods, and were aware that a vehicle was ahead. However, their attention lapsed, and this is thought to be what made them either collide with the vehicle ahead or feel they had almost collided with the vehicle ahead.

3.2. SIMULATION ANALYSIS

The degree of forward inattention involved when accidents occur was analyzed in light of the accident summary results and questionnaire survey results. In other words, although it appears inevitable for rear-end collisions to occur when inattention is prolonged for a considerable time due to diminished wakefulness or other such reasons, the large number of accidents makes it unlikely that so many drivers experience such prolonged inattention. The questionnaire survey also failed to uncover such a causal factor.

Consequently, simulation was used to explore the possibility that the accidents could occur due to short-term or momentary inattention.

272

It was supposed from the questionnaire results that the most frequent accident scene involved a driver following behind a vehicle moving ahead, and the vehicle moving ahead decelerated during momentary inattention by the driver.

The conditions were set as follows: The headway time between the vehicle moving ahead and the following vehicle (i.e., relative distance/speed of subject vehicle) was 1.5 seconds. The vehicle moving ahead would decelerate (deceleration of 0.3 G) at the moment when the driver of the subject vehicle was inattentive. After a set period of inattention, the driver would look ahead attentively and undergo recognition, judgment, and deceleration operation.

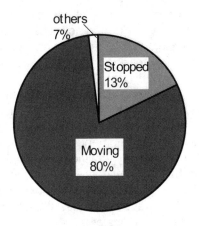

Fig.6 State of vehicle ahead during rear-end collision (according to questionnaire)

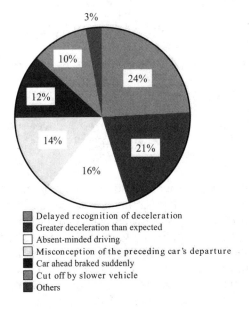

Fig.7 Factors involved in rear-end collisions (according to questionnaire)

During this process, the driver's response time (from initial attention to the initiation of deceleration) was set at 1.3 seconds[3][4], and the deceleration was set at 0.8 G during the deceleration operation. The simulation was conducted

with three different levels of driver inattention time at 1sec., 1.5sec., and 2.0sec.,respectively.

Fig. 8 shows the simulation results. The horizontal axis is the velocity of the following vehicle, and the vertical axis is the relative collision velocity. It is apparent from Fig. 8 that, although the length of inattention makes some difference, rear-end collisions take place when cruising at low to medium speeds while at higher speeds the driver decelerates so that rear-end collision does not take place. This result matches the tendency shown in Fig. 2 for more accidents to occur at low to medium speeds. The portion in Fig. 8 marked by bold line shows those cases where the vehicle moving ahead decelerates to a stop, after which the subject vehicle collides it. In other words, this shows that rear-end collisions with a stationary vehicle occurs at following vehicle speeds of 40 km/h and below.

In the accident statistics, such cases of rear-end collision with a cruising vehicle that has come to a stop are presented as rear-end collisions with a stationary vehicle. According to this simulation, a rear-end collision occurs while cruising at low to medium speed, and the status is collision with a stationary vehicle. These results, therefore, are in line with the tendencies shown above in Fig. 2 and Fig. 5. This also agrees with the tendency found in the questionnaire results, in which the scene of a rear-end collision occurring when cruising behind the vehicle ahead is found frequently. The results in Fig. 5 and Fig. 6 were earlier found to be in disagreement. In light of the above, however, this difference can be considered to arise from application of the statistics, and the actual accident scenes represented are similar.

In terms of the accident scene, the simulation was conducted on the assumption of a scene of rear-end collision with a stationary vehicle that had already come to a stop.

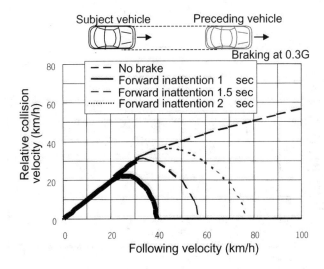

Fig.8 Results of simulation (while following)

In this case, the simulation takes the number of seconds by which the stationary vehicle is recognized in advance as one of its parameters. Fig. 9 shows the results on the condition that the driver, after recognizing the stationary vehicle, performs the deceleration operation with a specified response time (1.3 seconds, as discussed above). It is apparent from the results that rear-end collision does not occur when cruising at low to medium speeds if recognition occurs 2–2.5 seconds in advance of such collision. In this case, a driver must experience a relatively prolonged inattention regarding a vehicle that is already stationary, and must first recognize that vehicle approximately 2 seconds before collision. Otherwise a rear-end collision will not take place in the low to medium speed range. Consequently, it is very unlikely that there are many drivers to whom this would apply.

The above simulation results indicate that prolonged inattention alone is not necessarily a major factor in rear-end collisions with stationary vehicles. It was determined that rear-end collisions can occur instead due to momentary inattention while cruising at low to medium speeds.

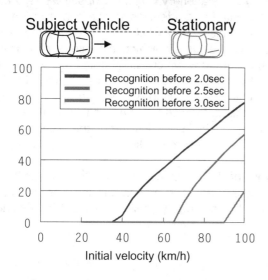

Fig.9 Results of simulation (with stationary vehicle)

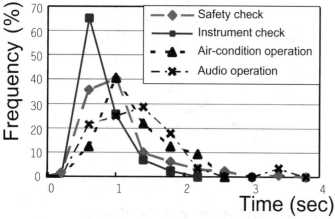

Fig.10 Forward inattention duration

3.3. DURATION OF INATTENTION WHILE DRIVING

An eye camera was then used to measure the extent of forward inattention that takes place during daily driving (Fig. 10). This measured the length of time that the driver's eyes moved away from the road ahead while the driver carried out ordinary activities such as checking for safety and operating audio equipment under cruising conditions. The horizontal axis shows the time of inattention and the vertical axis shows the frequency of inattention. Fig. 10 makes clear that momentary inattention about 1–2 seconds in duration occurs even during ordinary operations, such as operating the stereo or instrument check. In other words, the simulation showed that rear-end collisions can occur if the vehicle ahead decelerates during a period of forward inattention when the driver of the subject vehicle is engaging in everyday, frequently performed activities of this kind. This may be the reason that many rear-end collisions occur when cruising at low to medium speeds.

The objective of this research was defined, based on the above analysis, as development of a collision avoidance assist system that would be effective in averting rear-end collisions caused by momentary inattention such as occurs even during daily driving. The objective was to create a system that would not be conducive to unsafe driving and that would not invite driver over-confidence by making this collision avoidance assist system rather than an automatic avoidance system.

4. SYSTEM CONCEPTS

4.1. THE CONCEPT OF BRAKE CONTROL

In the case of a collision avoidance assist system, one first thinks of a warning system that assists driver recognition by providing a warning using sound or other such means at times of danger. Such a system must be made to time the warning to allow the driver a margin of time in which to make a judgment and carry out an operation after hearing the warning.

A driving simulator was used to study driver response times to warnings. The simulator was set up to make the driver look to the side while cruising behind another vehicle. At that point the forward vehicle would decelerate and a warning would be issued 2 seconds before the predicted collision. Fig. 11 shows the response time between the point when the driver hears the warning and begins to apply the brakes. This shows that a time of about 0.6–1.4 seconds is required for response. A warning system, therefore, must issue its warning with a timing that takes account of the time required for drivers themselves to decelerate their vehicle plus this anticipated response time.

When the time margin is made larger to accommodate a larger number of drivers, however, there is a possibility that warnings will be issued frequently when conditions are normal. There is concern, therefore, that warnings will lose their original significance as drivers grow accustomed to hearing them. On the other hand, if the timing of warnings is delayed, then there is a smaller margin of time for the driver to make a decision and then take action after hearing the warning. It is possible that, as a result, the warnings would not be fully effective.

Here, therefore, brake control was employed in order to make this a collision avoidance assist system that would be effective at times of danger without interfering with drivers under ordinary conditions.

In other words, the rationale was to time warnings so that they would interfere with the driver as little as possible, and in this way the warnings would not lose their original significance. The system was also structured to use brake control in order to make up for delays in driver judgment and operation.

4.2. STUDY OF THE CONTROL TIMING

In general, the distance required to avoid collision using braking alone is longer than the distance required to avoid collision by steering operation. The attempt to achieve certain collision avoidance using brake control alone may, therfore, interfere with the driver's avoidance steering operations.[5]

In considering the system operation timing, therefore, tests were conducted to measure the timing of driver avoidance steering operations. Fig. 12 presents the measurements of the distance from the vehicle ahead at the point that avoidance operation by steering is initiated when the subject vehicle is approaching the vehicle ahead at a certain relative velocity.

The horizontal axis shows the relative velocity, and the vertical axis shows the distance when avoidance operation is initiated. The various symbols in the graph indicates points at which drivers felt they were performing an ordinary avoidance operation, a somewhat dangerous avoidance operation, and considerably dangerous avoidance operation. The lines in the graph represent the margin of time before collision (i.e., relative distance/relative velocity).

Given the above findings, the following two points regarding the timing for warning issuance were considered for this system. First, the warnings should not sound so frequent that drivers end up becoming accustomed to them, and the warnings lose their original significance. Second, drivers should not be allowed to rely excessively on warnings to the extent that they think it is safe to be inattentive until a warning is issued. With this in mind, the system was set not to issue warnings frequently during ordinary driving, with a primary warning set to be issued approximately 3 seconds before interruption of an ordinary avoidance operation, and a secondary warning set to be issued approximately 2 seconds before a predicted collision, timed so that it would be issued to cause drivers to feel they were performing a somewhat dangerous avoidance operation. The timing for brake control (emergency braking) was set to correspond to the considerably dangerous avoidance operation so that it would cause almost no interference with driver operations. In other words, it was set at approximately 1 second before a predicted collision, and this timing would be experienced extremely infrequently during daily driving.

4.3. THE CONCEPT OF BRAKE CONTROL SETTINGS

The timing for issuance of the secondary warning is at the point that interrupts the somewhat dangerous avoidance operation. This is a point where the danger of rear-end collision has grown greater, so the warning should be definitely recognized by the driver.

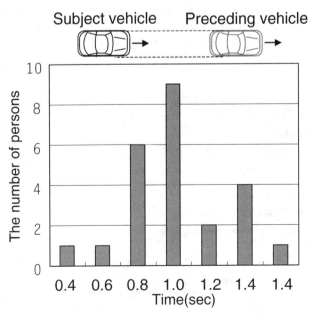

Fig.11 Driver's response time (by driving simulator)

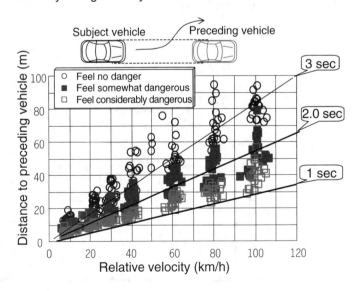

Fig.12 Avoidance operation timing test

Therefore, rather than issuing the warning by sound alone, the method of light braking is also used at the same time in order to convey the warning by physical channel.

This braking is termed alarm braking. After examination in various ways and using repeated testing, it was determined that the appropriate intensity (deceleration) for the alarm brake was about 0.1–0.2 G. This setting enables an enhanced warning effect to the driver in addition to the sound warning. This is also a setting that will not interfere with operations by drivers who attempt avoidance by steering at that point.

There was a notion that deceleration from the braking (termed emergency braking) that operates approximately 1 second before a predicted collision should be set at the maximum value for brake control because this was used in circumstances where a rear-end collision was predicted. It was confirmed through testing, however, that collisions could be avoided by using a combination of deceleration plus steering avoidance operation, to include recognition assistance from alarm braking. After repeated testing under a variety of conditions, it was decided that deceleration would be set within a control range with a maximum of 0.6 G.

The system is configured so that the amount of braking force applied by drivers who recognize danger and apply the brakes will be exerted in addition to the braking force exerted by the system's brake control.

The questionnaire results showed that when drivers caused rear-end collisions or felt they were about to cause rear-end collisions, they stepped on the brakes approximately 70% of the time (Fig. 13).

Therefore, the combination of driver operation and system brake control is likely to easily cause maximum deceleration. The use of brake control for deceleration also increases the margin of time before collision, and so increases the margin of time in which drivers can make a judgment and carry out an operation. This system can be expected to have the effect of facilitating braking operations even for drivers whose response times were previously too slow to allow them to make a judgment and carry out an operation. Fig. 14 shows the sequence of operational modes.

5. SYSTEM OVERVIEW

5.1. SYSTEM CONFIGURATION

Fig. 15 shows the system configuration. Millimeter wave radar sensor was employed as the sensor for forward vehicle detection because of its stable detection performance in different weather conditions. The adaptive cruise control (ACC) systems that various automobile manufacturers already have on the market mostly use relatively low-cost laser radar as the sensor for forward vehicle detection. It is characteristic of laser radar, however, that it is easily affected by rain, snow, fog, sunlight, soiling with mud, and so on, so that it can only be used in certain limited environments. The collision avoidance assist system developed here must be capable of stable operation in a variety of environments. Millimeter wave radar sensor was adopted, therefore, because it is typically less affected by rain, snow, fog, and other such conditions in the natural environment.

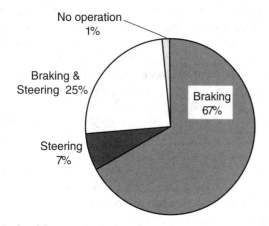

Fig.13 Avoidance operation during rear-end collision

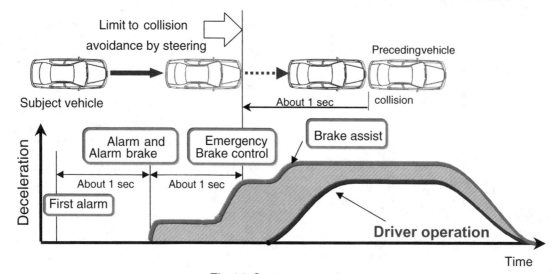

Fig.14 System operation

Table 1 shows the major specifications of the millimeter wave radar sensor used for the present system. In terms of ranging performance, this radar has the sensitivity to detect a passenger vehicle at 100 m or more. This makes it capable of detection approximately 3 seconds in advance when moving at a relative velocity of 100 km/h. With a detection area that ranges 16° to left and right, the sensor detects vehicles that are ahead on straight roads and gentle curves.

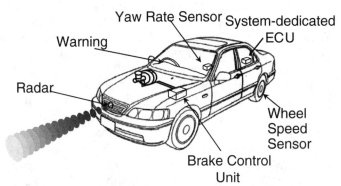

Fig.15 System configuration

Table.1 Major specifications of radar sensor

Frequency	76-77GHz
Waveform	FMCW
Detection Area	Horizontal : 16deg
	Vertical : 3.8deg
Range	4–120m
Range Accuracy	± 1.0m, ± 5%
Range Rate	±200km/h
Range Rate Accuracy	± 5%
Data Update Rate	0.1 sec
Target Tracking	8 Targets
Package Size	106(W)*8 8(H)*8 6(D)mm
Weight	0.65k

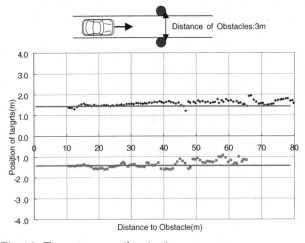

Fig.16 Target separation test

The system targets stationary vehicles in addition to moving vehicles. Therefore it must distinguish vehicles from many roadside objects such as signs, guardrails, and so on. A scanning type radar sensor is used so that the position of a stationary vehicle can be recognized accurately even in the kind of road environment described above. Fig. 16 shows the results of target separation testing of the radar sensor. Fig. 16 shows two obstacles position of detection by radar sensor, when the subject vehicle closes to center of two obstacles that located between 3meters. The vertical axis shows side position from subject vehicle to obstacle. The horizontal axis shows distance from subject vehicle to obstacle. Then the radar sensor detects two obstacles separately.

The subject vehicle is supposed to measure its own state quantities, and so it is equipped with a wheel speed sensor to measure the vehicle speed, a yaw rate sensor to measure turning, and a steering angle sensor to measure steering wheel operation. The subject vehicle estimates its course as shown in Fig. 17, and calculates its estimated lateral travel distance at the obstacle location.

It is also equipped with an alarm unit that provides warnings to the driver when the system is operating. A hydraulic actuator is also installed in order to carry out brake control. The hydraulic unit utilizes the brake actuator of the vehicle stability assist (VSA) system.[6] It is equipped with a pressure control valve to allow variation of the braking pressure.

5.2. CONTROL LOGIC

Fig. 18 shows the basic control flow. The system recognizes the vehicle ahead by radar sensor, and derives the relative distance (X1) from the subject vehicle, the relative velocity, and angle of direction. The subject vehicle's course is estimated from its state quantities, and the lateral travel distance is calculated. The system uses the obstacle location and estimated lateral travel distance of the subject vehicle to determine the amount of overlap of the subject vehicle and the obstacle. It then calculates the lateral travel distance (Y) required in order to avoid the obstacle by steering.

The system next calculates the relative distance (X2) required in order to accomplish that lateral travel. It judges the degree of danger from the calculated distance and from the relative distance, and controls the brakes accordingly. Here, brake control is initiated when X1 is smaller than X2.

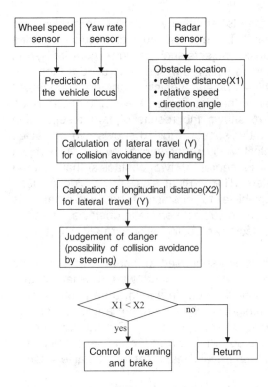

Fig.18 Block diagram of basic control

The location of the subject vehicle relative to the vehicle ahead may be fully overlapping when the subject vehicle is directly behind the forward vehicle, or it may be offset. These differ in the lateral travel distance to be accomplished by steering, and the system therefore makes its judgment by substituting the difference in lateral travel distances for the danger.

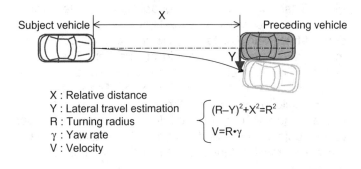

X : Relative distance
Y : Lateral travel estimation
R : Turning radius
γ : Yaw rate
V : Velocity

$$(R-Y)^2 + X^2 = R^2$$
$$V = R \cdot \gamma$$

Fig.17 Prediction of the vehicle locus

In other words, when the lateral travel distance is slight during an offset position, X2 can be set at a smaller value than calculated value that is when the position is fully overlapping (when the speeds are the same). This prevents interference with avoidance operations by the driver.

6. SYSTEM EFFECTIVENESS

6.1. EFFECTS IN SIMULATION

Simulation was used to examine the effects of the system in operation. The response time of 1 second was used for the interval from the time the driver hears the secondary warning until the brake is operated. This figure is the average of the response time results from the driving simulator shown in Fig. 11. The deceleration by emergency braking during system operation was set to a maximum of 0.6 G.

The simulation was carried out on the assumption that the vehicle ahead decelerates when the subject vehicle is cruising behind it. The test conditions were a headway time of 1.5 seconds when the subject vehicle is following, and a deceleration of 0.3 G by the vehicle moving ahead. The following state was fully overlapping, with the subject vehicle cruising directly behind the vehicle moving ahead. When the vehicle moving ahead decelerates, a relative velocity is produced between it and the subject vehicle, and their relative distance grows shorter. The system begins to operate when the danger of collision becomes great. The driver is considered to operate the brakes 1 second after the secondary warning is issued. The deceleration during this driver operation is set at 0.8 G. The results are shown in Fig. 19. The horizontal axis is the following velocity, and the vertical axis is the relative velocity at the time of collision. With this system, rear-end collisions are avoided when the subject vehicle is following behind at low to medium speeds, and it is clear that the collision velocity is reduced at the higher following speeds. Given these results and the results in Fig. 2, the system can be considered capable of greatly reducing rear-end collisions caused by momentary inattention.

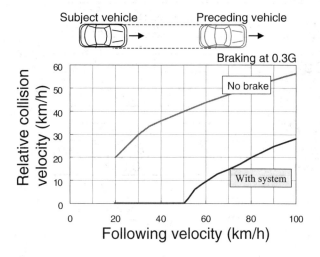

Fig.19 System effect (by simulation)

6.2. RESULTS OF TESTS WITH ACTUAL VEHICLE

Actual vehicle was next used to measure the reduction effect on collision velocity relative to a stationary object. The test was carried out by placing an obstacle representing a vehicle directly ahead of the subject vehicle.

The subject vehicle cruises toward it, and the system begins to operate when the danger of collision has become great. Here, again, the driver operates the brakes approximately 1 second after the secondary warning. Fig. 20 shows the test results. The horizontal axis is the velocity of the subject vehicle during its approach, and the vertical axis is the velocity at the time of collision. It is apparent from the results that collision is avoided at low to medium cruising speeds when the relative velocity is low, and the collision velocity is reduced at higher cruising speeds when the relative velocity is higher.

In either case, the low to medium speed range is the range within which the majority of accidents occur, as shown in Fig. 2. Therefore the system is thought to be useful.

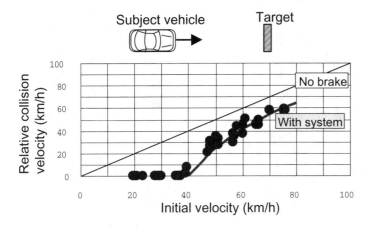

Fig.20 System effect (by actual test)

7. CONCLUSION

A system was developed that is effective in reducing the collision velocity during rear-end collisions by making up for driver delays in recognition, judgment, and operation by means of brake control and external sensing technology using millimeter wave radar sensor.

The following results were obtained:

1. We found that rear-end collisions can occur during ordinary driving as a result of momentary inattention.
2. A system that prevents interference with driver operation and assists in avoidance operations was created.
3. This system was determined to be effective in reducing the number of accidents at low to medium speeds, which make up the greater part of rear-end collisions.

In the case of a collision avoidance assist system such as this, changes in the consciousness of system users must also be taken into consideration. The present system is above all a collision avoidance assist system intended to make up for driver delays in deceleration. It is not an automatic avoidance system. Drivers must not be allowed to grow dependent on such systems, since they would then have the reverse effect of promoting unsafe driving.

Due to limits in sensing capability and other such factors, the present system is not able to prevent all types of rear-end collision. It is also necessary, therefore, to convey the system limits in clear, understandable terms. In this light, it is important for drivers to recognize the fundamental rule that driving is still their responsibility, as it always has been.

REFERENCES

1. Traffic statistics 1997,Institute for Traffic Accident Research and Data Analysis (ITARDA),May 1998
2. Verification of rear-end collision accidents, Signal, October 1997,pp.2-11
3. George T. Taoka: Brake Reaction Times of Unalerted Drivers, ITE Journal March 1989,Vol.59,No.3, pp.19-21
4. Nancy L. Broen, et al.: Braking response times for 100 driver in the avoidance of an unexpected obstacle as measured in a driving simulator: Proceedings of the human factors and ergonomics society 40th annual meeting, 1996,pp.900-904
5. Yasuhiko Fujita☐Kenji Akuzawa, Makoto Sato: Radar Brake System. Society of Automotive Engineers of Japan Conference Preprint Collection (9437845), No. 186, 1994
6. Horiuchi, Y., et al., VSA-Vehicle Stability Assist (in Japanese), Automotive Technology, May Vol. 47, No. 7, pp. 36-43 (1998)

Research on a Braking System for Reducing Collision Speed

Yoji Seto, Takayuki Watanabe, Teruo Kuga, Yoshinori Yamamura and Kenichi Watanabe
Nissan Motor Co., Ltd.

ABSTRACT

An investigation was made of the relationship between the driving speed at the time of impact and the injury levels suffered in accidents. The results showed that a 5 km/h or more reduction in collision speed tends to mitigate injury severity. Using sensors and brake actuators already in practical use, we have started to research a braking system aimed at reducing the collision speed by at least 5 km/h in rear-end collisions. The system estimates the risk of a collision with the vehicle ahead. If it judges there is a very high possibility of a collision, it applies the brakes.

INTRODUCTION

Progress has been made in recent years in the development of systems that measure the distance to a forward object and control the behavior of the host vehicle accordingly. Adaptive cruise control (ACC) systems that automatically control the distance to the vehicle ahead have already been implemented on production vehicles. A lot of work is also being done on the development of forward collision avoidance systems for preventing car accidents by avoiding collisions with objects in front of the host vehicle [1]. To accomplish collision avoidance, such systems require sensors for reliably detecting forward obstacles sufficiently in advance and actuators capable of producing the high deceleration needed to avoid a collision. However, collision avoidance systems have yet to be implemented on production vehicles owing to unresolved technical issues with respect to both sensing and actuation [2]. Therefore, we initiated a study aimed at the early deployment of a system for reducing even by a little the severity of injuries through the application of sensing technologies and actuators already in use.

In Japan, rear-end collisions and broadside collisions tend to account for a high percentage of all traffic accidents [3]. Technology for detecting a preceding vehicle has already been implemented in ACC systems and development work continues to move ahead. Accordingly, the system described here is aimed at reducing the severity of injuries in rear-end collisions that account for a high percentage of all traffic accidents.

We investigated the relationship between the collision speed and the severity of the injuries suffered. We wanted to predict what effect a certain reduction in vehicle velocity might have on reducing injury severity. As the velocity standard, we used the equivalent barrier collision speed and, as the standard for injury severity, we used the Japanese Abbreviated Injury Scale (JAIS).

The equivalent barrier collision speed is calculated by converting the vehicle weight and collision speed, which are different in every accident, into an equivalent speed for a collision with a barrier having an infinite mass. In a situation where vehicle A runs into the rear-end of vehicle B, the equivalent barrier collision speed of vehicle A, v_E, is given by the following equation [4]:

$$v_E = |v_A - v_B| \frac{M_B}{M_A + M_B} \qquad (1)$$

where

v_A, v_B : velocity of vehicle A and vehicle B

M_A, M_B : mass of vehicle A and vehicle B

As outlined in Table 1, JAIS defines the criteria for injury severities in vehicle accidents [4].

Table 1 JAIS (Japanese Abbreviated Injury Scale)

JAIS Code	Injury severity
0	Not injured
0.5 - 1.0	Minor injury
1.5 - 2.0	Moderate injury
2.5 - 3.0	Serious injury
4.0	Severe injury
5.0	Critical injury
6.0 - 9.0	Death
9.9	Unknown

The relationship between the collision speed and injury severity in vehicle collisions is shown in Fig. 1 [5]. The horizontal axis in this figure shows the average equivalent barrier collision speed and the vertical axis shows JAIS. Although the data show considerable scatter, a general tendency is seen for JAIS values to decrease by approximately one point when the collision speed is reduced by 5 km/h or more. In view of this tendency, the aim set for this system was to reduce the collision speed by at least 5 km/h.

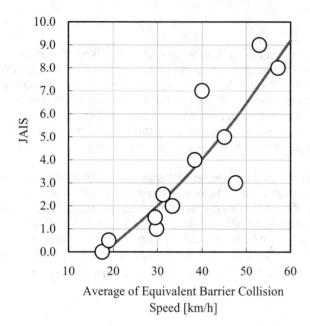

Fig. 1   Relationship Between Collision Speed and JAIS Values

SYSTEM CONFIGURATION

The system configuration is shown schematically in Fig. 2. It mainly consists of a laser radar, a vacuum-controlled brake actuator, wheel speed sensors and a controller, all of which are connected via a Controller Area Network (CAN). The laser radar and brake actuator have already been implemented on production vehicles as components of an ACC system.

Based on the headway distance and velocity of the host vehicle, the controller estimates the risk of a collision with the vehicle ahead. In the event it judges there is a very high possibility of a collision, the controller sends a brake pressure command signal to the brake controller to actuate the braking system for reducing the host vehicle's velocity.

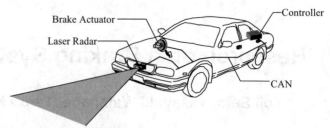

Fig. 2   System Configuration

BRAKING PATTERN AND BRAKE ACTUATION JUDGMENT

BRAKING PATTERN

This system actuates the brakes according to a pattern like that shown in Fig. 3 in order to obtain the desired speed reduction while giving utmost priority to the driver's intentions.

(1) If the system judges that the host vehicle is closing too rapidly on the vehicle ahead, it applies light braking to reduce the velocity by approximately 1 m/s^2. The driver is not likely to notice that level of deceleration. In the first part of the time line shown in Fig. 3, this light braking action is referred to as pre-braking. The purpose of pre-braking is to enhance the effective reduction of the collision speed by applying light braking force approximately one second before a judgment is made that there is a very high possibility of a collision with the vehicle ahead.

(2) The system applies braking force to reduce the host vehicle's velocity at a rate of approximately 2.5 m/s^2 only in the event that it judges there is a very high possibility of a collision with the vehicle ahead. In the latter part of the time line in Fig. 3, this is referred to as main braking. The magnitude of this main braking is equivalent to deceleration of 2.5 m/s^2, based on a consideration of the desired speed reduction in rear-end collisions, the capacity of the brake actuator and the results of a subjective evaluation made by a skilled driver.

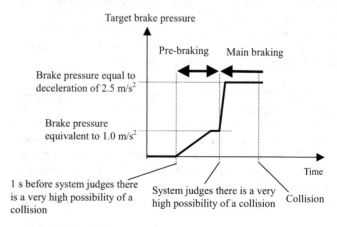

Fig. 3   Braking Pattern

CALCULATION OF THE DISTANCE FOR JUDGING THERE IS A VERY HIGH POSSIBILITY OF A COLLISION

This section explains the procedures for calculating the distance for judging that there is a very high possibility of a collision and the distance for judging a time of one second before there is such a possibility.

(1) Calculation of the distance for judging a collision is unavoidable by evasive steering

It is assumed that the driver will execute an evasive steering maneuver in t_{pre} s. In this case, the distance for judging that a rear-end collision is unavoidable, d_{str_avoid}, is given by the following equation:

$$d_{str\_avoid} = -\frac{1}{2}a_r(0)\cdot\left(t_{str}+t_{pre}\right)^2 - v_r(0)\cdot\left(t_{str}+t_{pre}\right) \quad (2)$$

Letting t_{pre} = 1.0 s, we calculate the distance one second before a rear-end collision becomes unavoidable by evasive steering, and when t_{pre} = 0.0 s, we calculate the distance at which a rear-end collision cannot be avoided by evasive steering.

where

$a_r(0)$: present relative acceleration

$v_r(0)$: present relative velocity

t_{str} : time needed to avoid a collision by evasive steering

The relative velocity is calculated with the following equation from the headway distance, and the relative acceleration is then calculated from the relative velocity.

$$v_r(s) = \frac{\omega_{vr}^2 s}{s^2 + 2\xi_{vr}\omega_{vr}s + \omega_{vr}^2}\cdot d_r(s) \quad (3)$$

$$a_r(s) = \frac{\omega_{ar}^2 s}{s^2 + 2\xi_{ar}\omega_{ar}s + \omega_{ar}^2}\cdot v_r(s) \quad (4)$$

The time needed to avoid a collision by evasive steering, t_{str}, is found with Eq. (5) using the driver's evasive steering characteristics [6-7] and the steering response characteristics of the host vehicle.

$$t_{str} = f_{str\_characteristic}\left(y_{avoid}\right) \quad (5)$$

where y_{avoid} is the amount of lateral movement needed to avoid a rear-end collision by evasive steering and is calculated with Eq. (6) based on the angular range, θ_1-θ_2, in which the laser radar detects the preceding

vehicle and the headway distance, d_r, as shown in Fig. 4.

$$y_{avoid} = \min\left(\frac{w_{body}}{2} - d_r\tan\theta_1 , \frac{w_{body}}{2} + d_r\tan\theta_2\right) \quad (6)$$

where w_{body} is the width of the host vehicle.

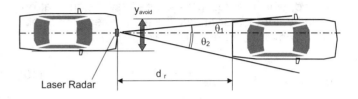

Fig. 4   Lateral Movement Needed to Avoid a Collision by Evasive Steering

(2) Calculation of the distance for judging a collision is unavoidable by braking action

It is assumed that the driver will execute evasive braking at a level of $\hat{a}_h$ in t_{pre} s. In this case, the distance for judging that a rear-end collision is unavoidable is found with the following expression:

$$d_{brk\_avoid} = \frac{v_r(t_{pre})}{2\hat{a}_r(t_{pre})} - \frac{1}{2}a_r(0)\cdot t_{pre}^2 - v_r(0)\cdot t_{pre} \quad (7)$$

Letting t_{pre} = 1.0 s, we calculate the distance one second before a rear-end collision becomes unavoidable by evasive braking, and when t_{pre} = 0.0 s, we calculate the distance at which a rear-end collision cannot be avoided by evasive braking.

where

$v_r(t_{pre})$: relative velocity at t_{pre}

$\hat{a}_r(t_{pre})$: estimated relative acceleration at t_{pre}

The relative velocity at t_{pre} s is given by

$$v_r(t_{pre}) = v_r + a_r(0)\cdot t_{pre} \quad (8)$$

The estimated relative acceleration at t_{pre}, assuming that the preceding vehicle continues to decelerate at the same rate of $a_t(0)$ and that the host vehicle decelerates at $\hat{a}_h$ in t_{pre} s, is found with the following equation:

$$\hat{a}_r(t_{pre}) = a_t(0) - \hat{a}_h$$
$$= a_r(0) + a_h(0) - \hat{a}_h \qquad (9)$$

where

$a_h(0)$: present acceleration rate of the host vehicle

From Eqs. (2) and (7), we obtain the following expression for the distance at which pre-braking is initiated, which represents the headway distance at one second before a rear-end collision becomes unavoidable either by evasive steering or braking.

$$d_{pre\_brk\_on} = \min\left(\left. d_{brk\_avoid} \right|_{t_{pre}=1.0} , \left. d_{str\_avoid} \right|_{t_{pre}=1.0} \right) \quad (10)$$

The distance for the onset of main braking, i.e., the distance at which a collision becomes unavoidable either by evasive steering or braking, is given by

$$d_{main\_brk\_on} = \min\left(\left. d_{brk\_avoid} \right|_{t_{pre}=0.0} , \left. d_{str\_avoid} \right|_{t_{pre}=0.0} \right) \quad (11)$$

Pre-braking is initiated in the event $d_r \le d_{pre\_brk\_on}$ and main braking is initiated if $d_r \le d_{main\_brk\_on}$.

SIMULATION RESULTS

SIMULATION MODEL

The control logic explained in the preceding section was incorporated into the simulation model shown in Fig. 5, and the effect on reducing the collision speed was calculated in two types of rear-end collision situations.

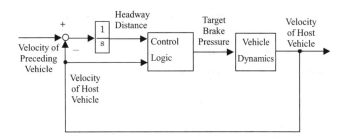

Fig. 5   Simulation Model

CALCULATED RESULTS FOR A REAR-END COLLISION SITUATION INVOLVING A VEHICLE AHEAD TRAVELING AT A CONSTANT VELOCITY

Figure 6 shows the calculated effect of the system on reducing the collision speed in a rear-end collision situation where the host vehicle, traveling at a velocity of 40 km/h, closes on a vehicle ahead that is traveling at a steady velocity of 10 km/h. The upper graph shows the velocity of the host vehicle. In the middle graph, the

boldface, dashed and solid lines show the headway distance, the distance for the onset of pre-braking and the distance for the onset of main braking, respectively. In the bottom graph, the solid line indicates the target brake pressure and the boldface line indicates the actual brake pressure.

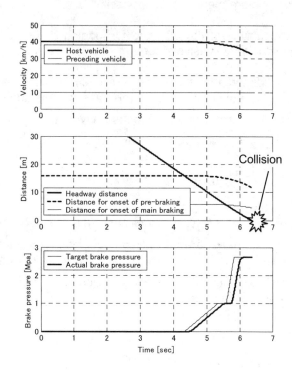

Fig. 6   Effect on Reducing Collision Speed with a Vehicle Ahead Traveling at a Constant Velocity

It is seen that pre-braking was applied initially at the point where the headway distance $\le$ the distance for the onset of pre-braking. Then, main braking was applied at the point where the headway distance $\le$ the distance for the onset of main braking, and ultimately a rear-end collision occurred when the headway distance = 0. The host vehicle's velocity at the time the headway distance = 0 was approximately 33 km/h. It is clear that the operation of the system had the effect of reducing the rear-end collision speed by approximately 7 km/h.

CALCULATED RESULTS FOR A REAR-END COLLISION SITUATION INVOLVING A DECELERATING VEHICLE AHEAD

Figure 7 shows the calculated results for a rear-end collision situation in which the vehicle ahead decelerated at a rate of 3 m/s^2 while the host vehicle was following it at a velocity of 40 km/h and at a constant headway distance. The host vehicle's velocity at the time the rear-end collision occurred was approximately 32 km/h. This indicates that the operation of the system was effective in reducing the rear-end collision speed by approximately 8 km/h.

284

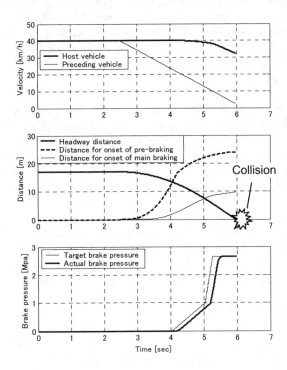

Fig. 7   Effect on Reducing Collision Speed with a Decelerating Vehicle Ahead

EXPERIMENTAL RESULTS

Experiments were conducted in which a sensor at the end of a bar attached to the host vehicle was crashed into a vehicle ahead to validate the speed reduction effect predicted in the simulations.

OVERVIEW OF EXPERIMENTAL PROCEDURE

The experiments were conducted with the setup shown schematically in Fig. 8 to simulate rear-end collision situations. A laser radar was positioned at the tip of a bar that was attached to the right side of the body of the host vehicle. When the laser radar was crashed into the vehicle ahead, the bar bent so as not to damage the struck vehicle.

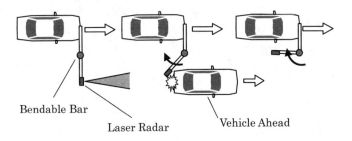

Fig. 8   Experimental Setup

EXPERIMENTAL RESULTS FOR A REAR-END COLLISION SITUATION INVOLVING A VEHICLE AHEAD TRAVELING AT A CONSTANT VELOCITY

Figure 9 shows the experimental results for a rear-end collision situation where the host vehicle was traveling at a velocity of 40 km/h and the vehicle ahead was traveling at a velocity of 10 km/h. The operation of the system reduced the velocity of the host vehicle from 40 km/h to approximately 32 km/h, for a reduction in rear-end collision speed of about 8 km/h.

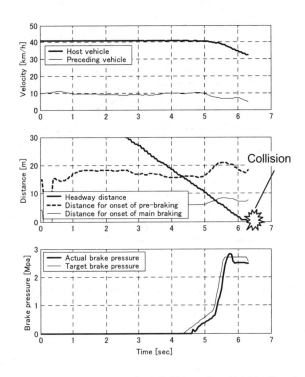

Fig. 9   Experimental Results for a Collision with a Vehicle Ahead Traveling at a Constant Velocity

EXPERIMENTAL RESULTS FOR A REAR-END COLLISION SITUATION INVOLVING A DECELERATING VEHICLE AHEAD

Figure 10 shows the experimental results for a rear-end collision situation in which the vehicle ahead decelerated at a rate of 3m/s^2 while the host vehicle was following it at a velocity of 42 km/h and at a constant headway distance. The system effectively reduced the velocity of the host vehicle from 42 km/h to approximately 34 km/h, for a reduction in rear-end collision speed of about 8 km/h.

However, even though the experiments were conducted under nearly the same conditions every time, there were a few cases where the collision speed was not reduced as expected. That was attributed to a delay in the onset of braking action because the sensor was slow in detecting the vehicle ahead. This variation in the detection performance of the sensor with respect to the obstacle in front of the host vehicle produced a similar variation in the effect of the system on reducing the collision speed. Further work must be done in the future to improve and stabilize the performance of the system.

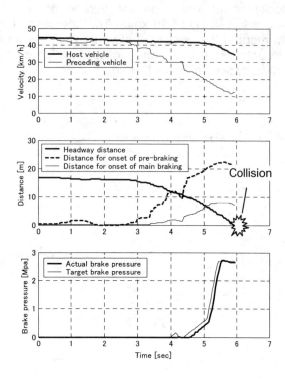

Fig. 10   Experimental Results for a Collision with a Decelerating Vehicle Ahead

SUMMARY

Research has been undertaken on a braking system aimed at reducing the collision speed in rear-end accidents by at least 5 km/h. One reason for undertaking this research is that rear-end collisions account for a high percentage of all traffic accidents. Another reason is that JAIS (Japanese Abbreviated Injury Scale) values tend to be reduced by one point when the collision speed is lowered by 5 km/h or more. To promote its early implementation, this braking system has been configured with sensor and actuator technologies already in use on production vehicles.

In order to give utmost priority to the driver's intentions, the system is designed to operate only when it judges that there is a very high possibility of a collision with the vehicle ahead based on the headway distance and velocity of the host vehicle.

It was confirmed in simulations that the system reduces the rear-end collision speed by approximately 7-8 km/h. In experiments, the collision speed was effectively reduced by approximately 8 km/h when the sensor

detected the vehicle ahead properly. However, there were also cases where the expected reduction of collision speed was not obtained because the sensor was slow to detect the vehicle ahead. In future work, the system will have to be improved further to stabilize its performance.

It will be noted that this system is intended to reduce the collision speed rather than to avoid collisions entirely. Additionally, there are limitations on the sensor's ability to detect objects in the surrounding environment. Accordingly, drivers will still have to pay constant attention to the driving environment as they operate their vehicles, just as they have done without this system.

In future work, efforts will be directed at reducing the severity of injury further by improving the performance of the sensor and actuators and at applying the system to other types of accidents besides rear-end collisions.

REFERENCES

1. Watanabe, K., et al., "Development of Forward Obstacle Collision Prevention Support System in ASV2," Proc. of JSAE, No. 58-01, 2001, pp. 9-12 (in Japanese).
2. Kodaka, K., et al., "Rear-end Collision Velocity Reduction System," Honda R&D Technical Review, Vol. 13, No. 1, 2001, pp. 159-166 (in Japanese).
3. Institute for Traffic Accident Research and Data Analysis, "Traffic Accident Statistics 1998," 1998 (in Japanese).
4. Sato, T., "Traffic Engineering 9, Traffic Accidents and Their Research," 1987 (in Japanese).
5. Regional Transport Bureau, Ministry of Transport of Japan, "FY 1989 Project Report: Research and Analysis Concerning Automobile Structures/Equipment and Occupant Injuries," 1989 (in Japanese).
6. Uno, H., et al., "Older Drivers' Avoidance Capabilities in Emergency Traffic Situations," Proc. of JSAE, No. 61-00, 2000, pp. 5-8 (in Japanese).
7. Tsutsui, S., et al., "Characteristics of Drivers' Avoidance Maneuvers," Toyota Technical Review, Vol. 45, No. 1, 1995, pp. 55-59 (in Japanese).

The Use of Fuzzy Controller for Optimizing the Brake Performance

Seyed Mohammad Reza Hashemi Mogaddam

Irankhodroo Car Manufacturing Company

ABSTRACT

Among various innovations on *Intelligent Transportation Systems ITS,* several researches are performing on implementation of *Adaptive Fuzzy Controllers* in braking systems.

The braking procedure consists of two basic consequent steps; the first step is the time interval (delay) which takes to reach an optimum deceleration, and the second is the time of actual braking. Though several efforts have been done to reduce the actual braking distance by increasing the effective contact of tires to the road surface (i.e. New generation of tiers, Anti-Lock Braking Systems,....), but this paper presents a fuzzy braking control system to enhance the performance of braking system by partial assistance of the driver in the first step of emergency braking, in this way the equipped vehicle starts to brake by a shorter delay.

INTRODUCTION

Advanced Driver Assistance Systems ADAS, involve some type of electronically assisted control in driving procedure while maintaining the driver as the best supervisor. Anti-Collision ASSISTance (Autonomous Support and Safety Intervention Sys) involve driver assistance by warning about front collision and emergency intervention on braking sys in dangerous situations on longitudinal motion.

Absorbing the kinetic energy of moving vehicles in the shortest duration of time in emergency braking is the main goal of braking system designers. Within the time of distinction of an obstacle till the efficient deceleration, there is a short but vital dead time without optimum braking; this time can be divided in two consecutive parts.

The first part is the time interval between obstacle detection (collision prediction) till brake pad activation, which is referred as *Driver Brake Reaction Time* "D.B.R.T ". Data from 321 drivers by G.Johanson and K.Rumar [2], showed that this time may vary from 0.4 to 2.7 Sec's with mean and deviation values of (1.01, 0.37). This means a traveling distance of about 30 meters for a vehicle of a speed of 118 km/h without any brake activation.

The second part is the time between brakepad touching till reaching to the nearly stable optimum and effective braking, this time is affected by several factors like environmental parameters, load parameters, brake pad activation intensity, braking system performance and response time, driver situation, and the adaptation level of driver to the braking system. This versatile time span may vary from one to a few seconds and may greatly reduce the braking performance and so increases the braking distance.

FUZZY CONTROL

Fuzzy Logic Control is a type of control, which is based on Fuzzy set theory and reasoning. In Classic set theory, an element either is or is not a member of a specific set, but in Fuzzy theory every element is related to each particular set with a *Membership Function* which can vary continuously from Zero to One. It is found that fuzzy reasoning is more likely near to human judgment style [1].

Fuzzy logic is an efficient way to put engineering expertise into technical solution, since it provides a convenient method for constructing nonlinear controllers via the use of heuristic information.

Fuzzy sys may be referred as a common language between Man-Machine [1], a Fuzzy control panel is the best choice for Human Machine Interface HMI. With this system the system engineer (or the driver) can control or tune the system behavior on his demand and his technical conceptualization.

The Fuzzy control is more user-friend, so it brings better feeling and more pleasant driving. In recent years fuzzy logic control techniques have been applied to a wide range of systems, David Elting and Mohammed Fennich in their research told that automotive systems realize superior characteristics through the use of fuzzy logic controllers [3] especially in nonlinear cases.

DRIVER EMERGENCY BRAKE REACTION

The driver normally uses the simplest type of visual information for controlling the car and brake activation. P.N Lee told that The *Time To Collision* " T.T.C " is the most important basic value, which determines the braking procedure of a normal driver [4].

Though " T.T.C " has some value in real world but the braking behavior of any driver is the result of his own conceptualization about *Collision Danger Level* "C.D.L ", for example with the same T.T.C and similar kinetic situations, the reaction and response of two different drivers are different, because two different drivers (i.e. one assured venturous, the other cautious) or even one driver on two different situations (i.e. vehicle, surface, brake performance, load, speed....) may feel two different *Collision Danger Levels*. Every driver has a specific cognition about the *Collision Danger Level* in any braking -more generally; driving- situation. This internal filling comes from personal history of each driver (i.e.: the personal collection of accident experiments). As a natural rule the driver pushes the brake pedal related to his or her own Fuzzy apperception of " C.D.L "; so a high level of " C.D " results a fast and intense brake activation.

THE CONTROL SYSTEM

The control system assists the driver by issuing rapid braking commands to the brake pad actuator. Fig 1 shows the fuzzy controller block diagram.

The processing unit evaluates the required *Deceleration Value,* this value after *Fuzzyfication* aggregates with *Fuzzy Adaptation Variables* in the *Inference Engine*. The final resultant value is the *Fuzzy Deceleration Value.* This value after *Defuzzification*, commands the actuating system (simply, a push rod connected to the brake pad) in a way to reach an acceptable Deceleration Value.

THE EVALUATION STAGE

This block is the original block, which performs the basic processing required for evaluation of brake activation signal.

TIME TO COLLISION

W.Van Winsaw and A.Heino stated that the basic parameter which evaluates the braking reaction of a driver [5] is *Time To Collision* "T.T.C"; it can be derived from several sources either computational or perceptual.

In modern vehicles with electronic distance measuring equipment, the computing system evaluates several parameters like absolute speed and acceleration, as well as other cinematic comparative parameters relative to the front vehicle or obstacle, by processing the data stream of measured values of position at any time. Though these values are sufficient for "T.T.C" calculation but it is more practical to calculate the *Required Deceleration "R.D"* value for collision avoidance.

Also it is possible to use broadcasting alarm signals from the front vehicle during braking, it can be ultrasonic, electromagnetic, infrared, or light signals transmitted from front vehicle during braking when its deceleration value is more than some threshold value. In this situation the system starts braking with a very small propagation delay after the braking of front vehicle and as a consecutive result the risk of sudden braking or collision decreases.

ADAPTATION STAGE

In this stage the evaluated value for braking is adapted to several peripheral parameters which affect the performance or efficiency of braking. Without this stage a desirable or pleasant driving or braking is impossible and the system may induce jerky deceleration [6].

VEHICLE SITUATION

Even in similar cinematic situations, safe and efficient braking may be quiet different; the optimum braking depends on based on both physical and environmental conditions such as:
-Surface friction condition as the result of tire and road interaction
-Temperature
-Road condition (downhill descending-uphill ascending-windy-...)
-Cornering situation (light cornering, heavy cornering, coming to corner, coming from corner,.....)
-Braking system sensitivity
-Accelerator pad position
-Gear position
-Speed
-Engine R.P.M
The control system can make these conditions into account by bringing them into a fuzzy block, which aggregates them to evaluate the resultant fuzzy parameter, which represents the vehicle situation effect on braking command.

DRIVING MODE

One important and essential property of each automotive control system is its good handling and derivability or its smoothness. A good automatic braking system must behave similar to the human driver and beyond that, it must act on driver desire and mimic to every individual driver regarding his mode of driving [7] at each time. Several driving modes are introduced at the following:

Full Race Mode

A driving mode for following the front vehicle with short following distance and rapid and intense brake activation when required, in this mode the driver can reduce his following distance.

Sport mode

A driving mode with moderate to high T.T.C values and rapid brake activation suit for fast drivers.

Confident mode

A driving mode with moderate T.T.C values and rapid brake activation to avoid any frontal accident.

Economic mode

A driving mode for attaining the least friction and wearing of moving parts. In this mode the best deceleration value must be derived from the wearing data of tires, brake pad shoes,... .

Comfort (Rest) mode

A driving mode in which the driver may be daisy or sleepy, so for improving his or her response, the system come into work without any consideration about *Over Reliance* avoidance [8]. This mode suggests longer brake activation duration.

Free mode

In this mode the system is brought to a situation with very smooth and loose response.

MODE SELLECTION

Since there is no exact borders between several driving modes a driver can select any mode on his demand; also a fuzzy reasoning system can diagnose the driving parameters and determine the fuzzy value of driving mode. For unprofessional use it is enough to pick only a few of above modes. For example wide throttle butterfly angles or high R.P.M and fast changes of steering wheel with fast brakes means *Sport mode*, or *in Full Race mode* one can observe fast widespread butterfly changes and rapid braking.

ACTUATING STAGE

Control of deceleration value is the final goal of braking control system. For the safety and security reasons the system is compatible with regular braking systems, so the brake activation is simply done through the brake pad with an auxiliary electromechanical module, which is connected to the pad and pushes it proportional to the control command. Thus the system only activates the brake a few seconds in advance compared with normal situation. This actuating module is an auxiliary module, which has no effect on the basic structure or performance of the braking system, so if it doesn't work well, the original system performs his duty without any noticeable trouble; only a small change in brake feeling may be sensed.

ACTUATOR

The simplest actuator module is one electromagnetic system concluding a solenoid and an iron core; the iron core pushes the brake pad proportional to its actuating current, in smooth braking the sys may cut in the engine power out.
In modern cars (A.B.S equipped vehicles), the system can control the slack value.

ACTUATOR CONTROL

Due to completely nonlinear and unstable time variant nature of conventional brake mechanisms and the deviation of it for each car, it is too difficult to implement a closed loop feedback system. Choosing a separate open loop actuating system makes it possible to adjust the actuating module for every individual braking system, also it makes it easier to tune the fuzzy controller respected to the physical parameter changes of brake due to aging and several other environmental factors. There are two common ways to determine the actuator situation; **_Direct_** and **_Indirect_** .In direct way one can measure the displacement or the force which come from brake actuator, though measuring the actuator force on brake pad may result in a better performance; but it is better to combine both force and displacement in open loop application, thus adjustment and tuning of the control parameters can be done manually or automatically based on supervised or unsupervised learning algorithms in tuning layer.

Indirect method implements the measured deceleration value induced by the actuator to adjust the actuation signal for the actuator.

CONCLUSION

The basic idea of presented paper about fast reacting to collision dangers, is a very useful idea, which can be implemented very easily. In the simplest form it can be composed of a simple electromagnetic actuator installed on a conventional brake, which reacts, according to accelerator fast releasing movement in emergency braking situations. The basic model can greatly improve the performance of booster free brakes. It reduces the length of free movement of brake pad, so it permits new design for master brake cylinder with reduced diameter and increased natural amplifying action based on Pascal's rule.

The use of this auxiliary system brings more safety, stability and comfort; also it reduces the driver workload and safe following distance. With reduced following distance, the utilization factor of streets and highways increases And the traffic flow improves significantly.

With the use off Fuzzy, the control system can react very smoothly and behaves mimic to natural drivers but increasing the safety. The fast developments of fuzzy systems will make them very popular and more economic in the near future [9]. Generalization of new fuzzy chips makes it possible to have the fuzzy controllers with the least price in the future.

REFERENCES

1. The Use of Self Tuning Fuzzy Controller for Impedance Matching, A.Adibi, H.Ghafoory, senior members IEEE; Seyed Mohammad Reza Hashemi Mogaddam, Amir kabir university, Iran IEEE international conference on control sys, Elm-o-sanat university, 1995, Tehran, Iran.
2. G.Johanson and K.Rumar. "Driver brake reaction times", Human Factors, No13, pp23-27, 1972.
3. David Elting, Mohammed Fennich, Robert Kowalczyk, Bert Hellenthal; "Fuzzy Anti-Lock Brake System", Intel Corporation, Automotive operation, Arizona.
4. D.N.Lee, "A theory of visual control braking based on information about time-to-collision", Perception,5,pp.437-459.
5. W.Van Winsam and A.Heino', "Choice of time headway in Car-Following and the role of Time-To-Collision information in braking", Ergonomics, No4,pp.579-592,1996.
6. J.Bernasch, K.Naab, "Adaptive Fuzzy Control for driver assistance", BMW AG, Munchen.
7. Johan Bengtsson, "Adaptive Cruise Control and Driver Modeling", Lund institute of technology.
8. J.Hitz, J.Koziol, A.Lam, "Safety evaluation results from the field operational test of an intelligent control sys".
9. Constantin Von Altrock, "Fuzzy Logic in Automotive Engineering", Circuit Cellar, Issue 1997. 88, Nov.
10. The Control of Reflected Signals from Body Tissue by a Fuzzy Controller, MS Thesis, 1995, Amirkabir University, Tehran, Iran.

CONTACT INFORMATION

Seyed Mohammad Reza Hashemi Mogaddam
Electronics engineer, MS. Degree.
He started his study on fuzzy sys from 1992.
Passenger car body production dept.
Irankhodroo car manufacturing company
It is the biggest car manufacturing company in Middle East.
Add: Km 14th Karaj road, Tehran, Iran.
Tel: 009821 4890 2331
Fax:009821 48905456
Email: m.hashemi@ikco.com
Personal address: No 55 Rangbarian Alley, Mellat Square, Hasan Bana Street, Shams Abad, Tehran, Iran, Po code 16738.

ADDITIONAL SOURSES

1. ImechE, Advanced Driver Assistance Systems, Seminar Publication, 1999-13, London, UK
2. NHTSA. Preliminary Assessment of Crash Avoidance Systems Benefits. 1996, NHTSA Benefits Working Group.
3. Tribe, R., Prynne, K., Wetswood, I. and Clarke, N. et al., 1995, Intelligent Driver Support. Proc. Of 2nd Word Congress on ITS. Yokohoma, Japan, 1995.

DEFINITIONS, ACRONYMS, ABBREVITIONS

A.B.S Anti-lock Braking Systems
A.C.A.S.I.S.T Anti Collision Advanced Support and Safety Intervention System
A.D.A.S Advanced Driver Assistance Systems
A.L.B Anti Lock Brake
A.S.I.S.T Advanced Support and Safety Intervention System
C.D.L Collision Danger Level
D.B.R.T Driver Brake Reaction Time
F.L.C Fuzzy Logic Controller
H.M.I Human Machine Interface
I.T.S Intelligent transport Systems
R.D Required Deceleration
R.P.M Revolution Per Minute
T.T.C Time To Collision

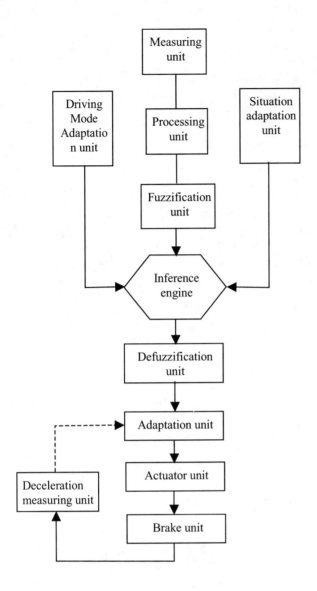

Figure 1. The controller block diagram

2002-01-0403

Decision Making for Collision Avoidance Systems

Jonas Jansson and Jonas Johansson
Volvo Car Corporation

Fredrik Gustafsson
Linköping University

ABSTRACT

Driver errors cause a majority of all car accidents. Forward collision avoidance systems aim at avoiding, or at least mitigating, host vehicle frontal collisions, of which rear-end collisions are one of the most common. This is done by either warning the driver or braking or steering away, respectively, where each action requires its own considerations and design. We here focus on forward collision by braking, and present a general method for calculating the risk for collision. A brake maneuver is activated to mitigate the accident when the probability of collision is one, taking all driver actions into considerations. We describe results from a simulation study using a large number of scenarios, created from extensive accident statistics. We also show some results from an implementation of a forward collision avoidance system in a Volvo V70. The system has been tested in real traffic, and in collision scenarios (with an inflatable car) showing promising results.

INTRODUCTION

It is well known that driver errors are the main cause, or contribute to increased severity, of most accidents. For instance, the Indiana Tri-level (Treat et al. 1979) found driver errors to be a cause or severity-increasing factor in 93% of the accidents. Furthermore, 27% of all accidents (USA 1997) were rear-end collisions. This shows the potential of forward collision avoidance (FCA) systems. The crucial part of the algorithm is the decision making, and the conflicting considerations are:

- Avoid all collisions

- Never do a faulty intervention

The design is a compromise between these mutually exclusive conditions. For many reasons, such as driver acceptance of the system and legal requirement that the system itself must not cause hazards, the second condition is the most important one when designing a FCA system.

The final responsibility must always be with the driver, and we stress that the system presented in this paper is driver assistance help.

A further consideration is that such an active system must not brake when the driver can still brake or steer to avoid an accident. This leads to mitigation rather than avoidance system, and in the sequel we refer to the system as Collision Mitigation by Braking (CMBB).

In a combined CMBB and collision mitigation by steering system, CMBB will mainly be activated at low speeds when braking is more efficient.

Algorithms previously proposed in literature for FCA by warning [7][17] and braking [2] are almost exclusively based on relative velocity and relative distance for decision making.

These metrics are easy to understand, but may not be sufficient information in more complex situations such as dense city traffic, situations where the driver switches lane or does hard handling maneuvers.

We here present a method to compute the risk for collision, taking into account measurement uncertainty and driver maneuvers. Decision making is then based on the probability density function for the relative position from the own vehicle to the most dangerous other object for the moment. Similar algorithms have been proposed and exist today for military aircraft, but have, as far as we know, not been studied for road based vehicles.

TARGET TRACKING

The information sources that are used for FCA systems come from one or more of the following sensors:

-Millimeter radar measuring bearing, range and range rate.

-IR Radar measuring bearing, elevation, range and range rate.

-Camera with image processing algorithms computing bearing and elevation.

If more than one sensor is used we have a sensor fusion problem, which also includes synchronization in time and space, which is elegantly handled in the Kalman filtering framework described below. The approach is model based using a state space model of the form:

$$X_{t+1} = A_t X_t + B_t v_t, Cov(v_t) = Q_t$$
$$Y_t = C_t X_t + e_t, Cov(e_t) = R_t$$

The model should be able to predict how the position of the tracked object evolves in time, and there are a variety of possible models available in the target tracking literature. The one we have chosen is based on the coordinated turn model, where the object is supposed to follow straight line segments and circle segments. This is a fairly good model of roads and typical driver maneuvers, and transitions between different segments are modeled as state noise v_t.

The state vector is:

$$X_t = \begin{pmatrix} x_t & y_t & v_{x,t} & v_{y,t} & \omega_t \end{pmatrix}^T$$

x_t , y_t = position coordinates in a ground fixed coordinate system at time t

$v_{x,t}$, $v_{y,t}$ = velocity at time t

ω_t = turn rate (yaw rate)

θ_t = heading angle

The state space matrices are given by

$$A_t(\omega) = \begin{pmatrix} 1 & 0 & \dfrac{a}{\omega_t} & \dfrac{b-1}{\omega_t} & 0 \\ 0 & 1 & \dfrac{1-b}{\omega_t} & \dfrac{a}{\omega_t} & 0 \\ 0 & 0 & b & -a & 0 \\ 0 & 0 & a & b & 0 \\ 0 & 0 & 0 & 0 & 1 \end{pmatrix}$$

Where:

$$a = \sin(\omega_t T)$$
$$b = \cos(\omega_t T)$$

$$B_t(v_x, v_y) = \begin{pmatrix} 0 & 0 & 0 & 0 & 0 \\ 0 & 0 & 0 & 0 & 0 \\ 0 & 0 & q_v \cos(\theta) & 0 & 0 \\ 0 & 0 & 0 & q_v \sin(\theta) & 0 \\ 0 & 0 & 0 & 0 & q_\omega \end{pmatrix}$$

Finally, the measurement equation incorporates the sensor information.

Sensor *i* gives

$$Y_i(t) = \begin{pmatrix} R & \dot{R} & \alpha \end{pmatrix}$$

R = Range

$\dot{R}$ = Range Rate

α = Bearing angle

There can be any number of sensors, and they need not be synchronized in time. The idea is to compute the a posteriori probability density function (PDF) of the state vector, given all sensor information up to time t. The model is then used to predict future positions and simulate how the PDF evolves in the near future. From this, we can compute the probability that the relative position belongs to a rectangle D, which is the size of the own vehicle. The posteriori probability density function (PDF) is provided by a model based filter, using the state space model. For linear models, the Kalman filter provides a finite dimensional algorithm to compute this. For non-linear models, such as the one we are using, the so called extended Kalman filter can be used to approximate the PDF, and this is what we have chosen.

Other alternatives include point mass filtering (deterministic numerical integration) and particle filtering (stochastic numerical integration). These may both be too complex for the computational platform we are using, though we have found them to be an interesting alternatives.

The Kalman filtering equations are summarized below:

$$\hat{X}_{t+1/t} = A_t \hat{X}_t$$
$$P_{t+1/t} = A_t P_{t/t} A_t^T + B_{v,t} Q_t B_{v,t}^T$$
$$\hat{X}_{t/t} = \hat{X}_{t/t-1} + P_{t/t-1} C_t^T (C_t P_{t/t-1} + R_t)^{-1}(Y_t - C_t \hat{X}_{t/t-1})$$
$$P_{t/t} = P_{t/t-1} - P_{t/t-1} C_t^T (C_t P_{t/t-1} C_t^T + R_t)^{-1} C_t P_{t/t-1}$$

Where:

$$C_t = \frac{Y_i(t)}{dX} = \frac{\left(R \quad \dot{R} \quad \alpha\right)}{dX}$$

$$= \frac{\left(R \quad \dfrac{\Delta X \cdot \Delta v_x + \Delta y \cdot \Delta v_y}{R} \quad a\tan(\dfrac{\Delta y}{\Delta x})\right)}{dX} =$$

$$= \begin{pmatrix} \dfrac{\Delta X}{R} & \dfrac{\Delta y}{R} & 0 & 0 & 0 \\[3mm] \dfrac{\Delta v_x}{R} - \dfrac{(\Delta y \cdot \Delta v_y + \Delta x \cdot \Delta v_x)\Delta x}{R^{\frac{3}{2}}} & \dfrac{\Delta v_y}{R} - \dfrac{(\Delta y \cdot \Delta v_y + \Delta x \cdot \Delta v_x)\Delta y}{R^{\frac{3}{2}}} & \dfrac{\Delta X}{R} & \dfrac{\Delta y}{R} & 0 \\[3mm] \dfrac{-\Delta y}{\Delta X \cdot \left(1 + \dfrac{\Delta y^2}{\Delta x^2}\right)} & \dfrac{1}{\Delta X \cdot \left(1 + \dfrac{\Delta y^2}{\Delta x^2}\right)} & 0 & 0 & 0 \end{pmatrix}$$

Here:

$$\Delta x = x_{obstacle} - x_{host}$$

$$\Delta y = y_{obstacle} - y_{host}$$

$$\Delta v_x = v_{x,obstacle} - v_{x,host}$$

$$\Delta v_y = v_{y,obstacle} - v_{y,host}$$

A thorough treatment of Kalman filtering and model choices is given in [16].

In a collision avoidance application you normally want to evaluate the probability of collision for several future positions. To do this one simply iterates the time update as many times as desired. For autonomous braking actuation we do not need to look further ahead in time than 1.5 seconds (because collisions become unavoidable when they are closer in time). This fact can be realized from calculating the minimum avoidance time from figure 5.

Other important issues for over all tracking performance such as how to handle multiple tracks (many potentially dangerous objects) and data association will not be discussed in this paper.

RISK ESTIMATION

The metric we propose for risk estimation is to calculate the probability of collision. This is done by forming the joint PDF of the host's and the other objects' position relative to each other. The probability is calculated by integrating the joint PDF over the area which corresponds to a collision (the area where the two objects physically overlap):

$$P(collision) = P(Pos_{Host} - Pos_{Object} \in D) =$$

$$= \iint_{Pos_{host} - Pos_{obj} \in D} f_{rel\_pos}(x, y)\,dx\,dy$$

Pos_{Host} = position of tracked object

Pos_{Object} = position of tracked object

D = the area which correspond to (geometrical) overlap between the host vehicle and the other object.

$f_{Rel\_pos}(x,y)$ = the PDF of the position of host and obstacle relative to each other

The joint PDF is formed from the PDF of the host's future position and the other obstacles future position. The PDF of the host's and other vehicles' future position can be obtained with one of the filtering methods mentioned above. An example of how the PDFs and the joint PDF look like for one situation is shown in Figure 1-3. Figure 1 below shows two vehicles meeting. In Figure 2 the PDF for the two vehicles is plotted at 4 different future time instances.

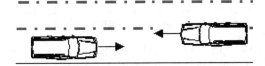

Figure 1. Two cars meeting.

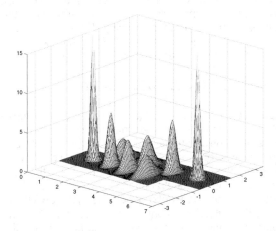

Figure 2. PDF's for the two vehicles in figure 1 at 4 time instances. The narrow peaks correspond to time 1 and the widest peaks correspond to time 4.

In figure 3 the joint probability of the vehicles' position relative to each other is shown.

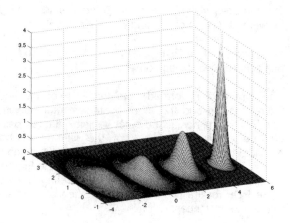

Figure 3. Joint PDF for vehicle's relative position. Again the narrow peak corresponds to time 1 and the widest peak corresponds to time 4.

Notice that in figure 3 we have switched the coordinate system to one that is fixed to the front of the car coming from the left. At time instance 3 there is a lot of density close to the origin. At this time instance the probability for collision is high.

In our case we get the Kalman filter estimate of X and covariance matrix P which can be used to compute a confidence region (ellipse) where the relative position will be with a certain probability. Furthermore, with a Gaussian assumption on the PDF, we can compute the risk for collision as described above.

COMPUTATIONAL COMPLEXITY

Apart from tracking accuracy, computational complexity of the algorithm is also an important issue. There are several considerations concerning the computational complexity of the algorithm. The algorithm has computational complexity of O(n) where n is the number of future time points evaluated. n is a function of how far ahead in time one wants to look and also how close you want the future positions to be in time (= how many future time instances that are calculated). As mentioned above for the CMBB application it is enough to look 1.5 seconds ahead in time. It remains to decide how many points one should calculate for these 1.5 seconds. This is a matter of computational power available, and for the best result one should space the points as close as possible in time. But there is of course a trade-off on how many objects you want to be able to track. The algorithm is O(m) where we let m be the number of tracks to be handled. Looking both at future time points and number of tracks the algorithm is O(n*m) in computational complexity.

THEORETICAL PERFORMANCE OF COLLISION AVOIDANCE BY BRAKING

We shall now specifically study what can be achieved by autonomous brake intervention. In order to avoid faulty interventions braking is only allowed when a collision becomes unavoidable i.e. the probability of collision is 1. To get a feeling for what performance that is possible to achieve we look at one specific scenario. The scenario studied is a head on collision with a stationary object (see fig. 4 below).

Figure 4. Car on collision course with a stationary obstacle

The braking distance and the distance needed to avoid the obstacle by steering away at different speeds is shown in figure 5 below

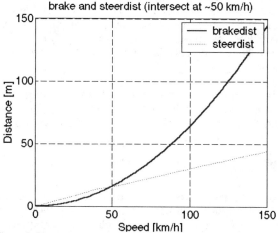

Figure 5. Distance needed to avoid a stationary obstacle with a width of 2 m by means of braking and steering.

We see in figure 5 that the braking distance (solid) and the steering away distance (dotted) cross somewhere around 50 km/h. This means that for low speeds, braking is the most efficient countermeasure and for high speeds steering away is more efficient. If full braking is applied at the point were collision becomes unavoidable (for low speeds at the braking distance and for high speeds at the steer away distance, then the collision speed as a function of the initial speed will be according to figure 6 below

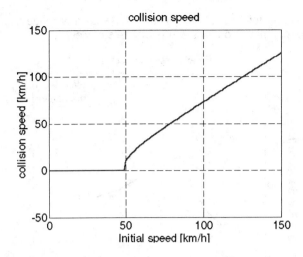

Figure 6. Collision speed when full braking is applied at the point where the collision becomes unavoidable.

SIMULATON RESULTS

To evaluate performance of the algorithm we have designed and simulated 30 different scenarios. The scenarios can be divided in to two groups. The first group of scenarios is designed to provoke the system to make a faulty intervention (= intervene when no collision occurs). The second group is designed to evaluate performance of the system in a situation where it should intervene i.e. when a collision occurs. Here we describe and present results from a selection of the simulated scenarios (all scenarios were simulated at speeds ranging from 10 –150 km/h):

Symbols:

: The CMBB vehicle

: The POV (Principal Other Vehicle)

GROUP ONE SCENARIOS:

These scenarios were designed to provoke a faulty intervention. The scenarios were simulated 10 times at each speed (speeds [10 20 30…. 150]). From the simulation results no faulty interventions were observed.

Scenario 1:

Head to Head, the CMBB vehicle turns right at the last moment.

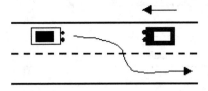

Figure 7. Scenario 1

Scenario 2:

Straight roadway, POV travelling in the same lane as the CMBB vehicle. Suddenly the POV brakes hard and then turns hard. The POV just clears the path of the CMBB vehicle.

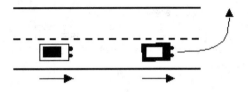

Figure 8. Scenario 2

Scenario 3:

The CMBB vehicle changes from right to left lane (both lanes in the same direction) at the same time as it meets another vehicle in the opposite lane.

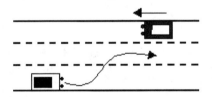

Figure 9. Scenario 3

GROUP 2 SCENARIOS:

The group 2 scenarios presented below are used to see what performance the system achieves.

Scenario 4:

Straight roadway, POV traveling in the same lane as the CMBB vehicle. Suddenly the POV brakes hard (deceleration 7 m/s^2). The headway is 15 m. The initial speed, prior to the POV brake maneuver, was the same for both vehicles.

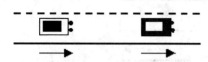

Figure 10. Scenario 4

Result: In figure 11 the relative speed at impact is plotted as a function of the initial speed. One can see that for low speeds the relative velocity at impact is reduced ~10-20 km/h. For higher speeds the system response time is too long to be able to reduce the collision speed.

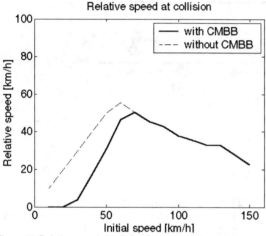

Figure 11. Relative speed at collision for scenario 4.

Scenario 5:

This scenario is the same as scenario 4 with the exception that there is a lateral offset of 0.5 m between the CMBB vehicle and the POV.

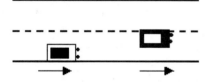

Figure 12. Scenario 5

Result: Again the relative speed of the vehicles at impact is plotted as a function of the initial speed (in figure 13). For comparison the results from the previous scenario has been plotted in figure 13. In the figure one can see that there is no intervention for low speeds. The

reason for this is that before the decision is made the target moves out of the sensors' field of view.

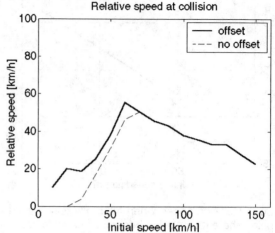

Figure 13. Relative speed at collision for scenario 5.

Scenario 6:

Object is on the side of the road and then suddenly "jumps" out 10 m in front of the CMBB vehicle.

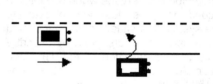

Figure 14. Scenario 6

The performance for scenario 6 is plotted as a phase diagram in figure 15. This shows impact speed and at what distance the system intervenes.

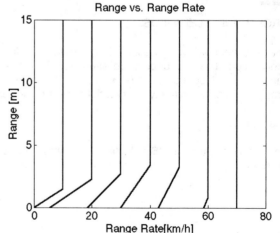

Figure 15. Range and Range rate between the host vehicle and the POV for scenario 6. The results are based on simulated data.

TEST RESULTS

The test vehicle is a Volvo V70 equipped with a millimeter wavelength radar and a laser radar. The sensor update rate is 10 Hz for both sensors and their field of view is ~8 degrees. The sensor fusion, data association and decision making algorithm execute on an on-board processing unit which is also connected to the vehicle's braking system. The purpose of having two sensors is to try to discriminate targets that are not valid. For example a millimeter wavelength radar can receive a strong echo from a tin can.

CORRECTNESS OF THE ALGORITM

To check if the algorithm makes faulty decision the prototype system has been driven in real traffic (urban and highway traffic) with braking disabled. We found that some faulty interventions do occur. These interventions mainly occur at low speeds when there are a lot of potential targets/obstacles close to the sensors (i.e. in front of the car). However we did not find any case where it was obvious that the algorithm made an erroneous decision based on a target that it had been tracking for some samples. All the faulty interventions seem to come from erroneous measurements, "false targets" or from bad initialization of obstacles.

PERFORMANCE OF THE ALGORITHM

To evaluate collision mitigation performance, collision tests against a stationary obstacle have been performed. As an obstacle, an inflatable car is used (fig 16).

Figure 16. Inflatable car for testing system performance when colliding.

Performance for the head-on scenario pictured in figure 4 is plotted in figure 17. In figure 17 we plot range between host vehicle and obstacle vs. range rate, the plotted result is an average from 5-20 collisions at each

speed. As can be seen in the figure, speed was reduced 10-17 km/h for initial speeds ranging from 30-70 km/h.

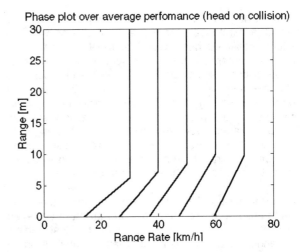

Figure 17. Range vs. Range Rate for the head on against stationary object scenario (figure 4). The result shown is the average performance from test drives with the V70 test vehicle.

CONCLUSION

A method for decision making in collision avoidance applications has been presented. The main advantages of the method are:

- The use of modern tracking theory makes it straight forward how to deal with measurement and process noise
- Motion in two dimensions is considered.

The prototype system presented in this paper significantly reduces the impact speed in frontal collisions. As can be seen in figure 17 interventions typically occur when the obstacle is closer than 20 m away from the CMBB vehicle (more than 90 % of all rear end collisions occur at relative speeds below 100 km/h [4]). A sensor with a shorter detection range but a larger field of view might be more appropriate for collision mitigation purposes.

Further work on the sensors and the sensorfusion is needed to have a system with 0 faulty interventions. It would be desirable to have a sensor with better target classification capability.

DISCUSSION

Differences between optimal performance (figure 5 and 6) and test result (figure 17) can be attributed to:

- Measurement uncertainties
- System response time (computational time, sensor measurement rate)
- System modeling errors

By improving these factors one can come closer to optimal performance. Some specific problems with the system presented here were that the laser radar and millimeter radar could not be synchronized. This of course causes some discrepancy between laser and radar measurement. Both sensor loses the target at close range (<10 m) because their narrow field of view. This cassias deteriorated performance at low speeds. The laser radar seem to have problem "seeing" the inflatable car which at some occasions caused missed interventions, since both sensors are required to detect the obstacle in order to have an intervention.

Another factor that limits the system performance is the time to build up brake pressure and maximum achieved pressure. The brake system used on the test vehicle here achieved decelerations between 5-7 m/s^2. A brake system that quickly gives a deceleration of 10 m/s^2 would give an additional speed reduction of ~10 km/h to the test results plotted in figure 17.

To design a good collision avoidance system we need to solve two issues. One is the risk estimation discussed in this paper. The other issue, that has not been addressed here, is that of object recognition. This is a matter of the sensing capabilities of the sensors but also a matter of how to do the sensor fusion. For correct decision making accurate target classification and feature (for example width) extraction is imperative.

REFERENCES

1. Minoru Tamura, Hideaki Inoue, Takayuki Watnabe and Naoki Maruko. Research on a Brake Assist System with a Preview Function. SAE Paper 2001-01-0357
2. Development of an Adaptive Cruise Control with Stop-and-Go Capability, SAE Paper 2001-01-0807
3. Berthold Ulmer. VITA II – Active Collision Avoidance in Real Traffic. 1994
4. Bo Zhu. Potential Effects on Accidents from Forward Collision Warning/Avoidance System. Master thesis at Linköping University, LITH-ITN-EX-150-SE, 2001.
5. Datta N. Godbole, Raja Sengupta, James Misener, Natasha Kourjanskaia, James B. Michael. (Transport Research record 1621,1998). Benefit Evaluation of Crash Avoidance System.
6. Farber E. I. Using the Reamacs Model to Compare the Effectiveness of Alternative Rear End Collision Warning Algorithms. Conference on Enhanced Safety of Vehicle, Munich, Germany, 1994.
7. CAMP (Crash Avoidance Metric Partnership), Final report, 1999?.
8. H. C. Joksch. Velocity Change and Fatality Risk in a Crash-A Rule of Thumb. Accident Analysis and Prevention, vol25, No. 1, 1993.
9. J. R. Treat, N. S. McDonald, D. Shinar, R. D. Hume, R. E. Mayer, R. L. Stansifer and N. J. Castellan. Tri-level Study of the Cause of Traffic Accidents. Report No. DOT-HS-805-085, NHTSA, 1977.
10. R. Wade Allen (Transportation Research Record 1059) Research Methodology for Crash Avoidance Studies.
11. R. Wade Allen (VTI Report 331 A) Crash Avoidance Models and Driver/Vehicle Handling.
12. Ribert M. Clark, Michael J. Goodman. Michael Perel, and Ronald R. Knipling (DOT/FHWA, 1996) Driver Performance and IVHS Collision Avoidance Systems: A Search for Design-Relevant Measurement Protocol.
13. Ronald R. Knipling. (1993). IVHS Technologies Applied to Collision Avoidance: Perspectives on Six Target Crash Types and Countermeasures.
14. Thomas A. Dingus, Steven K. Jahns, Avraham D. Horowitz, and Ronal Knipling (Human Factors in Intelligent Transportation System, 1997). Human Factors Design Issues for Crash Avoidance System.
15. William A. Leasure. Jr. (1994) NHTSA's IVHS Collision Avoidance Research Program: Strategic Plan and Status Update.
16. F. Gustafson. Adaptive Filtering and Change Detection. John Wiley, 2000.
17. P. Barber, N. Clarke. Advanced Collision Warning Systems. Industrial Automation and Control: Applications in the Automotive Industry (Digest No. 1998/234).

CONTACT

Jonas Jansson
Volvo Car Corporation
Dept 92421
SE-405 31 Göteborg, Sweden
Telephone: +46 31 595438
E-mail: jjansso1@volvocars.com

alt.

Linköping University
Division of Automatic Control
SE-581 83 Linköping, Sweden
Telephone: +46 13 286697
E-mail: jansson@isy.liu.se

A New Paradigm for Rear-end Crash Prevention Driving Performance

August L. Burgett and Robert J. Miller, Jr.
National Highway Traffic Safety Administration, Research and Development

ABSTRACT

This paper presents a new data analysis approach to describe driver performance in situations that have the potential of leading to a rear-end crash. The approach provides at least two key benefits. It provides a unified means of analyzing data from different sources such as simulators, test tracks, and instrumented vehicles. It may also provide a means of addressing the huge diversity of driver performance in pre-crash situations.

INTRODUCTION

This paper presents a new approach to analysis of data which describes driver performance in situations that often result in rear-end crashes. The analysis introduces the concept of a crash prevention boundary -- a theoretical, deterministic avoidance threshold that relates driver reaction to the dynamics between two vehicles in an impending crash. The crash prevention boundary provides two key benefits. It allows a unified means of analyzing data from different sources such as driving simulators, recorded naturalistic driving incidents, and controlled test track driving scenarios. It also provides a means of addressing the diversity of driver performance during the pre-crash situation.

The paper consists of a short review of past studies that sought to define driving conditions and driver behavior leading to rear-end crashes. Many of the studies also examined the modification of driver behavior by using a warning system to prevent crashes or mitigate crash severity. The review of previous work is followed by the definition of a deterministic relationship – a crash prevention boundary (CPB) – which becomes the framework for making comparisons of driver braking responses for different driving conditions. Driver response data from tests on the Iowa Driving Simulator are then presented in the CPB framework. The authors believe that this approach can be used to expand on previously published analysis of these experimental results. Analysis of these data demonstrates how the CPB can then be extended and applied to additional sets of similar driving data.

BACKGROUND

Recent data shows that drivers in the United States accumulate a total of more that 2.6 trillion miles of travel annually [1]. These same drivers experience more than 1.8 million crashes annually where one vehicle collides with the rear of another [2]. Thus, there is approximately one rear-end crash for every two million vehicle-miles of travel each year. Also, one study has found that any particular driver brakes about 50,000 times each year [3]. Most of these brake applications occur in routine stops and adjustments of speed in traffic; but each event has the potential to be a crash if the driver does not brake. This suggests that nationally, there are more than 10 trillion brake applications each year. Many of these, even if there is a relatively low level of deceleration, serve the purpose of preventing a collision. This leads to the question that underlies this paper, as well as a large body of other research; "What is different during those 1.8 million events were the driver could not, or did not, prevent a rear-end collision than during the other 10 trillion times that drivers braked and prevented a crash?"

A number of studies of rear-end crash dynamics examined the basis for warning drivers of potential rear-end crashes. Examples of such efforts from 1997 – 2000 include the following National Highway Transportation Safety Agency (NHTSA) contracts: Fostering Development, Evaluation, and Deployment of Forward Crash Avoidance Systems (FOCAS); Sensor Technologies & Systems analysis of rear-end warning system performance; Intelligent Cruise Control (ICC) Field Operational Test Evaluation; the Crash Avoidance Metrics Partnership (CAMP); University of Iowa Driving Simulator (IDS) Tests; and the Johns Hopkins University Applied Physics Laboratory (JHU/APL) Rear-end Collision Symposium. A synopsis of each is given below.

The goal of the FOCAS work [4] was to advance the development of sensors and systems for commercial use in assisting the forward crash-avoidance performance of drivers. To aid in progressing towards this goal, the program created tools, methodologies, and knowledge-bases to expedite the development of adaptive cruise control (ACC) systems as well as systems providing forward collision warning (FCW) alerts. The results, findings, and conclusions of the program are numerous. The program was evolutionary both in terms of hardware and software advancements and more importantly understanding the driver's role in the application of this new technology. Prototype systems were used by lay persons in naturalistic driving. The culmination of the project resulted in progress in five subject areas: (1) evaluation of ACC-with braking, (2) braking latency, (3) development of a NHTSA warning algorithm, (4) evaluation of three FCW algorithms, and (5) research of vigilance as it relates to deceleration authority of an ACC system. The ACC systems developed in this study were well liked by drivers, convenient to use, and did not present any clear safety concerns.

Sensor Technologies & Systems, formerly Frontier Engineering Sciences, Inc. [5], studied the improvement on driver behavior of a rear-end warning system as well as the effect of choice of headway values on the effectiveness of the warning system. Using the Iowa Driving Simulator, data was collected with and without the warning system for various driving conditions. This study concluded that a warning system is useful for shortest headway conditions tested and that drivers may be distracted or confused by collision warning information that is presented too early (nuisance alarm).

The ICC Field Operational Test Evaluation [6] collected data using instrumented vehicles with and without ACC. Prevailing tendencies of drivers in the choice of headway values as well as driving habits were studied. It was concluded that the ACC system that was tested provided a safety benefit for drivers.

In a study conducted by the Crash Avoidance Metrics Partnership [7] a series of controlled experiments were carried out on test tracks to determine driver response to several collision warning alert algorithms as part of an overall study to develop objective test procedures for rear-end collision warning systems. Useful data for driver braking behavior and response time were derived from this study.

The University of Iowa [8] studied the effect of a rear-end warning system on a distracted driver for varying driving conditions and settings for the warning system assumptions. It was concluded that the warning system that was tested reduces the chance of collision and that an early warning provided a greater benefit than a late warning.

NHTSA sponsored the APL symposium [9] that brought together a wide representation from industry and government on the subject of rear-end collision avoidance and ICC. Information was shared in presentations to promote synergism within the entire community of interested parties.

EFFECTIVENESS OF A CRASH WARNING SYSTEM

Each of the studies noted above has addressed specific aspects of rear-end crash avoidance analysis. Some of the studies addressed performance specifications, some addressed effectiveness of crash warning systems, some are based on naturalistic driving, while others included tests in driving simulators or on test tracks. A review of these studies points out that there is no common analytical framework for

comparing results. The work described in this paper is a first step toward development of such a framework.

A complete framework would cover all types of crashes and all subsets of each type. The framework developed in this paper is focused on the family of rear-end crashes that result from situations where two vehicles, that are initially traveling at the same speed, begin to close on each other due to deceleration of the lead-vehicle. At some point, the lead-vehicle will brake resulting in braking by the following-vehicle. The initial dynamic conditions of such a situation as well as the driver responses will lead to some crashes and some crash avoidances (no crashes). Proper countermeasures will help avoid many would-be crashes and lead to safer highways.

REAR-END CRASH DYNAMICS

Figure 1 illustrates a situation where two vehicles are initially traveling without any significant conflict. A driving conflict arises because the lead-vehicle brakes. The time at which the lead-vehicle begins to brake is used as a primary reference and is defined as $t = 0$. Also, the location of the front of the following-vehicle at $t = 0$ is defined to be zero distance. The driver of the following-vehicle notices the conflict due to brake lights, the perceived closing rate, other cues, or a warning at time, $t = t_w$. The driver then takes action at time, $t = t_b$ resulting in a total crash avoidance, a near crash, or a crash. If action is taken quickly enough with sufficient braking and/or steering, then a crash is avoided and the vehicles have a point of closest approach at $t = t_s$. If the driver is occupied or distracted with another task when the driving conflict arises, then a crash is more likely.

SCENARIO DEFINITION

The starting point (initial conditions) for this scenario definition is the time when the lead-vehicle begins to decelerate. Prior to this point, the two vehicles are traveling at a constant separation with no closing rate. After the starting point the vehicles are closing due to lead-vehicle deceleration.

Thus, the initial conditions at the starting point for this family of rear-end crash situations are the traveling speed (where both vehicles are initially traveling at the same speed), the initial separation between the two vehicles, and the level of deceleration of the lead-vehicle. Initial speed, V_o, and separation distance, R_o, may be combined to provide the value of headway, T_h. Headway is the amount of time it takes the following-vehicle to cover the distance, R_o, when traveling at speed V_o. The significance of headway is in its relationship to the response time of the following-vehicle driver. If the following-vehicle driver's brake response time is equal to the value of the initial headway and the following-vehicle driver applies the same deceleration profile as the lead-vehicle experiences, the two vehicles will come to a stop without a collision but will be bumper-to-bumper at the end of the event. If the following-driver responds more quickly, less braking is required; and if the following-driver responds less quickly, more braking is required to avoid a crash.

In such scenarios, the following-vehicle driver should notice the brake lights, higher closing rate, or other cues and react to

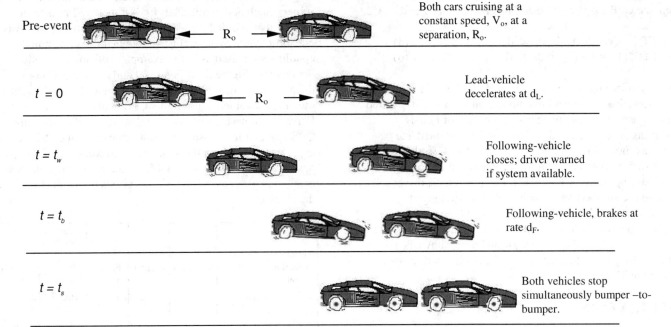

Pre-event		Both cars cruising at a constant speed, V_o, at a separation, R_o.
$t = 0$	R_o	Lead-vehicle decelerates at d_L.
$t = t_w$		Following-vehicle closes; driver warned if system available.
$t = t_b$		Following-vehicle, brakes at rate d_F.
$t = t_s$		Both vehicles stop simultaneously bumper –to-bumper.

them as the danger of a crash is perceived. The reaction should be to brake hard enough to slow or stop before a crash.

Figure 1 Typical Rear-end Driving Scenario

If the driver is distracted or does not perceive the lead-vehicle deceleration, an imminent crash warning can be given. Descriptions of algorithms for providing such a warning have been described in the literature [7, 11]. Assuming that only braking occurred, the two key variables that describe the following-vehicle driver's crash prevention response are:

1. t_b, the brake response time of the following-vehicle driver relative to the initial braking by the lead-vehicle, and
2. d_F, the level of deceleration of the following-vehicle.

The brake response time, t_b, is defined as the time span from start of lead-vehicle deceleration (initial conditions/starting point) until the initiation of braking by the driver of the following vehicle. The level of deceleration, d_F, of the following-vehicle is defined as the average deceleration over the time from the start of following-vehicle deceleration (braking) until the following-vehicle stops.

THE CRASH PREVENTION BOUNDARY

The underlying idea behind the analytical framework of a crash prevention boundary (CPB) is that for any given set of initial dynamic conditions, there is a subset of values of driver brake response time, t_b, and level of deceleration, d_F, which will result in crash avoidance. The corollary is that there is also a subset of values of these two variables that produce a crash. The CPB is a deterministic relationship that separates these two subsets of possibilities.

Thus, the CPB is an analytically derived expression that separates driver response values into those that provide crash avoidance and those that result in crashes. The CPB expression describes the limiting case between the two variables t_b and d_F. If a driver's brake response time and

deceleration satisfy the relationship, the two vehicles will have zero closing speed at the point of closest approach. Also, the point of closest approach will also be at zero range (i.e., the bumpers will be touching). The desired deterministic relationship for the CPB is a combination of logic criteria and algebraic relationships as shown in equations 1 and 2. The detailed development is provided in Appendix A.

$$t_b = R_0/V_0 + (V_0)[1/d_L - 1/d_F]/2 \quad \text{if } d_F < d_F^* \quad (1)$$

$$t_b = [(2V_0T_h)(1 - d_L/d_F)/d_L]^{1/2} \quad \text{if } d_F > d_F^* \quad (2)$$

where crossover deceleration, $d_F^* = d_L V_0^2/(V_0^2 - 2d_L R_0)$ (See Section A.1).

Both of the above equations assume that the lead-vehicle comes to a stop. In equation (1) the lead-vehicle and the following-vehicle are stopped at the point of closest approach. Equation (2) reflects the situation where both vehicles are moving at the point of closest approach. The value of crossover deceleration, d_F^*, is the separating criteria for these two situations. The derivation of the expression for following-vehicle crossover deceleration is given in Appendix A.1. Expressions (1) and (2) can be combined with expressions for the time-to-collision (TTC) at the beginning of the event to provide simplified expressions for the CPB. The expressions for time-to-collision are:

$$TTC = TTC1 = T_h + V_0/(2 d_L) \quad \text{if } d_L > d_L^*, \quad (3)$$

and

$$TTC = TTC2 = (2V_0T_h/d_L)^{1/2} \quad \text{if } d_L < d_L^* \quad (4)$$

where $d_L^* = V_0/(2T_h)$ and $T_h = R_0/V_0$ (See Section A.4).

Then, the CPB may be expressed in terms of TTC1 and TTC2,

$$t_b = TTC1 - V_0/(2d_F), \text{ if } d_F < d_F* \qquad (5)$$
$$t_b = TTC2(1 - d_L/d_F)^{1/2} \text{ if } d_F > d_F* \qquad (6)$$

Thus, the relationships (1) and (2) or (5) and (6) between t_b and d_F describe the Crash Prevention Boundary (CPB). Based on the equations given above and given a set of initial conditions of R_0, V_0, and d_L a CPB can be computed and plotted as shown in Figure 2. The value of TTC is also shown on this figure. It can be seen from equations 5 and 6 that the CPB is asymptotic to TTC. The following-vehicle driver's response is described by the point, (d_F, t_b). Braking at sufficient average level within the required time prevents a collision and plots below the CPB, while lighter braking with a greater delay will lead to a collision and plots above the CPB. Doing nothing after the initiation of the conflict will cause a collision at TTC.

APPLICATION

As an example of the application of the CPB approach, data from an experiment [10] using a driving simulator are presented in this format. The purpose of the experiment was to investigate how distracted drivers respond to imminent rear-end collision situations – both with a warning and without a warning. The experiment examined how variations in warning algorithm parameters affect the ability of a warning to aid distracted drivers. The derivation of this algorithm is described in [11].

Four sets of initial conditions were used in this experiment. Initial conditions included velocities of 35 and 55 mph, initial headway was either 1.7 or 2.5 seconds, and lead-vehicle decelerations were 0.40 and 0.55 g. Within each set of initial conditions, testing was performed using subjects with no warning (baseline), subjects aided by a short warning, and subjects aided by a long warning. In this context, long and short are used relative to the start of the driving scenario. Short warnings were based on the assumption that the following-vehicle driver would brake after a delay of 1.5 seconds after the warning at an average of 0.4g. Long warnings were based on the assumption that the driver would brake after 1.5 seconds at 0.75g. Long and short warning set points are also shown in Figure 2 to illustrate the relationship between the CPB and warning criteria used in this group of tests. In general, the "long" warnings occurred about 1 second later than the "short" warnings. Comparisons were made within each set between the baseline and short warning as well as comparison of baseline with long warning results. Drivers were distracted with a visually demanding number reading task. The simulator allowed drivers to follow a course, deliberately be distracted, observe a braking vehicle, and respond to the crash threat in a naturalistic way.

All test conditions are summarized in Table 1. Twenty subjects were tested for each of the 12 test conditions for a total of 240 tests in this experiment. The baseline driver performance data for one of the test conditions (IDS Test Condition 1) are shown in Figure 3. In this test, two of the 20

drivers chose to steer rather than brake. Thus, there are 18 subjects included in this analysis. Of these, 11 braked in a manner that avoided a crash while the performance of 7 was not sufficient to avoid a crash. Note that the same initial conditions are used as in the example collision prevention boundary of Figure 2. If Figures 2 and 3 are superimposed, the result is shown in Figure 4. This figure demonstrates a rather remarkable feature of the CPB analytical framework. Drivers who performed in a way that was predicted by the CPB to result in a crash, i.e. points above the line, did indeed experience a crash on the simulator. Conversely, drivers who performed in a way that the CPB predicted would avoid a crash did indeed avoid a crash, i.e. points below the line, on the simulator.

Initial Condition Set	Test Condition	Vo(mph)	d_L(g's)	T_h(sec)	Warning Algorithm dF(g's) Design Point	Warning
1	1	35	0.4	1.7	None	Baseline
	2	35	0.4	1.7	0.40	Short
	3	35	0.4	1.7	0.75	Long
2	4	35	0.55	2.5	None	Baseline
	5	35	0.55	2.5	0.40	Short
	6	35	0.55	2.5	0.75	Long
3	7	55	0.4	1.7	None	Baseline
	8	55	0.4	1.7	0.40	Short
	9	55	0.4	1.7	0.75	Long
4	10	55	0.55	2.5	None	Baseline
	11	55	0.55	2.5	0.40	Short
	12	55	0.55	2.5	0.75	Long

Table 1. IDS Test Design

The complete set of results from the driving simulator experiment are included in Appendix B. Each figure of Appendix B contains the data from a baseline condition in addition to a condition where there was a warning. Each figure also includes two other features. One is the crash prevention boundary that corresponds to the initial conditions for the particular IDS test condition. The other is a marker that identifies the set-point (assumed reaction time of a driver to the warning and level of deceleration) of the warning algorithm.

OBSERVATIONS

Perhaps the most noticeable result of comparing the experimental data with the corresponding CPBs is the additional insight that can be gained by having a graphical tool for quickly comparing experimental results with theoretical predictions. The CPB provides a quantitative and graphical means of describing the envelope of acceptable performance for specific dynamic situations. Experimental results of driver responses may then be compared to CPB. From the simulator experiment cited above, the ratio of the number of driver

Initial Conditions V$_o$=35mph, R$_o$=87.2 ft, d$_L$=0.4 g, T$_h$=1.7 sec

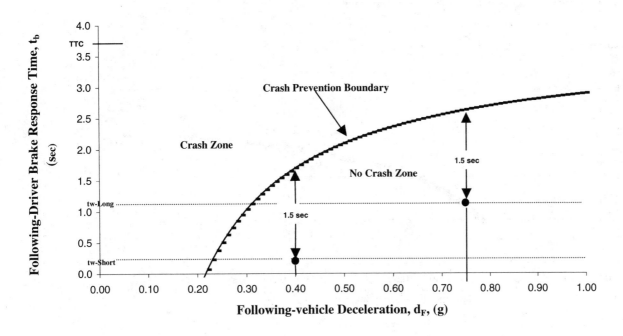

Figure 2. Sample Crash Prevention Boundary

IDS Test Condition 1
(V$_o$=35mph, R$_o$=87.2 ft, d$_L$=0.4 g, T$_h$=1.7 sec)

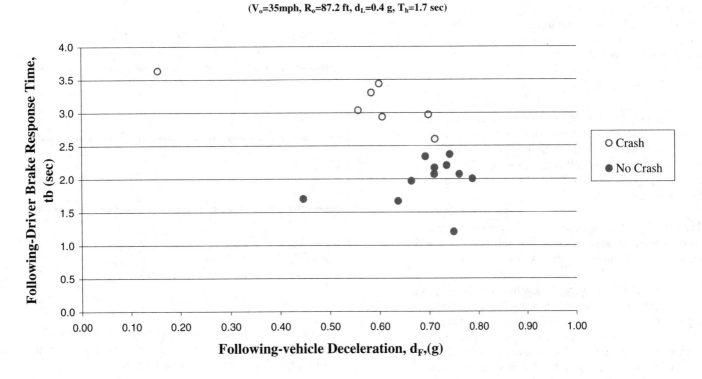

Figure 3. Driver Performance Experimental Results from Simulator Test Condition 1

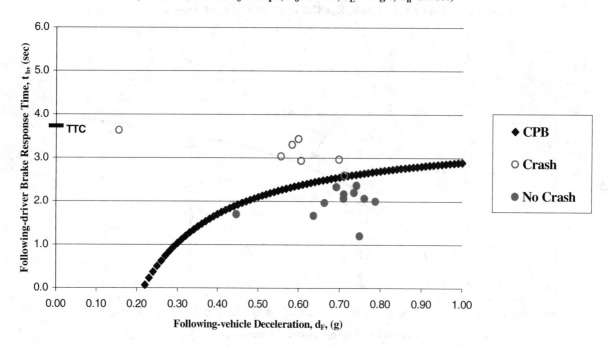

IDS Test Condition 1
(Initial Conditions V_o=35mph, R_o=87.2 ft, d_L=0.4 g's, T_h=1.7 sec)

Figure 4. Simulator Driving Results With CPB

responses above the CPB to the total number of driver responses for conditions with and without warnings can be obtained. These ratios are estimates of the crash probability for each set of conditions. Given the probability of a crash, the effectiveness, E, of a warning system may be computed as follows[12]:

$$E = (P_{c,w/o} - P_{c,w})/P_{c,w/o}$$

where $P_{c,w/o}$ is the probability of a crash without a warning and $P_{c,w}$ is the probability of a crash with a warning. The values of E for the four sets of initial conditions are given in Table 2. The values of crash probability are given in column 4. Drivers who steered instead of braking to avoid a crash are not included in these results.

When effectiveness is compared for short vs. long warning, it can be seen that the short warning is more effective than the long warning in eliminating crashes in all situations.

Effectiveness also is seen graphically by a comparison of the number of points on each side of the CPB thus gaining perspective on the significance of the estimates. One of these perspectives is the level of crash severity. A number of observers have noted that calculations of effectiveness such as those above do not include consideration of the relative importance of more severe crashes. Although, not included quantitatively in this paper, it can be shown that relative speed at the time of impact is related to the distance a point is from the CPB. Hence a combination of graphically based insights

and appropriate calculation procedures can provide additional estimates of the impact of a warning on overall crash-caused harm.

Warning Type	Test Condition	Total Tests	Crash Probability	Warning Effectiveness
Short (0.40g) ($P_{c,w}$)	2	19	0/19	1.00
	5	19	2/19	0.80
	8	19	1/19	0.88
	11	19	4/19	0.70
	Total	76	7/76	0.82
Long (0.75g) ($P_{c,w}$)	3	18	3/18	0.54
	6	19	5/19	0.52
	9	17	5/17	0.33
	12	16	5/16	0.55
	Total	70	18/70	0.50
No Warning ($P_{c,w/o}$)	1	18	7/18	
	4	18	10/18	
	7	16	7/16	
	10	17	12/17	
	Total	69	36/69	

Table 2. IDS Test Results

A third observation relates to the relative ease of identifying interesting features of experimental data. Two features of the driving simulator results are discussed here.

The first feature is the difference in baseline performance between the cases that started with long separation (i.e. test condition 10 which has an initial range of 201 feet) and the cases that started with shorter separation (between 88 and 137 feet for the other three test conditions). From the Figures B-7 and B-8 in Appendix B it can be seen that the cluster of points for test condition 10 (longer initial separation) is located substantially above the CPB while the cluster of points for test conditions 1, 4 and 7 are almost evenly divided on both sides of the CPB. This difference in location is also seen in crash probability for these (no warning) test conditions; test condition 10 has a crash probability of 0.7 while the probability of a crash for the other three is between 0.35 and 0.55.

Thus it appears that there is something fundamentally different about driver performance in baseline test condition 10 than in the other three baseline conditions. One possibility is that at the longer initial range, the drivers were not able to perceive that the lead vehicle was decelerating at a level that would produce an imminent crash. This lack of perception could be the result of limited graphical fidelity in the driving simulator or it could be a limitation in ability to perceive relative motion. The data from the experiment is not adequate to reach definite conclusions on this question. However, a quick review of capability of perceiving a looming object can provide some insight.

Figure B-9 in Appendix B shows the relationship between the rate of change of the subtended angle of the lead-vehicle as seen by the driver of the following-vehicle and the distance between the vehicles. This is consistent with previous research with regard to the perception of a "looming" object [7, p 157]. The reference paper suggested that a rate of change of 0.003 radians per second is a threshold below which subjects are not able to perceive a significant relative motion. Figure B-9 shows that this threshold is reached at a longer range and after a larger change in range for test condition 10 than the other three conditions. Thus, the graphical nature of the CPB presentation for analyzing experimental data suggests a difference in the driver's perceived level of threat; and points the way to an approach for investigating the issue.

A second significant driving simulator feature is the distribution of actual performance relative to the assumed performance that is the basis of the warning. For the three conditions with relatively short range at the beginning of the event, a summary of performance of the drivers is as follows.

For the short warning (assumed following-vehicle level of deceleration of 0.4g), the average reaction time (the time between a warning being given and application of the brakes) was 1.8 seconds, close to the assumed value of 1.5 seconds. For the long warnings (assumed level of deceleration of 0.75g) the average reaction time was 2.3 seconds, greater than the assumed value of 1.5 seconds. Similarly, the average deceleration for the short warning of 0.59g was greater than the assumed level of 0.4g; but for the long warning the average deceleration of 0.62g was closer to the assumed level of 0.75g. While these differences suggest that the drivers braked at the same level, it is not clear why on average their responses took longer for long warnings.

FUTURE WORK

Some future applications and extensions of CPBs include further analysis of rear-end crash conditions. This will include an analysis of naturalistic driving data from an intelligent cruise control field operational test, and data from other naturalistic driving experiments. It will also include derivation of CPB expressions for other families of rear-end crashes and for other types of crashes such as road departure. A third extension would lead to better understanding of the concept of nuisance warnings and near-crash conditions and is useful as a measure of "seriousness" of situations, i.e. it may be used as parameter in distribution of responses.

CONCLUSIONS

This paper has introduced the idea of an analytically derived deterministic crash prevention boundary and has shown its application to the analysis of rear-end crash data. The analysis of data from an experiment in a driving simulator led to additional insights into driver performance in situations where a rear-end crash was imminent. One insight is the possibility that limitations on driver's ability to perceive relative motion may have significant impact on crash prevention performance. Another insight is that extensions of the framework presented here may provide a better understanding of the relative severity of crashes. These insights may lead to additional testing or analysis to refine further the understanding of driver performance.

ACKNOWLEDGEMENTS

Many thanks to the NHTSA staff, especially Peter Martin of the Advanced Safety Systems Research Division as well as Wassim Najm of the Volpe National Transportation Systems Center for reviews, valuable comments, and contributions to this paper.

REFERENCES

1. Traffic Safety Facts, Overview 1998; U.S. Department of Transportation National Highway Traffic Safety Administration; DOT HS 808 956.
2. Wassim G. Najm, Christopher J. Wiacek, and August L. Burgett, Identification of Precrash Scenarios for Estimating the Safety Benefits of Rear-end Collision Avoidance Systems, Proceedings of the 5th World Congress on Intelligent Transport Systems, 12-16 October 1998, Seoul Korea.
3. Eugene I. Farber, Safety Improvements from Advanced Vehicle/Highway Technology, Proceedings of the 13th International Technical Conference on Experimental Safety Vehicles, Paris, France, Nov. 4-7, 1991, p 205.
4. P. Fancher, R. Ervin, Z. Bareket, S. Bogard, J. Sayer, J. Haugen, and M. Mefford, Fostering Development, Evaluation, and Deployment of Forward Crash Avoidance Systems (FOCAS) Final Report, June 2000, UMTRI-2000-27, DOT HS 808 437.

5. T. Wilson; Task 3 Interim Report: Test Results; Frontier Engineering, Inc., Advanced Programs Division; DOT HS 808 514.

6. P. Fancher, R. Ervin, Z. Bareket, S. Bogard, J. Sayer, J. Haugen, and M. Mefford, Intelligent Cruise Control Field Operational Test (Final Report), May 1998, UMTRI-98-17, DOT HS 808 437.

7. Forward Collision Warning Systems Final Report, Crash Avoidance Metrics Partnership (CAMP), March 12, 1999, NHTSA Cooperative Agreement Number DTNH22-95-R-07301, DOT HS 808 964.

8. J. D. Lee, D.V. McGehee, T.L. Brown, and M.L. Ries, Can Collision Warning Systems Mitigate Distraction Due to In-Vehicle Devices?, August 2000, NHTSA Sponsored Driver Distraction Internet Forum (http://www-nrd.nhtsa.dot.gov/driver-distraction/Welcome.htm).

9. National Highway Traffic Safety Administration Symposium on Rear-end Collision Avoidance Including Intelligent Cruise Control, The Johns Hopkins University Applied Physics Laboratory, October 22-23, 1998, DOT HS 808 809 012.

10. J. D. Lee, D.V. McGehee, T.L. Brown, and M.L. Ries, Driver Distraction, Warning Algorithm Parameters, and Driver Response to Imminent Rear-End Collisions in a High-Fidelity Simulator, July 27, 2000, Contract DTNH22-95-D-07168, University of Iowa Report IOQ No. Two(8-07633).

11. Burgett, A.; Carter, A.; Miller, R.; Najim, W.; and Smith, D.; A Collision Warning Algorithm for Rear-End Collisions, 16th International Technical Conference On the Enhanced Safety of Vehicles (ESV), Windsor, Canada, June 1-4, 1998, Paper No. 98-S2-P-31.

12. W.G. Najm and A.L. Burgett, Benefits Estimation for Selected Collision Avoidance Systems; 4th World Congress on Intelligent Transportation Systems, Berlin, Germany, October 1997.

CONTACT

The authors may be contacted at at the National Highway Traffic Safety Administration as follows: A. L. Burgett: august.burgett@nhtsa.dot.gov; R.J. Miller, Jr.: bmiller@nhtsa.dot.gov.

APPENDIX A – Derivation of Expressions for Crash Prevention Boundary

A.1 DERIVATION OF CROSSOVER DECELERATION, d_F*

Consider the case as seen in Figure 1 where the slowing vehicles just touch, bumper-to-bumper, without crashing. For this situation to occur, the following-vehicle deceleration, d_F, must be of a certain level that is dependent on the time of brake application. A subset of this case occurs when both vehicles stop at precisely the same moment. This will happen only at a single deceleration level, d_F*. If d_F is greater than d_F*, the lead vehicle will pull away after they touch. If d_F is less than d_F*, the lead-vehicle will stop before the following-vehicle (but they will eventually end up bumper-to-bumper). The term, d_F*, is defined as the crossover deceleration and is the basis of the Crash Prevention Boundary.

An expression for d_F* may be derived by considering the no-crash case when the vehicles stop simultaneously and are bumper-to-bumper. Suppose the initial conditions of range, R_o, velocity, V_o, and lead-vehicle deceleration, d_L, are given and t_b is the time of following-vehicle driver braking in reaction to a warning. The sequence of events is then: $t = 0$, t_b, and t_s.

Let $X_F(t_s)$ and $X_L(t_s)$ be the vehicle positions at time t_s. At time, t_s, the vehicles come to rest at and are just touching at which point both velocities are zero:

$$d(X_F(t_s))/dt = d(X_L(t_s))/dt = 0 \qquad (A-1)$$

and by definition,

$$d(X_L(t_s))/dt = V_0 - d_L t_s \qquad (A-2)$$

$$d(X_F(t_s))/dt = V_0 - d_F(t_s - t_b) \qquad (A-3)$$

in addition

$$t_s = V_0/d_L, \text{ and} \qquad (A-4)$$

$$t_s - t_b = V_0/d_F. \qquad (A-5)$$

The positions X_F and X_L of the vehicles then at time t_s are:

$$X_F(t_s) = t_b V_0 + V_0(t_s - t_b) - d_F(t_s - t_b)^2/2 \qquad (A-6)$$

$$X_L(t_s) = R_0 + t_s V_0 - d_L t_s^2/2 \qquad (A-7)$$

Equating A-6 and A-7 gives

$$V_0^2/d_L - V_0^2/2d_F = R_0 + V_0^2/2d_L \qquad (A-8)$$

$$V_0^2[1/d_L - 1/2d_F]/2 = R_0 + V_0^2/2d_L \qquad (A-9)$$

Solving for d_F we have the desired expression for d_F*:

$$\boxed{d_F = d_L V_0^2/(V_0^2 - 2d_L R_0) = d_F*} \qquad (A-10)$$

Thus, given a set of initial conditions of d_L, V_0, and R_0 when $d_F < d_F*$, the lead-vehicle stops before or at the same time as the following-vehicle. If the same initial conditions hold and $d_F > d_F*$ the following-vehicle stops before the lead-vehicle. For this second condition ($d_F > d_F*$) the bumper-to-bumper condition occurs while both vehicles are still moving.

A.2 BRAKE APPLICATION TIME IF THE LEAD-VEHICLE STOPS BEFORE THE FOLLOWING-VEHICLE.

In this scenario, the following-vehicle deceleration is less than d_F* (see Appendix A.1). At the end of the motion, the vehicles are stationary and bumper-to-bumper:

$$X_F(t_s) = X_L(t_s) \qquad (A-11)$$

And their final positions are

$$X_F(t_s) = V_0 t_b + V_0^2/(2d_F) \qquad (A-12)$$

$$X_L(t_s) = R_0 + V_0^2/(2d_L) \qquad (A-13)$$

Substituting into the first equation we have,

$$V_0 t_b + V_0^2/(2d_F) = R_0 + V_0^2/(2d_L) \qquad (A-14)$$

Solving for t_b gives the relationship

$$\boxed{t_b = R_0/V_0 + V_0[1/d_L - 1/d_F]/2} \quad \text{if } d_F < d_F* \qquad (A-15)$$

A.3 BRAKE APPLICATION TIME IF THE FOLLOWING-VEHICLE STOPS BEFORE THE LEAD-VEHICLE.

In this case the following-vehicle deceleration is greater than d_F*. The closest approach occurs while the two vehicles are still in motion so that they just touch at which time their velocities are equal. Thus, the <u>closest approach</u> of the two vehicles occurs at time t_c and requires the following relationships:

$$X_F(t_c) = X_L(t_c) \qquad (A-16)$$

$$dX_L(t_c)/dt = dX_F(t_c)/dt \qquad (A-17)$$

Also note that at closest approach the range rate, dR/dt, changes sign going from negative to positive, i.e.

$$dR/dt < 0 \qquad\qquad 0 < t < t_c \qquad (A-18)$$

$$dR/dt > 0 \qquad\qquad t > t_c \qquad (A-19)$$

The position of the two vehicles at the time of closest approach, t_c, is:

$$X_L(t_c) = R_0 + V_0 t_c - (d_L/2)t_c^2 \qquad (A-20)$$

$$X_F(t_c) = V_0 t_c - (d_F/2)(t_c - t_b)^2 \qquad (A-21)$$

Substituting into the position equation above,

$$V_0 t_c - (d_F/2)(t_c - t_b)^2 = R_0 + V_0 t_c - (d_L/2)t_c^2 \qquad \text{(A-22)}$$

Furthermore, the speed equations for the two vehicles at t_c may be written as,

$$dX_L(t_c)/dt = V_0 - d_L t_c \qquad \text{(A-23)}$$

$$dX_F(t_c)/dt = V_0 - d_F(t_c - t_b) \qquad \text{(A-24)}$$

which may be equated at the critical time, t_c:

$$V_0 - d_L t_c = V_0 - d_F(t_c - t_b) \qquad \text{(A-25)}$$

Rearranging equation A-25 gives,

$$t_b d_F = t_c(d_F - d_L) \qquad \text{(A-26)}$$

and solving for t_b yields:

$$t_b = [(d_F - d_L)/(d_F)]t_c \qquad \text{(A-27)}$$

Substituting this into equation A-25 above and simplifying to obtain an expression for t_c gives,

$$\begin{aligned} V_0 t_c - (d_F/2)[t_c - ((d_F - d_L)/(d_F))t_c]^2 = \\ R_0 + V_0 t_c - (d_L/2)t_c^2 \end{aligned} \qquad \text{(A-28)}$$

$$-(d_F/2)[d_L t_c/d_F]^2 = R_0 - (d_L/2)t_c^2 \qquad \text{(A-29)}$$

$$[d_L/2 - (d_F/2)(d_L/d_F)^2]t_c^2 = R_0 \qquad \text{(A-30)}$$

$$(d_L/2)[1 - d_L/d_F]t_c^2 = R_0 \qquad \text{(A-31)}$$

then $t_c = [2R_0/\{d_L(1 - d_L/d_F)\}]^{1/2} \qquad \text{(A-32)}$

Substituting into equation A-30 for t_b results in:

$$t_b = [(d_F - d_L)/d_F][2R_0/\{d_L(1 - d_L/d_F)\}]^{1/2} \qquad \text{(A-33)}$$

rearranging terms and simplifying gives the following expression:

$$\boxed{t_b = [(2V_0 T_h/d_L)(1 - d_L/d_F)]^{1/2}} \quad \text{if } d_F > d_F^* \qquad \text{(A-34)}$$

In summary then for both conditions:

$$t_b = R_0/V_0 + (V_0/2)[1/d_L - 1/d_F] \text{ if } d_F < d_F^* \qquad \text{(A-15)}$$

$$t_b = [(2V_0 T_h/d_L)(1 - d_L/d_F)]^{1/2} \quad \text{if } d_F > d_F^* \qquad \text{(A-34)}$$

A.4 TIME TO COLLISION EXPRESSIONS

As developed in Sections A.1 to A.3, given the condition of lead-vehicle braking, it is necessary to establish the governing mathematical relationships between d_F and t_b in relationship to the result of the conflict. For a specific rear-end driving scenario starting with initial velocity, V_0, initial range, R_0, and lead-vehicle deceleration level, d_L, there is a following-vehicle deceleration level, d_F, that determines a brake application time as described by the equations A-15 and A-34 which are the CPB equations.

It is often convenient to relate this expression to the time to collision, (TTC). TTC is the value in seconds at which collision will occur if the following-vehicle driver does not brake at all. TTC is obviously a function of lead-vehicle deceleration, d_L.

If d_L is relatively large, a collision will occur after the lead-vehicle has come to a stop. If d_L Is relatively small, the collision will occur before the lead-vehicle has come to a stop. Thus, for every initial value of R_0 and V_0 there is a value of d_L that separates these two collision conditions. That value of d_L, denoted by d_L^*, corresponds to the value for which the collision occurs at the instant that the lead-vehicle comes to a stop. The logic for development of the relationship for the time-to-collision (TTC) is similar to that in section A.1. However, the difference in the case discussed here from that of A.1 is that the following-vehicle takes no evasive braking action. To determine d_L^*, the lead-vehicle will take V_0/d_L seconds to come to a stop. During this time, the lead-vehicle will travel a distance of $V_0^2/2d_L$ and the following-vehicle will travel V_0^2/d_L. The locations for each vehicle after V_0/d_L seconds are $R_0 + V_0^2/2d_L$ and V_0^2/d_L, respectively for the lead and the following-vehicles. Since these locations must be the same, equating these expressions provides the relationship for d_L^*:

$$R_0 + V_0^2/2d_L = V_0^2/d_L \qquad \text{(A-35)}$$

And solving for d_L yields d_L^*:

$$d_L^* = V_0/(2T_h) \qquad \text{(A-36)}$$

In order to find expressions for TTC1 and TTC2, it will be sufficient to use equations already derived.

For the first condition using equation A-14 with the assumption that there is no braking by the following-vehicle and that t_b is defined as TTC1, gives:

$$V_0(\text{TTC1}) = R_0 + V_0^2/(2d_L) \qquad \text{(A-37)}$$

Solving for TTC1 and expressing the result in terms of T_h gives the espression

$$\text{TTC1} = T_h + V_0/2d_L \quad \text{if } d_L > d_L^* \qquad \text{(A-38)}$$

Then, for the second condition from equation A-31 where there is no braking by the following-vehicle and t_c is defined as TTC2, gives

$$\text{TTC2} = (2V_0 T_h/d_L)^{1/2} \quad \text{if } d_L < d_L^* \qquad \text{(A-39)}$$

Then in order to express the original equations in terms of TTC values, we have

$$t_b = \text{TTC1} - V_0/(2d_F), \quad \text{if } d_F < d_F^* \qquad \text{(A-40)}$$

$$t_b = \text{TTC2}(1 - d_L/d_F)^{1/2} \text{ if } d_F > d_F^* \qquad \text{(A-41)}$$

310

Thus, it is seen that d_F^* and d_L^* are analogous conditions that must hold simultaneously in relation to d_F and d_L for the proper expression of t_b in terms of TTC. Therefore, a hypothetical boundary of t_b vs. d_F can be formed for a set of initial conditions of R_0, d_L, and T_h which shall be termed the crash prevention boundary (CPB). The CPB may either be expressed in terms of equations A-15 and A-34 or by equations A-40 and A-41 with their attendant conditions.

APPENDIX B

SIMULATOR RESULTS

Tables 1 and 2, and the accompanying text, summarize a series of experiments that were run on the Iowa Driving Simulator. Figures B-1 through B-8 present details of the crash prevention performance for each subject in the experiment. Each figure contains performance data for a specific set of initial conditions in both the baseline condition, i.e. no warning was provided, and where a warning was provided. Each figure also includes the design point for the warning. The design point, or reference performance, for the "short" warning was a reaction time to the warning of 1.5 seconds and a braking level which produced a constant 0.4g deceleration. The design point for the "long" warning was a reaction time of 1.5 seconds and a constant deceleration of 0.75g. The crash prevention boundary that corresponds to the initial condition as well as the time-to-collision at the beginning of the event are also shown in each figure.

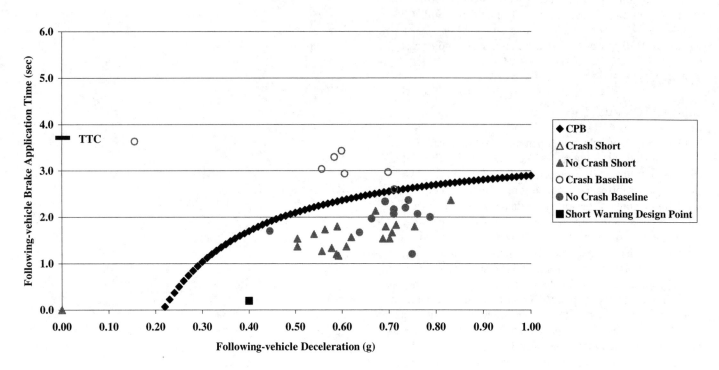

Figure B-1. IDS Test Conditions 1 and 2
(V=35mph, R=87.2 ft, dL=0.4 g's, Th=1.7 sec)

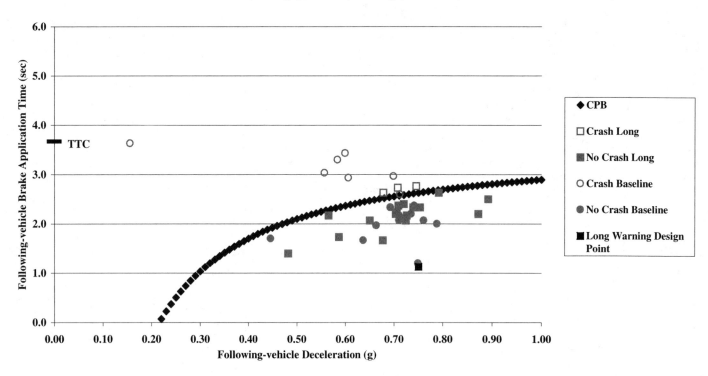

Figure B-2. IDS Test Conditions 1 and 3
(V=35mph, R=87.2 ft, dL=0.4 g's, Th=1.7 sec)

312

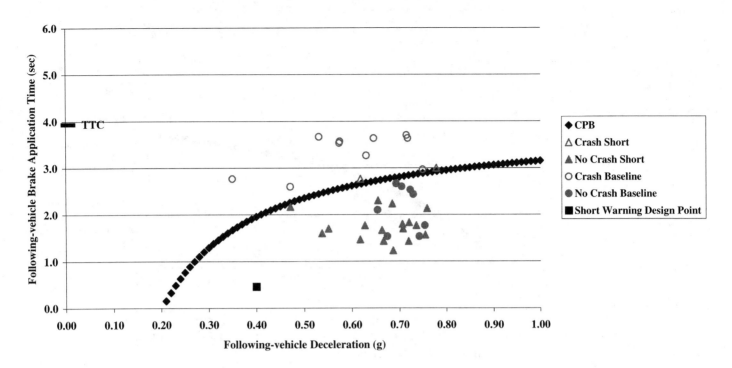

Figure B-3. IDS Test Conditions 4 and 5
(V=35mph, R=128.3 ft, dL=0.55 g's, Th=2.5 sec)

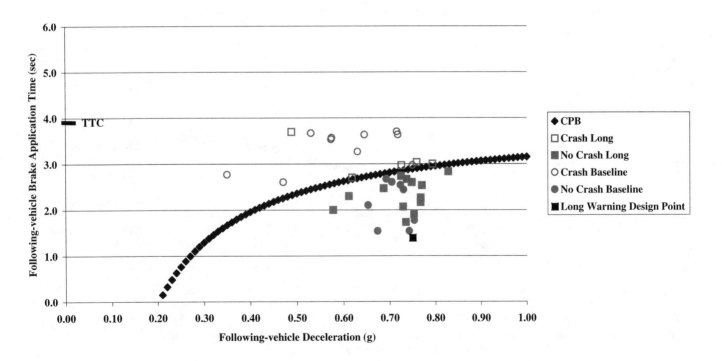

Figure B-4. IDS Test Conditions 4 and 6
(V=35mph, R=128.3 ft, dL=0.55 g's, Th=2.5 sec)

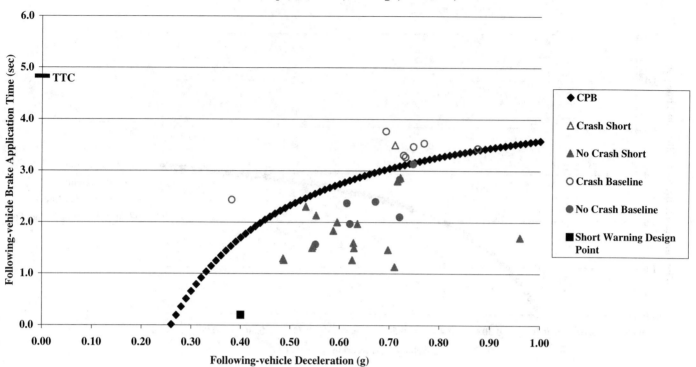

Figure B-5. IDS Test Conditions 7 and 8
(V=55 mph, R=137.1 ft, dL=0.4 g's, Th=1.7 sec)

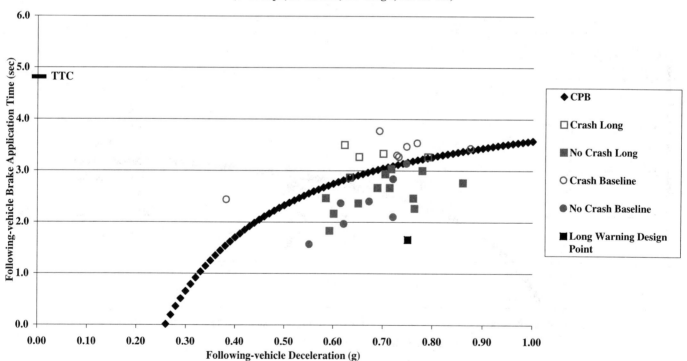

Figure B-6. IDS Test Conditions 7 and 9
(V=55 mph, R=137.1 ft, dL=0.4 g's, Th=1.7 sec)

314

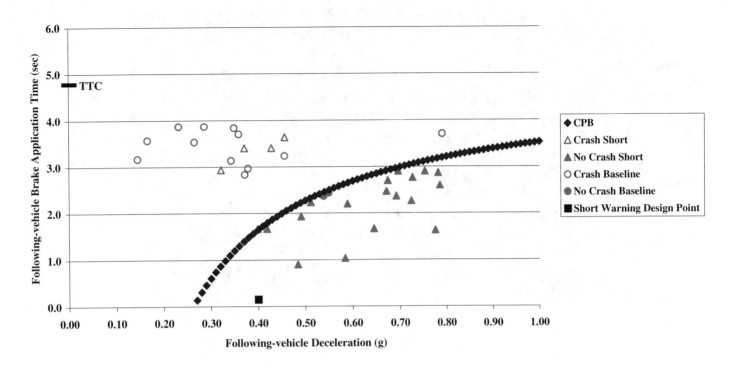

Figure B-7. IDS Test Conditons 10 and 11
(V=55 mph, R=201.7 ft, dL=0.55 g's, Th=2.5 sec)

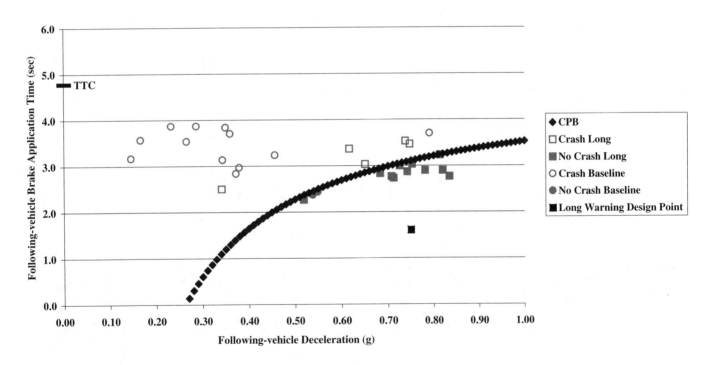

Figure B-8. IDS Test Conditons 10 and 12
(V=55 mph, R=201.7 ft, dL=0.55 g's, Th=2.5 sec)

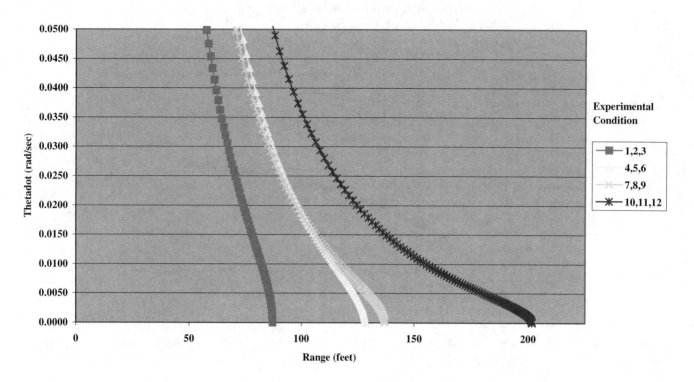

Figure B-9. Looming Effect for IDS Experiments

316

APPENDIX C. LIST OF FIGURES AND TABLES

LIST OF FIGURES

LIST OF TABLES

APPENDIX D. TERMS AND DEFINITIONS

ACC, Adaptive Cruise Control
CAMP, Crash Avoidance Metrics Partnership
CPB, Crash Prevention Boundary
d_F, average following-vehicle deceleration
d_L, average lead-vehicle deceleration
E, Effectiveness
FOCAS, Fostering Development, Evaluation, and Deployment
    of Forward Crash Avoidance Systems
IC, Initial Conditions
ICC, Intelligent Cruise Control
IDS, Iowa Driving Simulator
JHU/APL, Johns Hopkins University Applied Physics
Laboratory
NHTSA, National Highway Transportation Safety Agency
Range, Separation distance between two vehicles
R_0, Range at time of Initial Conditions
t_c, time of closest approach of two vehicles
T_h, headway at time of initial conditions
t_b, brake response time of following-vehicle
t_w, warning time
TTC, Time to Collision for following-vehicle
V_0, Velocity at time of Initial Conditions

2002-01-0396

Analysis of Off-Roadway Crash Countermeasures for Intelligent Vehicle Applications

Jonathan A. Koopmann and Wassim G. Najm
Volpe National Transportation Systems Center

ABSTRACT

This paper analyzes off-roadway crash countermeasure systems in support of the United States (U.S.) Department of Transportation's Intelligent Vehicle Initiative. Off-roadway crashes transpire when a moving vehicle departs the travel roadway and then experiences its first harmful event. This paper defines off-roadway crashes and describes their pre-crash scenarios and crash contributing factors. This information is then utilized to develop countermeasure concepts and concomitant functional requirements to warn drivers of imminent road edge crossing or vehicle control loss on straight or curved roadways. A technology survey follows to assess the status of state-of-the-art technologies within the categories of vehicle-based, infrastructure-based, or cooperative vehicle-infrastructure systems. This paper concludes with forecasts of the progression of future countermeasure systems towards the realm of cooperative technologies.

INTRODUCTION

This paper presents a review of off-roadway crash countermeasure concepts as part of the United States Department of Transportation's (U.S. DOT) Intelligent Vehicle Initiative (IVI). Off-roadway crashes are defined as those in which the first harmful event occurs off the roadway after a vehicle in transport departs the travel roadway. These crashes are one of seven focal areas in which the IVI encourages development and deployment of collision avoidance systems. The emphasis of the IVI programs is on enhancing the ability of the driver to solve traffic safety problems. This is accomplished primarily by vehicle-based countermeasure systems, but cooperative vehicle-infrastructure systems will play an increasingly large role as intelligent infrastructure is implemented [1]. Intelligent infrastructure is the necessary network of technologies, a communication and information backbone, which supports and unites key user services of the U.S. DOT's Intelligent Transportation System (ITS) program. The IVI explores possible vehicle-infrastructure cooperative systems that employ communications between the vehicle and the infrastructure, or among vehicles, to enhance the performance of vehicle-based systems or to enable the operation of some critical crash avoidance system functions.

Listed below are the steps that were undertaken to analyze promising countermeasure concepts for off-roadway crashes:
1. Identify common pre-crash scenarios and contributing factors
2. Devise crash countermeasure concepts
3. Develop countermeasure functional requirements
4. Review existing performance guidelines
5. Survey off-roadway crash avoidance technologies
6. Project the evolution of future systems

Next, this paper provides the results of Step 1. This is followed by the results of Steps 2-3. Step 4 summarizes and builds upon the efforts of a previous project that developed performance guidelines for run-off-road collision avoidance systems [2]. After, the results of Step 5 on the state-of-the-art technology survey are presented. Finally, this paper concludes with remarks related to Step 6.

OFF-ROADWAY CRASHES

Off-roadway crashes involved approximately 1,251,000 light vehicles (passenger cars, minivans, pickup trucks, and sports utility vehicles) in 1998 and 211,000 commercial vehicles (medium and heavy trucks) over a 3-year period from 1996 through 1998 in the U.S. based upon police accident reports from the General Estimates System (GES) crash database [3,4]. These vehicles comprised 11.5% and 11.3% respectively of all light and commercial vehicles involved in police reported crashes in the U.S.

PRE-CRASH SCENARIOS

Tables 1 and 2 provide the frequency of the six most common off-roadway pre-crash scenarios among light and commercial vehicles for freeways and non-freeways. The 1996-1998 GES databases do not contain any variable that directly identifies the roadway type. Therefore, freeways were defined as divided roadways with speed limits of 88.5 Km/h (55 mph) or greater. The identification of pre-crash scenarios was based on the

Table 1.
Distribution of Off-Roadway Pre-Crash Scenarios by
Roadway Type for Light Vehicles (1998 GES)

Pre-Crash Scenario	All Roadway		Non-Freeway		Freeway	
	Frequency	Relative Frequency	Frequency	Relative Frequency	Frequency	Relative Frequency
Going straight and departed road edge	327,000	36.6%	289,000	38.0%	38,000	29.0%
Going straight and lost control	210,000	23.5%	153,000	20.2%	57,000	43.1%
Negotiating a curve and lost control	153,000	17.1%	133,000	17.5%	19,000	14.8%
Negotiating a curve and departed road edge	104,000	11.6%	97,000	12.7%	7,000	5.4%
Initiating a maneuver and departed road edge	51,000	5.7%	49,000	6.4%	2,000	1.6%
Initiating a maneuver and lost control	48,000	5.4%	40,000	5.2%	8,000	6.0%
Total	892,000	100.0%	761,000	100.0%	132,000	100.0%

Numbers in cells rounded to the nearest 1,000

Table 2.
Distribution of Off-Roadway Pre-Crash Scenarios by
Roadway Type for Commercial Vehicles (1996-1998 GES)

Pre-Crash Scenario	All Roadway		Non-Freeway		Freeway	
	Frequency	Relative Frequency	Frequency	Relative Frequency	Frequency	Relative Frequency
Going straight and departed road edge	44,000	34.8%	36,000	34.7%	8,000	35.3%
Initiating a maneuver and departed road edge	37,000	29.3%	36,000	34.1%	1,000	6.2%
Going straight and lost control	18,000	14.4%	10,000	9.6%	8,000	37.2%
Negotiating a curve and lost control	12,000	9.5%	10,000	9.2%	2,000	11.0%
Negotiating a curve and departed road edge	10,000	7.7%	8,000	7.8%	2,000	7.0%
Initiating a maneuver and lost control	6,000	4.4%	5,000	4.6%	1,000	3.2%
Total	127,000	100.0%	105,000	100.0%	22,000	100.0%

GES variables of "critical event", "movement prior to critical event", and "roadway alignment". The critical event was broken down into two categories: control loss and road edge departure. Control loss scenarios involve excessive speed or speeding on poor road conditions. Road edge departure scenarios include vehicles traveling over the left or right edge of the roadway, or simply departing the end of a roadway (T-shape). The vehicle movement prior to the critical event was separated into traveling on a straight road, negotiating a curve, and initiating a certain maneuver. The last vehicle movement classification generally encompasses slow speed events such as turning at an intersection, changing lanes, parking, or slowing. The total of Tables 1 and 2 amounts to 71% and 60% of all light and commercial vehicles involved in off-roadway crashes, respectively. Excluded from these tables are off-roadway crashes in which one of the involved vehicles was backing up, had a component failure, performed an evasive maneuver to avoid hitting another vehicle, or had no impact (e.g., caught fire). These excluded crashes are either addressed by other areas of the IVI or are not part of the scope of the IVI.

CRASH CONTRIBUTING FACTORS

A priority scheme was devised to arrange and rank crash contributing factors in the following descending order:
- Alcohol or drugs
- Driver impairment
- Driver distraction
- Speeding
- Hit and run
- Other:
  - Light conditions
  - Atmospheric conditions
  - Roadway surface conditions

The analysis of crash contributing factors first determined the portion of target off-roadway crashes that involved alcohol or drugs. These crashes were removed from the data set once they were identified, and the remaining crashes were examined for the subsequent factor. Following this process of elimination, the next factor examined was the driver's physical impairment - illness, blackouts, drowsiness, fatigue, or impairment due to a previous injury. Subsequently, driver distraction factors such as passengers, vehicle instrument display, phone, other internal devices, other crash, or other external distractions were reviewed. It should however

Table 3.
Off-Roadway Crash Countermeasures by Roadway
Type for Light Vehicles (1998 GES)

Crash Countermeasures		Pre-Crash Scenarios	Contributing Factors*	Roadway Type					
				All Roadways		Non-Freeways		Freeways	
				Min.	Max.	Min.	Max.	Min.	Max.
Intoxicated Driver Monitor		S1 - S6	F1	19.7%		20.8%		13.2%	
Drowsy Driver Monitor		S1 - S6	F2	5.9%		5.5%		8.2%	
Excessive Speed Warning System:	Upcoming Curve	S3 & S4	(F4.a - F4.a &F4.b)	4.9%	10.2%	5.4%	10.5%	2.0%	8.0%
	Posted Speed Limit	S1 - S6	(F4.a - F4.a &F4.b)	11.4%	24.5%	11.7%	23.0%	9.4%	33.7%
	Roadway Surface Conditions	S1 - S6	(F4.a &F4.b - F4.a, F4.b, &F6)	24.5%	41.0%	23.0%	38.7%	33.7%	54.1%
Pavement Conditions Monitor		S1 - S6	(F6 - F6 & F4.b)	16.4%	29.6%	15.7%	27.0%	20.5%	44.8%
Vision Enhancement System:	Night Vision	S1 - S6	F5.a	10.3%		10.2%		10.4%	
	All Weather	S1 - S6	F5.a, F5.b, & F5.c	21.7%		20.9%		25.9%	
Road Edge Departure Warning System		S1, S3, & S5	(F3 & F5 - F3, F5, F1 & F2)[t]	22.0%	39.6%	23.3%	41.5%	14.7%	28.8%
Advanced Vehicle Controls		S2, S4, & S6	F4.a, F4.b, F5.a, F5.b, F5.c, F6.a, & F6.b	42.5%		38.7%		64.3%	

* ( ) indicates range of values

[t] Includes Non-Speeding - Day, Clear, Dry values

Pre-Crash Scenarios Key

S1 - Going straight and departed road edge
S2 - Going straight and lost control
S3 - Negotiating a curve and departed road edge
S4 - Negotiating a curve and lost control
S5 - Initiating a manuever and departed road
  edge (Turning, Parking, Changing Lanes)
S6 - Initiating a manuever and lost control
  (Turning , Changing Lanes, Slowing)

Contributing Factors Key

F1 - Alcohol or drugs
F2 - Driver impaired (excluding F1)
F3 - Driver distracted (excluding F1 & F2)
F4 - Speeding: (excluding F1, F2, & F3)
  F4.a - Dry roads
  F4.b - Slippery roads
F5 - Reduced Visibility: (excluding F1, F2, F3, F4, & Hit and Run)
  F5.a - Dark (Night, clear)
  F5.b - Adverse weather in daytime
  F5.c - Adverse weather in nighttime
F6 - Slippery surface (excluding F1, F2, F3, F4, & Hit and Run)
  F6.a - Clear weather
  F6.b - Adverse weather

Table 4.
Off-Roadway Crash Countermeasures by Roadway
Type for Commercial Vehicles (1996 - 1998 GES)

Crash Countermeasures		Pre-Crash Scenarios	Contributing Factors*	Roadway Type					
				All Roadways		Non-Freeways		Freeways	
				Min.	Max.	Min.	Max.	Min.	Max.
Intoxicated Driver Monitor		S1 - S6	F1	0.6%		0.4%		1.7%	
Drowsy Driver Monitor		S1 - S6	F2	1.7%		0.7%		7.1%	
Excessive Speed Warning System:	Upcoming Curve	S3 & S4	(F4.a - F4.a &F4.b)	3.8%	5.5%	4.1%	4.9%	2.6%	8.3%
	Posted Speed Limit	S1 - S6	(F4.a - F4.a &F4.b)	8.2%	14.8%	8.2%	12.6%	8.4%	25.9%
	Roadway Surface Conditions	S1 - S6	(F4.a &F4.b - F4.a, F4.b, &F6)	14.8%	28.6%	12.6%	26.3%	25.9%	40.2%
Pavement Conditions Monitor		S1 - S6	(F6 - F6 & F4.b)	13.8%	20.4%	13.7%	18.1%	14.3%	31.7%
Vision Enhancement System:	Night Vision	S1 - S6	F5.a	10.2%		8.4%		19.1%	
	All Weather	S1 - S6	F5.a, F5.b, & F5.c	18.9%		16.4%		31.5%	
Road Edge Departure Warning System		S1, S3, & S5	(F3 & F5 - F3, F5, F1 & F2)[t]	43.9%	45.7%	44.8%	45.5%	39.5%	46.8%
Advanced Vehicle Controls		S2, S4, & S6	F4.a, F4.b, F5.a, F5.b, F5.c, F6.a, & F6.b	21.1%		17.0%		42.3%	

321

be noted that distraction is rarely cited in police accident reports so the actual value could be significantly higher than the GES values of 3.9% for light vehicles and 5.0% for commercial vehicles; the 1997 crashworthiness data system (CDS) cites distraction as a cause of 18% of all light vehicle crashes. Following distraction, vehicle speed was investigated to determine if it played a factor in the crash. Next, the analysis classified "hit and run" crash cases in which the driver left the scene of the crash. If the crash fit none of the above factors, it was then categorized as "Other" and accordingly characterized by the environmental circumstances surrounding the crash. Light conditions were sorted as either light or dark. Dark conditions comprise dusk, dawn, and artificially lighted roadways. Atmospheric conditions were separated into clear and adverse weather. Adverse weather conditions include rain, sleet, snow, fog, and smog. Finally, roadway surface was grouped into dry or slippery. Slippery surfaces consist of wet, snowy, icy, or oily roadways.

OFF-ROADWAY CRASH COUNTERMEASURE CONCEPTS

Tables 3 and 4 list seven countermeasure concepts that were devised to address the target off-roadway pre-crash scenarios and concomitant contributing factors for light and commercial vehicles, respectively. Also shown in these tables is the percentage of crashes that could be averted, by roadway type, if the concepts were 100% effective. This paper will concentrate on road edge departure warning, excessive speed warning, pavement condition monitoring, and vision enhancement systems. The intoxicated and drowsy driver monitoring systems as well as advanced vehicle stability controls are not discussed here, since the first two systems are being more thoroughly investigated under other IVI programs while the third is available as products on current production vehicle models.

The excessive speed warning system concept may encompass three different functionalities with varying degree of complexity, each applying to different pre-crash scenarios and contributing factors. The first function alerts the driver if the speed of the vehicle exceeds the design safe speed of traversing an upcoming curve based on its geometrical configuration. The second function warns the driver if the speed of the vehicle surpasses the posted speed limit, and thus applies to all roadway alignments. The third function augments the first two functions by incorporating information about the conditions of the roadway surface ahead, and warns the driver if the speed of the vehicle is excessive for the existing roadway conditions even though it might fall below the design safe speed or posted speed limit. The pavement condition monitoring countermeasure simply functions as an advisor to the driver by providing information about an upcoming slippery surface. The driver in this case is responsible for making the decision to whether or not to slow down. The minimum values shown in Tables 3 and 4 encompass only non-speeding drivers while the maximum values

also include drivers who were speeding. Vision enhancement systems may consist of two types of possible countermeasures. The simpler of the two is night vision. These systems apply to reduced visibility situations at night but function properly only under clear weather conditions. On the other hand, the all weather vision enhancement system is more advanced and applies to all reduced visibility conditions, including inclement weather, both during the day and at night. Road edge departure warning countermeasures apply to an umbrella of contributing factors under the road edge departure pre-crash scenario category. The minimum values in Tables 3 and 4 were derived from addressing distraction, reduced visibility, as well as non-speeding – day, clear, dry conditions. The maximum values include every contributing factor except speeding with the belief the system would prevent some road departures of all types independent of driver state. Finally, advanced vehicle control systems target control loss crashes by affecting the dynamics of the vehicle to maintain the driver's intended course.

FUNCTIONAL AND PERFORMANCE REQUIREMENTS

Table 5 lists the minimum functional and performance requirements of the sensory element for two major off-roadway crash countermeasure concepts: road edge departure and excessive speed warning systems. Based on Tables 3 and 4, the road edge departure warning system has the potential to alleviate about 40% and 46% respectively of all light and commercial vehicles involved in off-roadway crashes on all roadway types. The excessive speed warning system, on the other hand, may mitigate about 41% and 29% respectively of all light and commercial vehicle involved in off-roadway crashes. The minimum performance requirements indicated in Table 5 were obtained from a previous project that developed performance specifications for single vehicle roadway departure crash avoidance systems [2]. This project built and tested an experimental testbed that performed lane drift warning and curve speed warning functionalities.

Both countermeasure systems share common sensory functions to:
1. Determine the lateral position of the vehicle within its lane of travel, and its orientation relative to the roadway;
2. Identify the geometric characteristics and traffic signs of the upcoming roadway segment;
3. Monitor the dynamic states of the vehicle; and
4. Predict driver intention.

The excessive speed warning system must also detect the pavement conditions of the upcoming roadway segment in addition to the sensory functions listed above.

Table 5.
Functional and Performance Requirements for the Sensory Element
of Selected Off-Roadway Crash Countermeasure Concepts

Functions	1	2	Performance
Determine vehicle position/orientation relative to roadway:			
a. Current travel lane	♦		
a. Lateral position within lane (accuracy)	♦		≤ 10 cm
b. Position relative to curve (accuracy)	♦	♦	≤ 5 m
i. Detect curve presence and characteristics		♦	≥ 200 m
Identify characteristics of upcoming road segments:			
a. Number of lanes	♦		
i. Lane width/Shoulder width (accuracy)	♦		≤ 10 cm
b. Roadway alignment (straight, curve right/left)	♦	♦	
c. Curvature of roadway segment (accuracy)	♦	♦	≤ 0.0005 m^{-1}
i. Operating radius	♦	♦	≥ 60 m.
ii. Change of curvature	♦		≤ 0.0004 m^{-2}
d. Superelevation (accuracy)	♦	♦	$\leq 3\%$
i. Functional superelevation		♦	$\leq 12\%$
e. Presence of exits or cross streets (detection distance)	♦	♦	≥ 200 m
f. Posted speed limit		♦	NA
Vehicle dynamic status:			
a. Velocity (accuracy)	♦	♦	≤ 3 mph
b. Longitudinal acceleration (accuracy)	♦	♦	≤ 0.03 g
c. Lateral acceleration (accuracy)	♦		NA
d. Yaw Rate (accuracy)	♦		≤ 1 deg./sec.
e. Vehicle path's radius of curvature	♦		NA
e. Stability Characteristics (roll stiffness, mass distribution, in vehicle load, tire condition, rollover potential)		♦	NA
Identify pavement characteristics of upcoming road segment:			
a. Side friction coefficient (accuracy)		♦	≤ 0.05
b. Longitudinal friction coefficient (accuracy)		♦	≤ 0.05
Predict driver's intent:			
a. Evasive manuever, turning at cross street, lane change, pulling off to side of roadway	♦		NA
b. Intended path of travel		♦	Vehicle's turn signals
c. Awareness of curve severity		♦	Curve negotiation behavior, brake pedal activation, accelerator pedal release

1: Road Edge Departure Warning System
2: Excessive Speed Warning System
NA: Not Available from [2]

SURVEY OF OFF-ROADWAY CRASH COUNTERMEASURE SYSTEMS AND ENABLING TECHNOLOGIES

A literature survey was conducted to evaluate the technical readiness of system and subsystem technologies capable of implementing the countermeasure concepts. Three system classes were considered: infrastructure-based systems, vehicle-based systems, and cooperative vehicle-infrastructure systems. The third class of systems includes cooperative infrastructure-to-vehicle systems and cooperative vehicle-to-vehicle systems. The inclusion of infrastructure-based systems in this survey helps to identify cooperative systems that would transmit remote sensor data from these infrastructure systems to vehicles via communications link. Such a link would enhance the performance of vehicle-based systems or enable the operation of additional critical countermeasure system functions.

ROAD EDGE DEPARTURE WARNING SYSTEMS

Road edge departure systems have been developed and tested using vehicle-based, infrastructure-based, and cooperative systems.

Vehicle-Based Systems: This class of systems generally uses a video camera together with on board real-time digital image processing to determine the lane boundaries and vehicle lateral position within the travel lane. Two such systems are currently commercially available: SafeTRAC and AutoVue. The SafeTRAC system consists of a camera and a dash-mounted display unit that contains the processor [5]. The processor is a 100 MHz 486 with maximum output data speed of 19,200 bps at 60 Hz over a RS232 connection. The forward-looking camera views the upcoming roadway and determines when the vehicle is making an unintended lane change. A feature unique to SafeTRAC is real-time driver alertness feedback. The system analyzes the driver's fatigue level based upon movement in relation to the lane and issues warnings when the driver is drowsy, impaired or inattentive. The AutoVue system is similar to the previous system except all the elements have been integrated into one unit [6]. This product is currently available as an option on some European trucks. Vehicle-based systems are at the conceptual and prototype level at many other companies and universities. The systems predominately use a single CMOS or CCD camera. A few more advanced systems use stereo cameras but this requires more processing power and sophisticated algorithms.

The major shortfall of all vision- or optical-based systems is the ability to recognize lane boundaries during adverse weather or on poor roadway conditions such as snow-covered roadways, night rain with oncoming headlights or streetlights, very low sun angles, shadows, inferior lane markings, or obscuration of markings by other traffic. Another common problem with this type of systems is the inability to see the roadway at farther distances, typically ≤ 30 m (98 ft), and the sensitivity to vehicle pitch and vibration as well as vertical road curvature [7]. The short look-ahead distance is caused mainly by the camera's low visibility angle from its mounting position within the vehicle. This is less of a problem in commercial vehicles because the camera can be mounted much higher and therefore see further down the road. Continuing advancements are being made in the camera and processing technology to overcome these physical and technological problems, but until they do these systems will only function in good visibility on well-marked roadways.

Another vehicle-based system still in the prototype stage is location finding using the Global Positioning System (GPS) combined with highly accurate digital maps. The accuracy of both GPS and maps are still not good enough however for these systems to be feasible for commercial use. Most map databases of roadways are no more precise than ±15 m (±49 ft) and coverage of all roadways is still not complete. In addition, even with the recent increase in accuracy of GPS with the removal of selective availability, it is still only able to find a location within ±3 m (±10 ft). For effective use in lane departure systems, GPS accuracy would have to be on the order of centimeters with accompanying maps of the same accuracy. A third common problem with GPS use is the

GPS signal dropout or loss of satellite lock due to the presence of large obstacles such as overpasses, trees, or tall buildings. When too few satellites are available, positioning isn't possible and GPS usually takes up to a minute to reacquire satellite lock and reestablish the location.

Infrastructure-Based Systems: Rumble strips are the main form of infrastructure road-departure warning systems available today. Milled-in rumble strips are the preferred type because of their effectiveness at issuing a warning, low installation cost, and their minimal negative impact to the roadway surface [8]. They have been widely deployed on highways and proven to reduce off-roadway crashes by 20-70% [9]. Like vehicle-based warning systems, rumble strips currently have limited applicability to most roadways.

Cooperative Vehicle-Infrastructure Systems: These systems involve elements of the vehicle interacting with intelligent infrastructure to alert the driver of an impending crash. Most systems utilize specially installed ferromagnetic road markers that can be easily and accurately identified by a sensing vehicle, including magnetic nails or tacks and magnetic tape [10]. The sensing vehicle travels the roadway tracking these lane markers and warns the driver if a lane departure is occurring. Cooperative systems offer the most accurate and consistent information; however, their effectiveness depends on the deployment of markers on roadways. The drawbacks of embedding magnets in the road include interference with other metals in the pavement and logistical problems such as cost, wide-scale deployment, and maintenance. Cooperative systems are also seen as the gateway toward the evolution of fully autonomous steering systems, however they are currently only at the prototype phase.

EXCESSIVE SPEED WARNING SYSTEMS

Excessive speed warning systems warn the driver of higher speed than is safe for curves, posted speed limits, or deteriorated roadway conditions. These systems are viewed to possess these three different functionalities with many overlapping components that could be easily integrated.

Vehicle-Based Systems: The warning of over speeding for an upcoming curve may be accomplished by the utilization of GPS with digital maps in order to locate the vehicle in relation to the approaching curve and to obtain the geometrical characteristics of the curve. The driver is then warned using information about the vehicle travel speed, distance to the curve, and recommended safe speed for the particular curve. The advantage of this system is that it does not require high accuracy GPS to issue warnings. The digital map information on the curve however must be accurate to the values listed previously in the functional and performance requirements section to issue warnings correctly. To limit crashes due to exceeding the speed limit, the maps would contain speed limit information for all roadways and could then be used to issue an alert when the limit is exceeded [11]. The two functionalities that warn the driver of over speeding for the upcoming curve and exceeding the posted speed limit could be easily integrated using digital map and vehicle speed information. The third functionality that warns the driver of speeding on poor roadway conditions is enabled by downward facing optical sensors to determine the coefficient of friction between the tires and the roadway. By focusing infrared light on the roadway, the sensors can detect the presence of snow, ice, or water and its thickness [12]. This information, combined with the vehicle dynamics, allows the system to calculate the safe travel speed and issue a warning if the current speed exceeds it.

Infrastructure-Based Systems: The use of roadside signs, coupled with radar and possibly weigh-in-motion sensors, provides information that allows warnings to be actively targeted to vehicles approaching the curve too fast. The radar measures the vehicle speed while the weigh-in-motion sensors determine the weight and type of the vehicle. Such systems have been implemented on a limited basis in California [13]. Active warning signs are particular useful for commercial vehicles on curves where rollovers often occur because the warning can be issued for their particular load circumstance. The main disadvantage of these systems is the cost and difficulty deploying them at more than a few key curves. Infrastructure-based systems also utilize radar to detect the vehicle speed and issue warnings to the driver via active signs based on a comparison with the posted speed limit. The disadvantage of active signs is that the driver may easily ignore them because they presumably know the posted speed limit from the roadside signs.

Cooperative Vehicle-Infrastructure Systems: These systems use combinations of the technologies mentioned previously and transmit information to the vehicle via communication link in order to more effectively and accurately warn the driver of over speeding for the upcoming curve or the posted speed limit. Cooperative systems relay information from surface sensors that detect poor roadway conditions to approaching vehicles. Other cooperative systems would transmit roadway information detected by a vehicle to other vehicles approaching that section of roadway via a cellular link. This would not be dependent on roadway sensors but vehicles would still receive advance warning of roadway conditions ahead.

PAVEMENT CONDITIONS MONITOR

This system only indicates to the driver when the roadway ahead is slippery. It does not process information about vehicle speed to warn the driver of excessive speed for the existing conditions. Thus, infrastructure-based systems are more feasible to implement such a system because no information is needed about the vehicle. Sensors could be placed in locations where crashes commonly occur due to poor roadway conditions, such as bridge overpasses, to advise drivers of the potential hazard ahead. In-vehicle systems would function essentially the same by advising drivers, but they would not need the computer logic necessary to assimilate all the various dynamic and vehicle speed information.

VISION ENHANCEMENT SYSTEM

Vision enhancement systems help the driver see the upcoming roadway better under poor visibility conditions.

Night Vision: These systems assist the driver to see in dark but clear weather. Night vision systems using in-vehicle thermal imaging have already been developed and put into production by General Motors. This system utilizes a thermal sensor mounted behind the grill of the vehicle to detect the upcoming roadway. The image is then displayed on the lower potion of the windshield below the line of sight of the driver. Night vision however only works well during clear conditions; heavy airborne moisture is difficult for the system to see through.

All Weather: This system improves the driver's view of the roadway in all poor visibility conditions. These are again in-vehicle devices that would likely display an image of the roadway ahead to the driver with the same type of head-up display as the night vision system. A prototype all weather system using active laser headlamps that illuminate the road with infrared light has been developed by DaimlerChrysler [14]. A camera records the reflected infrared light and displays the information to the driver. The major advantage of infrared light over conventional headlights is that it is nearly invisible to the human eye. This makes it possible for the lights and camera system to observe up to 152 m (500 ft) in front of the vehicle instead of the typical 40 m (130 ft) with conventional high beams, without blinding oncoming drivers. An alternative solution proposed by BMW is to measure the level of visibility and issue a warning if the vehicle is traveling too fast for the visibility conditions. There are also infrastructure improvements, which could reduce the amount of off-roadway crashes due to poor visibility. These include improving lane delineation via retro reflectors, roadway illumination, and shoulder improvements. Finally, cooperative systems using fluorescent paints and ultraviolet headlights have been explored in Europe at the research and demonstration levels.

DISCUSSION

This analysis attempted to project the path of development for off-roadway crash countermeasures by examining each countermeasure concept for existing prototype or experimental systems. The literature search conducted revealed predominantly vehicle-based systems; however, a few examples of infrastructure-based systems already implemented on a limited scale were found. Few truly cooperative vehicle-infrastructure systems were discussed, although this is viewed as the long-term direction of deployment for the IVI program as it accelerates deployment and increases safety benefits of crash avoidance technology.

The introduction of a new safety system depends on many factors such as system capability and maturity (mass production); liability and invasion of privacy concerns; and consumer acceptance, willingness to pay, and perceived benefits of the system. Introductions of these new systems could be constrained by the capability of the underlying technology and algorithms to meet desired performance objectives given vehicle size and operating characteristics, and to perform in a range of operating environments or conditions. If a new automotive product is ready for commercialization, its rollout will frequently be staged over a 2 to 5 year period and will also be limited to certain platforms or models and geographical markets. The cost ratio – the retail cost of the technology to the vehicle's base price – strongly influences the platforms and models on which these technologies are introduced. Given the high cost of many of these new emerging IVI systems, the tendency is to first offer them as options on the most expensive platforms and models. Although they might have already moved through the development and the testing stages, the commercialization of IVI systems would not occur until suppliers have succeeded in the creation of a centralized information control package for cost reasons. A limited consumer interest or the result of an ill-fated introduction of one IVI system would also delay the debut of these systems. Reliable estimates of IVI system safety benefits highly depend on the availability of data that describe driver performance with and without the assistance of an IVI system in real-world driving environment.

The current potential for deployment of off-roadway crash countermeasures varies based on the maturity and accuracy of the technology used. In-vehicle systems have, in general, shown the most large-scale deployment potential but their success to date still lacks the precision required to be viable countermeasures. Infrastructure-based systems, in contrast, have shown a high level of accuracy but a limited scope of deployment due their expensive nature, with the exception of rumble strips, and fixed location. These factors have led to most infrastructure-based countermeasures being deployed only in specifically targeted, high crash areas.

The development and installation of IVI systems should evolve into new vehicles in an integrated fashion. Many vehicle-based or cooperative vehicle-infrastructure systems based on GPS technologies will spawn as additions to existing telematics units already in vehicles. As telematics filter down to less expensive cars, systems such as excessive speed warning or roadway condition monitors become more feasible as options on all vehicles. The addition and integration of various vehicle subsystems that can share electronics, sensors, and software will increase system functionality and improve performance while minimizing complexity and cost.

CONCLUSION

An analysis was conducted to identify most promising countermeasures to off-roadway crashes. This analysis was based on a detailed definition of off-roadway crashes using 1996-1998 GES. Crash statistics were broken down by pre-crash scenario, roadway type, and contributing factor. Countermeasure concepts were considered for infrastructure-based, vehicle-based, and cooperative vehicle-infrastructure systems. For each proposed countermeasure system, this paper presented a summary of a literature survey about the technical readiness of available systems, sensors, and enabling technologies.

To expedite the development and deployment of IVI off-roadway crash avoidance systems, future research is recommended in the areas of performance specifications, objective test procedures, safety benefits estimation, and related driver performance data collection. This research must be technology independent and does not imply any particular implementation, either for sensors or for cooperative versus autonomous components. The technology implementation will be a deployment decision made by deployers based on trade-offs such as revenue and market issues, cost-benefit, and many other factors. Specifically, the performance specifications in Table 5 may be completed for the sensor, warning algorithm, and driver-vehicle interface elements of each concept. More naturalistic driving data will be needed to sufficiently understand the pre-crash kinematics problem and help to accurately fill in this table. The development of objective test procedures may begin for some concepts where the performance specifications are now adequate to solve a part of the crash problem. Finally, a field operational test may be initiated for those implementations that the deployers feel are ready for deployment, as a final pre-deployment decision gate.

ACKNOWLEDGMENT

The authors gratefully acknowledge the technical support that Dr. David L. Smith of the National Highway Traffic Safety Administration's Office of Vehicle Safety Research has contributed to this research effort.

REFERENCES

1. U.S. DOT ITS Joint Program Office, *National Intelligent Transportation Systems Program Plan Five-Year Horizon*. FHWA-OP-00-008, August 2000.
2. Pomerleau, D. and Everson, J., "Run-Off-Road Collision Avoidance Using IVHS Countermeasures – Final Report". DOT HS 809 170, December 1999.
3. Najm, W.G., Schimek, P.M., and Smith, D.L., "Definition of the Light Vehicle Off-Roadway Crash Problem for the Intelligent Vehicle Initiative". Paper No. 01-3194, TRB 80th Annual Meeting, Washington, D.C., January 2001.

4. Boyle, L. Ng and Najm, W.G., "Analysis of Off-Roadway Crashes for Intelligent Commercial Vehicle Applications". Paper No. 01TB-58, 2001 SAE International Truck & Bus Meeting & Exposition, Chicago, IL, November 2001.
5. Safetrac by Assistware: http://www.assistware.com
6. AutoVue by Iteris: http://www.iteris.com
7. Hamilton, L., Humm, L., Daniels, M., and Yen, H., "Vision Sensors and the Intelligent Vehicle". Automotive Engineering International, October 2001.
8. "Shoulder Rumble Strips: Effectiveness and Current Practice". Federal Highway Administration, Wyoming Division Office, April 2, 1998.
9. Hickey, J., "Shoulder Rumble Strip Effectiveness: Drift-Off-Road Accident Reductions on the Pennsylvania Turnpike". 76th Annual Meeting of the Transportation Research Board, Washington, D.C., January 1997.
10. Soma, H., Suzuki, K., Hiramatsu, K., and Ito, T., "Experimental Investigation of Dynamical Lateral Vehicle Position on Japanese Expressways for Design and Standardization of Lane Departure Warning System". Paper No. 3031, 6th World Congress on Intelligent Transport Systems, Toronto, Canada, November 1999.
11. AVV Transportation Research Center, Automated Vehicle Guidance with ADA Technology. February 2001.
12. Whipple, C.T., "Sensors Keep an Eye on Road Surfaces". Photonics Spectra, June 1996.
13. Wenham, R., "The Caltrans Advanced Curve Warning and Traffic Monitoring System". ITS Quarterly, Vol. 8, No. 4.
14. "Prototype infra-red night vision system". ITS International, May/June 2000.

National Transportation Safety Board Accident Investigations and Recommendations on Technologies to Prevent Rear-End Collisions

Jennifer H. Bishop and Lawrence E. Jackson
National Transportation Safety Board

ABSTRACT

Rear-end collisions account for over 1.7 million crashes[1] that occur on U.S. highways each year, and the number is growing. In the past 2 years, the National Transportation Safety Board (NTSB) investigated nine rear-end collisions in which 21 people died and 182 were injured. Technology exists today that could have saved these lives. Since 1995, the NTSB has recommended the testing and use of collision warning systems (CWS) to prevent or alleviate the severity of rear-end collisions. This report will describe the accidents that NTSB investigated in 1999 and 2000, how technology can help drivers prevent such accidents, and NTSB's position on these technologies that can save lives.

INTRODUCTION

Every year more than 6 million crashes occur on U.S. highways, killing over 41,000 people and injuring nearly 3.4 million more. Rear-end collisions account for almost one-third of these crashes, and between 1992 and 1998, the occurrence[2] of such collisions increased by 19 percent. Heavy trucks are involved in 40% of fatal rear-end collisions, even though they are only involved in 9% of all fatal crashes. In the past 2 years, the NTSB investigated nine rear-end collisions that took the lives of 21 people. Common to all nine accidents was the driver's degraded perception of traffic conditions ahead. Technologies that can alert drivers to the slowed or stopped traffic ahead can help prevent accidents like those the Safety Board Investigated.

ACCIDENTS

In the nine accidents investigated by the Safety Board, none of the striking vehicles had mechanical defects that

---

[1] According to the 1998 Fatal Analysis Reporting System, rear-end collisions accounted for 29.6 percent of all crashes that year.
[2] As a percentage of all collisions.

would have contributed to the accident, nor did alcohol or drugs degrade the driver's performance. Some of the collisions occurred because atmospheric conditions, such as sun glare, smoke, or fog, interfered with the driver's ability to see the traffic ahead. In other accidents, the driver did not notice that traffic had come to a halt due to congestion at work zones or due to other accidents. Still other accidents involved drivers who were distracted or fatigued. Below is a brief summary of the five major accidents that were investigated.

MORIARTY, NEW MEXICO – On January 14, 1999, about 7:45 a.m., a truck tractor semitrailer was traveling at about 120 kilometers per hour (kph) (the posted speed limit) eastbound on Interstate 40 near Moriarty, New Mexico. The truck struck the rear of a minivan and then a passenger car ahead of the minivan, both of which were traveling at about 40 to 50 kph. Two passengers in the minivan were fatally injured, one was seriously injured, and the remaining occupants of the minivan, the passenger car and the truck received minor injuries. Postaccident toxicological tests of the truckdriver were negative for drugs and alcohol. Visibility tests were conducted to determine whether the position of the sun would have had an effect on the drivers. In all three tests, the angle of the sun resulted in it being directly visible through the windshield, significantly impeding forward visibility and making it difficult to see the vehicles ahead.

SWEETWATER, TENNESSEE – On May 27, 1999, about 4:37 p.m., a minivan was stopped on southbound Interstate 75 near Sweetwater, Tennessee, due to congestion in a construction zone. A truck tractor semitrailer, traveling at about 70 to 80 kph, skidded approximately 18 meters and struck the minivan, fatally injuring all 4 occupants. The truckdriver received minor injuries and postaccident toxicological tests were negative for drugs and alcohol. The truck driver stated that before the accident, he glanced down at his speedometer, then observed the exit ramp ahead, and when he looked back at the road, saw traffic was stopped.

WELLBORN, FLORIDA - On March 8, 2000, about 7:49 a.m., a multivehicle collision, involving 24 vehicles (including 8 tractor semitrailers), and postcrash fire occurred on Interstate 10 near Wellborn, Florida, resulting in 3 fatalities, 17 serious injuries, and 12 minor injuries. Visibility at the time of the collision was reduced significantly as a result of smoke from local forest fires and fog in the area. The posted speed limit was 113 kph (70 miles per hour). The initial collision on the westbound side occurred when a truck tractor semitrailer began to slow to between 80 and 90 kph as the visibility on the highway diminished. The truck was struck from behind by another truck tractor semitrailer. Several subsequent drivers said they began to slow, but were unable to see the wreckage until it was too late to avoid colliding with the other vehicles. The initial collision on the eastbound side occurred when a vehicle stopped on the right shoulder was struck by a vehicle as it entered the smoke. The remainder of the collisions occurred when subsequent vehicles approached the accident and collided with it or the debris from it.

The accident in Wellborn, Florida, is very similar to one the Safety Board investigated in 1995 in Menifee, Arkansas. In both accidents, visibility was reduced to such an extent that drivers could not see slowed or stopped vehicles ahead. The Safety Board concluded in 1995 that "collision avoidance systems have the potential for avoidance or reduction of severity of low-visibility collision conditions such as in fog, snow, rain or darkness." This conclusion still holds true today. The Safety Board made recommendations at that time to the U.S. Department of Transportation (U.S. DOT) to begin fleet testing of CWS. The U.S. DOT's National Highway Traffic Safety Administration (NHTSA) recently began testing, in cooperation with Volvo and Eaton VORAD, of CWS and Adaptive Cruise Control (ACC) on commercial vehicles, more than 5 years after the Safety Board's initial recommendation.

TECHNOLOGIES

Several technologies currently exist that can help prevent rear-end collisions, or, at a minimum, alleviate the severity of such accidents. These technologies are either already available for deployment or are undergoing testing for deployment in the near future.

COLLISION WARNING SYSTEMS – Rear-end CWS can help drivers avoid accidents in which they strike the vehicle ahead, most likely when the driver is distracted and does not notice that the vehicle ahead is slowed or stopped or the degree to which the vehicle ahead is slowing. CWS alerts the driver to slowed or stopped objects in the vehicle's path ahead. The driver must then take evasive action such as braking or steering. CWS is not intended to relieve the driver from responsible and safe driving, but to aid drivers if for some reason (distraction or environmental factors) they do not

recognize a hazard ahead. The Eaton VORAD CWS, which is currently available on heavy trucks, uses radar to detect the objects ahead and uses visual and auditory alerts, which increase in intensity as the truck closes on the object, to warn the driver. The driver receives a visual alert at 107 meters from the obstacle ahead and another light is illuminated at a time-to-collision[3] of 3 seconds. The driver then receives an auditory alert (1 beep) at 2 seconds or 67 meters (whichever is less) and another auditory alert (double beep) at 1 second to collision. The system can distinguish objects that may not be a hazard, such as a guardrail in a curve. However, drivers report they still receive some false alerts from stationary objects (such as bridge abutments) and Eaton VORAD is improving the hardware and software to minimize these.

Most major truck manufacturers offer CWS as an option. Experience with CWS has been positive, with fleets experiencing accident reductions ranging from 35% to 100%.[4] The U.S. Army has evaluated CWS and is outfitting a portion of its fleet of trucks and transporters with CWS. In the Army's test, drivers reported avoiding 10 accidents in over 24,000 kilometers driven, and they expect a 30% decrease in convoy accidents and a savings of 15 soldier-lives per year with the deployment of CWS. A major U.S. trucking company, U.S. Xpress, reports a 75% decrease in rear-end accidents since the introduction of rear-end CWS and indicates that drivers like having the system on their trucks. NHTSA estimates that rear-end CWS can reduce all rear-end collisions by 48%, which would equate to a reduction of about 791,000 crashes each year.[5]

NHTSA is currently conducting two operational tests of CWS—one for passenger cars (with General Motors) and one for trucks (with Volvo). These are expected to be completed in 2003. NHTSA is also sponsoring pre-deployment enabling technology research, such as sensor design and human factors, in cooperation with GM and Ford. The Government of Australia is conducting research on CWS to determine whether the system will help train drivers to maintain longer headways (by alerting them when they are too close and there is a collision potential), thus increasing safety.

ADAPTIVE CRUISE CONTROL – While ACC is not advertised by manufacturers as a safety device (it is a

---

[3] Time-to-collision is the amount of time to impact if the driver does not take action to avoid a collision. It is based on the closing speed and the distance to the vehicle or obstacle ahead.

[4] This reduction may not be attributed completely to CWS because other technologies were implemented at the same time.

[5] U.S. Department of Transportation, National Highway Traffic Safety Administration, *Preliminary Assessment of Crash Avoidance Systems Benefits*, Benefits Working Group (October 1996).

convenience device), it can have safety implications. ACC is an enhancement to conventional cruise control in that it maintains the host vehicle at a constant speed, unless a slower-moving vehicle is detected ahead, at which point the ACC reduces the host vehicle's speed to a speed comparable to the slower vehicle. Once the slower vehicle moves out of the way, the host vehicle resumes its pre-set speed. ACC can only reduce the host vehicle's speed within limits (varying by manufacturer), generally by about 25%, through the application of engine control and limited braking (the industry maximum for braking is about 0.3g). Some manufacturers provide the driver with an alert if the ACC cannot slow the vehicle enough to prevent a collision. Current ACC systems are intended for use at highway speeds, and most will not operate well below 50 kph. Several manufacturers are working on "stop and go" systems that operate at low speeds and can detect a stopped vehicle ahead.

ACC is not intended to be used as a crash avoidance system, but as a convenience system to alleviate some of the stress of driving on high-speed roadways. According to one truck manufacturer, drivers report that ACC reduces their stress on the roadway by reducing the number to tasks the driver has to attend to at any one time. ACC allows drivers to focus on the task of lane-keeping and to pay more attention to the traffic around them while devoting less attention to speed control.

ACC was tested by NHTSA to evaluate how it operated in real-world conditions.[6] Drivers in this Field Operational Test used the ACC a majority of the time on high-speed roadways (77%) and drivers reported that they were comfortable with the system (98%). 56,000 kilometers were logged with the ACC engaged. Researchers anticipated no more than a 10% chance of a police-reported crash and none occurred. This is a crude data point supporting the safety of the system, and wider use needs to occur to support this data. Additionally, ACC lengthened drivers' headway times and led to a less aggressive driving style in many drivers. This is another benefit of the system. NHTSA is now sponsoring a Field Operational Test as part of its Intelligent Vehicle Initiative that includes ACC on trucks. (This is the test mentioned in with Volvo that includes CWS.) No further test data are currently available on the success or drawbacks of ACC.

Currently, ACC systems are available in Japan and Europe and in the United States on luxury class vehicles (Mercedes S- and CLK-class cars and the Lexus LS430). The Mercedes ACC retails for $2500 in the United States. Manufacturers expect the price to drop to around $500 within the next 6 months. Studies show that drivers are likely to purchase ACC when the prices are comparable to conventional cruise control. ACC is also available as an option on heavy trucks in the United States and Europe.

HOW TECHNOLOGIES CAN PREVENT ACCIDENTS

The Safety Board explored how some currently available technologies can be used to help prevent the rear-end collisions we investigated. In each accident, some form of CWS or ACC may have prevented or alleviated the severity of the collision.

COLLISION WARNING SYSTEMS – The accident in Moriarty, New Mexico, was simulated using a Human Vehicle Environment system (HVE2). An EDSMAC4 computer software program was used to simulate the truck/van/car accident dynamics. The EDVDS, EDVSM, and EDGEN software programs were used to model the vehicles' trajectory and rollover. First, the accident was simulated as it occurred. Then an Eaton VORAD CWS was incorporated into the simulated truck. Using the parameters of the system, the events were modeled to determine whether such a system would have helped the driver avoid the collision.[7]

In the accident, the driver did not see the slower moving vehicles ahead, probably because of the sun glare. If the Eaton VORAD CWS had been used, the driver would have received a visual alert at 107 meters. If the driver had noticed the visual alert when the truck was 107 meters from the slower-moving vehicles, she would have had adequate time (assuming a perception reaction time of 1.6 seconds)[8] to determine the potential hazard, apply the brakes, and steer to avoid the collision without needing to take extraordinary measures.

However, what would have happened if the driver was not looking at the display or if the bright environment washed out the lights on the system and she did not detect the visual alert? At a time-to-collision of 3 seconds, a second light would have been illuminated. Again, the driver may not have noticed it. As the driver continued to approach the vehicles, the CWS would have provided an auditory alert at a time-to-collision of 2 seconds or 67 meters, whichever is less (2 seconds in this case). Given that the typical driver reaction time ranges from 0.9 to 2.1 seconds, with the 95th percentile reaction time of 1.6 seconds, the truck driver would not

---

[6] U.S. Department of Transportation, *Intelligent Cruise Control Field Operational Test Final Report, May 1998*, DOT-HS-808-849 (Springfield, VA: NTIS).

[7] To view simulations, see http://www.ntsb.gov/Events/2001/MoriarityNM/mnm_video.htm

[8] Thomas A. Dingus, Steven K. Jahns, Abraham D. Horowitz, and Ronald Knipling, "Human Factors Design Issues for Crash Avoidance Systems," Woodrow Barfield and Thomas A. Dingus, eds., *Human Factors in ITS* (New Jersey: Lawrence Erlbaum & Associates, 1998).

have had enough time to slow the truck or swerve into the other lane and prevent the collision, if she responded to the auditory alert alone. This is due to the high difference in the rate of speed of the two vehicles. To avoid the collision, the driver would have had to react in about 0.73 seconds to the auditory alert to perform a severe lane change maneuver, placing the trailer on the verge of instability. Even the most well-trained and alert drivers probably would not have the ability to react so quickly and would likely need additional time to successfully avoid a collision.

Similarly, in the Sweetwater, Tennessee, accident, the driver reported that he was looking at the exit ramp and probably would not have noticed the visual alert. Thus, if he reacted only to the auditory alert, the driver would not have had any additional time for avoiding the accident (assuming a typical driver reaction time).

Accidents in which there are great speed differentials are generally the most severe. These examples illustrate the Safety Board's concern that, while CWS is beneficial and can prevent accidents, in extreme situations involving a great difference in vehicle speeds, the system may not provide sufficient audible warning. The Safety Board is also concerned that as more systems begin to be deployed, their operational characteristics may differ, thus leading to driver confusion. One system may give a warning that the driver is driving at speeds exceeding the capabilities of the system. Others may not. The driver may receive alerts at different times and the alerts may have different modalities (auditory, visual or tactile). Some systems may provide auditory warnings to address the above situations, while others may not. The variations may lead to driver confusion as drivers move from one vehicle to another. Thus, the Safety Board recommended that the U.S. DOT develop performance standards for CWS.[9]

The U.S. DOT has set a goal of equipping 25% of new trucks and 10% of new cars with Intelligent Transportation Systems (ITS) technologies by 2010. They have also set the goal of a 50% reduction in truck-related accident fatalities and a 20% reduction in all highway fatalities. The use of rear-end CWS can go a long way in helping the U.S. DOT meet its goals and increasing the safety of the traveling public. As evidenced by the experience of companies whose trucks are equipped with CWS and by the findings of Safety Board investigations, CWS can help reduce rear-end collisions. The Board recommended that, because of the severity of the accidents in which trucks are involved, heavy trucks be equipped with rear-end CWS. Because of the high cost of trucks, the addition of CWS does not add great cost proportionately; in passenger cars the cost differential is greater. We hope that in the future the prices will drop and CWS will become commonplace in

passenger cars. Rear-end CWS may have saved the lives of some of the people who were involved in the accidents we investigated.

ADAPTIVE CRUISE CONTROL – Although ACC is referred to by manufacturers as a convenience system, it can contribute to safety. Given the infancy of the ACC technology, the Safety Board has not yet investigated any accidents in which ACC was in use.

The Wellborn, Florida, collision illustrates how ACC can help drivers in certain situations. The accident occurred during smoke and fog on Interstate 10. The first truck to enter the smoky area began to slow because of limited visibility. The next truck did not slow its speed and ran into the first truck. This accident resulted in the trucks blocking both of the eastbound lanes, which led to a number of subsequent accidents. Had the second truck been equipped with and using an ACC system, it would have automatically begun to slow to a speed similar to that of the first truck and probably would not have struck the first truck. It would have formed a convoy with a 2- to 3-second headway (depending on the driver-selected setting) and followed the first truck through the smoke without colliding. The remainder of the vehicles, had they likewise been equipped with an ACC system, would have traveled through the smoke in a similar manner, adapting to the speed of the vehicle ahead. While it is most prudent for drivers not to enter situations in which visibility restriction are severe, in this case, the smoke appeared on the highway suddenly and the drivers were unable to divert their travel.

The primary concern that the Safety Board has about ACC use in poor visibility conditions is that drivers may not know whether their vehicles need to slow by more than the ACC capability. Some of the vehicles in Wellborn may have been traveling at speeds less than 25% of the speed at which other vehicles were entering the smoky area. Current ACC systems are designed only to slow the vehicle by about 25 percent, after which the driver must take action. If the driver is not aware that he must take action (for example in a low-visibility situation), he may not know he has to take additional action. Without a cue, the driver may not know whether the ACC is capable of slowing his vehicle sufficiently. The Mercedes passenger car ACC does provide the driver with a warning that the situation exceeds the ACC capabilities; however, the Jaguar system does not. Without performance standards for system operation and driver interaction with ACC, the usage of numerous and non-uniform systems may result in driver confusion, which may lead to inappropriate use of the system. Therefore, the Safety Board recommended that the U.S. DOT also develop performance standards for ACC for both heavy trucks and passenger cars.

---

[9] For a list of conclusions, see:
http://www.ntsb.gov/Publictn/2001/SIR0101.htm

CHALLENGES OF DEPLOYING TECHNOLOGIES

There are several challenges associated with the deployment of new technologies on passenger cars and commercial vehicles. These challenges include consumer acceptance and public perception and driver training.

CONSUMER ACCEPTANCE AND PUBLIC PERCEPTION – Although requiring CWS is critical, consumer acceptance of the technology is equally critical. For example, educating the public about the benefits of seatbelts has been as important as equipping the vehicles with or requiring the use of seatbelts. In a U.S. DOT study on consumer acceptance of various automotive technologies, drivers, particularly older drivers, were enthusiastic about ACC and CWS, but were wary about how they operated and their reliability. While only 43% of the drivers surveyed would purchase an ACC system, 98% of drivers who actually drove with an ACC in the U.S. DOT's operational test said they would purchase the system.

Some drivers may be wary of new technologies prior to using them. A similar situation existed when airbags were first deployed; people were initially apprehensive. To educate the public, the U.S. DOT and Allstate Insurance Company sponsored a demonstration of airbags using crash dummies.[10] The exhibit traveled to 100 cities in a three-year period beginning in 1990. The purpose of the exhibit, according to Allstate's Chairman and Chief Executive Officer was to "encourage consumers to purchase cars with air bags because we know they save lives and reduce injuries". A similar program could be developed to educate the public on the safety benefits of CWS. The average driver, whether a truck or passenger car driver, does not know what ITS technologies are available and has never experienced what it is like to drive with some of these technologies.[11] While industry predicts that the market for CWS and ACC will grow to over $3 billion by 2010,[12] steps need to be taken to reach that goal.

In discussing what the government can do to promote the implementation of technology, a trucking company representative said that the government could provide more information on the technologies so that the data coming from the manufacturers are not suspect (consumers may think the manufacturer is just trying to sell something).[13] He further stated that electronics in trucks are still relatively new and consumers are not yet completely comfortable with them. If the government publishes solid data on the benefit of certain technologies, and the on benefits of multiple technologies, the trucking industry may be more apt to adopt them. This is part of the impetus of the current Volvo operational test—to allow the U.S. DOT to gather data and form unbiased opinions and recommendations regarding CWS technologies. Transmitting this information to the public is crucial to the acceptance of ACC and CWS technologies. Therefore, the Safety Board recommended that the U.S. DOT, the Intelligent Transportation Society of America, and truck and automobile manufacturers develop and implement a program to promote the benefits, use, and effectiveness of collision warning systems among drivers.

TRAINING – The object of training is to ensure that specific skills or procedures are learned. Training can take the form of verbal instruction, demonstration and guidance, or practice[14] through the use of videos or computers. It is one of the standard methods used to aid people in acquiring safe behavioral practices.[15]

According to the president of U.S. Xpress, the company provides its drivers with extensive training on all the technologies that are on its trucks. For example, a driver receives orientation on ACC so that he understands what happens if the truck begins to slow down, why the truck is slowing (because there is a vehicle ahead), and how the driver must react. Recurrent training is also provided and is considered by U.S. Xpress to be necessary to help the driver be successful and to understand the technology.

Training has been provided in the operational tests that have been conducted to date with ACC or CWS. In the ACC operational test completed for NHTSA in 1998, the drivers received a limited introduction to the functions and capabilities of the system. This understanding allowed the drivers to use the ACC in the manner for which it was intended and made them aware of the necessity of intervening when harder braking was necessary.[16] The drivers surveyed during the U.S. Army field test believed that training was imperative because

[10] IIHS Status Report, Volume 25, Number 10. p. 11., Insurance Institute of Highway Safety. Arlington, VA, November 17, 1990.

[11] Michael A. Regan, Claes Tingvall, David Healy, and Laurie Williams. "Trial and Evaluation of Integrated In-Car ITS Technologies: Report on an Australian Research Program." 7th World Congress on Intelligent Transport Systems. Turin, Italy. November 5-9, 2000.

[12] http://www.tierone.com/accmtrexcerpt.html

[13] Public Hearing on Advanced Safety Technology Applications for Commercial Vehicles.

[14] Salvendy, Gavriel, ed. Handbook of Human Factors. John Wiley and Sons, Inc.. New York. 1987.

[15] Sanders, Mark S. and McCormick, Ernest J. Human Factors in Engineering and Design, 7th Edition. McGraw Hill, Inc. 1993.

[16] Intelligent Cruise Control Field Operational Test Final Report. May 1998. DOT HS 808 849. National Technical Information Service, Springfield, VA.

the systems were not intuitive.[17] Despite the U.S. DOT and Army experiences, according to Eaton VORAD, fleets have stated that the CWS is easy to learn without training. One of the parameters being explored in the Australian study is whether drivers are able to intuitively understand the operation of the CWS. Mercedes requires customers who purchase cars with ACC to watch a video on its operation and then receive a demonstration on their vehicle before they leave the lot.

The importance of training cannot be overstated, based on the experience of U.S. Xpress, the operational tests, and previous Safety Board investigations. Training is critical to the understanding of complex technical system functionalities so that drivers can respond adequately when the technology is in use. The Safety Board recommended that truck and automobile manufacturers develop training programs for drivers on the operation and capabilities of ACC and CWS. The Board also recommended that trucking associations encourage their members to provide or obtain training on ACC and CWS, once the systems are installed.

CONCLUSION

The Safety Board has recommended in the past that operational tests of CWS to prevent rear-end collisions be conducted. Because of delays by the U.S. DOT in conducting these tests and in beginning deployment, hundreds of lives may have been lost. The sooner these systems are deployed and the challenges are addressed, the safer our roadways will be. Leading the deployment should be heavy trucks because of their higher involvement in rear-end collisions and because of the costs of the vehicles. The Safety Board believes that performance standards need to be developed so that drivers understand how the systems work when they move from one vehicle to another. Drivers should also be educated on these safety-related systems so that they will begin to purchase the systems use them safely. Finally, once the systems are purchased, the drivers should be trained so that they completely understand what the systems can and cannot do and what any types of alerts or warnings mean. All of these steps can lead to safer highways and reduced fatalities due to rear-end collisions.

CONTACT

Jennifer H. Bishop is a project manager in the Office of Highway Safety at the National Transportation Safety Board. She can be contacted at bishopj@ntsb.gov.

DEFINITIONS, ACRONYMS, ABBREVIATIONS

ACC – Adaptive Cruise Control

CWS – Collision Warning Systems

EDGEN – Engineering Dynamics Corporation General Analysis Tool

EDSMAC4 – Engineering Dynamics Corporation Simulation Model of Automobile Collision, 4th Revision

EDVDS – Engineering Dynamics Corporation Vehicle Dynamics Simulator

EDVSM – Engineering Dynamics Corporation Vehicle Simulation Model

HVE2 – Human Vehicle Environment system version 2

ITS – Intelligent Transportation Systems

NHTSA – National Highway Traffic Safety Administration

NTSB – National Transportation Safety Board

U.S. DOT – United States Department of Transportation

---

[17] National Automotive Center Collision Warning Safety Convoy. K. Luckscheiter. U.S. Army Tank-automotive and Armaments Command. Warren, MI. September 1996.

2001-01-0357

Research on a Brake Assist System with a Preview Function

Minoru Tamura, Hideaki Inoue, Takayuki Watanabe and Naoki Maruko
Nissan Motor Co., Ltd.

ABSTRACT

Traffic accidents in Japan claim some 10,000 precious lives every year, and there is seemingly no end to the problem. In an effort to overcome this situation, vehicle manufacturers have been pushing ahead with the development of a variety of advanced safety technologies. Joint public-private sector projects related to Intelligent Transport Systems (ITS) are also proceeding vigorously.

Most accidents can be attributed to driver error in recognition, judgment or vehicle operation. This paper presents an analysis of driver behavior characteristics in emergency situations that lead to an accident, focusing in particular on operation of the brake pedal. Based on the insights gained so far, we have developed a Brake Assist System with a Preview Function (BAP) designed to prevent accidents by helping drivers with braking actions.

Experimental results have confirmed that BAP is effective in reducing the impact speed and the frequency of accidents in emergency situations.

ACCIDENT ANALYSIS

Figure 1 shows recent trends in the number of fatal accidents by types of violation. The number of accidents caused by drivers' pernicious behavior or lack of moral sense, such as disregarding traffic signals, drunken driving and excessive speed, has been decreasing year after year. On the other hand, the number of accidents caused by a slight error, such as operating mistakes, careless driving and distraction, has been increasing.

Figure 2 shows the number of fatal accidents by types of violation for different age brackets. The percentage of accidents caused by a slight error, such as operating mistakes, careless driving and distraction, increases with increasing age. We can estimate that in the near future the ratio of these accidents will increase more as the population continues to grow older.

Figure 3 shows recent trends in the ratio of traffic accidents by types. Rear-end and broadside collisions are the two types of accidents which occur most frequently in Japan and have been steadily increasing in recent years. Since 1990, rear-end and broadside collisions have accounted for over 50% of all traffic accidents.

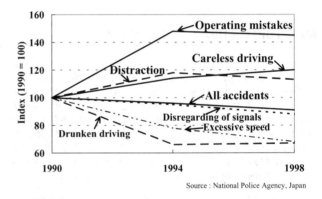

Source : National Police Agency, Japan

Fig 1. Trends in the number of fatal accidents by types of violation

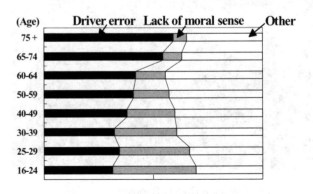

Fig 2. The number of fatal accidents by types of violation for different age brackets

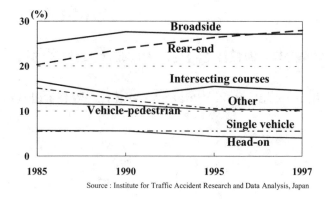

Source : Institute for Traffic Accident Research and Data Analysis, Japan

Fig 3. Trends in the ratio of traffic accidents by types

From the foregoing data, we can conclude that stepping up the development of technologies to assist drivers' recognition, judgment or vehicle operation under conditions with the potential for rear-end or broadside collisions will be a crucial factor in reducing the overall accident rate.

ANALYSIS OF DRIVER BEHAVIOR

Knowing how drivers respond and brake in emergency situations is essential in order to analyze the fundamental causes of accidents and determine the direction for technology development. Using the driving simulator shown in Figure 4, a model was developed of an emergency situation where another vehicle suddenly comes out of a side street at an intersection and one's own vehicle runs into it broadside; measurements were made of drivers' behavior and responses (Figure 5).

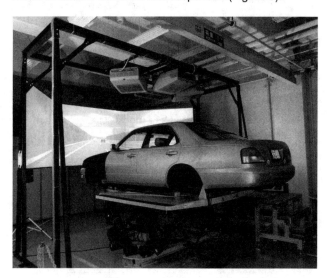

Fig 4. Nissan Driving Simulator

To prevent the test subjects from predicting an emergency situation, they were first instructed to stop at or pass through a number of intersections. Following those dummy tasks and at a point when it was thought the subjects were accustomed to the driving simulator and might be apt to drive carelessly, a vehicle was presented from a side street. In order to focus on braking behavior, a vehicle was parked in the opposite lane so that the subjects could not avoid the collision simply by steering.

Many subjects had depressed the accelerator pedal and accelerated the vehicle before the appearance of the other vehicle from the side street. After the other vehicle appeared and the subjects perceived the risk of a collision, histograms of their behavior show that they applied greater force to the footrest or gripped the steering wheel more strongly. After that, they soon released the accelerator pedal and pressed the brake pedal. (Figure 6)

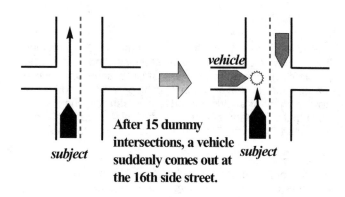

Fig 5. Simulated emergency situation

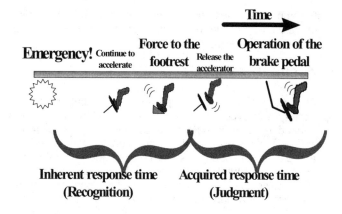

Fig 6. Time chart of drivers' responses and operations

Figure 7 shows that the inherent response time from a driver's perception of danger ahead to the application of force to the footrest or the steering wheel was approximately 0.62 second. The acquired response time from application of force to the footrest to operation of the brake pedal was 0.38 second, and the total response time was 1.00 second.

336

It was observed that some experienced subjects happened to be distracted when the danger arose ahead and were not able to avoid a collision. Little correlation is seen between the inherent response time and driving skill. However, the acquired response time shows little deviation and provides a criterion that readily indicates the difference in the driving skill of individual subjects.

To summarize this analysis of driver behavior, technologies to assist drivers in recognizing a vehicle ahead by using radar sensors would be effective in reducing accidents for all drivers.

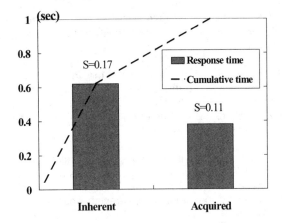

Fig 7. Response time of subjects

OBJECTIVE OF BAP

In order to reduce accidents, it would be desirable for one's own vehicle to detect the headway distance reliably and to apply suitable braking force to avoid a collision. However, further development of accurate headway detection capabilities and reliable deceleration control is needed to commercialize such a system.

As the first step toward future automatic braking technology, we are currently developing an emergency braking aid system called the Brake Assist System with a Preview Function (BAP). This system provides minute braking force when the sensor detects danger ahead. This braking force closes the gap between the brake rotor and the brake pad to improve braking response when the driver presses the brake pedal and thereby reduces the stopping distance and the impact speed.

An overview of a prototype vehicle equipped with BAP is shown in Figure 8. It has a millimeter-wave radar sensor on the grille to detect the distance and relative speed of a vehicle ahead, a throttle angle sensor to measure accelerator pedal, control units, and a brake actuator to provide minute braking force.

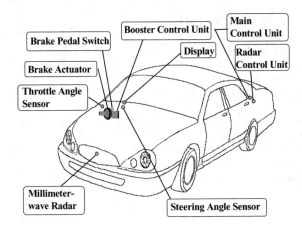

Fig 8. Overview of BAP-equipped vehicle

CONTROL ALGORITHM

The target deceleration for stopping without colliding with the vehicle ahead, Gp, is calculated as follows:

$$Gp = (Vm^2 - Vf^2) / (2 * ds)$$

where   Vf = velocity of the vehicle ahead

Vm = velocity of the host vehicle

ds = current distance between the two vehicles

Braking force begins to be provided when Gp exceeds a certain predetermined value, which is considered to indicate an emergency situation. A value of 5.88 m/s^2 (0.6 G) was used in this research.

The supplied brake pressure Pt is determined by the manner in which the host vehicle approaches the vehicle ahead. In this research, greater brake pressure was supplied when the host vehicle accelerated or the return speed of the accelerator pedal was faster than usual while approaching the vehicle ahead.

$$Pt = f1 (Am, Vm) * f2 (Pdot, Vm)$$

where   Am = host vehicle's acceleration

Pdot = return speed of the accelerator

The system judges that the emergency situation has passed on the basis of one of the following conditions.

- when the driver operates the accelerator pedal

- when the driver operates the brake pedal

- when a certain interval of time has elapsed without any operation of the accelerator pedal or the brake pedal

BENEFITS TO DRIVERS

The results of experiments conducted with the prototype vehicle show that the delay time from the operation of the brake pedal to the rise of the brake pressure was shortened by 100 msec with BAP (Figure 9).

Figure 10 shows the experimental results for the braking distance at an initial vehicle speed of 100 km/h. This test methods used was the "control braking method", which means that braking was done by an experienced driver. The distance was reduced by 3 meters or 6% with BAP.

That would translate into a 5 km/h reduction in impact speed compared with conventional vehicles in the most frequent accident scenarios in Japan, in which a driver becomes aware of a obstacle at a forward distance of 20 meters when traveling at 50 km/h. The impact speed of a BAP-equipped vehicle would be 17 km/h, or 5 km/h less than that the 22-km/h impact speed of a vehicle without this system (Figure 11).

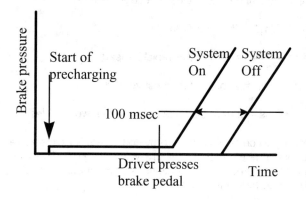

Fig 9. Comparison of brake pressure with system on and off

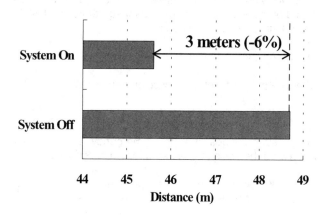

Fig 10. Comparison of stopping distance with system on and off

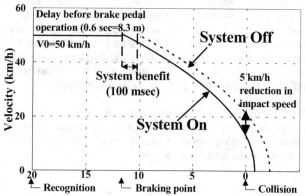

Fig 11. Comparison of impact speed with system on and off

Figure 12 shows the correlation between the vehicle speed and the distance when the drivers recognize potential danger ahead. These data were collected in accident scene investigations or are based on statements given by drivers and other persons involved.

The curved line in the figure indicates the physical limit at which the vehicle can stop without a collision as a result of maximum deceleration. BAP would enlarge the collision avoidance area by the amount between the dashed line and the solid line.

In the accident database, 12 of 157 cases were in this expanded area, so it can be estimated that approximately 2% of the rear-end collisions might have been avoided. This estimation is premised on the assumption that the sensor for detecting the headway distance can recognize the driving environment without fail. In actuality, however, the sensor sometimes loses the vehicle ahead, so the actual collision avoidance rate might be smaller than 2%.

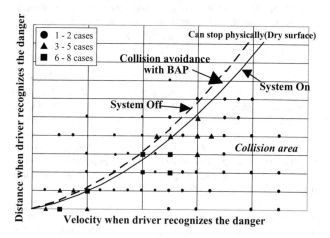

Fig 12. Estimation of collision avoidance

LIMITATIONS OF BAP

We have confirmed that BAP would be helpful in reducing the impact speed or frequency of accidents. However, this system is only aimed at improving the braking response when the driver operates the brake pedal; it is not an automatic braking system. Accordingly, drivers must press the brake pedal firmly just as in ordinary driving.

Although the performance of sensors for recognizing the driving environment has been improved significantly in recent years, sensors cannot detect, for example, objects that suddenly appear from crossroads or from behind some structures. It is expected that drivers will always try to drive carefully and not rely on the system excessively.

CONCLUSION

- The results of experiments and accident database analyses revealed the behavior and response of drivers in emergency situations.

- A prototype of a Brake Assist System with a Preview Function (BAP) was built which incorporates sensor technology for detecting a vehicle ahead. A control algorithm for this system has been developed that is effective in compensating for a driver's inattention or misperception in an emergency situation.

- Sensors for detecting the external environment and actuators for controlling vehicle motions are being vigorously developed by many organizations for use in ITS or with advanced safety technologies. Such fundamental technologies will be used in this research to commercialize the next generation of BAP as quickly as possible with the aim of contributing to a reduction of fatal accidents.

REFERENCES

1. 1997 Transportation Statistics: Institute for Traffic Accident Research and Data Analysis.
2. 1999 Traffic Green Paper: National Police Agency of Japan.
3. Web page of National Police Agency of Japan
4. Development of the Nissan Driving Simulator: Nissan Technical Review, No. 41 (1997)
5. 1996 Accidents Analysis Report: Institute for Traffic Accident Research and Data Analysis.
6. Statistical Methods in the Analysis of Road Accidents: Road Research Laboratory, U.K., 1969.

2001-01-0052

Driver-Vehicle Interface Requirements for a Transit Bus Collision Avoidance System

Stephen J. Reinach and Jeffrey H. Everson
Foster-Miller, Inc.

ABSTRACT

This paper discusses the development of driver-vehicle interface (DVI) requirements for an inner-city transit bus collision avoidance system (CAS). In 1998, there were over 23,000 transit bus collisions resulting in over 20,000 injuries. Using structured interviews with transit bus operators and naturalistic observation, the transit bus operating environment was characterized. Then, a set of CAS functional requirements was generated. Lastly, a set of human factors DVI requirements for a transit bus CAS was developed. The DVI requirements focused on the physical aspects of the display and the display's cognitive demands on the bus operator.

INTRODUCTION

In 1998 there were over 23,000 transit bus collisions resulting in over 100 fatalities and more than 20,000 personal injuries (Federal Transit Administration, 1999). The Department of Transportation recently launched the Intelligent Vehicle Initiative (IVI) in large part to increase the safety of our nation's transportation system. The Federal Transit Administration (FTA) is sponsoring several programs to address transit safety as part of the national IVI program, including the development of collision avoidance systems for transit buses. However, very little research currently exists to support researchers and designers in developing transit bus CAS systems or other transit technologies. In particular, very little research exists to support the design of a driver-vehicle interface (DVI) for transit bus operations.

As part of a recent Small Business Innovative Research (SBIR) program sponsored by the FTA, we explored some of the human factors issues involved in the preliminary design of a DVI for an inner-city transit bus CAS. This paper documents our effort at understanding the inner-city transit bus operational environment and developing DVI requirements for an inner-city transit bus CAS. A CAS that meets these requirements will be more likely to be accepted by its users, and therefore more likely to be effective in reducing the number of transit bus collisions.

OVERALL APPROACH

To develop the DVI requirements for a transit bus CAS, we used two primary sources of information: the experience and expertise of Massachusetts Bay Transit Authority (MBTA) driving instructors and bus operators, and past passenger and commercial vehicle collision avoidance research. MBTA instructors and bus operators served as subject matter experts and provided input into the transit bus operational environment, while past passenger and commercial vehicle CAS research (e.g., Young, Eberhard, and Moffa, 1995; Wilson, 1996; Kiefer, LeBlanc, Palmer, Salinger, Deering and Shulman, 1999) guided the development of the CAS functional requirements. Figure 1 illustrates the overall approach.

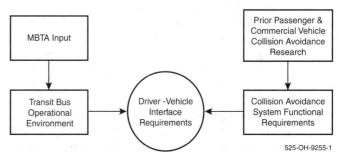

525-OH-9255-1

Figure 1. Overall approach to the development of transit bus DVI requirements

THE TRANSIT BUS OPERATIONAL ENVIRONMENT

To develop a set of transit bus DVI requirements, it was necessary to gain an appreciation and understanding of the transit bus operational environment. Data came from three sources:

1. Structured interviews with MBTA bus operator instructors.
2. Naturalistic observation (i.e., ride-alongs).
3. Review of the MBTA's transit bus operator training manual.

Information from these three sources was divided into three categories: characterization of the transit bus operating environment (i.e., the physical environment in which the bus operates), characterization of the operator working environment, including the operator's workstation and the bus as a whole, and delineation of operator performance requirements (i.e., what is the bus operator trained to do, or what must the operator do to negotiate and maneuver the bus in an urban environment).

THE TRANSIT BUS OPERATING ENVIRONMENT - To begin, the physical environment in which inner-city buses operate was identified and delineated, based on structured interviews with MBTA bus operator instructors. The following items characterize the inner-city transit bus operating environment.

Inner-city transit buses:

- Make frequent stops.
- Frequently merge into and out of traffic.
- Typically travel at relatively low speeds (less than 35 mph/56.3 kph).
- Operate in high density traffic.
- Require additional time to stop compared to passenger vehicles.
- Operate primarily in densely populated areas.
- Share the road with pedestrians and bicyclists.
- Share the road with double-parked vehicles (e.g., delivery truck).
- Have minimal side-to-side clearance.
- Frequently operate in construction areas, where roads are often narrower, turns are sharper, and views of the road are obstructed.
- Require large turning radiuses.
- Can weigh 29,000-36,000 lbs. (13,154-16,329 kg)
- Can extend over 40 ft (12.2 m) long (Novabus, Inc., 1999), or 60 ft (18.3 m) long if articulated.
- Can be 102 in (2.6 m) wide (Novabus, Inc., 1999).

THE TRANSIT BUS OPERATOR ENVIRONMENT - The next step in identifying the transit bus operational environment was to characterize the transit bus operator environment (i.e., the bus "cab" or workstation). Components of the bus operator environment include instrumentation (e.g., type of information, display modality, location); seating position; windshield characteristics (e.g., size, angle, field of view); mirror arrangements; and cabin characteristics (e.g., noise level, distractions). The MBTA made one of their transit buses available for the project to allow hands-on examination and measurement of the transit bus operator environment.

The instrument panel on the in-service TMC (now Novabus, Inc.) RTS 40 ft (12.2 m) bus was located approximately 45 degrees below the operator's line of sight (see Figure 2), and contained mostly primary displays and controls related to operating the bus: speedometer, battery voltage meter, the operator's Heating, Ventilation and Air Conditioning (HVAC) controls, air brake pressure displays and controls, and

vehicle status and telltale warning indicators. To the operator's left, right, and top, were secondary controls related to operating the bus. These are controls that are relevant to transit bus functioning, but not specifically to operating the bus in traffic. These secondary components were located within reach of the bus operator, and included a radio, fare box, front and rear door controls, inside light switches, and passenger HVAC controls.

Figure 2. MBTA TMC RTS 40 ft (12.2 m) bus instrument panel

One of the most distinctive aspects of the transit bus operator's workstation is the angle of the windshield with respect to the horizontal, the size of the windshield, and the vantage point offered by the combination of large windshield and high seated position. The windshield angle of the MBTA TMC RTS 40 ft (12.2 m) bus was approximately 70 degrees (see Figure 3). The front windshield measured approximately 87 in. (2.2 m) across and approximately 40 in. (1.0 m) high (see Figure 4). The distance between a seated bus operator and the forward windshield was approximately 30 in. (0.8 m).

The range of bus operator eye heights from the ground was also measured. Eye height measurements were based on manual measurements of the distance from the ground to the operator's seat (since the seat of the bus is adjustable, the measurements were taken at the lowest and highest possible settings) combined with sitting eye height data from NASA's *Anthropometric Source Book, Vol. 1: Anthropometry for Designers* (1978). Bus operator eye height from the ground ranged from 88.75 in (225.4 cm) for a 5th percentile female sitting at the lowest seat adjustment to 99.9 in (253.7 cm) for a 95th percentile male sitting at the highest seat adjustment. This compares to a range of 42-48.9 in (106.7-124.2 cm) for a 1994 Honda Accord LX (the height from the ground to the Honda's driver seat was approximately 15 inches, and the seat was not adjustable). Thus, bus operators sit about twice as high as many of those who drive automobiles.

Figure 4. Frontal view of MBTA RTS 40 ft (12.2 m) bus

Transit buses have large blind spots on all sides due to the dimensions of the bus and the placement of devices around the bus operator workstation. Most transit buses do not have rear windows (nor rear-view mirrors), so bus operators must rely on a pair of flat and/or convex mirrors on each side of the bus to assist them in detecting other vehicles and potential hazards around the bus. According to MBTA bus driving instructors, the convex mirrors in particular have enhanced bus operators' situation awareness around the bus. However, the low position of the mirrors on the left side of the bus produces a blind spot in front and to the left of the bus operators, where pedestrians rushing to catch the bus before it pulls away from the bus stop may be located. McNeil, Thorpe, and Mertz (2000) also note that in some buses, the dashboard, fare box, and right-hand side mirror produce additional blind spots for bus operators.

There are many differences between a transit bus workstation and a passenger vehicle workstation. Specifically, relative to a passenger vehicle, a transit bus:

Figure 3. Side view of MBTA RTS 40 ft (12.2 m) bus

- Has a larger forward vertical and horizontal field of view.
- Has a higher vantage point.
- Places the bus operator's physical position farther to the left, nearer the center stripe, due to the size of the bus.
- Has a larger steering wheel.
- Has larger accelerator and brake pedals.
- Places the seated operator in front of the turning wheels (rather than behind them).
- Has larger blind spots.
- Does not have a center-mounted rear-view mirror.
- Has a lower instrument panel (approximately 45 degrees below the operator's line of sight compared to 30 degrees for a typical automobile (Gish and Staplin, 1995)).
- Contains an internal source of nighttime glare due to the use of internal passenger lights.
- Contains a noise-rich internal environment. Sources of noise include the bus radio, rattling seats and fare

343

boxes, processing of fare box change, windshield wipers, brakes, various in-vehicle buzzers and alarms, HVAC blower, and passenger conversations.

- Operates in a noise-rich external environment, including sounds generated by emergency vehicles and construction equipment.

TRANSIT BUS OPERATOR REQUIREMENTS - Lastly, transit bus operator requirements for bus operation were identified. In addition to assembling information from the structured interviews with MBTA operator instructors and naturalistic observations, the MBTA's driver training program was reviewed. The driver training manual used by the MBTA was developed by the Department of Transportation-operated Transportation Safety Institute (TSI) in Oklahoma City, OK (Transportation Safety Institute, 1999). It is used by transit authorities across the United States.

Some of the driving "basics" that are characteristic of transit bus operation are:

- Right lane driving. Operators are taught to "protect the right" by driving in the right lane as frequently as possible to preclude the possibility of vehicles passing the bus on the right, where there is a large blind spot.
- Operators are taught not to jam the brakes in order to avoid an onboard injury, prevent skidding, reduce the likelihood of a vehicle striking the rear of the bus, and prevent overheating the brakes.
- Lane changes should be made smoothly to allow other vehicles to react appropriately to the maneuver.
- Operators are trained to maintain a 4 to 5 sec following distance. When it is not possible to maintain this separation, the operator is trained to remove his or her foot from the accelerator pedal and cover the brake pedal to reduce the brake response time.
- Operators are taught to cover the brake pedal with their foot when approaching and entering an intersection to reduce their brake response time.
- Operators are taught to scan their side mirrors every 3 to 5 sec and to constantly sweep their eyes back and forth across their forward view.
- Operators are taught to look ahead 12 sec when looking forward.
- Operators must scan bus stops for passengers.
- Operators must watch for bicyclists (e.g., couriers) darting into and out of traffic and between lanes. Bicyclists are particularly dangerous when they pass a bus on the right in the same lane.
- Operators must watch for taxis that may pull into a bus stop in front of the bus.
- Operators must interact with passengers, dispatchers, and other transit personnel.
- Often it is not possible to pull completely into a bus stop due to construction or obstructions (e.g., parked car). In such cases, the bus may have to stop in the

road, parallel to the bus stop, or the bus operator may try to pull part-way into the bus stop.

The sudden and sporadic nature of city driving requires transit bus operators to remain vigilant to avoid collisions with:

- Vehicles turning into and out of the traffic stream.
- Vehicles unsafely cutting in front of a bus.
- Double-parked vehicles (since transit buses often drive in the right lane to stop and pick up or drop off riders).
- Pedestrians walking directly in front of the bus (proximal pedestrian traffic).

A bus operator must be particularly cognizant of pedestrians who recently alighted the bus or who want to "catch" the bus before it leaves (distal pedestrian traffic). Bus operators also frequently must multitask while operating a bus. For example, a bus operator may have to look for potential passengers at an upcoming bus stop, answer a passenger's question, and attend to a passenger's request to stop, while driving safely at rush hour and given narrow side-to-side clearances.

Lastly, it should be noted that transit bus operators are a heterogeneous group, and vary in age, height, and driving ability. Each of these factors affects the bus operator's ability to operate a bus and interact with a DVI.

TRANSIT BUS COLLISION AVOIDANCE SYSTEM FUNCTIONAL REQUIREMENTS

To facilitate the development of DVI requirements for an inner-city transit bus CAS, it was necessary to generate a set of "straw man" functional requirements since none existed. The scope of the project was limited to frontal and side collision avoidance. Therefore, we used conversations with MBTA bus operator instructors to identify five types of driving maneuvers that were expected to benefit from a frontal and side transit bus CAS. They included driving straight, changing lanes or merging in traffic, entering and exiting a bus stop, and close clearance scenarios.

Based on these driving maneuvers, and review of past passenger and commercial vehicle collision avoidance research (e.g., Young et al., 1995, Wilson, 1996, Kiefer et al., 1999; McNeil et al., 2000), 16 crash scenarios were identified as countermeasure intervention opportunities for the transit bus CAS. These are collision scenarios that the transit bus CAS should be specifically designed to avoid. Based on this information, a set of basic CAS functional requirements was then developed. Briefly, the frontal and side CAS must be able to:

1. Detect the presence of, recognize, and identify vehicles, pedestrians, and roadside objects in the proximity of the CAS-equipped vehicle

2. Determine the likelihood of a collision involving the CAS-equipped vehicle
3. Given a certain (predetermined) likelihood of collision, warn the driver of the hazardous situation
4. If necessary, take temporary and limited control of the CAS-equipped vehicle to avoid a collision or mitigate its severity.

Reinach and Everson (2001, submitted) discuss the development of the transit bus CAS functional requirements in more detail.

TRANSIT BUS DRIVER-VEHICLE INTERFACE REQUIREMENTS

After the "straw man" functional requirements were delineated, it was possible to identify those features or aspects of a transit bus DVI that are required for the display, and the CAS as a whole, to be effective. That is, once the transit bus operational environment was characterized, and the CAS functional requirements were delineated, it was possible to focus on the interface between the two; i.e., the DVI. The following DVI requirements are based on the CAS functional requirements, information gleaned from multiple interviews and discussions with MBTA driving instructors, results from a focus group with MBTA bus operators, and a review of existing passenger and commercial vehicle CAS and human factors display design research:

- The display must be supportive of bus operator training requirements, such as lateral scanning and looking ahead several seconds. For example, the display must not encourage or require the bus operator's eyes to remain or dwell in one place for any substantial period of time.
- The display must minimize the potential to attract passenger attention.
- The display must be clearly conveyed and understood during low-light conditions, high-light conditions, and conditions of high glare due to internal passenger lights, external street lamps and car headlights.
- The display must be clearly conveyed and understood over the noise generated from the interior of the bus, including passenger conversations, rattling fare boxes and seats, two-way radio conversations, squealing brakes and windshield wipers, and internal alarms and buzzers (e.g., telltale indicators).
- The display must be capable of being displayed and understood under high mechanical vibration conditions.
- The display must not induce a startle response from the bus operator that results in hard braking, an inappropriate steering maneuver, or a delay in operator response.
- The display must be capable of conveying multiple levels of collision imminence (i.e., graded alerts and warnings).

- The display must be capable of being retrofitted into existing bus systems.
- The display must elicit the appropriate bus operator response to the crash avoidance situation (i.e., the display must provide congruent stimulus-response mapping).
- The display must indicate the direction of the hazard.
- The display must be well understood by a heterogeneous population.
- The display must indicate when the system has been initiated.
- The display must always indicate the status of the system, including normal functioning and when the system has failed or has become unreliable.
- The display must be able to be easily discriminated from other in-vehicle displays, and the different displays within the CAS must be capable of being discriminated from each another (e.g., a frontal warning vs. a side warning).
- The display must present warnings until the bus operator responds appropriately (adapted from Boff and Lincoln, 1988) or the hazard no longer exists.
- The display must minimize the bus operator visual load due to the high visual demands that are already exacted by the visual complexity of the inner-city operating environment, large field of view, significant blind spots that require visual scanning by the bus operator, and the need to look for passengers at wayside bus stops.
- The display must follow warning symbology and coding conventions, follow population stereotypes (e.g., the use of yellow to indicate caution and red to indicate a warning), and present information to the bus operator in a clear and concise manner.

In addition to these requirements, through conversations with MBTA instructors and bus operators, it was determined that it was not necessary for the transit bus DVI to convey the nature of the hazard (e.g., a pedestrian vs. a vehicle), but rather it is most important to convey to the bus operator the location of the hazard and the criticality of the situation.

DISCUSSION

Little research has been conducted and published on the transit bus operational environment. This paper begins to fill that gap by characterizing and documenting the inner-city transit bus operational environment. The transit bus operational environment is unique and poses multiple challenges to operating a bus in the inner-city. Designers should view these challenges not as insurmountable problems but rather as opportunities to develop interfaces that take into account and overcome these challenges. For example, the fact that bus operators must not suddenly stop nor suddenly apply the vehicle's brakes at any time (because it jeopardizes passengers' safety) can be seen as an opportunity to design a CAS that induces gradual deceleration through graded alerts and warnings. Further, given the multitude of vibrations, noises and light sources in the environment, a transit bus DVI should take advantage of multiple modalities (visual, auditory and haptic) to ensure successful transmission of the message to the bus operator. Reinach and Everson (2001, submitted) discuss and propose some preliminary transit bus CAS display designs that take into account this challenging and unique operational environment.

When considering a transit bus CAS and its DVI, it is also important to keep in mind that transit bus operators are highly trained. Designers may take advantage of this training to increase the acceptance and support the correct usage of the transit bus CAS, as well as other in-vehicle transit technologies.

Finally, although the DVI requirements developed as part of this project were developed for a transit bus CAS, they likely can be applied to other transit bus in-vehicle technologies that involve an interface with the operator. Further, even though the emphasis in this program was on the inner-city bus operational environment, the DVI requirements are also likely to be suitable to smaller city and suburban transit bus operational environments as well as cross-country bus and commercial vehicle (i.e., over-the-road truck) operational environments.

CONCLUSION

It is clear that the transit bus environment is distinct from both passenger and commercial vehicle operations and poses a multitude of challenges to the bus operator. A clear understanding of this environment is essential to designing an effective transit bus DVI. This paper documents the transit bus operational environment, and describes a set of DVI requirements that are considered essential to an effective transit bus CAS.

ACKNOWLEDGMENTS

The authors wish to thank Mr. Brian Cronin of the Federal Transit Administration for his guidance, input and support throughout this research program. The authors also wish to thank Mr. Ed Tighe of the MBTA for his invaluable suggestions, time, assistance and unwavering dedication to this project. Special thanks are also due to the MBTA bus operators who provided important feedback regarding the transit bus environment and DVI requirements. This research was conducted under Small Business Innovative Research Contract No. DTRS57-99-C-00083.

REFERENCES

1. Boff, K., and Lincoln, J. (1988). *Engineering Data Compendium: Human Perception and Performance*. Wright-Patterson Air Force Base, Ohio: AAMRL.
2. Gish, K. W., and Staplin, L. (1995). *Human Factors Aspects of Using Head Up Displays in Automobiles: A Review of the Literature*. (Report No. DOT HS 808 320). Washington, D.C.: National Highway Traffic Safety Administration, Office of Crash Avoidance Research.
3. Kiefer, R., LeBlanc, D., Palmer, M., Salinger, J., Deering, R., and Shulman, M. (1999). *Development and Validation of Functional Definitions and Evaluation Procedures for Collision Warning/Avoidance Systems*. (Report No. DOT HS 808 964). Washington, D.C.: U.S. Department of Transportation National Highway Traffic Safety Administration.
4. McNeil, S., Thorpe, C., and Mertz, C. (2000). A New Focus for Side Collision Warning Systems for Transit Buses. *Intelligent Transportation Society of America's Tenth Annual Meeting and Exposition*. Washington, D.C.: Intelligent Transportation Society of America.
5. NASA. (1978). *Anthropometric Source Book, Vol. 1: Anthropometry for Designers*. NASA Scientific and Technical Information Office. Houston, TX.
6. Novabus, Incorporated. (1999) Dimensions for RTS 40 ft bus. *RTS Brochure* (. http://www.novabus.com/english/products/rts/rts.pdf).
7. Reinach, S. and Everson, J. (Submitted for publication). *The Preliminary Design of a Driver-Vehicle Interface for a Transit Bus Collision Avoidance System*. Intelligent Transportation Society of America 11th Annual Meeting and Exposition, June, 2001. ITS America: Washington, DC.
8. U.S. Department of Transportation Transportation Safety Institute. (1999). *USDOT Transportation Safety Institute Vehicle Operations/Emergency Management/Customer Relations Participant's Manual*. Federal Transit Administration, Office of Safety and Security.
9. U.S. Department of Transportation Federal Transit Administration. (1999). *1998 National Transit Database Annual Report: Data Tables*. http://www.fta.dot.gov/ntl/database.html. Office of Program Management.
10. Wilson, T. (1996). *IVHS Countermeasures for Rear-End Collisions: Task 2-Functional Goals*. (Report No. DOT HS 808 513). Washington, D.C.: US Department of Transportation National Highway Traffic Safety Administration, Office of Crash Avoidance Research.
11. Young, S., Eberhard, C., and Moffa, P. (1995). *Development of Performance Specifications for Collision Avoidance Systems for Lane Change, Merging and Backing, Task 2 Interim Report: Functional Goals Establishment*. Washington, D.C.: TRW Space and Electronics Group for National Highway Traffic Safety Administration, Office of Collision Avoidance Research.

CONTACT

Mr. Stephen Reinach joined Foster-Miller in 1996 and is a Senior Engineer in the Human Performance and Operations Research Division at Foster-Miller, Inc. in Waltham, MA. Mr. Reinach has over six years of ground transportation safety and human factors research experience. He received his BA in Psychology from the University of Michigan in 1992 and his MS in Industrial Engineering (Human Factors Option) from the University of Iowa in 1996. Mr. Reinach is an Associate Human Factors Professional (Board of Certified Professional Ergonomists). He can be reached at sreinach@foster-miller.com .

Dr. Jeffrey Everson joined Foster-Miller in 1998 as a Senior Engineer to support program development in applications for Intelligent Transportation Systems and the DOT's Intelligent Vehicle Initiative. Dr. Everson served as the Principal Investigator for this program. He has extensive experience in collision avoidance systems, sensor technology, phenomenology, systems analysis, as well as program and department management. Dr. Everson received his Ph.D. in Physics from Boston College in 1976. He can be reached at jeverson@foster-miller.com.

Research and Design of Automobile Backward Automatic Ranging System

Hongwei Zhao, Xuebai Zang and Guihe Qin
Jilin University of Technology

ABSTRACT

In the paper, the research and design work of an automotive safety device is introduced. The device can automatically range, display the distance between the automobile and obstacle and sound alarm at a frequency varying with the distance. Equipped with the device, automobiles can avoid collision caused by drivers not noticing obstacles or not properly estimating the distance between automobile and obstacles when backing up. In the paper, the principle, the software, the hardware and the anti-interference technology of the device are concisely depicted. Test results showed that it is effective in collision avoidance.

INTRODUCTION

The measuring technology using laser, radar, infrared rays and ultrasound now is widely used in automobile collision avoidance and ranging. In collision avoidance, the area behind and between automobiles is usually monitored with laser or radar or infrared ray sensors. In ranging, ultrasonic sensors are usually used to monitor related area when automobiles change lane and obstacles when back up or parking.

The rear automatic ranging system introduced in the paper can measure the distance between automobile and obstacle, show the distance and give a audio and visible alarm, thereby increases automobile safety and decreasing possibility of trouble. Working automatically, the system needs no operation by drivers. The work introduced in the paper is a new approach and research method in collision avoidance.

ULTRASONIC RANGING PRINCIPLE

Ultrasonic ranging is a kind of contactless measuring method. It can not be affected by light ray, temperature, color and so on of the detected objects. Therefore, it has an excellent adaptability to environment. The property is very important for automobiles which work in various environments.

In ultrasound technology of continuous ranging, the widest used one is the ultrasonic pulse echo method (Sonar Method). The following is the principle of the method: The sender launches an ultrasonic pulse. The waves propagate in the medium. They are reflected backward when meeting an object. Depending on the time from sending to receiving and the propagating speed of ultrasound in the medium, the distance between the ultrasonic sensor and the object can be calculated.

If the distance between sensor and detected object is L, t is the time from sending to receiving and c is sound speed, then:

$$L = c * t \qquad (1)$$

When sender is separated from the receiver, if the range between them is 2a and the distance from sender to detected object is s, then:

$$s = c * t / 2 \qquad (2)$$

L can be calculated from:

$$L = \sqrt{s^2 - a^2} \qquad (3)$$

when $L \gg a$, $L \approx s$.

HARDWARE DESIGN

The hardware of the system is composed of ultrasonic sender and receiver circuits, signal processing circuit, single-chip computer circuit and display alarm circuit.

The single-chip computer controls the three sender/receiver circuits in turn through the multiplexer. The senders transmit 40KHz ultrasound. The receivers accept the reflected sound waves, convert them to rectangular waves through comparator circuit and send them to the signal processing circuit through the multiplexer. The signal processing circuit converts the time delay of the sent and received waves to a pulse width signal, and transmits it to the single-chip computer to process. The single-chip computer calculates the delay time, then calculates the distance from formula (3) and alarm with different frequencies depending on the distance.

The three groups of the sender/receiver are evenly installed on the back of automobile. They work in turn. The distance is displayed to three decimal places LED. The largest distance displayed is 9.99m. The power supply of the device comes from the backing-up light. Only when the automobile back up is the device turned on and begin to work. When back-up stops, the device is turned off.

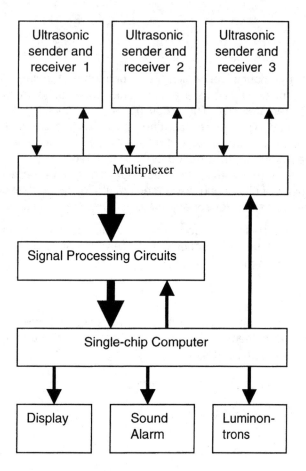

Figure 1. System Hardware Structure

SOFTWARE DESIGN

Software performs the following functions as needed by the system:

1. Data sampling
   The single-chip computer controls the sensors to work in turn and records the ultrasonic delay time of each channel when it is working.

2. Data processing
   The software completes such data processing work as digit filtering, correcting compensation, datum distinguishing and distance sectional detection.

3. Distance displaying
   The distance is changed into BCD form and sent to the LED through serial communication.

4. Audio-visible alarm
   The software also calculates the alarm frequency according to the distance and controls the audio and visible devices to alarm in the frequency.

Fig.2 is the diagram of the software.

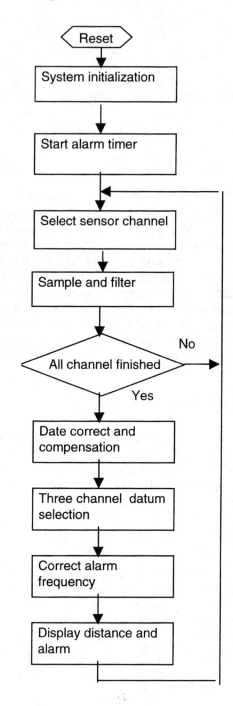

Figure 2. Software Diagram

ANTI-INTERFERENCE STRATEGIES

A blind zone and a distance limitation exist with an ultrasonic ranging method. When the amplitude of the received signal is larger then the specified threshold, the signal is considered valid. This causes the distance limitation, for the larger the distance between sensor and detected object, the weaker the reflected signal. The threshold must be larger than noise amplitude. Otherwise, it is difficult to differentiate the needed signal from the noise.

The system mainly employs the following software anti-jamming measures:

1. System error compensation.
   The main time delay error includes circuit time lag, transmission delay in cable and delay caused by coupling layer, and sound transmission mask of the receiver. When the device, sender/receiver, and cable are set, these time delays become fixed. They can be processed as a kind of system error.

2. Sending pulse modulation technique.
   The number of pulse sent in each sample phase varies with the distance. If the distance is longer, the number is larger, and vice versa.

3. Delayed receiving technique.
   Many substances exist in air which can cause scattering, for example, dust and micro drip. The reflecting face is also not smooth. All these can disturb the received signal and make the signal have many stray waves to cause false receptions. To decrease the disturbance, for an interval after the ultrasound is sent the receiver ignores any arriving signal.

4. Digit filter technique.
   In the software, elimination of faulty data, arithmetic mean and sliding filtering methods are used in data processing.

5. Time delay canceling after-vibration of sending.
   The software has a time delay after each sample to cancel the disturbance of after-vibration to next sample.

In fact, there are many stochastic interferences affecting measure precision and range. The following methods can be employed in increasing measurement precision and range:

1. Sound speed correction
   Sound speed can be affected by the temperature, composition and pressure intensity of air. When calculating the distance the system can correct the sound speed with the factors.

2. Amending focalizer
   Focalizers can adjust squint to increase the detected distance.

REFERENCES

1. Wang Wutong, "Functions and Requirement of Energy Transduce of Ultrasonic Detection of Defects", Nondestructive Detection, Vol.19, No.3.

2. Zhao Yufang, "Acoustic Principle", National Defence Publishing House, 1981.

3. Acoustic Research Dept. of Tongji University, Ultrasonic Industry Measure Technology, Shanghai Renmin Publishing House, 1977.

4. Fang Peimin, "New Edition of Sensor principle, Application and Circuit Explication", Electronic Industry Publishing House, 1995.

5. Zhao Hongwei, Zhu Hongwen, Zang Xuebai, "Microcomputer Interface Technology", Jilin University Publishing House, 1998.

6. Yuan Bo, "Amplification Circuit of Selecting Frequence Circuit Network & Ultronsic Signal in High Precission Measure", Electronic Technology, No. 4,1997.

7. Zhang Zhaofu, "Foreign Automotive Electronic Information", Automotive Magazine, No. 6, 1996.

1999-01-1290

Driver Crash Avoidance Behavior with ABS in an Intersection Incursion Scenario on the Iowa Driving Simulator

Elizabeth N. Mazzae
National Highway Traffic Safety Administration

G. H. Scott Baldwin
Transportation Research Center Inc.

Daniel V. McGehee
University of Iowa

ABSTRACT

The National Highway Traffic Safety Administration (NHTSA) has developed its Light Vehicle Antilock Brake Systems (ABS) Research Program in an effort to determine the cause(s) of the apparent increase in fatal single-vehicle run-off-road crashes as vehicles undergo a transition from conventional brakes to ABS. As part of this program, NHTSA conducted research examining driver crash avoidance behavior and the effects of ABS on drivers' ability to avoid a collision in a crash-imminent situation. The study described here was conducted on the Iowa Driving Simulator and examined the effects of ABS versus conventional brakes, speed limit, ABS instruction, and time-to-intersection (TTI) on driver behavior and crash avoidance performance. This study found that average, alert drivers do tend to brake and steer in realistic crash avoidance situations and that excessive steering can occur. However, this behavior did not result in a significant number of road departures.

INTRODUCTION

Antilock brake systems (ABS) have been introduced on many passenger car and light truck make/models in recent years. In general, ABS appears to be a promising safety device when evaluated on a test track. Under many pavement conditions antilock brake systems allow the driver to stop a vehicle more rapidly while maintaining steering control even during situations of extreme, panic braking. Brake experts anticipated that the introduction of ABS on passenger vehicles would reduce both the number and severity of crashes. However, a number of crash data analyses have been performed in recent years by NHTSA, automotive manufacturers, and others which indicate that the introduction of ABS has not resulted in a reduction in the number of crashes to the anticipated extent.

CRASH DATA – Kahane [1] found that, with the introduction of ABS, involvements in fatal multi-vehicle crashes on wet roads were significantly reduced by 24 percent, and nonfatal crashes by 14 percent. However, these reductions were offset by a statistically significant increase in the frequency of single-vehicle, run-off-road crashes, as compared to cars without ABS. Fatal run-off-road crashes were up by 28 percent and nonfatal crashes by 19 percent.

A later, 1998 study by Hertz, Hilton, and Johnson [2, 3]. This study is similar to an earlier study by the same authors [4] except that it is based on more recent (1995 - 96) crash data. The effects found by the 1998 study were generally similar to the findings of the earlier study except that ABS now appears to be decreasing one particular subtype of single-vehicle road departure crashes, frontal impacts with fixed objects, rather than increasing their numbers.

NHTSA'S LIGHT VEHICLE ABS RESEARCH PROGRAM – In an effort to investigate possible causes of the crash rate phenomena, NHTSA developed its Light Vehicle ABS Research Program. This program contains nine separate tasks addressing such issues as ABS hardware performance, examination of ABS crash reports, and assessment of driver behavior with ABS (as outlined in [5]). To date, NHTSA research has found no systematic hardware deficiencies in its examination of ABS hardware performance, except for known degradations in stopping distances on gravel (as documented in [6]). It is unknown, however, to what extent the increase in run-off-road crashes may be due to drivers' incorrect usage of ABS, incorrect response to ABS activation, incorrect instinctive driver response (e.g., oversteering), changes in driver behavior (e.g., behavioral adaptation) as a result of ABS use, or some other factor.

TASK 5: HUMAN FACTORS STUDIES OF DRIVER CRASH AVOIDANCE BEHAVIOR

To determine whether some aspect of driver behavior in a crash-imminent situation may be counteracting the potential benefits of ABS, NHTSA embarked on a series of human factors studies. These studies, which compose Task 5 of the research program, focus on the examination of driver crash avoidance behavior as a function of brake system and various other factors.

One of the theories Task 5 sought to address was whether the apparent increase in single-vehicle crashes involving ABS-equipped vehicles may be due to characteristics of driver steering and braking behavior in crash-imminent situations. According to this theory, in situations of extreme, panic braking, drivers may have a tendency to brake hard and make large steering inputs to avoid a crash. Without four-wheel ABS, aggressive braking may lock the front wheels of the vehicle, eliminating directional control capability, rendering the driver's steering behavior irrelevant. With four-wheel ABS, the vehicle's wheels do not lock; therefore, the vehicle does not lose directional control capability during hard braking, allowing drivers' steering inputs to be effective in directing the vehicle's motion. This directional control could result in drivers avoiding multi-vehicle crashes by driving off the road and experiencing single-vehicle crashes.

To investigate this theory, Task 5 sought to determine whether:

- Drivers tend to both brake and steer (as opposed to only braking or only steering) during crash avoidance maneuvers;
- Drivers tend to make large, potentially excessive, steering inputs during crash avoidance maneuvers;
- Drivers' crash avoidance maneuvers in ABS-equipped vehicles result in road departures more often than in conventionally braked vehicles;
- Drivers avoid more crashes in ABS-equipped vehicles than in conventionally braked vehicles; and
- Speed limit has an effect on whether drivers avoid more crashes in ABS-equipped vehicles than in conventionally braked vehicles.

Task 5 of NHTSA's Light Vehicle ABS Research Program included three studies. Two studies were conducted on a test track (one on dry pavement, one on wet pavement) and one on the University of Iowa's Iowa Driving Simulator (IDS).

These studies used a right-side intersection incursion scenario to elicit a crash avoidance response from human subjects. This scenario was chosen because it was likely to induce steering and obstacle avoidance behavior and had the potential for induce the subjects to drive the vehicle off of the road. This intersection-related obstacle avoidance scenario is obviously not responsible for all run-off-road crashes and results may not be representative of driver behavior in all situations leading to vehicle road departure. Many run-off-road crashes occur when drivers are unable to maneuver through a curve in the roadway or when they are drowsy or under the influence of alcohol. However, it is believed that the results of this study will be useful in determining not only the extent to which drivers are able to maneuver a vehicle, but also drivers' physical capacity to supply control inputs to the vehicle. Insight into drivers' ability to maintain vehicle control during a panic maneuver and ability to avoid a collision can also be gained from this research.

Although the same scenario was involved in each of these experimental venues, advantages to both test track and simulator means of observing driver behavior were present. The test track experiments allowed driver behavior to be examined in a realistic environment at moderate speeds in real vehicles with simulated obstacles on both dry and wet pavement. The IDS study allowed for driver behavior to be examined using a highly repeatable test method in a simulated environment at higher travel speeds and with no chance of actual physical collision. This paper discusses the method and results of the study conducted on the Iowa Driving Simulator.

METHOD

APPARATUS

Iowa Driving Simulator (IDS) – The Iowa Driving Simulator incorporates recent technological advances to create a highly realistic automobile simulator. IDS uses four multi-synch projectors to create a 190 degree forward field-of-view and a 60 degree rear view. Motion cues are produced by a six-degree of freedom motion base. Inside the simulator dome is a fully instrumented vehicle cab. The vehicle cab used in this study was a 1993 Saturn SL2. However, both the vehicle dynamics simulation and the antilock brake system modeled were of a Ford Taurus, a typical mid-sized American car. The Ford Taurus vehicle dynamics model used in this study was that developed by NHTSA for use with the National Advanced Driving Simulator (NADS).

Hardware-in-the-Loop Antilock Brake System (ABS) – To facilitate testing of driver behavior with ABS in this study a hardware-in-the-loop antilock brake system was developed by NHTSA's Vehicle Research and Test Center for implementation on the IDS. Details of this hardware-in-the-loop ABS are will be provided in the final NHTSA report for this project. The ABS operates as it would on an actual vehicle by providing both haptic brake pedal feedback and auditory feedback. This hardware-in-the-loop system has the capability of being enabled/disabled by an operator switch so both ABS and non-ABS conditions could be run efficiently without any hardware changes.

Instrumentation – In addition to the objective data quantifying the subjects' vehicle control inputs, four video cameras were also used to record the events on video tape for analysis of driver behavior, response timing, and reaction to the incursion event. One camera focused on the throttle and brake pedals. Another focused on the driver's face. A third focused on the driver's hands on the steering wheel. The fourth camera recorded the forward view of the road scene. Both sensor data and video data were collected at a rate of 30 Hz.

SUBJECTS – Sixty males and 60 females between the ages of 25 and 55 years were selected for participation in this study. Each participant was required to hold a valid driver's license and be able to pass a general health screening. In general, subjects placed in the conventional brake system condition had conventional brakes on their primary personal vehicle, while those in the ABS condition had ABS on their primary vehicle.

EXPERIMENTAL DESIGN – The experimental design used in this study was a 2 (brake system; ABS, conventional brakes) x 2 (ABS instruction; video, no video) x 2 (speed limit; 45, 55 miles per hour) x 2 (time-to-intersection, TTI; 2.5, 3.0 seconds) x 2 (gender) within subjects partial factorial design. This design produced twelve experimental conditions. This paper focuses on the results for brake system, ABS instruction, speed limit, and gender.

To address whether drivers may be more likely to crash in an ABS-equipped vehicle due to lack of knowledge about ABS, ABS instruction was included as an independent variable in this study. Of the subjects in the ABS condition, half received ABS instruction and the other half received no ABS instruction. ABS instruction consisted of an initial segment describing the Iowa Driving Simulator and procedures for entering and exiting the simulator and what to do in the event that the subjects experienced signs of motion sickness, and a latter segment which illustrated ABS operation and was use taken from an OEM video [7] designed to be given to a buyer with the purchase of a new vehicle. Subjects in the conventional brake system condition received a shortened version of the same video containing only the segment describing the nature of the IDS and related safety precautions.

TTI was defined as the time it took the subject vehicle to reach the intersection at their current velocity as measured at a "trigger" point in the road. The purpose of this independent variable was to examine whether subjects altered their collision avoidance strategy based on the time available to respond to the event.

To assess whether drivers are more likely to have unsuccessful crash avoidance maneuvers in ABS-equipped vehicles while traveling at higher speeds, a speed limit independent variable was included in this study. The speeds chosen were 45 and 55 miles per hour (mph). Results for the 45 mph condition could be compared to results for the dry test track study for the same speed.

For safety reasons, speed limits in the test track studies were kept to 45 mph on dry pavement and 35 mph on wet pavement.

A counterbalance scheme was used to ensure that each condition accommodated for differences in days, the time of day, and gender differences. An equal number of males and females participated both in the morning and in the afternoon on each day of testing.

PROCEDURE – To help ensure that subjects would not anticipate the intersection incursion event, subjects were informed that they would be driving for approximately 30 minutes. In actuality, the drive was approximately 15 minutes in length. In addition, subjects were told that their task was to assess the looks and feel of the simulator and that they would be given a questionnaire to collect their impressions on this topic after their drive.

Upon entering the simulator, an in-vehicle experimenter instructed the subject to adjust the seat and mirrors. The subject was then told to begin driving and was given time to get comfortable with the feel of the vehicle. The in-vehicle experimenter instructed the subject to get a feel for the steering, braking, and acceleration, during an initial portion of the drive in which no data were collected. No mention was made by the in-vehicle experimenter of the presence of ABS, where applicable, and no encouragement was given for the subjects to practice activating the ABS. Approximately five minutes into the drive, subjects were asked to begin driving normally and assess the simulator.

During their drive, subjects experienced a slow-moving semi tractor-trailer on a hill. This truck required them to reduce speed to approximately 25 mph. Oncoming traffic was spaced such that passing the truck was not an option. Once the subject had crested the hill, the truck pulled over and stopped on the shoulder of the roadway and the subject was able to return to driving at the posted speed limit.

Shortly thereafter, another vehicle, called the "lead vehicle," appeared in the distance ahead of the subject vehicle. This vehicle maintained a six second headway with respect to the subject vehicle. As the subject vehicle approached the intersection where the incursion would take place, the lead vehicle could be seen by the subject driving through the intersection without stopping or braking. The purpose of the lead vehicle was to encourage the subject to feel that there was no need to slow down or stop at the intersection and that it was safe to continue on their path through the intersection.

The intersection at which the incursion event took place was the only intersection present in the route. As depicted in Figure 1, two vehicles were positioned on the perpendicular roadway at the intersection. One vehicle was stopped at a stop sign on the left side of the intersection (a light truck) and another vehicle stopped at a stop sign on the right side of the intersection (a Buick Regal). At the time of the incursion event, no oncoming traffic was present. At the specified TTI the vehicle on the right side of the intersection drove into the intersection and stopped with its front bumper at the center of the subject's lane of travel, causing subjects to perform evasive maneuvers to avoid collision. Subjects reactions were captured by sensor and video data. Upon the completion of the intersection incursion event, the subject's drive was over. Each subject experienced the incursion event only once.

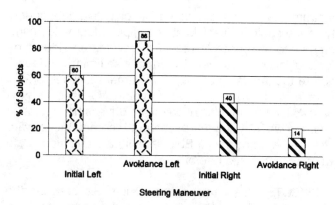

Figure 2. Steering response directions during incursion scenario.

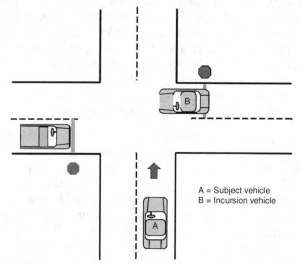

Figure 1. Illustration of intersection and position of vehicles during the incursion event.

RESULTS

CRASH AVOIDANCE STRATEGY -- OVERALL – All 120 subjects attempted both steering and braking inputs in an attempt to avoid colliding with the scenario vehicle as it encroached into their lane. Seventy-nine percent of the subjects applied the brakes before steering as their initial response. Four percent of the subjects initiated braking and steering inputs simultaneously as an initial response. Seventeen percent of the subjects steered before applying the brakes.

As the incursion vehicle began its motion into the intersection, 60 percent of the subjects chose to steer left as an initial steering input (defined as the first steering input of magnitude greater than six degrees which the subject made after the initiation of the incursion vehicle's motion), and 40 percent chose to steer right.

An "avoidance steering input" was defined as the steering input which a subject made that was intended to maneuver the subject vehicle around the crossing vehicle. This input was not necessarily the subject's first steering input in response to the incursion. During the collision avoidance maneuver, 86 percent of the subjects chose to try to steer left of the encroaching vehicle and 14 percent made the decision to steer right to avoid a collision (see Figure 2). Thirty-six percent of the subjects who steered left crashed, while 29 percent who steered right crashed.

STEERING BEHAVIOR -- OVERALL – The average magnitude of avoidance steering input observed was 148 degrees. The highest observed steering input from an individual subject in this study was 540 degrees during the avoidance maneuver.

The average maximum steering rate obtained during the avoidance maneuver was 514 degrees per second. The highest observed steering rate achieved by a subject in this study was 1416 degrees per second. Ninety-five percent of steering rates observed were less than 981 degrees per second.

BRAKING BEHAVIOR -- OVERALL – The overall average maximum brake pedal force obtained was 90 pounds. The highest observed brake pedal force input generated by a subject in this study was 278 pounds. Ninety-six percent of the subjects either activated ABS or locked the vehicle's wheels with conventional brakes during the avoidance maneuver.

ROAD DEPARTURES -- OVERALL – Eight people fully departed the roadway during the collision avoidance maneuver. Six subjects made steering inputs severe enough to cause yaw rates resulting in some degree of vehicle spin. In 4 of these 6 cases, the vehicle spun off the road.

CRASHES -- OVERALL – During the intersection incursion event, 35 percent of the subjects collided with the scenario vehicle as it encroached into their lane.

ABS VS. CONVENTIONAL

Overall, 80 subjects were assigned the ABS condition and 40 were assigned the conventional brake system case.

CRASH AVOIDANCE STRATEGY BY BRAKE SYSTEM – For those who braked then steered during the avoidance maneuver, the delay time from when they initiated braking to when they began to steer did not differ significantly by brake system. Those with ABS waited 0.70 seconds after braking to initiate steering while those with conventional brakes waited 0.62 seconds.

STEERING BEHAVIOR BY BRAKE SYSTEM – Table 1 summarizes characteristics of observed steering behavior according to brake system.

Results for the magnitude of avoidance steering inputs by brake system are given in Figure 3. The average magnitude of the avoidance steering input for subjects in the ABS condition was 125 degrees. The average magnitude of the avoidance steering input for subjects in the conventional brake system condition was significantly larger at 192 degrees [p = 0.0064].

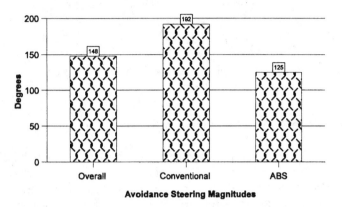

Figure 3. Magnitude of avoidance steering inputs by brake system.

As stated previously, a majority of subjects chose to steer left for the avoidance maneuver. On average, those with ABS steered 121 degrees to the left; whereas, the conventional group steered 176 degrees to the left. For those who steered left during the avoidance maneuver, those with ABS steered with a significantly smaller magnitude than those with conventional brakes [p = 0.0334].

Table 1. Characteristics of observed steering behavior by brake system (* These values are significantly different).

Steering Input Characteristics	Brake System	Value
Average magnitude of avoidance steering input (degrees)	Overall	148
	Conventional	192*
	ABS	125*
Average maximum steering input rate (degrees per second)	Overall	514
	Conventional	595*
	ABS	473*
Time to maximum steering input (seconds)	Conventional	3.69*
	ABS	3.24*

The average maximum steering rate of the avoidance maneuver for subjects in the ABS condition was 473 degrees per second. The average maximum steering rate of the avoidance maneuver for subjects in the conventional brake system case was significantly higher at 595 degrees per second [p = 0.0524](Figure 4).

For subjects in the ABS condition, the time to maximum steering input was significantly less [p = 0.0052] (3.24

seconds) than that observed for those with conventional brakes (3.69 seconds).

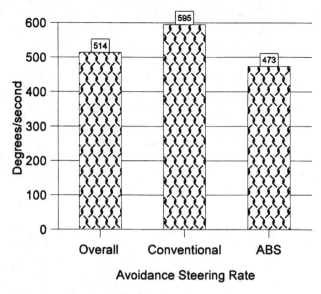

Figure 4. Rates of avoidance steering inputs by brake system.

BRAKING BEHAVIOR BY BRAKE SYSTEM – The average maximum brake pedal force observed for subjects in the ABS condition during the avoidance maneuver was 86 pounds. For subjects in the conventional brake system case, the average was 98 pounds (see Figure 5). This difference was not statistically significant [p = 0.0944] at the p = 0.05 level.

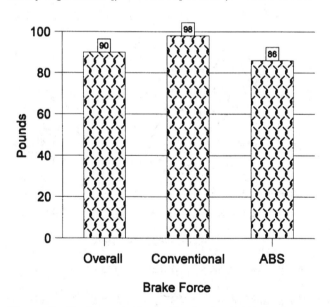

Figure 5. Brake pedal forces by brake system.

One might expect that observed brake pedal application durations should be longer for ABS if drivers were using the ABS properly. However, the average brake pedal application duration observed during the crash avoidance maneuver in this study was significantly longer [p = 0.0435] for conventional brakes (3.12 seconds) than for ABS (2.69 seconds). One also might expect that subjects receiving ABS instruction might have longer brake pedal application durations as a result of being told not to "pump" the brake pedal with ABS. However, subjects receiving ABS instruction had an average brake pedal application duration (2.84 seconds) which was not significantly longer than for those with ABS who received no instruction (2.54 seconds).

ROAD DEPARTURES BY BRAKE SYSTEM – Four subjects out of the 80, or 5 percent, who were assigned the ABS condition drove completely off the road (all four wheels) during the avoidance maneuver as shown in Figure 6. In each of these four cases the ABS was activated during the crash avoidance maneuver. Four of the 40 subjects, or 40 percent, who had conventional brakes also drove completely off the road during the avoidance maneuver.

All of the instances of four-wheel road departure with ABS were at the 45 mph speed limit. Each of the road departures for conventional brakes was in the 55 mph speed limit condition. Unfortunately, due to the small number of road departures observed in this test, it is difficult to determine whether there is a significant brake system by speed limit interaction.

Six partial (two-wheel) road departures were also observed in this study. One of the cases involved a subject driving with ABS, while the other five involved conventional brakes.

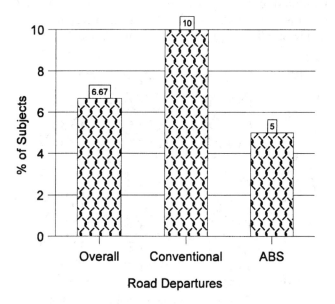

Figure 6. Percent road departures by brake system.

CRASHES BY BRAKE SYSTEM – Figure 7 illustrates results for the number of crashes by brake system. Subjects with ABS crashed less, 31 percent of 80 subjects, than those in the conventional brake system condition, 43 percent of 40 subjects. However, this difference was not statistically significant.

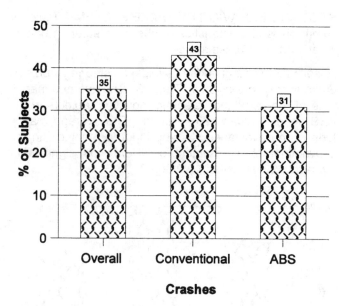

Figure 7. Percent crashes by brake system.

When considering only the sub-sample of subjects who were assigned the ABS condition but were not provided with ABS instruction, these subjects did experience fewer crashes (30 percent crashed) than the conventional group (43 percent crashed). Additional details regarding the effects of ABS instruction on driver behavior and crashes will be provided in the NHTSA final report on this project.

SPEED LIMIT

CRASH AVOIDANCE STRATEGY BY SPEED LIMIT – Regardless of speed limit, subjects tended to brake first and then steer in an attempt to avoid colliding with the crossing vehicle. Of the 20 subjects who steered before applying the brakes, 50 percent were in the 45 mph speed limit condition and 50 percent were in the 55 mph speed limit condition.

STEERING BEHAVIOR BY SPEED LIMIT – Table 2 summarizes subjects' steering behavior as a function of speed limit and brake system.

The average magnitude of the avoidance steering input for subjects in the 55 mph condition was 132 degrees. The average magnitude of the avoidance steering input for those in the 45 mph condition was higher at 163 degrees [p = 0.1055](see Figure 8).

The average maximum steering input rate during the crash avoidance maneuver for subjects in the 55 mph speed limit condition was 507 degrees per second. The average maximum steering rate of the avoidance maneuver for subjects in the 45 mph speed limit condition was similar at 520 degrees per second as shown in Figure 9.

Table 2. Characteristics of observed steering behavior by speed limit and brake system.

Steering Input Characteristics	Speed Limit	Brake System	Value
Average magnitude of avoidance steering input (degrees)	45	Overall	163
		Conventional	215
		ABS	139
	55	Overall	132
		Conventional	172
		ABS	112
Average maximum steering input rate (degrees per second)	45	Overall	520
		Conventional	569
		ABS	496
	55	Overall	507
		Conventional	621
		ABS	450

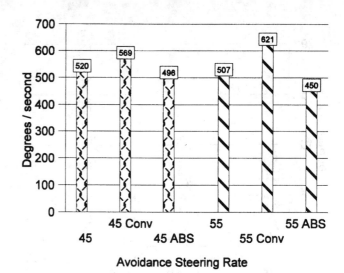

Figure 9. Avoidance steering rates by speed limit and brake system.

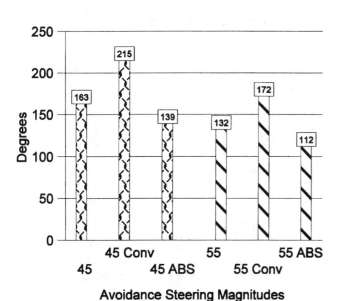

Figure 8. Magnitude of avoidance steering inputs by speed limit and brake system.

BRAKING BEHAVIOR BY SPEED LIMIT – For subjects in the 55 mph speed limit condition, the average maximum brake pedal force obtained during the avoidance maneuver was 98 pounds. The average maximum brake pedal force observed was 82 pounds for the subjects in the 45 mph speed limit condition [p = 0.0981] (Figure 10).

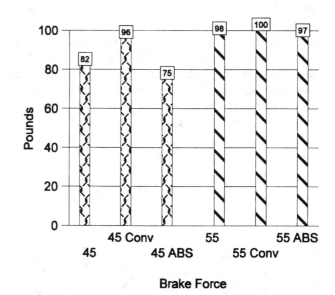

Figure 10. Brake forces by speed limit and brake system.

ROAD DEPARTURES BY SPEED LIMIT – As stated earlier, four of the instances of road departure were observed for the 45 mph speed limit and all of these involved the ABS-equipped condition. Each of the four road departures for the 55 mph case involved subjects in the conventional brake condition (Figure 11).

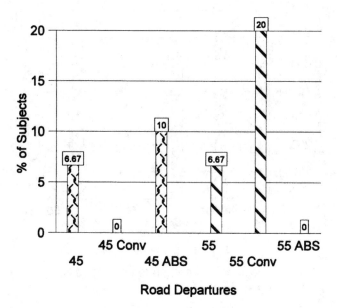

Figure 11. Road Departures by speed limit and brake system.

CRASHES BY SPEED LIMIT – For half of the 120 subjects, the posted speed limit on the roadway was 55 mph and for the other half the posted speed limit was 45 mph. Forty-two percent (40 percent ABS, 45 percent non-ABS) of those at the 55 mph speed limit collided with the encroaching vehicle. Only 28 percent (23 percent ABS, 40 percent conventional) crashed in the 45 mph speed limit condition [p = 0.126] as shown in Figure 12.

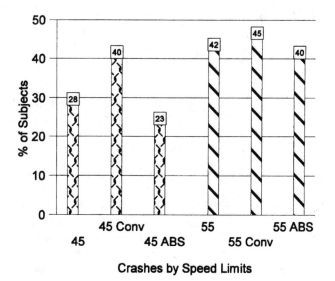

Figure 12. Crashes by speed limit and brake system.

GENDER

STEERING BEHAVIOR BY GENDER – The average magnitude of the avoidance steering input for females was 142 degrees and for males was 154 degrees as shown in Figure 13. The average maximum steering rate in any direction for females was 454 degrees per second

and for males was 573 degrees per second as illustrated in the incursion vehicle motion to the time of initial steering input nor the time from initiation of incursion vehicle motion to the time of maximum steering input varied significantly by gender.

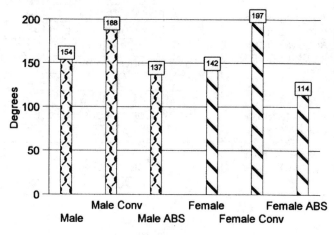

Figure 13. Magnitude of avoidance steering inputs by gender and brake system.

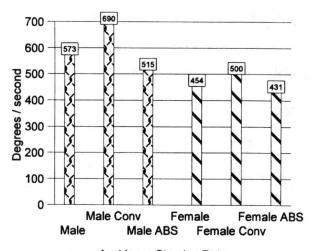

Figure 14. Avoidance steering rates by gender and brake system.

BRAKING BEHAVIOR BY GENDER – The average maximum brake pedal force was 86 pounds for females and 93 pounds for males as shown in Figure 15. This difference was not statistically significant.

In general, braking behavior did not differ according to gender. Males and females produced similar results in terms of the time from the initiation of the incursion vehicle motion to throttle release and maximum braking input. However, reaction time to initial brake application did differ by a nearly significant level [p = 0.06] with males applying the brakes within an average of 1.10 seconds and females doing so within 1.17 seconds.

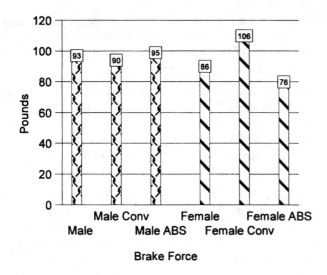

Figure 15. Brake forces by gender and brake system.

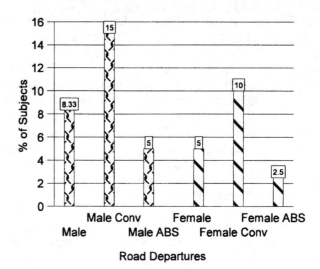

Figure 16. Road departures by gender and brake system.

ROAD DEPARTURES BY GENDER – Of the eight subjects who drove completely off the road to avoid a crash, five were males and three were females (see Figure 16).

CRASHES BY GENDER – Figure 17 illustrates crash rates observed in this study by gender and brake system. Thirty-two percent of female subjects crashed into the conflict vehicle during the intersection incursion scenario. Twenty-three percent of the female subjects in the ABS condition crashed, while 50 percent in the conventional case crashed. These data correspond to a statistically significant effect of brake system on crash rates for females, wherein, females in the ABS condition crashed significantly less [p = 0.031] than those with conventional brakes.

Thirty-eight percent of males crashed during the intersection incursion scenario. Forty percent of the males in the ABS condition crashed and 35 percent of the males with conventional brakes crashed.

DISCUSSION

DO DRIVERS TEND TO BOTH BRAKE AND STEER DURING CRASH AVOIDANCE MANEUVERS? – All subjects in this study both braked and steered in an attempt to avoid colliding with the incursion vehicle. Seventy-nine percent of subjects braked before steering during their collision avoidance maneuver. The delay time between when subjects initiated braking to when they made their first steering input did not vary significantly as a function of brake system.

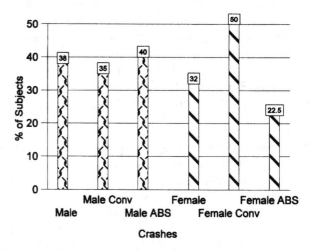

Figure 17. Crashes by gender and brake system.

DO DRIVERS TEND TO MAKE LARGE, POTENTIALLY EXCESSIVE STEERING INPUTS DURING CRASH AVOIDANCE MANEUVERS? – In general, steering inputs exhibited by subjects in this IDS study were larger and quicker than those observed in the related test track studies [8]. This difference is believed to be attributable to the lack of "road feel" present on the IDS as well as the limited range of travel of the simulator motion base. Both of these qualities of the IDS are believed to have contributed to subjects' perception that they were driving on a wet road, although the roadway coefficient of friction in this study was stipulated at 0.8. Despite the larger magnitudes and rates of steering inputs observed in this study, the significance of effects corresponds very well to findings obtained in the wet test track study.

Steering inputs characterized by large magnitudes and high rates of application were observed in this study. However, drivers appeared to alter their steering behavior based on the degree to which they felt the steering inputs were affecting the motion of the vehicle in the desired direction. Subjects with ABS made smaller steering inputs and used lower application rates than subjects with conventional brakes. The reason for this is believed to be that subjects made increasingly large steering inputs with conventional brakes since, with locked wheels, their steering inputs were not effective in directing the vehicle's motion.

DO PEOPLE EXPERIENCE MORE ROAD DEPARTURES IN ABS-EQUIPPED VEHICLES THAN IN VEHICLES WITH CONVENTIONAL BRAKES? – Overall, results from this study indicate that although subjects were observed making large steering inputs at high rates, these aggressive steering inputs did not result in a significant number of road departures. This conclusion is true for both conventional brakes and ABS for this study. An equal number of subjects, four per condition, experienced full road departures in both the conventional brakes system and ABS conditions. Six partial road departures were observed in this study, only one of which involved ABS.

DO PEOPLE CRASH LESS FREQUENTLY IN ABS-EQUIPPED VEHICLES THAN IN VEHICLES EQUIPPED WITH CONVENTIONAL BRAKES? – Overall in this study, subjects driving with ABS did not crash significantly less than those with conventional brakes. However, females in the ABS condition did crash significantly less than those in the conventional brake system condition. Males in this study crashed approximately the same amount with ABS as they did with conventional brakes.

At the 55 mph posted speed limit, the number of crashes observed by brake system did not vary greatly. However, for the 45 mph speed limit, only 20 percent of subjects with ABS crashed as opposed to 40 percent of subjects with conventional brakes. This finding appears to contradict the opinion of some experts that if ABS is found to be associated with an increase in crashes, this increase is likely to only be associated with vehicle traveling at high rates of speed.

CONCLUSIONS

An experiment was conducted in which drivers' collision avoidance behavior in a simulated right-side intersection incursion scenario was examined as a function of vehicle brake system (ABS versus conventional brakes), time-to-intersection (TTI), instruction, and travel speed (45 mph or 55 mph). Drivers' reactions in terms of steering and braking and their success in avoiding the incursion vehicle were recorded.

Subjects in this study were alert and sober. These subjects did demonstrate the capability to make aggressive steering and braking inputs. However, despite the high

magnitudes and rates of many inputs observed, few road departures were observed. Road departures which were observed could not be judged attributable to ABS performance or driver interaction with ABS.

Overall, ABS was not found to be associated with significantly fewer crashes in this study as compared to conventional brakes. However, females were found to crash significantly less with ABS than with conventional brakes. More crashes were observed in this study in the 55 mph speed limit condition than at 45 mph.

The results of this study do not appear to indicate that a problem exists due to the crash avoidance of alert, sober drivers. However, an examination of the behavior of sleepy drivers or drivers under the influence of alcohol may produce different results. This study revealed no indications that driver interaction with ABS may be contributing to the apparent increase in fatal single-vehicle road departure crashes that has been identified in conjunction with vehicles transitioning from conventional to antilock brake systems.

Results from this study will be examined in conjunction with the results of other tasks included in NHTSA's Light Vehicle ABS Research Program to determine whether the collective results viewed as a whole provide some insight into the cause(s) of the increase in fatal single-vehicle crashes observed in conjunction with the implementation of ABS.

ACKNOWLEDGMENTS

The authors acknowledge the contributions of Peter Grant of Ford Motor Company and Garrick Forkenbrock of Transportation Research Center Inc. toward the development of the hardware-in-the-loop ABS for implementation on the IDS for use in this study.

REFERENCES

1. Kahane, C. (December, 1994). Preliminary Evaluation of the Effectiveness of Antilock Brake Systems for Passenger Cars, National Highway Traffic Safety Administration, Technical Report.

2. Hertz, E., Hilton, J., and Johnson, D. M. (1998). "Analysis of the Crash Experience of Vehicles Equipped with Antilock Brake Systems (ABS) - An Update." ESV Paper No. 98-S2-O-07. Enhanced Safety of Vehicles Conference, February, 1998.

3. Hertz, E., Hilton, J., and Johnson, D. (1998). Analysis of the Crash Experiences of Vehicles Equipped with Antilock Braking Systems -- An Update. NHTSA Technical Report No. 808 758. Washington, DC: National Highway Traffic Safety Administration.

4. Hertz, E., Hilton, J., and Johnson, D. (1995). An Analysis of the Crash Experience of Passenger Cars Equipped with Antilock Braking Systems. NHTSA Technical Report No. 808 279. Washington, DC: National Highway Traffic Safety Administration.

5. Garrott, W. R. and Mazzae, E. N. (1999). An Overview of the National Highway Traffic Safety Administration's Light Vehicle Antilock Brake Systems Research Program. SAE Paper No. 1999-01-1286. Warrendale, PA: Society of Automotive Engineers.

6. Forkenbrock, G., Flick, Mark, and Garrott, W. G. (1999). NHTSA Light Vehicle Antilock Brake Systems Research Program Task 4: A Test Track Study of Light Vehicle ABS Performance Over a Broad Range of Surfaces and Maneuvers. Washington, DC: USDOT.

7. General Motors Corporation (1994). Your New Vehicle's Antilock Brake System; How It Works For You. Part No. 15709859.

8. Mazzae, E. N., Barickman, F. S., Baldwin, G. H. S., and Forkenbrock, G. (1999). Driver Crash Avoidance Behavior with ABS in an Intersection Incursion Scenario on Wet Versus Dry Pavement. SAE Paper No. 1999-01-1288. Warrendale, PA: Society of Automotive Engineers.

ADDITIONAL SOURCES

Additional information regarding Task 5 of NHTSA's Light Vehicle ABS Research Program can be found in an upcoming NHTSA report titled, "Examination of Drivers' Collision Avoidance Behavior Using Conventional and Antilock Brake Systems."

1999-01-0822

Improved Crash Avoidance Using Performance-Based Brake Test Results

S.J. Shaffer and A.-C. Christiaen
Battelle Memorial Institute

ABSTRACT

Current crash avoidance systems combine vehicle speed with knowledge of position (and change in position) of potential obstacles in front of the vehicle to trigger alarms warning of impending collisions. The various alarm levels are triggered using a simple set of minimum time delays. Although knowledge of on-board vehicle braking capability is not currently incorporated into these systems, such knowledge can improve the effectiveness of crash avoidance systems. A round robin test series of performance-based brake testers (PBBTs) was conducted in which the brake forces on several configurations of control vehicles were measured. Using the PBBT-reported brake forces and vehicle weights, combined with knowledge of limiting tire/road coefficient of friction, the maximum deceleration potential can be determined and incorporated into on-board crash avoidance systems. This paper presents the results of the PBBT round robin tests and shows how the information obtained with a PBBT can be incorporated in the algorithms used to warn of impending collisions and minimize the risk of unstable braking.

INTRODUCTION AND BACKGROUND

The Intelligent Transportation System (ITS) uses modern computational and communications technology to improve the safety of highway and roadway travel, as well as to expedite the transport of commerce and passengers. One of the safety advantages of the ITS is the ability of on-board computers to perform real-time analysis of data to assist the driver in avoiding collisions or in maintaining vehicle stability. Inputs such as vehicle speed, lateral and longitudinal accelerations, road conditions, as well as the position and speed of other vehicles are among those required for such safety-enhancing systems.

ON-BOARD SYSTEMS FOR IMPROVED SAFETY – One example of an on-board safety-enhancing system is a front and "lane-change" collision warning system in which the speed and position of the vehicle and objects near the vehicle are collected and analyzed. As a result of the analysis, collision alarm warnings can be provided to the driver via visual or audible signals. At present, these warnings are based on fixed time-delays to the instant of the impact, for example three-, two-, and one-second. For such time-based alarms, the real-time braking capability of the vehicle is not considered.

The braking performance of commercial vehicles can vary widely over the course of a day, as a result of changes in loading or road conditions. In addition, the effectiveness of pneumatic, s-cam design brake systems can also degrade severely within a few weeks when proper maintenance is neglected. In more sophisticated future collision warning systems, knowledge of the current vehicle braking capability, along with the limiting road/tire coefficient of friction (COF), can be incorporated into the alarm triggers. It is envisioned that these future systems could take over control of the acceleration, braking, and steering of the vehicle to avoid a collision when, for example, the driver has become unconscious. Such an automated braking system would complement an already demonstrated automated steering system [1].

As a specific example, ITS-based on-board computers which store current data on the available brake forces and loads on a wheel-by-wheel basis can also be used to warn of changes in vehicle load distribution or road conditions which may lead to unstable stops, including trailer swing-outs and jack-knifes.

In these examples, the safety of the vehicle can be improved through use of the ITS combined with the knowledge of the current braking capability of the vehicle. Knowledge of the current braking capability of commercial vehicles can be assessed using data provided by performance-based brake testers (PBBTs). The future generation of computer controlled PBBTs could export braking capability data for use by the ITS, including on-board computers.

Since PBBTs are relatively new to North America, particularly the use of portable models, a PBBT round robin test program was conducted to ensure that the PBBT measurements are truly representative of the braking capability of the tested vehicle. This paper presents the results of the round robin and shows how the information obtained with a PBBT can be incorporated in the algorithms used to warn of impending collisions and to minimize the risk of unstable braking.

ROUND ROBIN SUMMARY

While PBBTs have been in common use in Europe for more than 20 years, performance-based brake testing is only recently becoming more common in North America, with interests being primarily focussed on portable versions. An initial evaluation of the capabilities of PBBTs was conducted by investigators at the National Highway Traffic Safety Administration's (NHTSA's) Vehicle Research and Test Center (VRTC) [2 – 6]. Subsequently, the feasibility of use for roadside enforcement was evaluated through a field-study program sponsored by the U.S. DOT's Federal Highway Administration's (FHWA's) Office of Motor Carriers (OMC) [7 – 9]. However, the ability of PBBTs to accurately predict a vehicle's stopping performance has only been investigated recently. This investigation, accomplished through round robin testing of ten different PBBTs, was initiated for the purpose of determining whether the PBBT assessments of a given vehicle's braking performance were accurate and consistent, and for confirming that the proposed criteria for enforceable regulations were reasonable.

DESCRIPTION OF ROUND ROBIN – The test vehicles were specifically configured to assess the ability of the participating PBBTs to:

1. Identify vehicles that are unsafe as a result of weak brakes, in a valid and consistent way,
2. Measure brake forces accurately and consistently, and
3. Measure wheel weights accurately and consistently.

In addition, the influence of environmental conditions on the measured brake forces was investigated.

A total of ten PBBT stations were included in the round robin conducted at NHTSA's VRTC:

1. Four portable Roller Dynamometers (RDs),
2. Two in-ground RDs,
3. Two Flat Plate brake testers (FPs),
4. One Breakaway Torque Tester (BTT), and
5. One 32.2 km/hr road stop.

The principles of operation of these machines have been described elsewhere [3, 9 – 12].

As over 80% of the axle configurations of commercial vehicles on the road include three-axle tractor, two-axle trailer combinations (3-S2) and 2-axle straight trucks, one of each of these vehicles was utilized during the round robin. Loading the two test vehicles to their gross vehicle weight ratings (33,000 lbs. and 80,000 lbs. for the straight truck and the 3-S2, respectively) was accomplished by placing concrete blocks on the flatbeds. Each vehicle was instrumented with a fifth wheel speed sensor from which stopping distances and decelerations were derived. In addition, both vehicles were equipped with a Labeco on-board computer which, tied to a switch placed on the brake pedal, assisted in computing stopping distances. On the 3-S2 vehicle, individual speed sensors were placed on selected wheels, a torque wheel was fitted on the left wheel of the second drive axle (wheel 5) and air pressure was monitored using transducers at the glad hand, at each of the six tractor wheel air chambers and at the trailer distribution valve. All data were collected at a frequency of 100 Hz. Table 1 describes the tests conducted during the round robin and the corresponding vehicle configurations.

For the four originally planned tests, the two vehicles were set up with specific weak brakes and overall target deceleration of 0.4g (equivalent to a ratio of total brake force (BF_{tot}) to gross vehicle weight (GVW) of 0.4), under two loading conditions, unladen and fully laden.

Table 1. Test matrix for the PBBT round robin

Test No.	1	2	3	4	5	6	7	8	9
Vehicle Type	3-S2	2	3-S2	2	2	2	2	2	2
Loading Conditions	Fully Laden		Unladen		Empty	1/3 Laden	2/3 Laden	2/3 Laden	Empty
Test Surface Conditions	Dry	Dry	Dry	Dry	Dry	Dry	Dry	Wet	Wet
Brakes Conditions	Weak Brakes				Fully Adjusted, Strong Brakes				
Number of Tests	3, Separate				3, Sequential				

In addition, target brake force to wheel load ratios (BF/WL) of 0.25 and 0.35 were set for a steer axle wheel and a non-steer axle wheel, respectively. These BF/WL ratios and overall BF_{tot}/GVW requirements are consistent with those under consideration for enforceable performance criteria to be used for on-road brake inspections [13].

The additional tests 5 – 9 were conducted on the 2-axle vehicle with fully adjusted strong brakes, under unladen and partially laden conditions, and under dry and wet conditions. Under unladen conditions, the rear axle brakes were capable of locking the wheels as the ratio of BF/WL under full brake application pressure was greater than the COF between the test surface and the vehicle tires.

SUMMARY OF TEST RESULTS

Accuracy – It is critical for prediction of vehicle stopping capability that data obtained using PBBTs be accurate. For example, a machine that over predicts the deceleration of a vehicle, i.e. which outputs a ratio of BF_{tot}/GVW higher than the true available deceleration, presents a safety issue. In this case the ITS calculated time-delays will be incorrect and the warnings or corrections may not be implemented in a timely manner to prevent collisions. Conversely, from a safety standpoint, a PBBT which under predicts the deceleration of a vehicle, i.e. which outputs a ratio BF_{tot}/GVW lower than the true deceleration, is conservative. However, it presents economic difficulties for fleet owners if vehicles with adequate stopping capability are unfairly placed out of service.

The accuracy of PBBT measurements was assessed through the comparison of PBBT results with known quantities, either from certified instrument measurements or derived from independently measured quantities. The accuracy of the ratio of BF_{tot}/GVW reported by the PBBTs was determined by comparison with the deceleration measured in the 32.2 km/hr stopping tests. The accuracy of PBBT-measured BF was determined through comparison with calibrated torque wheel data. The accuracy of the PBBT measurements of WL was determined through comparison with certified scale measurements.

Accuracy of PBBT-reported BF_{tot}/GVW – The deviation of BF_{tot} calculated from the standard is plotted against the deviation of the GVW measured from the standard weight in Figure 1. The diagonal bands indicate the 10 percent acceptable range of variation on the derived deceleration (BF_{tot}/GVW). Results showed that most PBBTs measured vehicle weights to within 5% of the standard weights. For weakly braked vehicles, measured BFs were often slightly higher than the standard values. For strongly braked vehicles, both low and high BFs with respect to standard brake forces were reported. Brake forces higher than on-road forces are expected if the

COF between the tire and the test surface is higher than that between the tire and the road, and if the brake output is sufficient to lock up the tire on the high COF surface. Brake forces lower than the on-road values are expected in instances of premature test termination, insufficient PBBT capacity or low tire/test surface COF.

Accuracy of PBBT-reported BFs – The deviations of PBBT-reported BFs from the torque wheel data are illustrated in Figure 2, for the laden and unladen configurations, respectively. The PBBT-measured BFs showed larger deviations in the unladen configuration than in the laden configuration. Because the algorithms used by the PBBT vendors were not known, three methods of determination of the maximum BF from the torque wheel data are plotted. Method 2, the average of all data points which are within 80% of the maximum, appeared to be the most appropriate method for most PBBTs.

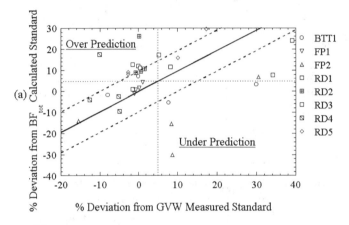

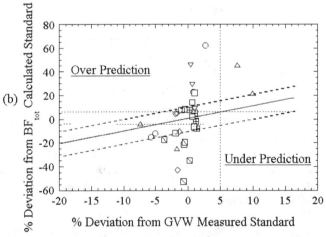

Figure 1. Plots showing position of BF_{tot} and GVW for the average of three repeats of each vehicle condition. The data for vehicles with weak brakes (Tests 1 – 4) are shown in (a) while the data for the strongly braked two-axle vehicle (Tests 5 – 9) are shown in (b). Note that the BF_{tot} standard is calculated from the 32.2 km/hr deceleration multiplied by the measured GVW.

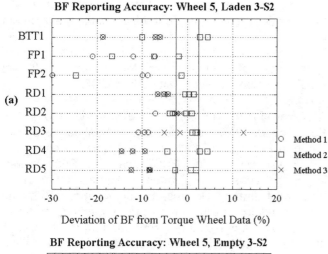

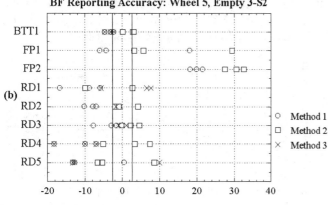

Figure 2. Deviations of PBBT reported BFs from BFs computed from torque wheel data collected on the 3-S2 (a) in the laden condition and (b) in the unladen condition.

Although the PBBT-reported BF values showed some deviation from those computed from the torque wheel data using method 2, the brake force versus time data reported by the PBBT vendors (not shown in this paper) was seen to match the data collected by the torque wheel. As such, the variations must be explained by the choice of differing algorithms used to determine BFs from time history data (BFs as a function of time). The round robin showed that a common procedure is required so that the reported BF is, for each type of PBBT, PBBT-vendor independent.

Accuracy of PBBT-reported WLs – PBBT results for axle load measurements are shown in Figure 3. The measurement of axle weights of the 2-axle vehicle by the PBBTs were accurate (~ 3%), and no axle-specific or loading conditions-specific variations could be identified. For the 3-S2 vehicle, variations from standard values were larger for axles 2 and 4 for all PBBTs except one FP and one in-ground RD. These large variations were attributed to the four-spring suspension system, to the posi-

tioning of the truck on ramps, as well as to the vehicle alignment with the machine. Other suspension systems, such as air-bag, are not expected to exhibit such variations in loads. In addition, longer ramps and more accurate positioning are expected to resolve the problem.

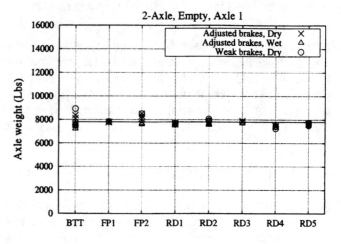

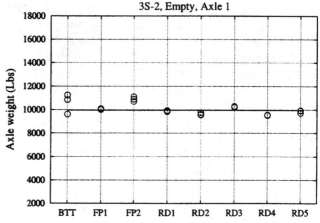

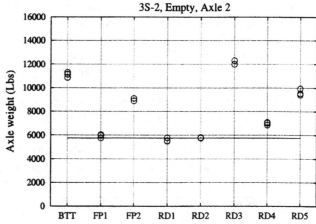

Figure 3. Axle load measurements for (a) axle 1 of the unladen 2-axle vehicle, (b) axle 1 and (c) axle 2 of the unladen 3S-2 vehicle.

Repeatability – Three replications of each test were conducted in order to evaluate the repeatability of the PBBT measurements. Three types of uncertainties are associated with measurements made on PBBTs:

1. Real life braking variations, i.e. variations in the brake torque output or deceleration during several given stops.

2. Uncertainty in the data measurements due to transducer accuracy and/or data reduction or manipulation. This range of acceptable accuracy of PBBTs under controlled conditions is proposed in the functional specifications [14].

3. Variation in test conditions, either geometrical parameters (truck characteristics such as total number of axles, position of the axles or type of suspensions) or variability in driver/operator conduction of the test.

Combining (1) and (2) defines the acceptable repeatability range (ARR), as listed in Table 2.

Table 2. Summary of acceptable range for repeatability for PBBTs.

	Expected Real Life Variations, %	Accuracy Allowance, %	Acceptable Range for Repeatability (ARR), %
Brake Force	10	± 2.5	15
Wheel Load or GVW	1	± 2.5	6
BF/WL or BF_{tot}/GVW	10	± 5.0	20

For determination of the repeatability in identifying weak brakes, the three replications were not conducted sequentially, rather they were performed in a circuit from station to station. Two wheels on each truck were set up with low BFs, resulting in a total of 8 cases (192 total measurements) of weak brake forces to be identified. The repeatability was acceptable in 94 percent of the measurements. For determination of the repeatability in characterizing fully adjusted brakes, only the 2-axle truck was utilized and all 3 repeats were conducted consecutively in each machine. 96 percent of these 540 measurements were within the ARR for all participating PBBTs.

Finally, the assessed overall vehicle performance or measured BF_{tot}/GVW (quantity used as stopping capability input for incorporation in the ITS-based collision warning systems) were within the ARR for 95 percent of the 234 measurements.

For those measurements that fell outside of the ARR, the causes could be attributed to operator error, variations in driver brake application, vehicle misalignment, erroneous tests or premature test terminations.

CONCLUSIONS FROM THE PBBT ROUND ROBIN – It was concluded from the results that the accuracy and repeatability of all PBBTs were very close to being acceptable, if not already acceptable, for use in enforcement and for use in predicting stopping capability from 32.2 km/hr. The deceleration calculated with PBBT results (BFs and WLs) correlated well with the average deceleration of a 32.2 km/hr on-road stop, for all participating PBBTs, regardless of the principle of operation.

Weight measurements were better on a straight 2-axle vehicle than on the 3-S2 vehicle with the four-spring tandem axle suspensions. Since preliminary calibration demonstrated that PBBT weighing systems were accurate, deviations were attributed to specific characteristics of the vehicle. For strong brakes, with a suitable test surface/tire COF, lock up occurs and an incorrect WL will have no consequence. When brakes are unable to lock up, then use of the BF/WL ratio might lead to an unwarranted failure if the PBBT-reported WL is high.

Although PBBT-measured BFs matched the torque wheel data, in some cases, the PBBT-reported BFs at the end of the test showed deviations from standard or known values. The probable causes are specific to the type of PBBT and are expected to be corrected after recognition of the causes of the deviations. These were typically software-related.

Accurate measurements of BFs and WLs are critical in assessing the overall deceleration capability of a vehicle. It was recognized that the enforceable criterion needs to take into account allowable variations as determined during the round robin.

The COF between the test surface and the vehicle tires must be close to that of the road and tires. Indeed, this value defines the maximum limit for BF/WL that can be imparted to the road. When the test surface/tire COF is higher than that of the road/tire, and when the brakes are strong enough to lock up the wheel, then a limiting value must be used to avoid over-prediction of stopping capability. It is important to note that test surface/tire COF must be near that of the tire/road in wet conditions as well in order to simulate real life braking.

Recommendations following the round robin include the alteration of the test surface of some portable RDs to increase the COF in order to meet the road/tire COF requirements, standardization of test methods, reporting and analysis procedures to provide PBBT-independent results, and upgrading the driving capacity of some roller drives. Although a tremendous step was taken during the PBBT round robin, it was just a beginning. The next one, after the above-mentioned modifications are completed, is to compare PBBT measurements to 96.6 km/hr on-road stops.

USE OF PBBT RESULTS FOR CRASH AVOIDANCE

INCORPORATION OF BRAKE FORCES INTO ALGORITHMS – If the individual brake forces of a vehicle are known, then algorithms for both maximum deceleration and dynamic vehicle stability during stopping or other high demand maneuvers can be refined to take these limiting values into account. An additional limiting value required for improved predictions is the COF of the tire/road interface. It is anticipated that a method for real-time assessment of this quantity will be available in the future for input to the ITS crash avoidance algorithms.

<u>Overall Braking Capability (minimum deceleration)</u> – Computation of stopping distance from a given speed requires some assumptions about the deceleration profile. An example of a deceleration profile is shown in Figure 4, in which T_0 is the driver's reaction time, and T_1 is the time for full energization of the brake system to maximum constant deceleration.

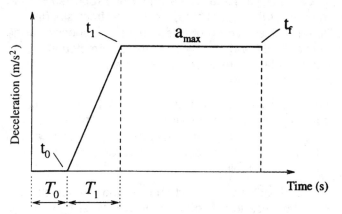

Figure 4.   Assumed deceleration profile of a vehicle.

The corresponding equations to predict stopping time, t_f, and stopping distance, x_f, using this assumed profile, in which α_{max} is given by BF_{tot}/WL from a PBBT measurement, are:

$$t_f = \frac{v_i}{a_{max}} + T_0 + \frac{T_1}{2} \tag{1}$$

$$x_f = v_i\left(T_0 + \frac{T_1}{2}\right) + \frac{v_i^2}{2\,a_{max}} - a_{max}\,\frac{T_1^2}{24} \tag{2}$$

where v_i and α_{max} are the initial vehicle speed and the constant average deceleration during the stop, respectively.

If the object is stationary and ahead of the vehicle, Equations 1 and 2 provide the required input for the warning alarms. Clearly, a warning alarm algorithm would require factors of safety on T_0 and x_f for use in any other capacity than complete automated takeover of vehicle braking control.

For objects in motion ahead of the vehicle, additional modifications to the algorithms would be required. If the object is moving in the same direction as the vehicle, then the equations for time (t_c) and distance (x_c) to collision can be obtained by solving the following equations:

$$-\left[\frac{a_{max} - a_{max}^{obj}}{2}\right]t_c^2$$
$$+\left[\frac{a_{max}(2T_0 + T_1)}{2} + (v_i - v_i^{obj})\right]t_c$$
$$-\left[a_{max}\left(\frac{T_0(T_0 + T_1)}{2} + \frac{(T_1)^2}{6}\right) + x_i^{obj}\right] = 0 \tag{3}$$

$$x(t_c) =$$
$$v_i\,t_c - a_{max}\left[\frac{T_1^2}{6} + \frac{(t_c - (T_1 + T_0)(t_c - T_0)}{2}\right] \tag{4}$$

where x_i^{obj}, v_i^{obj}, and a_{max}^{obj} are the initial distance between, velocity of, and rate of deceleration of the object in front of the vehicle. These solutions can be easily obtained using a computer.

As described in [15], the brake forces and wheel loads measured by the PBBTs can be used to compute the vehicle's deceleration potential as simply BF_{tot}/GVW. Note that while accurate representation of actual vehicle performance was shown for most PBBTs during the round robin for stops from 32.2 km/hr, it was not demonstrated for 96.6 km/hr stops. For such higher energy stops, if the foundation brakes exhibit in-stop fade as shown in Figure 5, a modification to the deceleration profile will likely be required. Note that not all brake materials and brake systems will behave alike. If a vehicle's brakes do not exhibit any significant variation in brake force output with speed, then modifications to the above algorithms are not required.

For algorithms that determine whether a vehicle can maintain travel within a 3.66 meter lane width during a stop, additional factors are required such as, for example, individual wheel brake forces and loads, axle spacings, position of the center of gravity of both tractor and trailer, tire lateral stability properties and extent of dynamic load shift. Many of these factors can also be incorporated into the on-board computers.

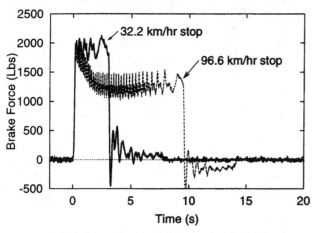

Figure 5. Variation of an individual wheel's brake force during 32.2 km/hr and 96.6 km/hr stops, collected by an instrumented torque wheel.

CONCLUDING REMARKS

While it remains considerable work to achieve a significant reduction in vehicle collisions, the sensors, measurement devices, method of input to on-board computers, and algorithms required are presently available. Incorporating the knowledge of current braking capacity in such advanced systems is reasonably straight forward. In the case of relatively constant load, hydraulically braked vehicles, an assumed braking capability or minimum deceleration potential is adequate. However for commercial vehicles, the deceleration capability is highly variable, even on a given vehicle. With up-to-date knowledge of the current deceleration capability of commercial vehicles possible using PBBTs, sophisticated collision-avoidance systems could be implemented with the help of ITS for assimilation into the commercial vehicle equipment market.

REFERENCES

1. Pomerleau, Dean, "RALPH: Rapidly adapting lateral position handler", IEEE Symposium on Intelligent Vehicles, Detroit, MI (September 1995).

2. M. A. Flick, "Evaluation of Hunter Heavy Duty Plate Brake Tester", Interim Final Report, DOT HS 808 275 (February 1995).

3. M. A. Flick, "An Overview of Heavy Vehicle Brake System Test Methods", SAE Paper 962215, Society of Automotive Engineers, Detroit, MI (October 1996).

4. M. A. Flick, "Tests to Evaluate B&G Brake Torque Tester", Interim Final Report , DOT HS 808 468 (September 1996).

5. M. A. Flick, "Tests to evaluate Renstar Infrared Heavy Vehicle Brake Temperature Measurement System", Interim Final Report, DOT HS 808 567 (April 1997).

6. R. W. Radlinski, M. A. Flick, and G. S. Clark, "Enhancing the Roller Brake Tester", SAE Paper 922444, Society of Automotive Engineers, Warrendale, PA (1992).

7. S. J. Shaffer and G. H. Alexander, "Evaluation of Performance-Based Brake Testing Technologies", Interim Report, FHWA-MC-96-004 (December, 1995).

8. S. J. Shaffer and P. A. Gaydos, "Development, Evaluation and Application of Performance-Based Brake Testing Technologies", Final report submitted to FHWA (April, 1998).

9. "Development, Evaluation, and Application of Performance-Based Brake Testing Technologies", FHWA-OMC Publication No. FHWA-MCRT-98-001 (July 1998).

10. "Use of Brake Performance Measurements to Assess Safe Stopping Capability", US DOT/FHWA-OMC Publication No. FHWA-MC-98-030.

11. S.J. Shaffer and J.W. Kannel, "Understanding the Portable Roller Brake Dynamometer", SAE Paper 982829, Society of Automotive Engineers, Warrendale, PA (1998).

12. Robert Bosch GmbH, "Automotive Brake Systems", Distributed by SAE, pp. 188 - 189 (1995).

13. "Public Meeting to Discuss the Development of In-Service Brake Performance Standards for Commercial Vehicles Inspected With Performance-Based Brake Testers", Federal Register, Vol. 63, No. 166 (1998).

14. "Development of Functional Specifications for Performance-Based Brake Testers Used to Inspect Commercial Motor Vehicles", Federal Register Publication, FHWA-1998-3611-1, Federal Register, Vol. 63, No. 108 (1998).

15. S. J. Shaffer and A.-C. Christiaen, "Judging the Stopping Capability of Commercial Vehicles using the Results of a Performance-Based Brake Force Measurement", SAE Paper 982830, Society of Automotive Engineers, Warrendale, PA (November, 1998).

The Application of State Space Boundaries in the Safety Evaluation of Collision Avoidance Systems

Joseph S. Koziol, Jr. and Andy Lam
Volpe National Transportation Systems Center

ABSTRACT

This paper describes the concept of using state space boundaries to evaluate the safety effects of longitudinal collision avoidance systems from data produced in field operational tests. The boundaries are represented in terms of the relative range and range rate between a lead vehicle and the vehicle hosting the collision avoidance system. Phase plane diagrams are used to illustrate the state space boundaries. Parameters of curves representing the boundaries were selected such that the boundaries would be fairly well distributed over the range vs. range-rate space with the ones closer to the horizontal axis (range = 0) being indicative of a relatively higher hazard potential. The application of these state space boundaries is examined with data available from a recently completed field operational test sponsored by the National Highway Traffic Safety Administration.

INTRODUCTION

PURPOSE – The purpose of this paper is to present and describe a set of state space boundaries that can be used to evaluate the potential benefits of longitudinal collision avoidance systems from field operational test data. In the absence of collisions during the field tests, surrogate measures must be established that can give an indication of the relative safety effectiveness of new Intelligent Transportation Systems (ITS) devices that are coming on the market. The surrogate measures need to be robust enough to discriminate safety effects between comparable devices as well as between different road types, driver ages, driving experiences and environmental conditions. Other safety surrogates such as headways, velocities, and deceleration levels, have been used extensively in past operational tests and evaluations [1, 2], and have provided valuable information on the potential safety effectiveness of the tested device. Further, the phase plane method has been used by researchers [3] as a method for presenting field operational test data. In this paper, the state space boundary concept is introduced. As will be illustrated below, the state space boundaries have the advantage of integrating a number of important accepted measures of safety in a manner that can be related to specific driving scenarios. The measures that are integrated include range, range rate, and deceleration level. As such, more direct safety inferences may be drawn from operational test data, particularly as relates to pre-crash scenarios, for which an abundance of collision data has been recently accumulating.

BACKGROUND – The National Highway Traffic Safety Administration (NHTSA) is involved in a multi-year program [4] aimed at developing an improved understanding of the causes of crashes on today's highways, the potential for reducing these crashes through the application of advanced technologies, and to use this knowledge base to encourage and facilitate industry efforts in developing and introducing effective collision avoidance products. Emphasis has been recently shifting to projects which address issues of system capability, usability and benefits [5]. Operational tests and demonstrations of collision avoidance systems are being tailored to provide broad exposure of these systems to the driving public. In order to evaluate the potential benefits achievable from the collision avoidance systems and to determine the success of the tests and demonstrations, a comprehensive, meaningful, and transferable set of measures of effectiveness will be needed. This will entail the usage of surrogate measures, since sufficient collision data, the ultimate measure of safety, will most likely not be available for discriminating effects that are statistically significant.

STATE SPACE BOUNDARY DEFINITIONS

This section defines and describes the state space boundaries. The state space boundaries are intended to be used to evaluate the potential safety benefits of longitudinal collision avoidance systems from field operational test data.

PHASE PLANE REPRESENTATION – A set of state space boundaries is shown in Figure 1. The boundaries are a set of curves in the range vs. range rate state space or phase plane, that are spread somewhat uniformly above the abscissa. The closer a driving situation (i.e., actual range, range rate data) is to the abscissa (for R_{dot} < 0), the closer that driving situation is to a collision. Thus, the degree to which actual driving situations or data cross the lower boundaries provides an indication of a relatively higher hazard potential for that situation.

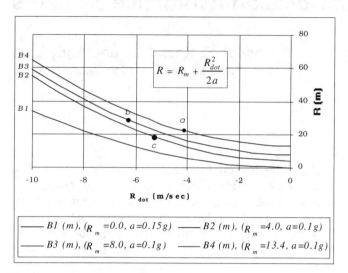

Figure 1.   State space boundaries

These boundaries can be thought of as representing the initial range and range rate conditions required to bring a following or host vehicle, closing in on a lead vehicle at a constant rate and then braking at a constant deceleration level, to the range indicated by the intercept of that curve with the ordinate. Furthermore, the initial conditions may be interpreted as initial conditions for potentially hazardous driving scenarios. The scenarios may be a particular driving situation or condition that suddenly confronts the driver of the host vehicle. It may be a cut-in or a realization of an in-lane situation involving a lead vehicle traveling at a lower constant velocity. The curves would indicate the initial conditions required at a given and immediate braking level, to just avoid a collision (those curves that pass through the origin in the phase plane) or for a near miss (those curves that intercept the ordinate above the origin).

The equation for the set of curves is as follows:

$$R = R_m + (R_{dot})^2 / 2a$$

Where
R_m = minimum range separation
R = range between lead and host vehicles
R_{dot} = range rate between lead and host vehicles
a = host driver braking level

The equation and a set of parameters for the curves are also shown in the figure. It is the form of the equation that allows physical interpretation to be given to the curves. The equation and parameters may be chosen to suit the needs of the research. The number of boundaries will govern the degree of courseness of the analysis. The placement of the boundaries should be in state space region of interest where the analysis is to be focused. For purposes of the application given in this paper, namely, the evaluation of an ICC system, four boundaries were

found to be adequate to discriminate between driving mode effects. These boundaries, as indicated in Figure 1 cover a region bounded by a 60 meter range and a 10 meter per second closing range. Only moderate deceleration levels (see discussion below) were found to be necessary (the application here focused on interstate driving). Other studies may require extended state space regions or higher deceleration levels (0.7 g may be used as an upper bound, 0.25 g was found to be a high level for interstate driving). Some experimentation may be required to adjust the number and location of the boundaries in order to achieve maximum benefit in discriminating effects.

The state space boundaries in Figure 1 thus integrate three variables that may be considered separate measures of safety: range, range rate, and deceleration. The integration of the three variables through the equation given in the figure establishes a single measure. This measure, with its interpretation defined above, not only has the advantage of reducing the number of measures, but can resolve potential conflicts of individual measures. For example, in Figure 1, consider points a and b. Using the measures separately, the range indicated by point a is shorter than the range indicated by point b. Yet the closing range indicated by point b is greater than that for point a. Which point presents the greater hazard? The state space boundaries resolve this conflict. Point b presents the greater hazard since, at the same deceleration level, it will lead to a closer minimum headway.

As another example, consider points b and c. Using the measures separately, the range indicated by point c is shorter than the range indicated by point b. Yet the closing range indicated by point b is greater than that for point c. Which point presents the greater hazard? With the use of the state space boundaries, point c presents the greater hazard since, at the same deceleration level, it will lead to a closer minimum headway.

It should be pointed out that the state space boundaries given by the equation in Figure 1 not only represent the initial conditions for the above mentioned scenarios, but also their theoretical time tracks in the state space as well. (The time variable is implicit in the state space.) From a boundary perspective, any point on any curve represents the initial conditions for a constant velocity closing situation, that would result in the indicated minimum range and for the indicated deceleration level. The boundary condition can be also thought of as addressing the question "How far back of the lead vehicle (initial range) must the host vehicle be to result in the desired minimum range, for a given initial range rate and constant deceleration by the host vehicle?"

From a time track perspective, points on any single curve represent motion in time and space with the motion proceeding downwards in the state space (upper left quadrant). From any point on any curve (initial conditions) in Figure 1, the motion will proceed to another point lower on that curve until reaching the minimum range.

The metric for the state space boundaries is the relative occurrence of exceeding or violating (falling below) each of the boundaries. The occurrence can be either the number of times a boundary is crossed or the time spent below a boundary. The resulting data would be normalized with respect to driving exposure. This metric is readily calculable from the range and range rate data that is obtained with the use of instrumented vehicles.

TIME DOMAIN REPRESENTATION – The driving scenarios mentioned above (*cut-in*, *approach*, *lane change*), that correspond to the safety boundaries shown in Figure 1, are described and illustrated in this section. There is no driver response time in these scenarios. The scenarios begin with the driver immediately responding by braking to a given range and closing range rate condition. The lead vehicle can either be moving at a constant speed or stationary (stopped) throughout the scenario. The host vehicle is initially moving at a constant speed. The driver brakes at a constant and uniform level, *a,* to bring the vehicle to just within a minimum acceptable separation (range), given by R_m.

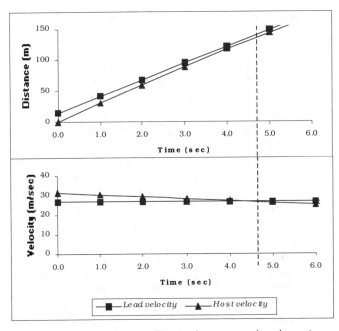

Figure 2. Time response illustrating scenario where two vehicles are initially closing at constant velocities, and the host vehicle brakes at a constant deceleration level of 0.10g.

This type of scenario is illustrated in Figure 2. The figure shows that, for the selected parameters, (lead vehicle initial and constant speed = 26.8 m/sec, host vehicle initial speed = 31.4 m/sec, initial range between the two vehicles = 14.6 m, and driver braking level = 0.1g), the host vehicle will come to within 4.0 m of the lead vehicle (closing rate = 0 when range = 4.0 m). This occurs at approximately 4.7 seconds after the scenario begins. The initial range separation (14.6 m), and the initial closing range rate (4.6 m/sec), represent in effect, a data point on state space boundary *B2* in Figure 1.

STATE SPACE BOUNDARY APPLICATION

The application of these state space boundaries is examined next in the evaluation of an Intelligent Cruise Control (ICC) system. The particular ICC system here was operationally tested in southeastern Michigan during 1997 and 1998 under a cooperative agreement between the National Highway Traffic Safety Administration (NHTSA) and the University of Michigan Transportation Research Institute (UMTRI) [6]. While not a safety system, the ICC incorporated a forward looking sensor and performed headway control by modulating the throttle and downshifting [7]. This control, particularly for close headway situations where downshifting would be necessitated, could serve as a cue or warning to the driver that braking may be necessary to avoid a collision.

ALL DRIVING – INTERSTATES – The results of the state space boundary violations from the ICC field operational test for all driving on interstates are indicated in Figure 3. The results are shown in terms of boundary violations per kilometer for the different driving modes: Manual, Conventional Cruise Control (CCC) and ICC [8]. For ICC, the driver had a choice of three headway settings: 1.0 second, 1.4 seconds, and 2.0 seconds.

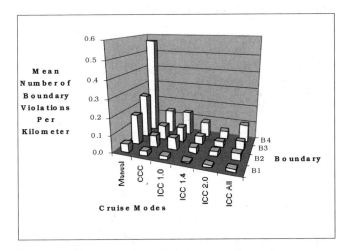

Figure 3. Mean number of state space boundary violations per kilometer, as a function of cruise mode and state space boundary

The robustness of the state space boundary metric is indicated in the figure by its ability to discriminate potential safety effects between the different driving modes and ICC headway settings. Boundary violations occurred significantly less than once per kilometer for all driving. Most apparent was the high relative occurrence of exceeding the state space boundaries for manual driving compared to driving with CCC or ICC. Furthermore, when aggregated together, the violation rate for ICC was less than that for CCC for each boundary.

For the ICC mode, the violation rate was inversely related to the time headway setting. It was highest for the 1.0 second setting and lowest for the 2.0 second setting. Overall, there were few occurrences of exceeding the most critical or hazardous boundary, i.e., boundary *B1*. To the extent that this boundary was exceeded, only moderate braking levels would be required in most cases to prevent a collision. This was born out by the test data and is examined further below for some of the "worst case" scenarios encountered during the test.

CUT-IN SCENARIO – INTERSTATES – Data from the ICC field operational test were disaggregated by driving scenario [8]. Four types of pre-crash scenarios (*cut-in, approach, lane change, and lead vehicle decelerates*) were categorized. The remaining data were also categorized by type (following, separating, and cruising - no lead vehicle present). This section utilizes the state space boundary concept to compare and analyze ICC driver behavior during *cut-in* scenarios. Here, to illustrate the added utility of the state space boundary concept, actual driving scenarios are compared directly with respect to the boundaries rather than in terms of the violation rate of the boundaries.

Figure 4 shows the specific state space boundaries used in this study to examine *cut-in* scenarios. As a pre-crash scenario and in the current study, *cut-ins* are defined as situations where a lead vehicle changes lanes in front of a host vehicle going straight and the host vehicle proceeds to close on the lead vehicle.

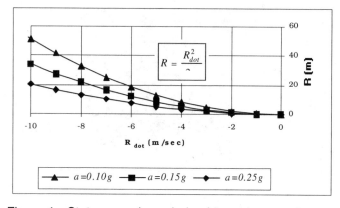

Figure 4.   State space boundaries for *cut-in* scenario

The state space boundaries represented in the figure are similar to the ones in Figure 1 in that they represent the initial conditions for constant velocity, closing situations between the lead and host vehicles. The deceleration levels of the host vehicle in Figure 4 are 0.10 g, 0.15 g, and 0.25 g, as indicated, and all the curves pass through the origin in the phase plane. Each of these boundaries thus represents the initial conditions to just avoid a collision, assuming the driver in the host vehicle braked immediately at the constant level indicated by the curves.

Specific *cut-in* scenarios were drawn from the field operational test and superimposed on these *cut-in* state space boundaries. The results are shown in Figure 5. The scenarios selected here were amongst the "worst case" *cut-in* scenarios observed on interstates during the field operational test. The "worst case" *cut-ins* were those that produced the highest level of braking on the part of the

host vehicle driver, or the nearest encounters. It is felt that these 'worst case" scenarios, including the host driver response, are a prelude to actual collisions and thus a study of these extreme cases would provide more insight into the potential safety effects of ICC driving. For a complete analysis, one would need to factor in the exposure to the "worst case" scenarios as well.

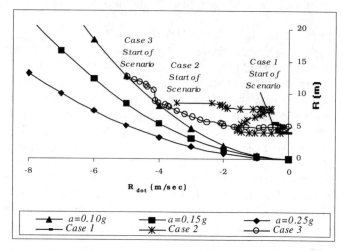

Figure 5.   *Cut-in* state space boundaries and ICC scenarios

Figure 5 shows three of the "worst" *cut-ins* that occurred during ICC driving. The three *cut-ins* represent the time tracks for each of the cut-ins. It is to be noted that the start of a time track coincides with the start of that scenario. It is also to be noted that the time tracks are not shown in the right half plane. The right half plane represents a separating situation and the pre-crash scenarios are considered to be ended in this study when their time tracks cross into the right half plane. As mentioned above, these *cut-ins* were selected from among the "worst case" *cut-in* scenarios observed on interstates during the field operational test. Thus the time tracks in Figure 5 not only describe the driver responses in each case but also provide a sense of "worst case" situations confronting the host vehicle driver in terms of the initial conditions, namely range and range rate at the start of the *cut-in* scenario.

All of the *cut-in* cases were at or above the boundaries indicating that the driver would not need to brake above 0.10 g to just avoid a collision. None of the situations developed where the driver actually ended up in the scenario braking at 0.25 g or higher. All three drivers were already braking when these scenarios started. The minimum headways in all three cases were above 4 meters. It is to be noted from the test data that none of the time tracks for these scenarios indicated a response delay by the host driver, from the onset of the *cut-in* maneuver. The response delay would have been indicated by a near vertical initial time track. This may have been due largely to the fact that the sensor would not pick up the lead vehicle until it completely or near completely crossed into the lane ahead of the host driver, and after, presumably, the driver was responding to the situation developing in the adjacent lane.

LANE CHANGE SCENARIO - INTERSTATES – This section utilizes the state space boundary concept to compare and analyze ICC driver behavior during *lane change* scenarios. As a pre-crash scenario and in the current study, *lane changes* are defined as situations where the host vehicle changes lanes behind a lead vehicle going straight and the host vehicle <u>proceeds to close on the lead vehicle</u>.

Specific *lane change* scenarios were drawn from the field operational test and superimposed on the same state space boundaries used in the previous section for *cut-ins*. The results are shown in Figure 6. Again, the scenarios selected here were amongst the "worst case" *lane change* scenarios observed on interstates during the field operational test.

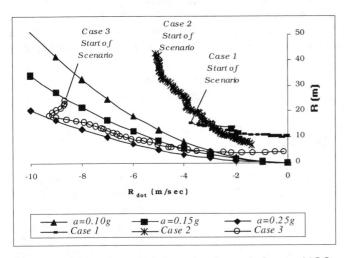

Figure 6. *Lane change* state space boundaries and ICC Scenarios

Two of the lane change cases (Case 1 and Case 2) were above the boundaries indicating that the driver would not need to brake above 0.10 g to just avoid a collision.

Case 1 was initiated at a range of about 13 meters with the host vehicle traveling at about 120 km/h. The initial closing rate was 4 meters per second. The ICC system responded and resolved the closing situation by decelerating the host vehicle at 0.06 g. The driver of the host vehicle did not brake and the nearest approach was 10 meters.

In Case 2, the lane change was initiated at about 45 meters, the braking level by the host driver was a relatively low 0.03 g, and the nearest approach was more than 8 meters. The maneuver ended with another lane change by the host driver. The braking occurred right at the start of the lane change, thus disengaging the ICC.

Case 3 started above and reached the lowest boundary. The boundary condition thus indicates that the driver would have to brake at a constant level of 0.25 g at that instant to just avoid a collision. In actuality, the driver braked at 0.26 g, less than one second after the lane

change. This initial braking lasted for about three seconds and was followed by a more severe braking (0.30 g) for about one second. The entire lane change maneuver lasted about 7 seconds, ending as the host vehicle slowed to the same speed as the lead vehicle. The minimum approach was about 4 meters. Note that, at the beginning of the scenario, there was a short time delay by the host driver and a slight deceleration (.05 g) by the lead vehicle. There was no perceptible response of the ICC system to this maneuver during the one second period before the driver braked.

APPROACH SCENARIO – INTERSTATES – This section utilizes the state space boundary concept to compare and analyze ICC driver behavior during *approach* scenarios. As a pre-crash scenario and in the current study, approaches are defined as situations where the host vehicle is going straight and <u>encounters a lead vehicle that is going straight at a lower speed</u>. (Near and far regions were defined in front of the host vehicle in order to distinguish *approach* scenarios from *cut-in* and *lane change* scenarios.)

Specific *approach* scenarios were drawn from the field operational test and superimposed on the same state space boundaries used in the previous sections. It should be noted that the start of the *approach* scenarios was defined as the time that the lead vehicle first appeared ahead of the host vehicle. In some cases this was the range limit of the sensor, in other cases this was the initial sensing of the lead vehicle in a curve or on a hill, and in still other cases this was the initial range sensed in the far region, e.g., after a cut-in maneuver. The results are shown in Figure 7. Again, the scenarios selected here were amongst the "worst case" *approach* scenarios observed on interstates during the field operational test.

Two of the approach cases (Case 1 and Case 2) started and remained substantially above the boundaries indicating that the driver would not have to brake above 0.10 g to just avoid a collision. Case 3 crossed the top boundary but not the middle boundary indicating that a braking level between 0.10 g and 0.15 g would be needed to just avoid a collision.

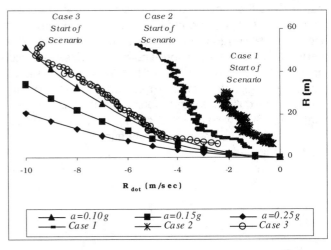

Figure 7. *Approach* state space boundaries and ICC scenarios

In Case 1, the approach was gradual. The initial closing rate was 2.0 meters per second at a range of 32 meters. The ICC system managed the approach by decelerating the host vehicle at 0.03 g. Near the end of the scenario, the lead vehicle suddenly braked (at 9 meters) and the host vehicle driver reacted quickly (about 0.5 second later) and also braked. The braking levels of both vehicles were extremely mild and the braking durations were short. The minimum approach was 6 meters. The driver of the host vehicle may have waited to see if the ICC system would resolve the situation.

In Case 2, the *approach* developed as a lead vehicle changed lanes (cut-out) and another vehicle was encountered traveling in the same lane at a lower speed. The initial conditions for the *approach* scenario were a range of 50 meters and a closing range rate of 5.5 meters per second. The ICC system managed the situation for 15 seconds decelerating the vehicle at 0.04 g. At a range of 7.5 meters the driver braked at 0.5 g and the lead vehicle accelerated away to end the scenario. The minimum approach was 4.5 meters. The driver of the host vehicle may have waited to see to what extent the ICC system would resolve the situation.

In Case 3 the driver braked at 0.30 g 7 seconds after the start of the *approach* scenario. (The scenario started as the sensor initially detected the vehicle on a curve at a range of 52 meters.) After an initial delay of less than a second, the ICC responded at 0.06 g until the driver braked. The minimum approach was 6 meters. The driver of the host vehicle may have waited to see if the ICC system would resolve the situation.

CONCLUDING REMARKS

A safety metric has been introduced that can be used in the evaluation of Intelligent Transportation Systems (ITS) such as collision avoidance systems from field operational test data. The state space boundary concept integrates a number of important and accepted measures of safety in longitudinal driving situations and thereby provides an enhanced measure of relative hazardousness. By relating to actual driving scenarios and incorporating parameters that govern the relative hazard of the condition, the state space boundary metric allows more direct safety inferences to be drawn from operational test data, particularly as relates to pre-crash scenarios, for which an abundance of collision data has been recently accumulating. The safety metric may serve as a surrogate measure of collisions, and would limit the size of tests that would otherwise be required to provide sufficient collision data for discriminating safety effects that are statistically significant.

An illustration of the application of the state space boundaries was given based on data from a recently completed field operational test of an Intelligent Cruise Control system. The robustness of the state space boundary metric was indicated by its ability to discriminate potential safety effects between the different driving modes and ICC headway settings.

An illustration was also given of the utilization of the state space boundary concept for examining three specific pre-crash scenario types, namely *cut-ins, lane changes,* and *approaches.* The boundaries for these scenario types integrate three variables that are commonly related to safety: range, range rate, and deceleration. The integration of the three variables establishes a single measure involving the three variables.

The state space boundaries were then used as a basis for examining actual "worst case" scenarios from the ICC field operational test. The variables associated with each scenario were reviewed and analyzed. Future work calls for the incorporation of driver response time into the state space boundaries, the application of the state space boundary concept to the *lead vehicle decelerates* scenario type, and the examination of violation rates of the specific state space boundaries for the pre-crash scenarios.

REFERENCES

1. Perez, William, A. et al, *TravTek Evaluation Safety Study*, Publication No. FHWA-RD-95-188, February, 1996.

2. Fancher, Paul, et al, *Fostering Development, Evaluation, and Deployment of Forward Crash Avoidance Systems (FOCAS)*, Final Report DOT HS 808 437, May, 1995.

3. Fancher, Paul, et al, *Tests Characterizing Performance of an Adaptive Cruise Control System* Paper presented at 1997 SAE International Congress and Exhibition, Detroit, Michigan, February 24-27, 1997.

4. National Highway Traffic Safety Administration, IVHS Plan. June 12, 1992.

5. National Highway Traffic Safety Administration, ITS Plan (1997-2002).

6. National Highway Traffic Safety Administration (1995). Cooperative Agreement Number DTNH22-95-H-07428.

7. Fancher, et al, *Intelligent Cruise Control Operational Test*, Interim Report, DOT HS 808 622, August, 1997.

8. Inman, Vaughan, et al, *Evaluation of the Intelligent Cruise Control System*, DOT Final Report, To Be Published, 1999.

1999-01-0816

Threat Detection System for Intersection Collision Avoidance - Real-Time System Performance

Edward H. Jocoy
Consultant

Herbert A. Pirson
Veridian, Calspan Operations

ABSTRACT

Veridian, Calspan Operations is currently developing an on-vehicle threat detection system for intersection collision avoidance (ICA) as part of its ICA program with the National Highway Traffic Safety Administration.

This paper briefly reviews the system design and describes recent efforts that include the development of a multi-radar collision avoidance (C/A) system. Results from in-traffic experiments utilizing two radars to simultaneously observe crossroads traffic from both directions will be presented for the first time. Warning functions were applied to the data in real time and for non-real-time signal processing. These functions warn the ICA vehicle against entering the intersection when targets are predicted to be present.

This research demonstrates the feasibility of preventing many intersection collisions.

INTRODUCTION

As part of its Intersection Collision Avoidance (ICA) program with the National Highway Transportation Safety Administration [1,2], Veridian, Calspan Operations is developing an on-vehicle threat detection system for intersection collision avoidance. In February of 1998 we presented, at this conference, simulation data and single-beam radar data from in-traffic experiments for two intersection scenarios [3].

In this paper we will present new results of in-traffic collision avoidance experiments and discuss radar/collision warning system performance. We will:

- briefly update the radar and signal processing design;
- discuss the concept of using three existing headway radars in a three-beam configuration to accommodate several important traffic scenarios; and

- present results from in-traffic experiments using a multi-beam radar configuration.

Intersection collisions constitute approximately twenty-six percent of all accidents in the United States. This translates into 1.7 million crashes annually. Because of their complexity and demands on the perceptual and decision making abilities of the driver, intersections present an increased risk of collisions between automobiles. A program to develop systems to prevent these collisions is being sponsored by the National Highway Traffic Safety Administration Office of Crash Avoidance Research (NHTSA OCAR). The Intersection Collision Avoidance Using IVHS Countermeasures Program is a five year effort that proceeded from an analysis of accident data to quantify the problem, to the fabrication and testing of a candidate countermeasure system. Previous papers and reports have described the use of accident data to determine intersection crash characteristics [2] in 4 crash scenarios. Table 1 (from [3]) summarizes the percentages of accidents associated with each scenario. Previous papers have also described a candidate threat detection system including the radar sensor and a simulation of the sensor, tracker and C/A algorithm, along with simulation and experimental results [3,4].

Table 1. Intersection Collision Scenarios

Scenario	Characteristics	Percentage
1	Parallel Path - Left Turn Across Path	23.8
2	Perpendicular Path - Inadequate Gap	30.2
3	Perpendicular Path - Violation of TCD	43.9
4	Premature Intersection Entry	2.1
	Total	100.0

Many variants of the four crash scenarios exist. The purpose of the ICA system is to prevent or reduce the severity of intersection crashes. Further discussion of the scenarios can be found in the Task 5 Interim Report [1] for this program, which also provides details of the components of the ICA Countermeasure. This paper will focus on the continuing development of the collision warning algorithm and multi-radar performance. System performance in Scenario 2 will be addressed. The system consists of a three-radar assembly that will monitor vehicle traffic along the approaching cross lanes and opposing lanes of traffic. A metric of gap time based on predicted time of arrival at the intersection is described. The use of the gap time metric to provide a warning to the driver will be illustrated. Both post processing of recorded data and in-traffic, real-time performance will be used to evaluate the system.

THREAT DETECTION SYSTEM

Figure 1 shows some of the basic components of the ICA threat detection/tracking/collision warning system. As indicated, the intersection collision avoidance system utilizes a Geographic Information System (GIS) and the Global Positioning System (GPS) to complement the on-board sensors. The GIS consists of a digital map database which contains information regarding roadways and traffic control devices (TCD) within a given area.

The GPS/GIS allows the countermeasure to identify the roadway on which it is traveling and the approach of an intersection. The intersection countermeasure utilizes the GIS map datafile to determine the configuration of the approaching intersection and angles of adjoining roadways. This feature allows the threat detection system to scan the intersection's highest risk areas during the target vehicles' approach to or traversal of the intersection.

Signal processing operates on targets detected within threat regions. The "size" of the intersection provided by the GIS will be used in developing a metric called gap time which utilizes the time predicted by the tracker for vehicles to enter and exit the intersection. The tracker uses the target's range, range rate and heading to develop tracks. The collision warning algorithm develops the gap time metric and issues warnings to the subject (ICA) vehicle if the gap time is insufficient for the ICA vehicle to execute a maneuver such as crossing an intersection. Other information such as the ICA vehicle's distance to the intersection, speed and intended turn actions are also used (it takes longer for the radar vehicle to cross the intersection than to make a right turn into the intersection).

Figure 2 shows an example of how approaching traffic on crossroads and opposing lanes will be observed by a 3-radar system for a four-lane intersection (maximum radar range - 400 ft.). For two-lane roads (one lane in each direction), it is unlikely that much scanning (if any) will be required. However, for multiple lane roads, a small (10°-20°) sector scan about a nominal pointing angle will be

required to improve angular coverage, particularly for the crossroads. The pointing angles, scan sectors and the relatively slow (~20°/sec) scan rates are completely computer controlled from within the ICA vehicle. The pointing angles can be varied with the position of the ICA vehicle as it approaches an intersection.

While the GPS/GIS system is absolutely required for an operational system, this paper will concentrate on the radars, trackers and warning systems. Intersection dimensions will be assumed or actually measured. Figure 3 illustrates three scenarios and the distances to be computed using range, range rate and bearing from a three-radar system currently implemented on the Calspan Instrumented Vehicle. Three independent trackers/Kalman filters are implemented for the three radars. Only in the collision warning system are the warnings that are developed from each sensor/tracker merged. In this paper, samples of data and results from two radars will be given.

REAL-TIME PROCESSING

In the 1998 SAE paper [3], in-traffic data and results were presented for both Scenario 1 and 2 traffic configurations. The radar data, while recorded in real-time, were post processed (with no tracker) and verified using video of the test. In addition, results were presented from a comprehensive computer simulation of the ICA tracker and collision avoidance algorithm written in MATLAB®.

In this paper we report on radar data that has been collected and processed with the ICA tracker and collision avoidance algorithm in real-time. A new tracker with audible warnings was developed [4] for the real-time processing system. The real-time system is based on the MATLAB® simulation, optimized for real-time execution. The new system also outputs data files containing both unprocessed radar data and processed tracker and warning outputs. The computer simulation software has been modified to accept these unprocessed radar data files from the real-time system. Comparison of the real-time tracker and warning data with the simulation output using the radar data files shows them to be equivalent. The data presented herein were post processed by the MATLAB® implementation of the tracker and collision warning algorithm.

The real-time processing system is hosted on a Pentium based computer. Multitasking software allows the three radars to be processed simultaneously and their warning outputs combined. The real-time system implements a separate tracker and collision avoidance processor for each radar. The resulting three warnings are merged into a single driver warning, which is currently implemented as an audible tone. Future improvements to the warning system, such as a heads up display and active brake control, are currently under investigation.

The real-time processing system also controls the three radar scanning platforms and the GPS. The scan platforms use gear-driven 12 Vdc servo motors commanded by a closed loop floating point DSP-based four axis motion controller. The GPS system combines dead reckoning with differential GPS to provide a continuous and accurate vehicle position. The dead reckoning algorithm uses a fiber optic gyro and odometer input. The system provides a location accuracy of about 10 ft.

A separate computer tracks the vehicle's position using the GPS data and a GIS map database. Approaching intersection information is provided to the tracker and collision warning processors in real time.

COLLISION WARNING SYSTEM EXPERIMENTAL RESULTS

The first multi-radar experiment was conducted with the ICA vehicle parked in Calspan's "East Gate" driveway which exits onto a heavily traveled six-lane highway (three lanes each way). Figure 4a is a photo of the two radars mounted on the roof of the ICA vehicle. The mountings were temporary and the pointing angles adjusted manually (no scanning platforms). Figure 4b is a sketch of the test scenario. The third radar for detecting oncoming targets (see Figures 2 and 3) is mounted on the bumper and is not shown in Figure 4a.

Figure 5 shows the tracker-derived range (as opposed to the raw radar range) from ICA vehicle to target for the roof mounted radar that was pointed to the right. While traffic was heavy during the nearly two minutes of data collection, a 10 second segment was selected that had, for clarity, only three targets. From these data, target vehicle state variables of position (X,Y) relative to intersection center, speed and acceleration were computed by the tracker/Kalman filter (see [4] for details). Targets with opening range rates, detected from the near lanes, were eliminated in the signal processing. While no actual intersection exists at this location, the on-board computer can easily represent a six-lane intersection (12 ft. lanes) and develop predicted times for targets to enter and exit the intersection (see Figure 3). These predicted times are shown in Figure 6 for the three targets approaching from the right. Also shown in Figure 6 are the times that the ICA vehicle would occupy the "intersection" if it tried to cross the six lanes of traffic from stand-still and with a 0.2g acceleration profile. Since the ICA vehicle was positioned several feet from the nearest roadway, it takes a little less than a second to reach the simulated intersection and accounts for the fact that the times the ICA vehicle would occupy the intersection don't start at zero. A sketch defining the predicted times accompanies Figure 6. When target and ICA vehicle are predicted to simultaneously occupy the intersection, an audible warning is issued. This is shown graphically on Figure 7 for each of the targets approaching from the right. The composite warning for targets approaching from the right is shown at the bottom of Figure 7.

Figure 8 shows the tracker-derived ranges from ICA vehicle to targets approaching from the left. Two targets were observed by radar #1 (pointed left) during the same time interval as was selected from radar #2 (pointed right). The predicted times for these targets to enter and exit the intersection are shown in Figure 9 along with the estimate for the ICA vehicle to enter and exit the intersection (the latter is the same as in Figure 6, of course). The warnings derived from radar #1 are shown in Figure 10. The composite warnings from radars 1 and 2 from Figures 7 and 10 are combined in Figure 11 and it is clear that there is no gap in the warnings during which the ICA vehicle could cross the intersection.

SUMMARY

As part of its Intersection Collision Avoidance (ICA) program for the National Highway Traffic Safety Administration, Office of Crash Avoidance Research, Veridian, Calspan Operations is developing an on-vehicle threat detection system. This system utilizes three off-the-shelf headway radars, modified for the ICA task and integrated into a multitarget collision avoidance system. Warning functions were developed which warn the ICA vehicle when it and any target are predicted to occupy an intersection simultaneously. Last year, at this forum, we presented simulation and single-beam radar data from in-traffic experiments but post processed in non-real time. This year, multiple beam radar data and collision warnings were developed in real time. Sample results from in-traffic experiments using two radars (simultaneously) were presented. The real-time signal processing requirements were briefly reviewed.

The system evaluation effort is continuing. Evaluations of warning activation and deactivation times for all three radars are currently underway on the Calspan instrumented track and intersection. Multiple-target tests for all traffic situations are also being conducted on the test track. Evaluations of the GPS/GIS system, used in conjunction with the radar system, are underway. Finally, in-traffic tests will be performed at a wide variety of intersections.

ACKNOWLEDGEMENTS

This work is being performed for the Office of Crash Avoidance Research, National Highway Traffic Safety Administration. The authors would like to acknowledge Mr. Art Carter and Dr. August Burgett of OCAR for their support and direction. The authors would also like to acknowledge Mr. John Pierowicz, project engineer at Calspan, for his guidance.

REFERENCES

1. J.A. Pierowicz, E.H. Jocoy, M.M. Lloyd, O. Kelly, A. Bittner, *Intersection Collision Avoidance Using IVHS Countermeasures, Task 5: Design of Testbed System,* Calspan SRL Corporation, Buffalo, 1996.

2. J.A. Pierowicz, "Development of an In-Vehicle Intersection Collision Countermeasure," SPIE Paper No. 2902-34, *SPIE Proceedings, Transportation Sensors and Controls: Collision Avoidance, Traffic Management, and ITS,* Volume 2902, pp. 284-292, Boston, 1996.

3. E.H. Jocoy, J.A. Pierowicz, "Threat Detection System for Intersection Collision Avoidance," SAE Paper No. 980851, International Congress and Exposition, Detroit, MI, Feb. 1998.

4. E.H. Jocoy, J.R. Knight, "Adapting Radar and Tracking Technology to an On-Board Automotive Collision Warning System," presented at 17th Digital Avionics System Conference, IEEE, Oct. 1998.

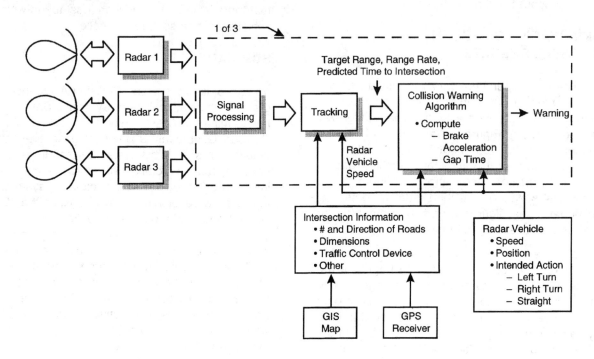

Figure 1. ICA Threat Detection System Showing Detailed Use of Intersection Information

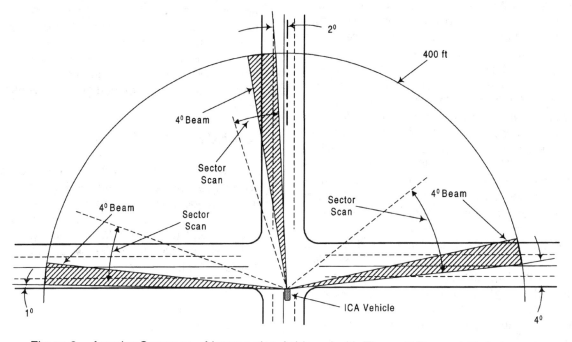

Figure 2. Angular Coverage of Intersection Achieved with Three 4° Beamwidth Antennas

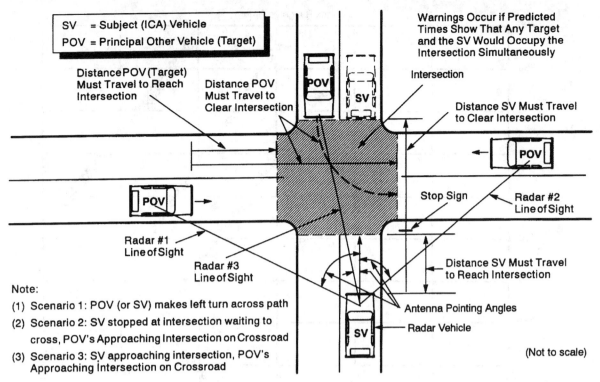

Figure 3. Scenario 1, 2, and 3 Showing Distances Used to Compute Predicted Times and Warnings

(a) ICA Vehicle

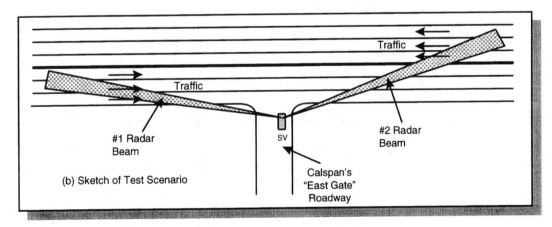

(b) Temporary Mounting of Radar on ICA Vehicle and Plan View of Test Scenario

Figure 4.

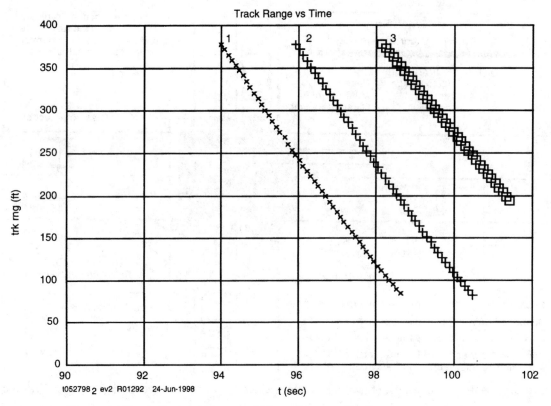

Figure 5. Track-Derived Range of Three Targets, Radar 2 (Pointed Right)

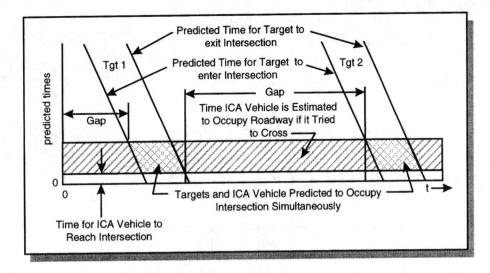

Sketch of Predicted Times

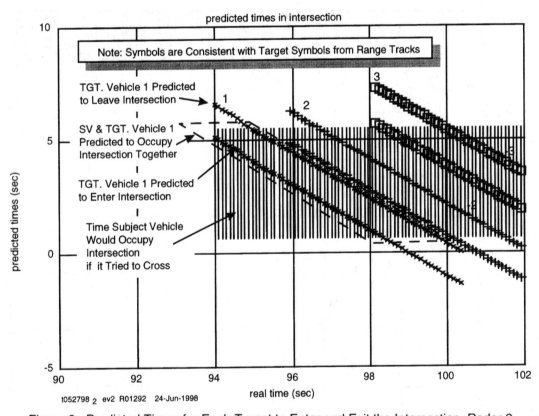

Figure 6. Predicted Times for Each Target to Enter and Exit the Intersection, Radar 2

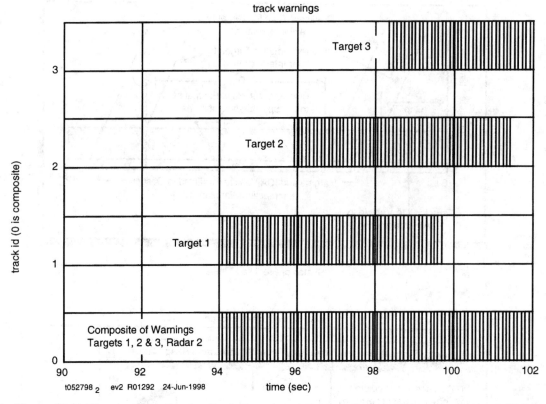

Figure 7. Binary (0-1) Warning Function for Each Target and Composite for all Targets, Radar 2 (Pointed Right)

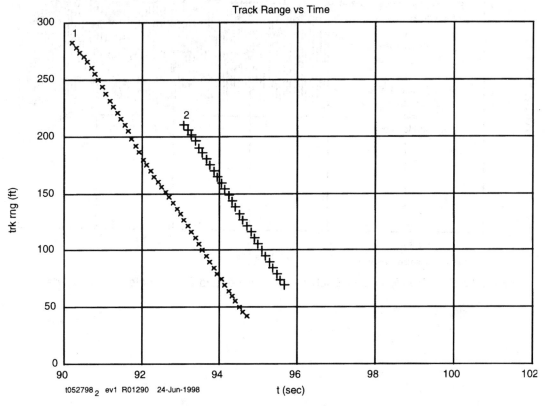

Figure 8. Track-Derived Range of Two Targets, Radar 1 (Pointed Left)

386

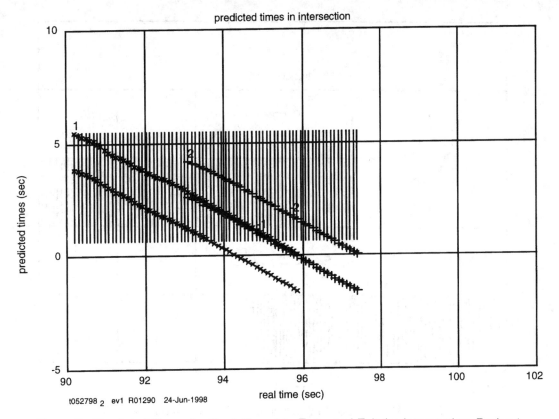

Figure 9. Predicted Times for Each Target to Enter and Exit the Intersection, Radar 1

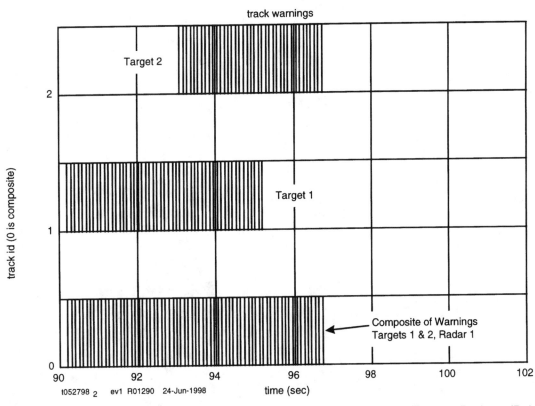

Figure 10. Binary (0-1) Warning Function for Each Target and Composite for all Targets, Radar 1 (Pointed Left)

387

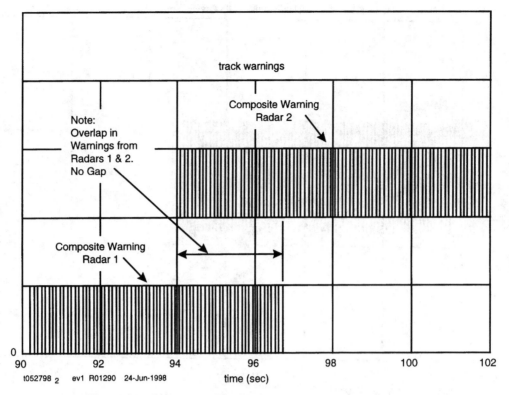

Figure 11. Composite Warning from Radar 1 and Radar 2

388

1999-01-0493

Side Collision Avoidance Systems: Better Agreement Between Effectiveness Predictions and Real-world Data

R. Steven Hackney
A.L.I.R.T. Advanced Technology Products

ABSTRACT

Considerable effort has been invested in the development of models to predict the effectiveness of side collision avoidance systems ("SCAS"). These estimates, based on reliability theory, indicate that SCAS can produce a measurable improvement in safety , but that safety improvement is sensitive to the method of sensor use.

The support of real-world data for these models is inconclusive. Objectively measured "right-clear" data show varying improvement with the use of SCAS, yet professional drivers of large vehicles (buses, heavy trucks) report favourable responses to the idea of SCAS use. Unscientific surveys of the general driving public support this favorable reaction, and also indicate a significant perceived cost to near miss incidents.

This paper proposes a mathematical model of SCAS reliability that takes the above factors into consideration. This model is heavily based on earlier work in the field.

The model predicts decreases in accident rates from 20% to 95%, depending on style of usage of the SCAS. Reduction in occurrence of near miss incidents is predicted to be between 5% and 80%, depending on usage style.

Usage style (i.e. use of the sensor as a substitute for or as an addition to direct observation) is determined to be the critical variable affecting overall performance.

Sensitivity analysis indicates that the assumed proportion of lane changes with potential conflict is somwhat important. The model is not generally sensitive to other factors

INTRODUCTION

As part of efforts to develop the intelligent transportation system, considerable effort has been invested in development of collision avoidance systems. These efforts have included work on predicting and measuring the effectiveness of these systems in informing the drive of obstacles, or in reducing actual accidents.

Mazzae et. al., for example, have attempted to measure changes in accuracy of driver answers to questions about obstacles – other vehicles – beside the host vehicle. These tests did not indicate significant improvements in accuracy [3,4,5,8]. However, researchers noted measurable differences in the accuracy of driver judgements when making actual lane changes, as opposed to answering researcher queries, suggesting that "Right clear?" questioning may not correlate completely with actual obstacle detection performance.

By comparison, fleet data collected from passenger buses reportedly showed measurable improvements in accident rates [7]. Focus groups of professional drivers of large vehicles (buses, heavy trucks) report favourable responses to the idea of SCAS use: unscientific surveys of the general driving public produce a similarly favourable response.

EXISTING RELIABILITY MODEL PREDICTIONS

In another approach to this problem, Tijerina and Garrott presented a mathematical model of SCAS-driver behaviour which attempted to predict the reduction in accidents which could be expected to result from different driver usage patterns for SCAS [1].

This model was based on calculation of reliability of individual elements of the lane change decision-making system. For example, estimated or derived values were used for reliability of driver perception of hazards, reliability of sensor detection of hazards, etc.

These individual estimates were combined according to different decision making structures: sole reliance on the driver; sole reliance on SCAS; Boolean AND; and Boolean OR combinations of warnings. These provided an overall prediction of the reduction in accidents which would occur as a result of the use of SCAS.

Sensitivity testing using this model showed effectiveness estimates ranging from −250% through +70%, depending primarily on accuracy of detection of the sensor, and proportion of lane change maneuvers with conflict incidence [2].

This modeling also indicated a strong dependency on driver usage method. Maximum safety improvement is predicted if lane change maneuvers are postponed on detection of a hazard by *either* of the driver or the CAS (Boolean "OR" of warnings), as opposed to reliance solely on the driver or the CAS, or in the worst case, proceeding with a lane change if either driver or CAS indicated a non-hazard condition (Boolean "AND" of warnings).

INPUTS AND ASSUMPTIONS IN EXISTING MATHEMATICAL MODEL

Tijerina and Garrott's model was based on a number of assumptions. These can be summarized as follows:

- 244,000 reported lane change accidents per annum ("NACC")
- Average one lane change attempt per mile driven ("LM")
- 2.35×10^{12} miles drive per annum ("M")
- 78% of lane change accidents can be addressed by SCAS ("PSCAS"). The remaining 22% cannot be prevented by SCAS, because of ?????.
- Conflict Incident of 2% ("CI" – proportion of contemplated lane changes that feature another vehicle in a hazard location)

From these assumptions, the probability (PH) of a driver detecting a hazard is calculated to be 0.999715. This is calculated as follows:

PH=1-(PSCAS x NACC / (LM x M)) ^ 0.5

$= 1-(0.78 \times 244,000 / 1.0 \times 2.35 \times 10^{12})$ ^ 0.5

Four decision making models were developed using a probability tree approach. These models are:

- **Driver-Only**: this model assumes that the driver ignores the CAS, depending on personal observation. It corresponds to the "base case" of no CAS availability.

- **CAS-Only**: this model assumes that the driver ignores personal observation, depending solely on warnings issued by the CAS.
- **Driver in Series** (Boolean AND of warnings): this model assumes that the driver will postpone a lane change only on a warning of the CAS that is confirmed by personal observation.
- **Driver in Parallel** (Boolean OR of warnings): this model assumes that the driver will postpone a lane change as a result of either personal observation or CAS warning.

Results predicted by these models were combined into four scenarios. These were formed by taking the average of the four decision models, weighted as shown in Table 1.

FACTORS EXCLUDED FROM EXISTING MODEL

This model has excluded a number of factors in estimating reliability.

NEAR MISS INCIDENTS – Tijerina and Garrott have implicitly considered only two outcomes from a lane change attempt: it can either end in a collision, or safely.

A safe outcome can result be the result of either a lane change, or a postponement of a contemplated lane change. In the case of a lane change, one form of safe outcome occurs when the driver of the principal vehicle overlooks a hazard, and a collision is avoided by later evasive action. For example, the driver of the principal other vehicle may take evasive action, or use the vehicle horn to cause the host vehicle driver to take evasive action. Informal discussions with the general public provide anecdotal evidence of a large number of these near miss incidents ("NMI").

These discussions suggest that there is potential value in avoiding not just collisions, but also NMI. This value is largely psychological, and consequently more difficult to quantify than death, injury, or material damage resulting from accidents. Despite this, it is reasonable to infer that this value is not zero.

Decision Model	Cautious Skepticism	Cautious Acceptance	Ideal Acceptance	Complacent Acceptance
Driver Only	0.750	0.250	0.000	0.125
CAS Only	0.000	0.000	0.000	0.125
Driver in Series	0.000	0.000	0.000	0.125
Driver in Parallel	0.250	0.750	1.000	0.625

Table 1: Weights used for combining decision models into usage scenarios

Tijerina and Garrott include this value implicitly. Calculations of driver reliability include errors made by driver of both the host vehicle, and the principal other vehicle. Essentially, an error made by the "host" driver will become a NMI based on either (i) evasion by the driver of the other vehicle, or (ii) late realization by the host driver that a hazard condition exists.

ERROR IN DETECTING CAS WARNINGS – The model proposed by Tijerina and Garrott does not consider a separate value for the probability of detection of the CAS alarm by the driver. This value is either included in the estimate of CAS reliability, or implicitly assigned a value of 1.0.

However, there is merit in considering this value separately, particularly since a considerable amount of research effort is currently dedicated to the ergonomic design of CAS warning alarms.

A NEW MATHEMATICAL MODEL

Based on the general approach outlined by Tijerina and Garrott, probability trees were developed for the four CAS decision models. The probability trees for these models are shown in Figures 1 through 4.

These models differ from calculations presented by Tijerina and Garrott in several respects:

1. Explicit values are included for probability of driver detection of CAS warning.

2. Consideration has been given to NMI rates, as distinct from accident rates.

3. Probability of Driver Detection of Hazard (PH) has been recalculated based on slightly different assumptions.

4. In developing NMI rates, explicit consideration has been given to the probability of evading an accident because of "late realization of hazard" by the host driver, or because of evasive tactics taken by the driver of the principal other vehicle (POV). Further, the SCAS system is assumed to affect this "late realization" probability, since the system will present the driver with an additional warning.

INPUT FACTORS AND CALCULATIONS

PROBABILITY OF DRIVER DETECTION OF HAZARD – Probability of driver detection of hazard (PH) has been calculated based on overall error rates for lane change maneuvers, as developed from accident data.

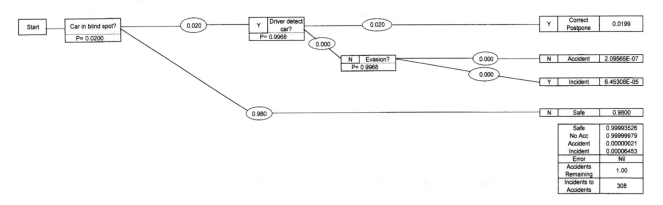

Figure 1. Probability Tree for Driver-Only Decision Model

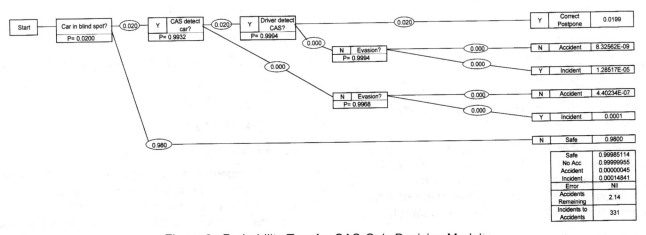

Figure 2. Probability Tree for CAS-Only Decision Model

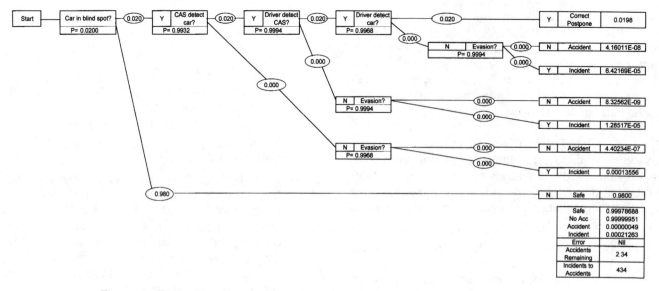

Figure 3. Probability Tree for Series Decision Model (Boolean AND of warnings)

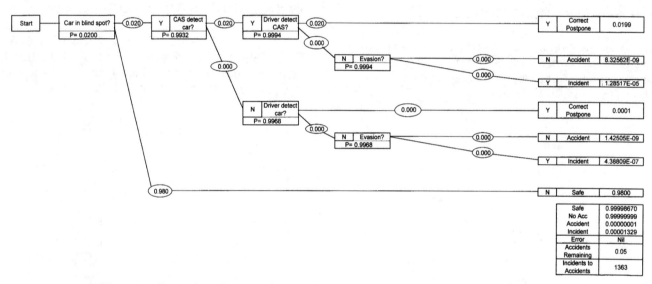

Figure 4. Probability Tree for Parallel Decision Model (Boolean OR of warnings)

Using Tijerina and Garrott's assumptions for lane change and conflict incidence, the total number of attempted lane changes with conflict (TLCC) can be calculated as:

TLCC = LM x M x CI

= 1.0 x 2.35 x 10^{12} x .02

= 4.70 x 10^{10}

In other words, if drivers gathered no data about adjacent vehicles, 4.7 x 10^{10} lane change accidents would occur annually.

The actual number of lane change accidents (NACC) occurring was 630,000. This is the sum of the 244,000 reported annual lane change accidents and the 386,000 estimated unreported. Of these, 78% were deemed to be addressable by collision avoidance systems [1].

The error rate for drivers making the lane change maneuver can therefore be calculated as:

Error = (PSCAS x NACC / (LM x M x CI))

= (.78 x 630,000) / (1.0 x 2.35 x 10^{12} x .02)

= (491,400 / 4.70 x 10^{10})

= 1.04455 x 10^{-5}

and the overall reliability of drivers making lane changes is:

Reliability = 1 - Error

= 1 - 1.04468 x 10^{-5}

= 0.99998954

For an accident to occur, the drivers of both the subject and principal other vehicles must fail to evade the accident – there must be a failure on the part of each of the two drivers. If the probability of failure is equally distributed between the two drivers, then error rate for each driver is the square root of the overall reliability:

$$PH = 1 - (Error ^\wedge 0.5)$$

$$= 1 - 0.0032335$$

$$= 0.9967665$$

Note that this is not a general probability of driver detection of hazards around a vehicle. Rather, it is a specific value applying only to detection of vehicles presenting a hazard during lane change maneuvers.

It should also be noted that the assumption that drivers of both subject (SV) and principal other vehicles (POV) have equal PH is suspect. In many instances, the POV will be located in the "blind spot" slightly behind the SV, making it more likely that the SV driver will fail to detect the hazard. It is difficult to quantify this effect and no objective data were found relating to it: consequently, sensitivity analysis was performed directly on this factor.

ADDITIONAL ASSUMPTIONS – The following additional factors were assumed in development of probability trees:

- **Reliability of CAS detection of hazard.** This value was 0.9932, taken from testing of preproduction and production SCAS sensors [8].

- **Probability of successful evasion of an accident, after driver failure to detect hazard.** This value is initially assumed to be PH(POV), in cases when no CAS warning is present. This value is the same as PH(SV) for the base case. However, in cases where a CAS warning is present, this value can be expected to increase. This assumes that the ongoing CAS

warning is likely to assist the host vehicle driver in detecting the hazard, even if detection is late. Values were expressed as a decrease in probability of error: this factor was initially set to 5.0. No data were available to support this assumption, so this value was chosen arbitrarily, and sensitivity analysis was performed directly on this variable.

- **Probability driver detection of hazard.** As noted above, PH is probably not equal between SV and POV drivers. In performing sensitivity analysis, a value of PH(SV) was set such that the SV driver was twice as likely to err as the POV driver, with the same overall reliability for the lane change "system." No data were available to support this assumption, so this value was chosen arbitrarily, and sensitivity analysis was performed directly on this variable.

- **Reliability of Driver detection of CAS warning.** This value was included explicitly in calculations. It was estimated that a driver would be 5.0 times less likely to miss a well-designed CAS warning than to err in detecting a hazardous principal other vehicle. No data were available to support this assumption, so this value was chosen arbitrarily, and sensitivity analysis was performed directly on this variable.

MODEL PREDICTIONS

INITIAL ASSUMPTIONS – For the initial case of assumed inputs, the probability calculations provide the results shown in figure 5.

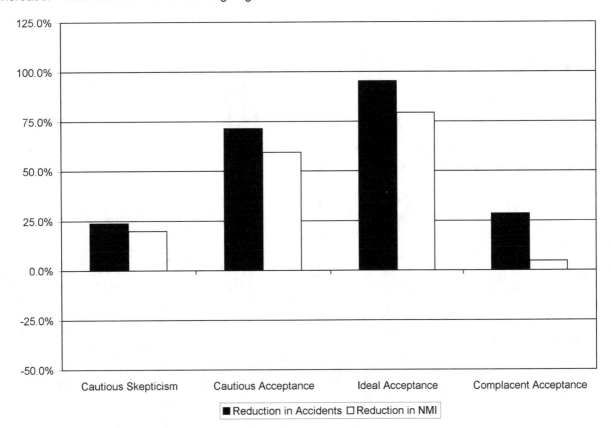

Figure 5. Predicted Reduction in Accidents And NMI For Base Case Assumptions

SENSITIVITY ANALYSIS – The model was evaluated for several different values for the input factors noted below:

- Conflict Incidence – the proportion of lane change attempts which have a principal other vehicle ("POV") in a hazardous location.

- Detectability of CAS warning – expressed as a multiple of the detectability of the hazardous POV. In essence, it is assumed to be "X" times as difficult to overlook the flashing red light or buzzer as it is to overlook the POV itself.

- Probability of Late Detection of Hazard – the increase in likelihood that the driver will detect the hazardous POV using SCAS after initiating a lane change (i.e. turn a potential accident into a NMI). In the absence of a SCAS alarm, the probability of avoiding an accident is equal to the reliability of the POV driver detecting the hazard. However, if a SCAS alarm is present, it is assumed that the SV driver may recognize this alarm and take evasive action, even after initiating the lane change. This potential improvement accident avoidance is expressed as a ratio between error rates with and without the SCAS.

- Distribution of errors between SV and POV drivers. This value is expressed as a ratio between SV and PV error rates (the SV driver is assumed to be more likely to fail to detect the hazard).

Table 2 shows the input values used for sensitivity analysis.

Case	Conflict Incidence	Detectability of CAS warning	Probability of Late Detection of Hazard	Ratio Between SV and POV Error Rates
Base	0.02	5.0	5.0	1.0
1	0.05	5.0	5.0	1.0
2	0.02	2.0	5.0	1.0
3	0.02	5.0	2.0	1.0
4	0.02	5.0	5.0	2.0

Table 2: Input Factors For Sensitivity Analysis

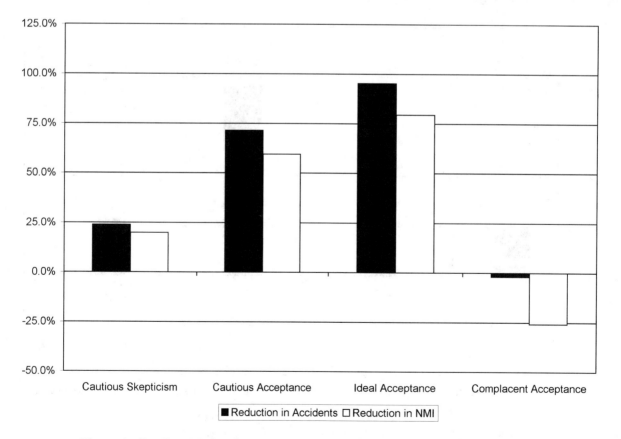

Figure 6. Predicted Reduction in Accidents And NMI For Conflict Incidence 0.05

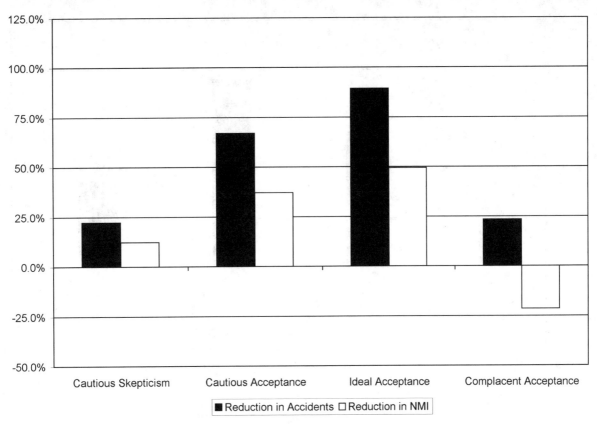

Figure 7. Predicted Reduction in Accidents And NMI For SCAS Warning Detectability 2.0

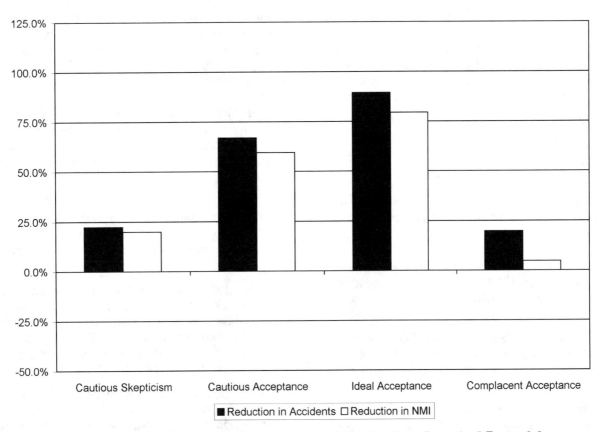

Figure 8. Predicted Reduction in Accidents And NMI For "Late Detection" Factor 2.0

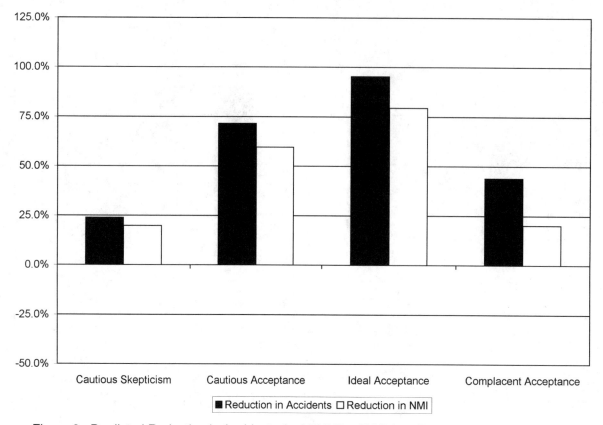

Figure 9. Predicted Reduction in Accidents And NMI For SV Driver Errors 2X POV Driver Error

In performing the sensitivity analysis, one of the factors varied was CI. Because CI was used in calculating the PH – the probability that a driver will detect a hazardous principal other vehicle – the change in CI necessitated a recalculation of PH. A change in CI from 0.02 to 0.05 resulted in a change in PH from 0.996763 to 0.997953.

Results of the sensitivity analysis are presented graphically in figures 6 to 8.

DISCUSSION

ACCIDENTS – In many respects, agreement with the predictions of Tijerina and Garrott is quite good. In particular, there is considerable sensitivity to "usage scenario." This is understandable: usage in parallel represents a "double check" which will invariably improve reliability. The greater the weighting of this decision mode, the more that accidents and incidents will be reduced.

As expected, with reasonably reliable SCAS sensors, accident rates are likely to increase only under the most adverse circumstances. This was found to correlate most strongly with CI.

However, this should not be interpreted as an indication that sensors are potentially more "dangerous" in heavy traffic situations. Because the model derives new values for driver reliability PH as the assumed value of CI increases, a change in CI represents a different description of driver behaviour in addition to a different driving environment. Essentially, increases in CI suggest that

drivers must be increasingly reliable at detecting hazards, in order to produce the known number of accidents. Therefore, "Complacent Acceptance" of an SCAS with fixed reliability must therefore degrade performance to a greater extent.

NEAR MISS INCIDENTS – For the "Base Case" – a decision mode where the driver ignores the SCAS – the model predicts that NMI will occur three hundred to five hundred times as often as actual accidents. Although no data are available, this value seems reasonable, based on the author's unscientific observations.

For most scenarios, the model predicts a decrease in NMI approximately proportionate to the decrease in accidents. However, this relationship degrades under two circumstances.

Firstly, under the "Complacent Acceptance" scenario, NMI are likely to increase, even as accidents decrease. In this case, the driver is assumed to begin a lane change based on information from the SCAS. Two sorts of errors may occur: either SCAS failure to detect the hazard, or driver failure to detect the SCAS warning.

The accident can be averted by evasive action on the part of the driver of the principal other vehicle, or on "late recognition" of the SCAS warning by the driver of the host vehicle. In either case, a NMI will occur.

Secondly, the proportion of NMI reduction to accident reduction decreases when the SCAS warning is more difficult to detect.

At worst case it was assumed that the SCAS warning would be twice as easy to detect as the actual hazard. Although there is still an overall decrease in accidents for most decision scenarios, a portion of the decrease results from "late recognition" of the alarm, resulting in a NMI.

These results suggest that the emphasis on ergonomic design of SCAS warnings is well placed.

CONCLUSIONS

A new mathematical model was used to predict reduction of lane change accidents and incidents with the use of a side collision avoidance sensor.

Reductions in accident rates of up to 95% are predicted, with optimal use of the sensor.

Reductions are predicted to be as low as 20% with less-than-optimal usage patterns, such as overly heavy dependence on the SCAS.

Reduction in near miss incidents is likely to be less than reduction in accidents. A maximum reduction of 80% is predicted, with an *increase* of 5% predicted with excess driver reliance on the sensor.

Performance is predicted to depend greatly on the quality of the SCAS warning. Both accident and incident rates can be expected to drop as the warning becomes easier to detect or recognize. This suggests that a more aggressive warning, such as an audible warning inter-locked with the vehicle turn signal, may be advisable.

Predictions are also sensitive to reliability of driver detection of hazards, as calculated from fleet accident rates and assumed conflict incidence values.

REFERENCES

1. Tijerina, L. and Garrott, W.R.: A Reliability Theory Approach to Estimate the Potential Effectiveness of a Crash Avoidance System to Support Lane Change Decisions. SAE Technical Paper 970454, Society of Automotive Engineers, 1997.

2. Tijerina, L: Sensitivity Analysis of a Reliability Model of the Potential Effectiveness of a Crash Avoidance System to Support Lane Change Decisions. SAE Technical Paper 980854, Society of Automotive Engineers, 1998.

3. Garrott, W.R., Flick, M.A. and Mazzae, E.N.: Hardware Evaluation of Heavy Truck Side and Rear Object Detection Systems. SAE Technical Paper 951010, 1995, Society of Automotive Engineers

4. Mazzae, E.N. and Garrott, W.R.: Human Performance Evaluation of Heavy Truck Side Object Detection Systems. SAE Technical Paper 951011, 1995, Society of Automotive Engineers.

5. Mazzae, E.N., Garrott, W.R. and Flick, M.A.: Human Factors Evaluation of Existing Side Collision Avoidance System Driver Interfaces. SAE Technical Paper 952659, 1995, Society of Automotive Engineers.

6. Woll, J.D.: Vehicle Collision Warning System with Data Recording Capability. SAE Technical Paper 952619, 1995, Society of Automotive Engineers.

7. Mazzae, E.N., Garrott, W.R. and Cacioppo, A.J.: Utility Assessment of Side Object Detection Systems for Heavy Trucks. Proceedings of the Human Factors and Ergonomics Society 38th Annual Meeting, 1994, Human Factors and Ergonomics Society.

8. Hackney, S.: Performance Assessment of a Side Collision Avoidance Sensor. SAE Technical Paper 1999-01-1235, 1999, Society of Automotive Engineers.

CONTACT

ALIRT Advanced Technology Products
33 Grovepark Street
Richmond Hill, Ontario
Canada L4E 3L5

Phone: (905) 773-2257
Fax: (905) 773-3437
Email: eng@alirt.com
Internet: www.alirt.com

FUTURE TRENDS AND NEEDS

Future Driver Assistance: Basing Safety Features on What the Driver Should Do!

Helge Schäfer, Jürgen Rataj and Mark Vollrath
German Aerospace Center, Institute for Transportation Systems

ABSTRACT

A systematic approach to the design, the development and the evaluation of new generations of advanced driver assistance systems is outlined. In contrast to the predominant technology-driven developments we propose a task-oriented method as a basic procedure for future developments that is rooted to two common societal aims: safety and efficiency. These aims are the framework that defines an aspired **target-strategy**, which will be derived by a systematic process: the **target-strategy-identification**. To keep the driver-vehicle system within the limits of appropriate and safe traffic participation the target-strategy has to be implemented into the technical system to provide a database for necessary and adequate assistance. For the system's decisions about the required assistance the actual driver-strategy has to be compared with the target-strategy. The driver-strategy will be estimated from data collected by special methods of driver-monitoring based on psychological behavior models.

INTRODUCTION

In order to develop novel driver assistance systems a variety of approaches is recently followed. Most of these approaches are simply driven by the availability of new technologies (sensors and actors, computational resources, software engineering techniques etc.). After that, another group of approaches adapts the system limits to human behavior patterns e.g. to avoid false reactions in emergency braking. A next step towards human centered design considers appropriate man machine interfaces.

However, the crucial question in the design and development of assistance systems lies beyond the system's surface. Following Johannsen (1993) the allocation of functions between man and machine and the structure of the remaining human tasks are the primary keys to an effective technical assistance of the human driver. In addition: system designers have to

reflect upon the users' mental models of assistance systems to assure that users understand

- how the technical system works,
- what its capabilities are and
- where its limits are set.

Färber et al. (1998) included this view in their model for the human-process-communication when they emphasized the importance of the consistency of the operator's mental model of the system and the system itself to ensure a secure and efficient operation. Closely related to this idea is the concept of situation awareness (SA) (Endsley, 1995a, 1995b, 2000) which includes the perception, the comprehension and the projection of the state of a man-machine-system into the future while operating the system in a more or less complex situation. SA has proven to be an effective construct in the analysis of aviation accidents and a guideline for the development of assistance systems in aviation as pointed out by Schulte (2002).

The essence of the ideas sketched above is applied to the field of driving in this paper. The aim is to design and develop new assistance functions that enhance driving safety and driving efficiency by enabling the driver to correctly perceive and comprehend traffic situations and to anticipate the necessary near-future actions. This is achieved by developing a situation awareness within the technical system including the following situation-elements: driver, vehicle and the surrounding traffic situation.

In the context of this paper situation-awareness is based on the target-strategy. The target-strategy summarizes those subsets of completed driving tasks that result in a safe and effective driver vehicle system performance. Therefore we establish a model of driving behavior which is represented by combinations of relevant subtasks that have to be completed in a specific traffic situation.

The target-strategy has to be related to sets of driving actions. In this case driving actions include every necessary action to drive along a road, e.g. looking out for known deer passes. The successful completion of sets of driving actions is an indication for the fulfillment of driving tasks. The observed driving actions - either successful or unsuccessful completed - are used to estimate the strategy realized by the driver: the driver-strategy. To realize the required driver monitoring we use our test vehicle: the DLR-ViewCar.

A technical implementation of the target-strategy compared with the monitored driver action indicates possible starting points for driver assistance by the identification of the driving task which is under execution. Thus, this target-strategy plays a key role for the design of new driver assistance systems.

To establish the target-strategy and the means to estimate the fulfillment of the driving tasks, the following questions have to be answered:

1. how to **identify** and delimit specific **traffic situations**

2. how to **identify situation-independent driving tasks** as a basis for the target-strategy

3. how to **identify** the appropriate (safe and efficient) **target-strategy** for different situations,

4. how to implement the target-strategy into a technical system (**technical-implementation**),

5. how to estimate the **driver-strategy** through driver monitoring.

In further steps the difference between target-strategy and driver-strategy will be used to implement situation- and driver-adequate assistance.

ESTABLISHING THE TARGET-STRATEGY

IDENTIFYING DELIMITED TRAFFIC SITUATIONS AND SITUATION-INDEPENDENT DRIVING TASKS

Fastenmeier (1995) suggested a definition of the traffic situation as a helpful concept for the analysis and evaluation of driver behavior. It allows for a precise description of all significant situation elements. These elements are divided into static and dynamic factors influencing driver's behavior. The static factors including constructional (intersection, curve ratio, etc.) and operational (road signs, toll regulations, etc.) elements are nearly time invariant and attached to locations. The dynamic factors change with time: weather conditions, lighting, composition of other road users etc.

Besides its relevance for a successive description of driver behavior the decomposition of trips into well-defined units - the traffic situations - based on place, time and behavior (cf. Erke & Gstalter, 1985) is the main requirement for a situation-adaptive driver-assistance. The classification system for traffic situation originally developed by v. Benda et al. (1983) is adapted by Fastenmeyer et al. (1995) to be appropriate for

1. developing new assistance functions,
2. arranging the experimental design for the assessment of driver assistance systems and
3. the evaluation of safety effects.

These classes of traffic situations are summarized to clusters of comparable situations e.g. all situations concerning the departure of a street to the right. In the following these clusters are denoted as generic traffic situations. In this way the number of traffic situations is strongly reduced.

Based on the situation classification codes by Fastenmeyer (1995) the manifold situation dependent driving tasks are determined which are applicable to these clusters of described situations. From this starting point we accumulate the situation dependent driving tasks to more abstract tasks that depend only on the generic traffic situations. This will be done for all clusters.

In the next step the identified tasks of every generic traffic situation will be analyzed to find similar or concurrent tasks which can be modified to fit in every generic traffic situation. The final output are situation independent tasks, in the following denoted as generic tasks.

IDENTIFYING THE TARGET-STRATEGY

To successfully pass a generic traffic situation a defined set of generic tasks has to be performed. More explicitly: successfully passing a generic traffic situation means that the driver-vehicle system only takes system-states in which no self-inflicted accident occurs.

Transitions between system-states are invoked by the completion of a member of a subset of generic tasks which is valid for the regarded generic traffic situation. In addition the sequence of the completion is determined.

To validate the derived generic tasks and their connectivity we propose an additional empirical task analysis that applies different psychological methods. The methodology is comparable to the Man-Technology-Organization (MTO) Approach described by Strohm and Ulich (1997) which summarizes a framework for the analysis of jobs in organizations. In transferring the MTO-approach to the field of road traffic the tools applied to different driving tasks are adapted accordingly: monitoring (by human surveyors or technical means), intrusive surveys (incl. on-task-interview), questionnaires, expert opinion polls (e.g. accident researchers), document analysis (e.g. traffic rules, driving education manuals). This approach benefits from the understanding of traffic as a multi-element system.

This empirical task analysis procedure results in data about the validity of the determined generic tasks and their connectivity. Furthermore the sequences of actions to fulfill these subtasks, the frequencies and lengths of actions will be derived. This result can also be used to identify which generic driving task the driver is about to perform.

The work-package will be carried out using our specially equipped test vehicle: the DLR-ViewCar. With this test vehicle we are able to answer the above mentioned two work-packages: validating the generic tasks and determining the sequences of actions to identify the task execution.

TECHNICAL-IMPLEMENTATION OF THE TARGET-STRATEGY

Driver assistance - as we have defined it above- means: keeping the driver-vehicle-system within the safe and efficient target-strategy. Therefore the sketched variety of data - achieved from task analysis as well as empirical studies - has to be transformed to generate a model of target driver behavior that may be processed by a technical system. To this aim, objective and replicable methods to appropriately modeling the target-strategy have to be chosen. Prospective methods for the data treatment are:

- qualitative scripts of action sequences,
- mappings (incl. logical deductions),
- formal representations,
- petri-nets (describing the feasible states of the driver-vehicle-systems and including parameters that limit the transitions between system-states)

In a first iteration we will establish a behavior space similar to the knowledge spaces proposed by Falmagne, Koppen, Villano, Doignon and Johannesen (1990). In assuming all generic tasks in a delimited traffic situation to equal the items in a knowledge space - which might either be solved or unsolved - all possible combinations of solved and unsolved generic tasks will be created.

A behavior space as considered here will take into account that solving one task might be a prerequisite for the fulfillment of another task. In constructing an and/or-graph by formal methods all accessible behavior states can be described. A behavior state is defined here as a set of solved generic tasks and the behavior space as the set of all accessible behavior states. The super set of safe behavior states represents the target-strategy.

To enable the assistance systems to process the knowledge incorporated in the target-strategy and the

dramatically increasing amount of data for driver monitoring and situation interpretation (sensory data, static data like digital maps and car performance models etc.) new approaches to the technical design of the system architecture of vehicle electronics will be established. Cognitive system architectures (COSA) as described by Putzer and Onken (2002) will replace today's arrangement in independent control units. These cognitive architectures have to admit for establishing knowledge about delimited traffic situations and the target-strategy appropriate to handle these. With the knowledge about the situation and the measurements a situation interpretation is executed. In the next step the result of the situation interpretation is compared to the target-strategy. This process is called the situation diagnosis. On the basis of the difference between target-strategy and the estimated driver-strategy the system will derive its decisions about the preferable assistance.

DRIVER MONITORING: DETECTING THE DRIVER-STRATEGY

In our aim to design and develop new driver assistance functions basing on the target-strategy a critical milestone is the implementation and further development of powerful means for the detection of the actual driver-strategy: the driver monitoring methods.

Driver monitoring is the basis for the system's understanding of what actions the driver is about to perform. The level of detail of action-description particularly depends on the availability of highly accurate and reliable monitoring methods. From this detected actions the driver's engagement in specific driving tasks will be estimated just as the probability of his task completion.

While specific actions are connected with a delimited set of generic tasks this connection is confounded with behavior variances within or between subjects. This variance is partly caused by individual differences in information processing styles, the drivers' skills and abilities, their driving experience and their actual psychological and physiological state etc. The consideration of these variables will be the main content of another DLR-research project. This includes a broader consideration of physiological measurement methods which provide further features to the detection of driver states.

An additional factor influencing the load of the perception modalities and the required mental resources of the driver are the considered traffic situations with the inherent generic task sets. In accordance the composition of the affected mental processes varies as well. Figure 1 shows the effects of task variables on mental resources and the evoked mental processes. As pointed out by Gstalter (1988) and Schäfer (2002) the demands on mental resources vary with the task level and the type of decisions that are involved in fulfilling the task.

Hence, the technical system has to cope with various driver behavior patterns whether they vary within one individual or between individuals.

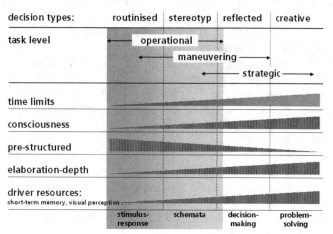

Figure 1: Demands of the driving tasks (c.f. Schäfer, 2002)

As will be pointed out in the description of our test vehicle, some of the necessary means to detect the driver-strategy are already implemented in mass-produce cars, e.g. the CAN-busses. Additional information will be provided by assistance systems that have just entered the market, like the adaptive cruise controls. But some very important monitoring devices - for example gaze-tracking - have just been implemented into test vehicles for research activities.

ADAPTING THE ASSISTANCE SERVICE

Several developmental goals have been proclaimed for future driver assistance systems. Besides an overall orientation on human factors, the main requirement for the future of in-car technology is to design systems that are:

- driver-adaptive and
- situation-adaptive.

Driver adaptation must refer to both: detecting the driver's state and understanding the driver's intentions. The former information is used to adapt the system to counteract dangerous or unpleasant states of the driver (strain, fatigue). The latter information - used to predict the intended course of action - enables the system to support the driver at the decisive points of the task-completion with the appropriate and necessary information or actions. For the technical system this means the adaptation of the system's functionality or the choice of the modal resources for the interaction with the driver. This is realized by implementing an assistance-strategy into the system that bases on the target-strategy described above. Only systems that have a model of successful behavior in a definite traffic situation

will be able to show safe and efficient performance and a high level of adaptation to driver and situation.

THE DLR-VIEWCAR: OUR MEANS OF INVESTIGATION

The multivariate actions shown by drivers and the analysis of the complex traffic situations in which their performance is accomplished requires a differentiated technical instrumentation. For this aim we have realized our test vehicle, the DLR-ViewCar. It is equipped with measurement means for

1. the estimation of the driver's sensation,
2. the direct observation of his motor behavior and the resulting performance,
3. the recording of the vehicle state,
4. and the description of traffic-situations with a high level of detail.

The current technical configuration can be summarized by

1. the signals of all CAN-busses,
2. a static video system capturing the traffic situation ahead,
3. a video-based gaze-tracking device
4. a dynamic camera platform pursuing the driver's head rotations
5. a laser scanner for objective distance measurements of the traffic situation
6. an on-board data recording system
7. and an off-board workplace for data analyses.

The Car Area Network bus

The Car Area Network busses (CAN) are implemented in every current passenger car. The measurement and recording of the signals transferred in this network provides specific information concerning the power train, the vehicle's electrical system state, the fail state of the vehicle's systems, the states of all driver controls including those of the multimedia-systems and parts of the vehicle dynamics. The recently implemented sensory devices like short-range radar (for example as part of the adaptive cruise control (ACC)) even add limited information about the surrounding traffic situation.

To a great deal these data enables us to accurately capture, analyze and describe driving behavior concerning the control of the vehicle: steering, acceleration, braking-behavior, preferred longitudinal and lateral accelerations and resulting yaw rates.

Furthermore the recording of all driver inputs on the vehicle controls e.g. the indicators, the wipers or the car radio allows for a real-time appraisal of driver intends and of the objects in the focus of his attention.

By this means we gather direct and indirect information about the driver's actions, the vehicle state and to some extent about the traffic situation. These data provided by CAN-network also enable us to estimate pursued driver actions and on basis of these actions the particularly intended generic tasks.

The video based gaze-tracking device

As an enhancement to the data received from CAN-busses which bases primarily on executed driver actions we use a video-based gaze-tracking device (see Figure 2) to gather knowledge about driver's visual search patterns. From these, an estimation of the driver's sensation is possible. This may be used, for example, to verify and refine the driver actions concerning in-vehicle situation elements estimated by means of CAN-bus data. For example, the head-down time is one critical measure for the safety of in-car systems and can only be measured by gaze-tracking. The CAN-bus data gives only information whether an input or output occurred.

Figure 2: Video-based gaze-tracking system

The implemented gaze-tracking system analyzes head and gaze direction basing on a stereo-camera system that captures the drivers head with a frequency of 60 Hz. The cameras have been installed on top of the cockpit to provide most adequate input to the image processing system when the driver looks straight through the windshield or scans the dashboard behind the steering wheel.

Every driver's head is modeled in an interactive process between user and system to achieve an exact as possible tracking performance. The image processing relies on specific face features that the system extracts with a time accuracy of approx. 30ms. By usage of its infrared detection the system is also applicable under low light conditions. We simultaneously use the cameras' video stream for driver observation to verify the systems accuracy and reliability and as a basis for the analysis of the driver state, e.g. to detect surprise or anger from driver mimics.

Based on the measured gaze-behavior and visual search patterns the dynamic camera platform of the test vehicle is controlled.

The on-board video-system

To gather knowledge about the external situation we use a video-based system and as described below a laser-scanner. The video-system is split into two parts:

a) a wide-angle camera to capture the complete driving scenery ahead which can be extended to a surrounding view and

b) a high-resolution camera on a motion platform (Figure 3) which follows the driver's head direction measured by the gaze-tracking device within the car.

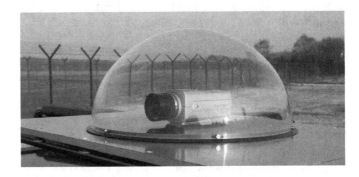

Figure 3: High-resolution camera on motion base

With the wide-angle camera we are able to qualitatively identify the situation elements of the traffic situation ahead. The high-resolution camera which follows the head direction of the driver records in much more detail the situation elements in the focus of the driver's attention. Additionally placing a graphical gaze-direction indicator into the appropriate place of the video-stream of the high-resolution camera shows the section of the traffic situation where the driver primarily searches for information at this moment.

With this information a connection between stimuli in the real world and the driver's reactions, determined via the CAN-bus, can be drawn. This allows the description of the duration between stimulus sensation and performing a driving reaction. The expected reaction can be derived from the focussed feature in the traffic situation. For example while entering the city limits via a country-side road the driver reduced his velocity in congruence with the position of the place name sign. Thus, the verification and refinement of the estimated driver actions concerning the complete traffic situation are possible.

To summarize: In order to estimate which driving task the driver is about to fulfill it is in most cases necessary to answer:

1. where the driver searches for information,
2. what is at the point of his visual focus and
3. how he reacts to specific stimuli.

This will be achieved by the system described so far.

The quantitative distance measurement

To achieve a more objective description of the actual surrounding traffic situation the car is also equipped with a laser-scanner that is able to detect the relative positions of situation elements (e.g. pedestrians, cars and an approximation of the road's boundaries).

With this exact quantitative data we can achieve derivatives like time-to-collision which are necessary to describe stimulus-response-connections including variations in reaction time and execution strength.

LEAVING THE HIGHWAY: A DEMANDING DRIVING SITUATION

In order to evaluate the approach described above, a first empirical study is in process focusing on one driving situation which is stressful for the driver and potentially dangerous: leaving the German highway ('Autobahn'). In a first step through a document analysis (traffic rules, driving education manuals) a working-version of the target-strategy is established. To refine this target-strategy additional information is gathered from traffic experts. Therefore structured interviews with a group of 5 driving teachers are carried out to model the range of safe and efficient driving behavior in highway exit-situations more appropriately.

In an empirical part of the study (results will be available during the oral presentation) 10 subjects drive on the highways A2 and A39 in the region of Braunschweig (Lower-Saxony) and perform several highway departures. The effect of traffic density, constructional factors (e.g., curve ratio) and of the presence of other drivers is controlled.

The goal of this study is to gather empirical data that serve as a basis for real world driver-strategies while leaving the highway. These data are additionally used to put forth indicators and measurement-methods for the fulfillment of the target-strategy.

FIRST RESULTS

During the approach to a departure point, several phases are distinguished each of which is initiated by a preceding traffic sign (the following is valid for road design following the German regulations).

The first announcement of the departure point is usually displayed on the sign posting of the antecedent departure point, in this study approximately 6-8 km before the departure takes place. The driver doesn't have to react immediately but has to consider the distance to the departure point in all of his driving decisions, e.g. whether he starts another overtaking manoeuvre.

A second phase is initiated by the so called "1000m-sign" which explicitly points at the nearing departure point. In this phase, the driver has to change to the right lane (if still necessary) to prepare for his departure.

Thirdly - 500 m before departing - the driver should be at the right lane and has to survey the preceding and following traffic to draft a realistic plan of upcoming driving actions (Figure 4).

Figure 4: 500m before exit point

A triplet of signal bakes precedes the lane-change to the exit lane (300/200/100 meters before departing). At this point, the driver should signal his departure. Afterwards, he should check the rear-view mirror and / or turn around to be sure that no other car is overtaking him at the right side. Drivers are not allowed to drive at the right side of the right lane - an emergency lane - but sometimes do so; see Figure 5. The driver also has to make sure that he doesn't have a significant speed difference compared to other cars that might eventually have stopped on the exit lane.

Figure 5: In the right rear mirror - a dangerous situation element

Figure 7: One example of the exits' curves,

At the start of the exit lane (see Fig. 6) the driver has to fulfil the lane-change-manoeuvre and to decelerate or brake (depending on the speed before departure). At the beginning of the curve, a safe speed - depending on the curve ratio and the vehicle dynamics - has to be reached.

Figure 6: The discontinuous line signals the start of the exit lane

While driving through the exit-curve (see Fig. 7) a safe speed should not be exceeded. The curve ratios in this study vary to a certain degree and more important in some cases drivers are not able to preview the roads curvature due to noise absorbing walls. At the end of the curve, the driving situation 'highway departure' ends.

From these empirical observations, a detailed description of highway departure is achieved that contains required actions which will be assessed as being performed or not. Moreover, information about the critical order of these actions is included. The quality of the performance will be scored using the video observation. Finally, from the behavioural and the CAN-bus data, a quantitative description of these actions is done.

CONCLUSION

This paper presents the outline of our methodology for deriving a model for safe and efficient driving behavior. In our view the implementation of knowledge about driver behavior into driver assistance systems will be one of the main steps to generate assistance functions that will simultaneously be driver- and situation-adaptive.

We started the necessary fundamental research on different levels of analysis: task-oriented and action-based. First results concerning the task structure have been qualitatively described in the "first results" section.

During the empirical studies - which yet haven't been completely analysed - we have found out that another necessary step to accomplish an utilizable driver model is to gain much more accurate information about the vehicle's position relative to the lane and to other situation elements, e.g. other vehicles. With regard to this challenge we are constantly developing the ViewCar further, soon implementing a lane-detection system and an inertial system. Thus, additional and highly accurate information will be gained like the acceleration into all directions, absolute velocity vectors, position relative to the lane etc.

The proposed approach to identifying, modeling and implementing the target-strategy is still under development, since we have found out that our ViewCar needs to be upgraded as sketched above. Also we have to expand our studies to a broader variety of situations in order to be in the position to describe driving on the task level.

During the presentation video material will demonstrate crucial driver actions and hazardous situation factors for the highway departure situation.

ACKNOWLEDGEMENTS

We thank Reiner Suikat, Martin Ruschinzik, Mark Schröder and Henning Mosebach who have pushed the technical development of our test vehicle and thus enabled us to conduct these studies.

REFERENCES

1. v. Benda, H., Graf Hoyos, C., &Schaible-Rapp, A. (1983). Klassifikation und Gefährlichkeit von Straßenverkehrssituationen. Bericht zum Forschungsprojekt 7320. Bundesanstalt für Straßenwesen: Bergisch Gladbach.

2. Endsley, M.R. (1995a). Measurement of Situation Awareness in Dynamic Systems. *Human Factors,* **37(1)**, 65-84.

3. Endsley, M.R. (1995b). Toward A Theory of Situation Awareness in Dynamic Systems. *Human Factors,* **37(1)**, 32-64.

4. Endsley, M.R. (2000). *Theoretical Underpinnings of Situation Awareness: A Critical Review.* In: Endsley, M.R., & Garland, D.J. (Ed.). Situation awareness analysis and measurement. pp.: 3-32. Mahwah, NJ: Lawrence Erlbaum Associates.

5. Erke, H., Gstalter, H. (1985). Verkehrskonflikttechnik: handbuch für die Durchführung und von Erhebungen. Unfall- und Sicherheitsforschung Straßenverkehr, Heft 52. Bundesanstalt für Straßenwesen: Bergisch-Gladbach.

6. Falmagne, J.-C., Koppen, M., Villano, M., Doignon, J.-P., &Johannesen, L. (1990). Introduction to Knowledge Spaces: How to Build, Test, and Search Them. Psychological Review, 97(2), pp. 221-224.

7. Färber, G., Görke, W., Polke, M., &Steusloff, H. (1998). Mensch-Prozess-Kommunikation: Aufgaben, Mensch, Technik - Status und Trends. In: G. Färber, W. Görke, M. Polke, & H. Steusloff (Ed.). Mensch-Prozess-Kommunikation: Vorträge der GMA-Tachtagung anläßlich des VDE-Kongresses '98. VDE-Verlag GmbH: Berlin.

8. Fastenmeyer, W. (1995). Die Verkehrssituation als Analyseeinheit im Verkehrssystem. In: W. Fastenmeyer. (Ed.). Autofahrer und Verkehrssituation. Deutscher Psychologen-Verlag: Bonn.

9. Gstalter, H. (1988). Transport und Verkehr. In: D. Frey, C. Graf Hoyos & D. Stahlberg (Ed.). Angewandte Psychologie - Ein Lehrbuch. Psychologie Verlags Union: Weinheim.

10. Johannsen (1993). Mensch-Maschine-Systeme. Springer: Berlin.

11. Putzer, H., &Onken, R. (2002). COSA und COSYflight. Ein generischer Ansatz für kognitive Systeme und seine Anwendung. CCG Seminar TV 3.02. DLR Braunschweig. Source: http://www.unibw-muenchen.de/campus/LRT/LRT13/deutsch/artikel/CCG2002_COSA.pdf.

12. Schäfer, H. (2002). *Fahrerassistenzsysteme: Zuweisung von Entscheidungsprozessen und Gestaltungsspielräume zur Fahrereinbindung.* In: Zentrum für Verkehr der Technischen Universität Braunschweig e.V. (Hrsg.). Automatisierungs- und Assistenzsysteme für Transportmittel. Möglichkeiten, Grenzen, Risiken. Fortschr.-Ber. VDI Reihe 12, Nr. 185, pp.: 269-279. Düsseldorf: VDI Verlag.

13. Schulte, A. (2002). Verbesserung des Situationsbewusstseins durch Automation von Missionsmanagementaufgaben im Kampfflugzeug-cockpit. In: M. Grandt und K.P. Gärtner (Hrsg.). Situation Awareness in der Fahrzeug- und Prozess-führung. DGLR-Bericht 2002-04. Bonn: DGLR e.V.

14. Strohm, O., &Ulich, E. (Ed.). (1997). Unternehmen arbeitspsychologisch bewerten. Ein Mehr-Ebenen-Ansatz unter besonderer Berücksichtigung von Mensch, Technik und Organisation. vdf Hochschulverlag AG an der ETH Zürich: Zürich.

CONTACT

Dipl.-Psych. Helge Schäfer

Deutsches Zentrum für Luft- und Raumfahrt

Institut für Verkehrsführung und Fahrzeugsteuerung

Lilienthalplatz 7

38108 Braunschweig

Tel.: 0531 / 295-3435

Fax.: 0531 / 295-3402

e-Mail: Helge.Schaefer@dlr.de

The Driving Need for Human Factors in the Car of the Future

David J. Wheatley

Motorola Labs, User Centered Research Group

ABSTRACT

The car of the future will be very different from that which we know today – no longer just a mechanical transportation device, but a mobile communications and entertainment platform providing many new functions and services to the occupants.

The use of cellular telephones whilst driving is already a source of major concern and further legislation regarding phone usage will undoubtedly be forthcoming. The future will also add display based navigation, internet access, e-commerce and other location dependent services which will increase the drivers cognitive load. We will also see driver assistance systems such as adaptive cruise control, collision avoidance warnings and vision enhancement systems which will change the driving task and could potentially reduce cognitive load almost to the level of autopilot monitoring.

As the nature of the in-vehicle activity therefore evolves from being a perceptual-motor driving task to that of a communications and vigilance task (set against a background of infotainment services) then we can not only anticipate emergent behaviors but will also see a new need for the management of multimodal information presentation to the driver according to his driving circumstances and indeed, personal preferences. Consequently driving could be as different from what we know today, as today is from the time of leather helmets and driving goggles.

The aim of this presentation is to provide an overview of the major human factors and safety issues which these new systems are raising and which will be fundamental to their effective and safe use as well as to their commercial viability and success.

INTRODUCTION

There was a time when driving a car involved wrestling with heavy and unresponsive steering, using strong leg muscles to operate heavy brakes, paying attention to avoid potholes in the road and double de-clutching. But that of course was a long time ago when life and cars were both much simpler than they are today. Back then cars were relatively simple mechanical devices whose primary purpose was that of transporting people and their luggage from A to B – what some classic car enthusiasts and purists might call "the good old days".

But both life and cars are far more complex today and the latter at least are on the threshold of becoming much more complex still – and will likely change the nature of driving much more than we have seen thus far. We now have abs, traction control, cruise control, adaptive cruise control – in the upmarket models additionally active suspension, night vision, parking assistance and so on - and of course just about everything is power assisted. Though we are still a very long way from actually automating the driving task, many of these technological developments (whose safety benefits incidentally have been very substantial and worthwhile) have gradually changed the nature of the driving task. The in-vehicle environment has also changed substantially – built in video to keep the kids entertained, navigation systems to help avoid getting lost, emergency call and breakdown for peace of mind, information services to the vehicle, cellular telephones to keep in touch and of course, the cup holders necessitated by the drive-thru concept.

But both the evolutionary changes to the driving task and the changes to the in-vehicle environment have at least one thing in common – they both act to place barriers between the driver (and of course the passenger) and the external world. In the "good old days" when driving was more of a seat of the pants experience – the noise level, the wind in the hair, the bumps in the road – this was how the driver sensed the outside world. Not today – he is coccooned in a comfortable, smooth and quiet air-conditioned environment where any sensing of the external environment has to occur through some sort of interface such as the speedometer, thermometer and various warning lights for low traction etc. The importance of these various interfaces is becoming greater at the same time as they are becoming more complex. Just as the technical developments in the vehicle have made it much easier to drive, increases in traffic density and road system complexity have made it more difficult. As a consequence the drivers attention is directed more towards the external vehicle environment and less towards the internal environment.

But there is one thing that has changed little if at all over this period – and that is this driver, the fallible human being whose perceptual and cognitive capacity is not increasing year on year to keep up to speed with the information with which he is bombarded. Though there is still a cognitive demand involved in driving there are different things which are now demanding the drivers attention and this is creating the driving need to pay serious attention not just to traditional physical ergonomics in the vehicle but increasingly to the psychological and cognitive issues with which the driver will have to contend in this information rich future mobile environment.

THE USE OF CELLULAR TELEPHONES IN THE VEHICLE

The problems that these new types of in-vehicle activity will bring are not imagined – we are witnessing right now the response to safety questions raised about the use of mobile telephones whilst driving [8]. Applying a physical ergonomics type of approach to this issue would clearly tell us that there is a finite limit to what a driver can do – he only has a finite number of hands, fingers and feet with which to operate controls. And one of the first major studies into the use of cellular telephones in the vehicle carried out in Japan in 1996 [1] confirmed that this was the case. This study investigated 129 motor vehicle crashes in which use of a cellular telephone was involved and found that 16% of the drivers were conversing on the phone at the time of the crash, 32% were dialling, 5.4% were hanging up the phone and 42% were responding to a call. The large proportion of crashes related to handling the telephone (74%) may be a reflection of the fact that, at that time, 94% of the telephones sold in Japan were hand-held models. The conclusion therefore was that the major safety risk was due to the physical demands of reaching for the phone in the glove box or bag, and physically opening the flip cover and dialling the number.

At the time of this study, handsfree telephone systems were not common and the mobile phones in use were often flip phones without repertory dialling facilities which were often not cradled within reach of the driver. It was a relatively straightforward and logical step to conclude from these results that a handsfree interface to a cradled telephone would solve many, if not all of these problems and mitigate the potential safety hazards.

But this was not the case. The more recent study carried out by Redelmeier and Tibshirani in Toronto during 1994 and 1995 [2] concluded that there were still possible safety issues arising from the use of cellular telephones which were not due solely to the physical handling of the device. They concluded that talking on a cellular telephone quadrupled the risk of a collision and that it

was the high cognitive load of being involved in conversation which was a significant hazard. They concluded also that the increased likelihood of accidents persisted even after the call had been completed. In other words the drivers attention was still, to an extent, focused on the verbal interaction and not on the task of driving. The use of handsfree systems might address the physical elements but not these cognitive elements.

It is the same sort of logic which is creating the widely held belief that verbally interacting with the in-vehicle functions by means of a voice interface will adequately address the concerns over cognitive demands on the driver. Unfortunately this is not entirely the case either. This has been shown by a recent study carried out at the University of Iowa [3] in which subjects were required to listen to email messages whilst driving in a simulator. The cognitive demand of listening to the verbalised information led to an increase of 30% in the reaction time to the brake lights of a leading vehicle. Even if the use of a voice interface were to be the best solution to the problem of accessing information functions in the vehicle – it is clear that the technology, particularly speech recognition, is not yet sufficiently robust and flexible to deliver a reliable and easy to use interface – by itself it is increasing the cognitive load on the driver rather than reducing it.

DESKTOP COMPUTING IN THE VEHICLE

The use of cellular telephones in the vehicle is just the beginning of an information revolution in the vehicle and the issues that have been raised by this will be even more critical in the context of the additional functions which are starting to be commonplace in high end vehicles. Functions such as navigation, email and internet access, various information functions as well as those valuable functions clearly targeted at increasing safety and security, such as ecall, breakdown call and automatic notification of airbag deployment. Though attention is being paid to the user interfaces of these devices most, if not all of those currently on the market, derive their user interface principles largely from human factors efforts in the desktop computing environment. The visual displays are often multi-colored, visually rich and frequently use softkeys for menu selection – this is fine for the desktop but not the best approach for using a system in the vehicle where such interaction is not the primary task. Quite simply, many of them represent a visual distraction and one thing that most seem to have in common today is that their usability is generally extremely poor. Perhaps we should consider whether there is a place here for a completely different approach to the user interaction with such systems which is specifically tailored to the unique situation of in-vehicle system interaction.

THE COMPUTER SCREEN AS AN IN-VEHICLE USER INTERFACE

This does not suggest that we therefore ignore all of the learning that has been accrued from research and evaluation of desktop systems, but there are certainly many differences with future in-vehicle systems which need to be considered;

1] Fundamentally, as I have previously stated, the interaction with in-vehicle information systems is a secondary task – driving is the primary task and one which is also safety critical. Though this can be addressed to some extent by "locking out" some or all functions and services when the vehicle is moving or in drive – this approach has the side-effect of seriously limiting the potential value of those functions.

2] Secondly, the user population is very broad, encompassing a very wide range of age, language, educational and driving experience. Though this is also true of desktop computer users there is more time to learn a system or application and the consequences of making a mistake are not so great.

3] Thirdly, it is not valid to assume that potential users have experience with or indeed any interest in computer technology.

4] Fourthly drivers are likely to expect any in-vehicle system to be as intuitive and easy to use as the vehicle itself. The car is not the sort of place to store instruction manuals and even if it were, most people would not read them. It is also unreasonable to expect users to take part in any training course to operate their infotainment system.

What this is suggesting is that the user interface between the driver and the variety of navigation, infotainment and driver assistance systems which are likely to be available in the coming years, is a matter which should receive special attention. And if this transition from a perceptual motor task to one which is more akin to a vigilance and monitoring task is going to be successful and viable then the motor manufacturers, system suppliers and content providers need to collaborate to a much greater extent than hitherto.

THE HUMAN FACTORS CHALLENGES AHEAD

There are a number of fundamental human factors challenges which need to be addressed in the coming years for in vehicle information, navigation, collision avoidance and other such systems to be safe, simple and easy to use. As well as the technological and legal challenges along the way, these human factors issues also need to be seriously addressed to successfully manage the transition from the vehicle as a mechanical transportation device to that of a mobile information and communications platform.

FUNCTIONALITY

Once wideband wireless communications are available to an in-vehicle device there is the potential for a vast number of new functions and services to be delivered. Marketing forces will undoubtedly come into play to exploit these to their maximum – almost certainly, in some cases without consideration being given to the user environment and the primary task demands in that environment. A degree of responsibility will need to be exercised, with the support of reliable human factors data, regarding the appropriateness of various functions and activities when associated with the driving task. For example; verbally presented turn by turn directions are clearly information which supports and should enhance safe driving. At the other extreme there are potential location based events such as trivial and intrusive advertisements when passing close to certain stores or restaurants or movie viewing. Inbetween these extremes are a range of possible functions whose benefits are less clearly defined and whose impact on driving safety must be established.

Many functions and services which pose hazards by unnecessarily distracting the driver could be perfectly acceptable, even desirable for front seat passengers. There are major challenges not only in defining what and when functions should be available but also to whom and how these are presented.

SYSTEMS INTEGRATION

As these different systems start to proliferate, there will additionally be a variety of auditory alerts and alarms as well as visual and probably haptic information presented to the driver. These could be alerting him to anything from an incoming email message to an imminent and life threatening forward collision. In order to ensure that these signals are correctly perceived, understood and acted upon, systems integration is needed within a vehicle. The driver/system interaction must be considered as a whole to avoid the presentation of competing, confusing or non-prioritised information.

INTUITIVE IN USE

The car driver does not expect to, nor does he read an instruction manual neither to learn how to operate the vehicle or the radio, entertainment system or HVAC. The expectation (but not always the reality) is that this is all obvious and sufficiently consistent between different manufacturers that he can work out any small differences. The same will be expected of future in-vehicle systems. Many navigation systems currently available are notoriously difficult and complex to use, are expensive and often give little advantage over using a traditional paper map [4]. Not only does this affect customer take up but can also be a significant safety hazard by adding to rather than alleviating the cognitive load on the driver.

EMERGENT BEHAVIORS

Emergent behaviors are new and potentially unforeseen human behavior patterns which result from the introduction of new technology or systems or which arise from the interaction effects of existing systems. For example some research has suggested that the use of cruise control leads to later braking when approaching behind a slowing vehicle – possibly due to a slight reluctance to go through the process of turning it off temporarily only to turn it on again. A potential example related to Adaptive Cruise Control (ACC) could be an even greater reluctance to brake when approaching behind a slowing vehicle or not braking at all if the driver believes that the ACC will do it for him. If the system happens to be turned off or not available, for instance in a rental car, then clearly this can be very dangerous. Long term experimental studies of driver assistance and other systems will be essential for the prediction of these kinds of emergent behaviors.

MULTIMODAL INFORMATION PRESENTATION

As the amount of information presented to the driver increases other modes of presentation will be utilised. There are already a number of chimes and dings to tell him about open doors and safety restraints. There are likely to be still more indicating upcoming turn manouvre, insufficient headway, vehicle in blind spot, lane deviation and any number of events such as incoming traffic information or email message. In order to avoid overloading the drivers visual capability the auditory and proprioceptive channels are starting to be used. Research needs to be done to determine the associative meaning of auditory icons, spoken commands or haptic events. Particularly for those events which are infrequent, for example an auditory alarm for forward collision warning; the alarm stimulus must be perceived, correctly interpreted, and must elicit the appropriate response within an extremely short time span. For an FCW warning which statistically could only occur once every 25 years or so [5] the driver needs to immediately and instinctively know what it means – there is no time to think about or interpret the meaning of the alarm and there will have been no opportunity to reinforce the associative meaning without increasing the frequency of false alarms and thereby reducing the effectiveness and credibility of the function.

Similarly cross modal information presentation will add a new dimension of complexity. Information may be presented simultaneously through more than one sensory channel and any apparent discrepancy between the associative meaning of multiple simultaneous signals will greatly increase reaction time and reduce effectiveness.

COGNITIVE LOAD MANAGEMENT

Although the concerns about the use of cellular telephones whilst driving center on the high cognitive load of the interaction, the driving task is in fact subject to both extremes of cognitive load. There are also times when the cognitive load is too low – typified by the drowsy driver who lacks situational awareness and may even fall asleep at the wheel. An integrated in-vehicle system could provide some opportunities to moderate these extremes by managing events temporally so as to even out the demand and maintain a required cognitive level closer to the optimum. Lack of integration of potentially competing systems cannot enable this to occur [6]. An example of this might be an incoming telephone call coinciding with and masking, a verbal navigation instruction at the same time as an alert regarding lane deviation.

PERSONALISATION AND PREFERENCES

Once the systems within the vehicle begin to be digital, many hitherto impossible functions and features start to be possible. For example personalisation and customisation of the car has long been a tradition though this has largely been applied to the exterior rather than the interior and has not generally touched upon the dashboard or on the information presented to the driver. It is also a one-time customisation and does not permit changes to be made frequently or in response to changes in preferences. Integrated in vehicle systems have the potential to enable personalisation, such as choosing, within limits, what information is provided to the driver by the vehicle, in what form it is presented and even when it is presented. Family members who share use of the same vehicle could well have different preferences. There are also significant cost benefits to the auto manufacturers who have additional opportunities for personalisation at the point of sale as well as the potential for a more rapid rollout of model updates for instance by the release of software updates to increase the features and functions of a vehicle – these could even be downloaded wirelessly.

BETWEEN VENDOR STANDARDISATION

I have referred to several potential future in vehicle systems - clearly those which are most crucial are those which directly affect safety. Most of us will have had the experience, usually in rental cars, of being totally unable to find the trunk release or the lightswitch or flashing the lights when we intended to turn on the windshield wipers. These are not new functions, they have been in cars for many years yet standardisation of the location and operation of many driving controls still appears to be almost non-existent beyond the steering wheel and pedals. Perhaps it is only mechanical necessity that is driving this consistency and there will be attempts to move these as soon as robust drive by wire systems permit. Similarly, if there is inconsistency between the incoming email alert between different makes and models of cars then the result could be annoyance or even inconvenience. The result is more likely to be injury or death if there is a lack of standardisation of collision warning alarms for example. Standardisation and consistency will therefore be an essential not a desirable element of safety related systems.

INDUSTRY COLLABORATION

The word convergence has frequently been used to allude to the coming together of computing with telecommunications. What will be happening in this future is the three way convergence of computing and telecommunications within the automobile. For the business opportunities that this convergence offers to be successfully exploited, will require a huge degree of collaboration not only within the auto industry but also between players in the computing and communications industry with wireless service providers, content providers, internet companies and many more. In fact a market report recently issued in the UK reported that "Partnering, collaboration and product positioning will be key to addressing the telematics opportunity" [7]

THE WAY FORWARD

The aim of this presentation has been to provide an overview of the major human factors and safety issues which new driver information and assistance systems are generating. These factors will be fundamental to their effective and safe use and to their commercial viability and long term success. The fundamental factors which should drive the way forward and make these systems both safe and acceptable are ;

• Driving a vehicle is not the same as operating a desktop computer so do not assume that information should be presented in the same way.

• Consider the whole driver/system interface – if subsystems are developed independently and separately there could be similar or confusing or competing alerts and alarms.

• Use real human factors research to better understand how, when and through which sensory modality they should be presented.

• Determine the instinctive or intuitive meaning associated with various sounds and exploit the potential of voice interaction with the vehicle.

• The presentation of any safety critical information in the vehicle should be consistent within and between manufacturers, consistent over time and ideally internationally.

• New relationships and collaborations will need to be established between companies who have not traditionally worked together if the true potential of these future systems are to be delivered to the man in the street, or in his car.

CONCLUSIONS

Many new driver assistance and information systems will be appearing in the cars of the future and there is much that we can potentially learn from experiences in the aircraft industry – about glass cockpits and speech synthesis, about haptic displays and vibrating sticks. What I personally hope we will not learn too much about from this is about automation, autopilots and the removal of control from the driver. The automobile is not quite like other industrial products but is one with which we human beings tend to have a special relationship unlike any other manufactured artifact. The task of driving is also one which is perceived quite differently from other tasks. As the nature of the in-vehicle environment evolves into a more information rich and electronic future, it is important to ensure that some of the pleasures of driving and some of the pleasures of car ownership remain whilst fully exploiting the safety benefits and business opportunities which these future systems can offer.

REFERENCES

[1] "An Investigation of the Safety Implications of Wireless Communications in Vehicles" *NHTSA Report*
[2] "Association between Cellular Telephone calls and Motor Vehicle Collisions". Redelmeier, D.A. & Tibshirani, R.J. *The New England Journal of Medicine* 336, 2, 1997, pp. 453 – 458.
[3] "Does a Speech-Based Interface for an In-vehicle Computer Distract Drivers ?" Lee J.D., Brown T.L., Caven B., Haake S., Schmidt K, Dept. of Industrial Eng., University of Iowa – unpublished report – May 2000.
[4] "When you Absolutely, Positively Have to get There... Eventually*" Autoweek,* Jan 3, 2000 pp 15 – 21.
[5] "In Vehicle Collision Avoidance Support under adverse Visibility Conditions – Warnings" in Y Ian Noy – Ergonomics and Safety of Intelligent Driver Interfaces 1997 (Mahwah; Lawrence Erlbaum) pp 221-230
[6] "Beyond the Desktop – and into your Vehicle" Wheatley D.J. *CHI2000 Development Consortium,* Den Haag, Netherlands, April 1-7 2000 - Extended Abstracts pp 43–44.
[7] "In Car Telematics Terminals Market 2000-2006" Strategy Analytics, Strategy Advisory Service, London, Feb 2000
[8] "Death by Distraction" , Kobe G. – Automotive Industries, May 2000 pp 30-37.

A Vision of the Future of Automotive Electronics

Daniel K. Ward and Harold L. Fields
Delphi Delco Electronics Systems

ABSTRACT

The Future of new electronic systems application in the automobile is extremely promising. New systems such as mobile multimedia, control-by-wire systems, advanced safety interiors, and collision avoidance coupled with smart sensors and actuators, in a potentially new integrated vehicle wiring system, means significantly more electronic content in the automobile of the future.

The challenge of the automotive electronics industry is to develop a vision of these future products, and then follow a defining process to develop the technologies necessary to offer timely, reliable, and cost effective products to the automotive consumer.

This paper presents a methodology by which global market trends, market considerations, and engineering developments are combined to create a product vision. This vision is used to generate technology roadmaps that spawn technology development projects. Technology projects in turn help to create product strategies that result in new marketable products.

Delphi Automotive System's vision of "Next Century Winners," which are stimulating new product development and technology projects that enabled new product developments, are examined in this paper.

INTRODUCTION

What is a reasonable vision of the type of automotive electronics consumers will demand in the future? Automobiles have had enormous impact on our global society and environment. Automotive production is the engine that has propelled western economies, especially that of the United States of America, for more than 70 years. However, if you look at the current state of our planet there are also negative contributions made by the automobile: air and noise pollution, depletion of the earth's natural resources, accidents causing injury and property damage, urban sprawl, traffic jams, and drivers lost in a maze of roads and ramps are just a few examples. Consumers continue to demand solutions to these negative impacts of automobile use, as well as options and amenities to provide comfort, entertainment, connectivity, and ease of operation.

Simply stated, we as consumers want cars that are non-polluting, fuel efficient, safe, efficient in getting us from point to point, comfortable, with entertainment features, connected to phone and computer, easy to operate, and, most important, affordable to purchase and use.

The vast majority of automotive improvements desired by consumers are enabled by electronics. This is great news for the future of automotive electronics industry. However, the real challenge is to produce electronic systems that are desirable, marketable, and manufacturable.

VISIONING PROCESS

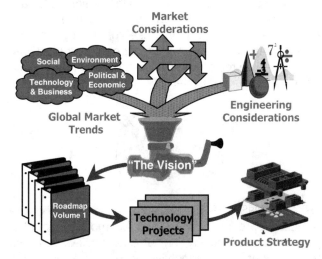

Figure 1.   The Visioning Process.

Dictionaries define vision as both "the ability to anticipate and make provisions for future events: foresight" and "insight: imagination." To have a vision is one thing to act on it is all together different. In order to make a vision into reality there must be a process.

Figure 1 illustrates an example of a model that provides a structure for turning visionary product concepts into functioning hardware. It combines global market trends, market considerations, and engineering considerations into product and technology roadmaps, which in-turn create technology projects and subsequent product strategy.

GLOBAL MARKET TRENDS – Global market trends consist of social demographic, technology, business, political, economic, and environmental issues.

Social demographic issues include longer life expectance, generational preferences such as generation-X with cell phones and Internet access, growing population centers, increased global automobile demand, and social and economic infrastructures.

Technology and Business factors include the current explosion of computer technology, communications and technology developments made in consumer electronics, after sale services becoming moneymakers and differentiators.

Political and economic trends that must be considered are globalization of the automotive industry (engineering, manufacturing, and sourcing), reductions in product life cycles and new competitors from other electronic markets.

Environmental issues concerning the automobile are many. Reduced pollution into the environment during manufacture and operation, use of energy saving recycled materials in the manufacture of new products, recycling of electronic products at end of their useful life are a few of the major issues.

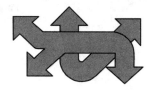

MARKET CONSIDERATIONS – The expectations of the marketplace are many. First and foremost the market expects improved quality, higher functionality, and less time to market at a competitive price. The market also expects globalization of products, but with regional differentiation. New products must be integrated into systems as modules that can varied to fit individual applications. Technology is another major discriminator. Technologies must be proven reliable, but new state-of-the-art technologies are preferred. Electronic hardware must be simulated with software then validated through controlled conditions to assure proper functionality.

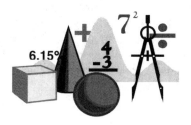

ENGINEERING CONSIDERATIONS – Many engineering and design challenges face the automotive electronic product designer as he/she deals with the complexity of future designs. Cost per function will be the major hurdle. New systems will be very complex with many different functions. Design will demand a level of integration far above current levels. Minimum size, mass, and volume will become major market discriminators.

As products are miniaturized, power and management design issues become harder to solve. Producability, manufacturability, and testability will become significant challenges.

Computer power and speed and software will play an increasing important role in these new product. Power and speed of automotive computers will approach those of office applications, and software will become orders of magnitude more complex than used in automobiles today.

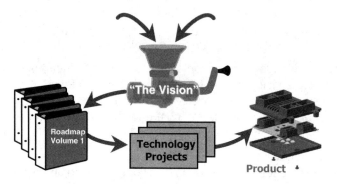

THE VISION

Determining a vision or direction for product development is the responsibility of all the major disciplines in an electronics manufacturing company. Marketing, sales, advanced engineering, product design, purchasing, manufacturing engineering, and operations, each have a specific role to play in the development of a new product.

Marketing and sales have the responsibility to determine what products are wanted or needed by consumers and automotive original equipment manufactures. These want and needs are then translated into product development plans called product roadmaps which define product requirements. Engineering then determines the technologies, which have to be developed or refined to meet these requirements then originates technology development projects. A key step in this process is to merge product roadmaps and technology projects to determine fit and timing so gaps can be resolved

As technology projects progress, development activities are supported by cross-functional teams which include product design, purchasing, manufacturing engineering, and operations. These groups provide needed inputs to the project. Each product line also learns of the specific technology issues in areas of responsibility that will enable successful implementation of specific technologies. New technologies should not be implemented into production without a coordinated multi-functional team with representation from all disciplines.

It is important to note that as new technologies are developed for specific products these technological advances also become available for use on existing products. Therefore, technology projects stimulate the development of new product strategies. Technology visioning is the engine that propels new innovative products. But it also can move new technologies into existing products for future applications.

DELPHI NEXT CENTURY WINNERS

Delphi Automotive Systems has a vision of marketable products, or winners, for the next century. [1]

1. Advanced Thermal Comfort Systems
2. X-by-Wire Control Systems
3. Modular Chassis Systems
4. Collision Avoidance
5. Mobile Multi-Media
6. Advanced Safety Interiors
7. Advanced Energy Systems
8. Smart Sensors and Actuators
9. Integrated Vehicle E/E System

This paper will focus on three of these so-called winners: x-by-wire control systems, collision avoidance, and advanced energy systems.

X-BY-WIRE CONTROL SYSTEMS – X-by-wire systems can be used to control steering, throttle, braking, and suspension. The advantages of these type systems are the elimination of mechanical links from the driver's controls to the control actuator. Mechanical links are expensive to manufacture, they add weight, complicate and add labor to the assembly of the vehicle, and require excessive maintenance over the life of the automobile.

In addition to the obvious saving that can be realized on conventional automobiles, X-by-wire systems are enablers for the automated highways of the future. Electronic systems can be controlled by central computers and ultimately will drive themselves in complete harmony with other automobiles in dense traffic. Automated travel reduces energy consumption, improves travel time, and makes highways more efficient thus reducing the need for more expensive highway construction.

COLLISION AVOIDANCE – Collision avoidance systems consist of radar and vision sensors, warning displays, brake, throttle, and steering control systems (X-by-wire), and processors and software. These systems are first and foremost designed to inform the driver of impending danger of either a collision or an out of control situation that could lead to a dangerous consequence (e.g., rollover). The next level of system design would be to take control of the automobile and make corrective action to advert danger in parallel with a warning to the driver.

Like X-by-wire control, collision avoidance systems, such as radar and vision sensors, are enablers of automated highways of the future. Radar and vision sensors will be the eyes of the computer that controls the automobile.

ADVANCED ENERGY SYSTEMS – Operation of the highly automated vehicles of the future will take more onboard energy. In order to power the additional electronic options being considered today, higher voltages (such as 42 Volts) are being considered. This requires new energy generation, storage, and control systems.

As automobile engines move toward hybrid and fully electric designs new motors/generators, converters/inverters, and storage batteries need to be developed. These systems will provide the power and range the automobile needs with far less energy consumed.

TESTING THE PROCESS: A CASE STUDY ON-ENGINE CONTROLLER

GLOBAL MARKET TRENDS – There is a strong market pull by automobile manufacturers to mount engine control modules on-engine. This pull is generated by engine manufacturers who want to assemble and test engines as a mechanical and electrical system, prior to shipment to automobile assembly plants. This assures "known good engine systems" are assembled into automobiles. It also reduces assembly cost and disruption of the assembly line, when compared to today's marriage of controller to engine during the final assembly of the automobile.

MARKET CONSIDERATIONS – In addition to the new mounting location the market also demanded additional features and functions. Improved emissions control, enhanced diagnostics, integration of engine and transmission controls, up integration of EMS functionality, and electron throttle control are expected in the new controller designs. Figure 2 shows the increasing controller I/O growth by year of introduction.

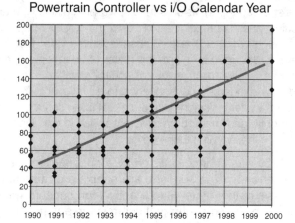

Functional Content:
Powertrain Controller vs i/O Calendar Year

Calendar Year

Figure 2.   Market Drivers.

ENGINEERING CONSIDERATIONS – Moving the mounting location from under hood to on- engine dramatically changes the operating environment of the controller. Operational temperature increase while vibrational concerns become more pronounced. Figure 3 illustrates the increasing severity of environmental conditions expected by moving the controller location from underhood to on engine. [2]

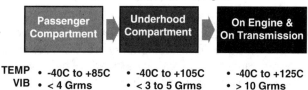

Market Demand Shift from Passenger Compartment to Underhood and On Engine

Passenger Compartment	Underhood Compartment	On Engine & On Transmission

TEMP	• -40C to +85C	• -40C to +105C	• -40C to +125C
VIB	• < 4 Grms	• < 3 to 5 Grms	• > 10 Grms

Figure 3.   Engineering Considerations

Size and weight of the controller designs are also expected to be reduced. Form factor reduction facilitates ease of mounting on-engine, and ease of final vehicle assembly

THE VISION/THE PLAN – Case examples of the described visioning processes have been executed at Delphi Delco Electronics Systems. In order to meet the size and weight specifications of the on-engine controller with a similar IC set, market and engineering considerations dictated that the majority of the IC's would have to mounted as bare die to save substrate space.

Higher component assembly density and advanced materials technologies can results in complications of design (especially thermal effects) and increased costs. Several design and cost studies were made. Flip chip bare die mounted to thin, high-density laminate with heat sinking directly to the flip chips was identified as the most robust and cost effective design for engine control applications. Three technology projects were initiated and completed as a result of this design decision: flip chip on laminate development, [3] flip chip capability in high-volume surface mount assembly operations, [4] and backside thermal heat sinking of flip chips. [5]

Figure 4 illustrates the final assembled circuit and the key technology elements developed for the on-engine controller.

Flip Chip IC Packaging
• Solder Bumping
• Known Good Die
Substrate Technology
• Fine Geometry's
• 0.062" Thick Structure
Thermal Management
• Through Substrate Heat Removal
Manufacturing Process
• Placement and Soldering
• Underfill Dispense and Cure
• Wave Soldering

Figure 4.   Key elements of flip chip on laminate assembly for engine control application.

PRODUCT STRATEGY – The new technologies developed for this specific on- engine product have benefited other product lines. For example, an advanced brake controller design subsequently utilized the flip chip on laminate technologies. The ABS application required thicker substrate and greater current carrying capabilities than the engine control application. Also, conventional thru-board thermal transfer techniques are required. Figure 5 presents key elements of advanced brake controller Figure 5. Key elements of flip chip on laminate assembly for brake control application. design.

Flip Chip IC Packaging
• Solder Bumping
• Known Good Die
Substrate Technology
• Fine Geometry's
• 0.062" Thick Structure
Thermal Management
• Through Substrate Heat Removal
Manufacturing Process
• Placement and Soldering
• Underfill Dispense and Cure
• Wave Soldering

Figure 5.   Key elements of flip chip on laminate assembly for brake control application.

Ceramic hybrid products have also benefited from the developments of the on engine laminate controller. Improved underfill materials, larger IC die size and I/O functionality, new design guideline, common assembly processes with laminate, and fine line substrates are a few of the enhancements which have been realized for ceramic substrates applications. Figure 6 shows a ceramic product with these improvements.

Flip Chip IC Packaging
- Solder Bumping
- Known Good Die

Substrate Technology
- Ceramic Thick Film
- Fine Geometries

Thermal Management
- Through Substrate Heat Removal

Manufacturing Process
- Placement and Soldering
- Underfill Dispense and Cure
- Robust Assembly Encapsulation

Figure 6. Key elements of flip chip on ceramic assembly for automotive application.

CONCLUSIONS

It is extremely important for progressive electronics automotive companies to have a clear vision of future products. This vision, through the use of a defining process, is used to generate technology roadmaps that identify technology development projects. These projects enable product vision realization through innovative product designs.

Technologies developed can be used, or reused, on other new or existing product designs, which results in expanded benefits from the new design rules and manufacturing processes developed.

ACKNOWLEDGEMENTS

The authors of this paper would like to thank Dr. Robert Schumacher, Product Line Executive for Mobile Multi-Media at Delphi Delco Electronics, and Dr. Richard Lind, Director of Advanced Engineering at Delphi Delco Electronics, for their inputs and support in writing this paper.

REFERENCE

1. R. Schumacher, "Next Century Winners in Automotive Electronics,"

2. G. DeVos and D. Helton, "Migration of Powertrain Electronics to On-Engine and On-Transmission," *SAE Intl. Conf. & Expo.*, (SAE, 1999).

3. M. Witty et al., "Flip Chip Assembly on Rigid Organic Laminates: A Production Ready Process for Automotive Electronics," *Intl. Conf. & Expo. on MCM and High Density Pack.* (IEEE Press, ISBN 0-7803-4850-8, 1998), pp. 77-82.

4. P. Jones, S. Davidson, and C. Delheimer, "Adding Flip Chip Placement Capabilities to a SMT Manufacturing Process is NOT Difficult, Even in High Volume," *Intl. Conf. on Elect. Assy.: Materials and Process Challenges* (IEEE Press, 1998), pp. 39-48.

5. S. Brandenburg and J. Daanen, "Thermal Reliability Characteristics of Top-Side Flip Chip Heat Sinking Method," *IMAPS Intl. Sys. Packaging Symp.*, (IMAPS 1999), pp.139-145.